SECOND EDITION

RULES OF THUMB FOR CHEMICAL ENGINEERS

A manual of quick, accurate solutions to everyday process engineering problems

CARL R. BRANAN, EDITOR

Gulf Publishing Company
Houston, Texas

To my four grandchildren:
Alex Matthew
Katherine Richard

RULES OF THUMB FOR CHEMICAL ENGINEERS

SECOND EDITION

Gulf Publishing Company
Book Division
P.O. Box 2608 □ Houston, Texas 77252-2608

10 9 8 7 6 5 4 3 2

Printed on Acid-Free Paper (∞)

Library of Congress Cataloging-in-Publication Data

Rules of thumb for chemical engineers: a manual of quick, accurate solutions to everyday process engineering problems / Carl R. Branan, editor. —2nd ed.
p. cm.
Includes bibliographical references and index.
ISBN 0-88415-788-1 (alk. paper)
1. Chemical engineering—Handbooks, manuals, etc.
I. Branan, Carl.
TP151.R85 1998
660—dc21 98-26530
CIP

Contents

13: Vacuum Systems, 190

14: Pneumatic Conveying, 198

15: Blending, 202

SECTION THREE
Plant Design 207

16: Process Evaluation, 209

17: Reliability, 243

18: Metallurgy, 246

19: Safety, 269

20: Controls, 286

SECTION FOUR
Operations 293

21: Troubleshooting, 294

SECTION ONE
Equipment Design

1
Fluid Flow

Velocity Head

Two of the most useful and basic equations are

$$\Delta h = \frac{u^2}{2g} \quad (1)$$

$$\Delta P\,(V) + \frac{\Delta u^2}{2g} + \Delta Z + E = 0 \quad (2)$$

where Δh = Head loss in feet of flowing fluid
u = Velocity in ft/sec
g = 32.2 ft/sec^2
P = Pressure in lb/ft^2
V = Specific volume in ft^3/lb
Z = Elevation in feet
E = Head loss due to friction in feet of flowing fluid

In Equation 1 Δh is called the "velocity head." This expression has a wide range of utility not appreciated by many. It is used "as is" for

1. Sizing the holes in a sparger
2. Calculating leakage through a small hole
3. Sizing a restriction orifice
4. Calculating the flow with a pitot tube

With a coefficient it is used for

1. Orifice calculations
2. Relating fitting losses, etc.

For a sparger consisting of a large pipe having small holes drilled along its length Equation 1 applies directly. This is because the hole diameter and the length of fluid travel passing through the hole are similar dimensions. An orifice on the other hand needs a coefficient in Equation 1 because hole diameter is a much larger dimension than length of travel (say ⅛ in. for many orifices). Orifices will be discussed under "Metering" in this chapter.

For compressible fluids one must be careful that when sonic or "choking" velocity is reached, further decreases in downstream pressure do not produce additional flow. This occurs at an upstream to downstream absolute pressure ratio of about 2:1. Critical flow due to sonic velocity has practically no application to liquids. The speed of sound in liquids is very high. See "Sonic Velocity" later in this chapter.

Still more mileage can be gotten out of $\Delta h = u^2/2g$ when using it with Equation 2, which is the famous Bernoulli equation. The terms are

1. The PV change
2. The kinetic energy change or "velocity head"
3. The elevation change
4. The friction loss

These contribute to the flowing head loss in a pipe. However, there are many situations where by chance, or on purpose, $u^2/2g$ head is converted to PV or vice versa.

We purposely change $u^2/2g$ to PV gradually in the following situations:

1. Entering phase separator drums to cut down turbulence and promote separation
2. Entering vacuum condensers to cut down pressure drop

We build up PV and convert it in a controlled manner to $u^2/2g$ in a form of tank blender. These examples are discussed under appropriate sections.

Source

Branan, C. R. *The Process Engineer's Pocket Handbook,* Vol. 1, Gulf Publishing Co., Houston, Texas, p. 1.

Piping Pressure Drop

A handy relationship for turbulent flow in commercial steel pipes is:

$$\Delta P_F = W^{1.8}\mu^{0.2}/20{,}000d^{4.8}\rho$$

where:

ΔP_F = Frictional pressure loss, psi/100 equivalent ft of pipe
W = Flow rate, lb/hr
μ = Viscosity, cp
ρ = Density, lb/ft^3
d = Internal pipe diameter, in.

This relationship holds for a Reynolds number range of 2,100 to 10^6. For smooth tubes (assumed for heat exchanger tubeside pressure drop calculations), a constant of 23,000 should be used instead of 20,000.

Source

Branan, Carl R. "Estimating Pressure Drop," *Chemical Engineering,* August 28, 1978.

Equivalent Length

The following table gives equivalent lengths of pipe for various fittings.

Table 1
Equivalent Length of Valves and Fittings in Feet

Nominal Pipe size in.	Globe valve or ball check valve	Angle valve	Swing check valve	Plug cock	Gate or ball valve	45° ell	Short rad. ell	Long rad. ell	Hard T.	Soft T.	90° miter bends			Enlargement					Contraction				
														Sudden			Std. red.		Sudden			Std. red.	
														Equiv. L in terms of small d									
						Weld thrd	Weld thrd	Weld thrd	Weld thrd	Weld thrd	2 miter	3 miter	4 miter	d/D = ¼	d/D = ½	d/D = ¾	d/D = ½	d/D = ¾	d/D = ¼	d/D = ½	d/D = ¾	d/D = ½	d/D = ¾
1½	55	26	13	7	1	1 2	3 5	2 3	8 9	2 3				5	3	1	4	1	3	2	1	1	—
2	70	33	17	14	2	2 3	4 5	3 4	10 11	3 4				7	4	1	5	1	3	3	1	1	—
2½	80	40	20	11	2	2 ..	5 ..	3 ..	12	3 ..				8	5	2	6	2	4	3	2	2	—
3	100	50	25	17	2	2	6	4	14	4				10	6	2	8	2	5	4	2	2	—
4	130	65	32	30	3	3	7	5	19	5				12	8	3	10	3	6	5	3	3	—
6	200	100	48	70	4	4	11	8	28	8				18	12	4	14	4	9	7	4	4	1
8	260	125	64	120	6	6	15	9	37	9				25	16	5	19	5	12	9	5	5	2
10	330	160	80	170	7	7	18	12	47	12				31	20	7	24	7	15	12	6	6	2
12	400	190	95	170	9	9	22	14	55	14	28	21	20	37	24	8	28	8	18	14	7	7	2
14	450	210	105	80	10	10	26	16	62	16	32	24	22	42	26	9	—	—	20	16	8	—	—
16	500	240	120	145	11	11	29	18	72	18	38	27	24	47	30	10	—	—	24	18	9	—	—
18	550	280	140	160	12	12	33	20	82	20	42	30	28	53	35	11	—	—	26	20	10	—	—
20	650	300	155	210	14	14	36	23	90	23	46	33	32	60	38	13	—	—	30	23	11	—	—
22	688	335	170	225	15	15	40	25	100	25	52	36	34	65	42	14	—	—	32	25	12	—	—
24	750	370	185	254	16	16	44	27	110	27	56	39	36	70	46	15	—	—	35	27	13	—	—
30	—	—	—	312	21	21	55	40	140	40	70	51	44										
36	—	—	—		25	25	66	47	170	47	84	60	52										
42	—	—	—		30	30	77	55	200	55	98	69	64										
48	—	—	—		35	35	88	65	220	65	112	81	72										
54	—	—	—		40	40	99	70	250	70	126	90	80										
60	—	—	—		45	45	110	80	260	80	190	99	92										

Sources

1. *GPSA Engineering Data Book,* Gas Processors Suppliers Association, 10th Ed. 1987.
2. Branan, C. R., *The Process Engineer's Pocket Handbook, Vol. 1,* Gulf Publishing Co., p. 6.

Recommended Velocities

Here are various recommended flows, velocities, and pressure drops for various piping services.

Sizing Steam Piping in New Plants
Maximum Allowable Flow and Pressure Drop

	Laterals			Mains		
Pressure, PSIG	600	175	30	600	175	30
Density, #/CF	0.91	0.41	0.106	0.91	1.41	0.106
ΔP, PSI/100′	1.0	0.70	0.50	0.70	0.40	0.30
Nominal Pipe Size, In.	**Maximum Lb/Hr × 10^{-3}**					
3	7.5	3.6	1.2	6.2	2.7	0.9
4	15	7.5	3.2	12	5.7	2.5
6	40	21	8.5	33	16	6.6
8	76	42	18	63	32	14
10	130	76	32	108	58	25
12	190	115	50	158	87	39
14	260	155	70	217	117	54
16	360	220	100	300	166	78
18	...	300	130	...	227	101
20	...	...	170	...	...	132

Note:
(1) 600 PSIG steam is at 750°F., 175 PSIG and 30 PSIG are saturated.
(2) On 600 PSIG flow ratings, internal pipe sizes for larger nominal diameters were taken as follows: 18/16.5″, 14/12.8″, 12/11.6″, 10/9.75″.
(3) If other actual I.D. pipe sizes are used, or if local superheat exists on 175 PSIG or 20 PSIG systems, the allowable pressure drop shall be the governing design criterion.

Sizing Cooling Water Piping in New Plants
Maximum Allowable Flow, Velocity and Pressure Drop

	LATERALS			MAINS		
Pipe Size in.	**Flow GPM**	**Vel. ft/sec.**	**ΔP ft/100′**	**Flow GPM**	**Vel. ft/sec.**	**ΔP ft/100′**
3	100	4.34	4.47	70	3.04	2.31
4	200	5.05	4.29	140	3.53	2.22
6	500	5.56	3.19	380	4.22	1.92
8	900	5.77	2.48	650	4.17	1.36
10	1,500	6.10	2.11	1,100	4.48	1.19
12	2,400	6.81	2.10	1,800	5.11	1.23
14	3,100	7.20	2.10	2,200	5.13	1.14
16	4,500	7.91	2.09	3,300	5.90	1.16
18	6,000	8.31	1.99	4,500	6.23	1.17
20	....	....	....	6,000	6.67	1.17
24	....	....	....	11,000	7.82	1.19
30	....	....	....	19,000	8.67	1.11

Sizing Piping for Miscellaneous Fluids

Dry Gas	100 ft/sec
Wet Gas	60 ft/sec
High Pressure Steam	150 ft/sec
Low Pressure Steam	100 ft/sec
Air	100 ft/sec
Vapor Lines General	Max. velocity 0.3 mach 0.5 psi/100 ft
Light Volatile Liquid Near Bubble Pt. Pump Suction	0.5 ft head total suction line
Pump Discharge, Tower Reflux	3–5 psi/100 ft
Hot Oil Headers	1.5 psi/100 ft
Vacuum Vapor Lines below 50 MM Absolute Pressure	Allow max. of 5% absolute pressure for friction loss

Suggested Fluid Velocities in Pipe and Tubing
(Liquids, Gases, and Vapors at Low Pressures to 50 psig and 50°F–100°F)

The velocities are suggestive only and are to be used to approximate line size as a starting point for pressure drop calculations. The final line size should be such as to give an economical balance between pressure drop and reasonable velocity.

Fluid	Suggested Trial Velocity	Pipe Material
Acetylene (Observe pressure limitations)	4000 fpm	Steel
Air, 0 to 30 psig	4000 fpm	Steel
Ammonia		
Liquid	6 fps	Steel
Gas	6000 fpm	Steel
Benzene	6 fps	Steel
Bromine		
Liquid	4 fps	Glass
Gas	2000 fpm	Glass
Calcium Chloride	4 fps	Steel
Carbon Tetrachloride	6 fps	Steel
Chlorine (Dry)		
Liquid	5 fps	Steel, Sch. 80
Gas	2000–5000 fpm	Steel, Sch. 80
Chloroform		
Liquid	6 fps	Copper & Steel
Gas	2000 fpm	Copper & Steel
Ethylene Gas	6000 fpm	Steel
Ethylene Dibromide	4 fps	Glass
Ethylene Dichloride	6 fps	Steel
Ethylene Glycol	6 fps	Steel
Hydrogen	4000 fpm	Steel
Hydrochloric Acid		
Liquid	5 fps	Rubber Lined
Gas	4000 fpm	R. L., Saran, Haveg
Methyl Chloride		
Liquid	6 fps	Steel
Gas	4000 fpm	Steel
Natural Gas	6000 fpm	Steel
Oils, lubricating	6 fps	Steel
Oxygen	1800 fpm Max.	Steel (300 psig Max.)
(ambient temp.)	4000 fpm	Type 304 SS
(Low temp.)		
Propylene Glycol	5 fps	Steel

Fluid	Suggested Trial Velocity	Pipe Material
Sodium Hydroxide		
0–30 Percent	6 fps	Steel
30–50 Percent	5 fps	and
50–73 Percent	4	Nickel
Sodium Chloride Sol'n.		
No Solids	5 fps	Steel
With Solids	(6 Min.–	
	15 Max.)	Monel or nickel
	7.5 fps	
Perchlorethylene	6 fps	Steel
Steam		
0–30 psi Saturated*	4000–6000 fpm	Steel
30–150 psi Saturated or superheated*	6000–10000 fpm	
150 psi up superheated	6500–15000 fpm	
*Short lines	15,000 fpm (max.)	
Sulfuric Acid		
88–93 Percent	4 fps	S. S.–316, Lead
93–100 Percent	4 fps	Cast Iron & Steel, Sch. 80
Sulfur Dioxide	4000 fpm	Steel
Styrene	6 fps	Steel
Trichlorethylene	6 fps	Steel
Vinyl Chloride	6 fps	Steel
Vinylidene Chloride	6 fps	Steel
Water		
Average service	3–8 (avg. 6) fps	Steel
Boiler feed	4–12 fps	Steel
Pump suction lines	1–5 fps	Steel
Maximum economical (usual)	7–10 fps	Steel
Sea and brackish water, lined pipe	5–8 fps } 3	R. L., concrete, asphalt-line, saran-
Concrete	5–12 fps } (Min.)	lined, transite

Note: R. L. = Rubber-lined steel.

Typical Design Vapor Velocities* (ft./sec.)

Fluid	Line Sizes ≤6″	8″–12″	≥14″
Saturated Vapor			
0 to 50 psig	30–115	50–125	60–145
Gas or Superheated Vapor			
0 to 10 psig	50–140	90–190	110–250
11 to 100 psig	40–115	75–165	95–225
101 to 900 psig	30–85	60–150	85–165

**Values listed are guides, and final line sizes and flow velocities must be determined by appropriate calculations to suit circumstances. Vacuum lines are not included in the table, but usually tolerate higher velocities. High vacuum conditions require careful pressure drop evaluation.*

Typical Design* Velocities for Process System Applications

Service	Velocity, ft./sec.
Average liquid process	4–6.5
Pump suction (except boiling)	1–5
Pump suction, (boiling)	0.5–3
Boiler feed water (disch., pressure)	4–8
Drain lines	1.5–4
Liquid to reboiler (no pump)	2–7
Vapor-liquid mixture out reboiler	15–30
Vapor to condenser	15–80
Gravity separator flows	0.5–1.5

**To be used as guide, pressure drop and system environment govern final selection of pipe size.*
For heavy and viscous fluids, velocities should be reduced to about ½ values shown.
Fluids not to contain suspended solid particles.

Usual Allowable Velocities for Duct and Piping Systems*

Service/Application	Velocity, ft./min.
Forced draft ducts	2,500–3,500
Induced-draft flues and breeching	2,000–3,000
Chimneys and stacks	2,000
Water lines (max.)	600
High pressure steam lines	10,000
Low pressure steam lines	12,000–15,000
Vacuum steam lines	25,000
Compressed air lines	2,000
Refrigerant vapor lines	
High pressure	1,000–3,000
Low pressure	2,000–5,000
Refrigerant liquid	200
Brine lines	400
Ventilating ducts	1,200–3,000
Register grilles	500

**By permission,* Chemical Engineer's Handbook, *3rd Ed., p. 1642, McGraw-Hill Book Co., New York, N.Y.*

Suggested Steam Pipe Velocities in Pipe Connecting to Steam Turbines

Service—Steam	Typical range, ft./sec.
Inlet to turbine	100–150
Exhaust, non-condensing	175–200
Exhaust, condensing	400–500

Sources

1. Branan, C. R., *The Process Engineer's Pocket Handbook, Vol. 1,* Gulf Publishing Co.
2. Ludwig, E. E., *Applied Process Design for Chemical and Petrochemical Plants, 2nd Ed.,* Gulf Publishing Co.
3. Perry, R. H., *Chemical Engineer's Handbook, 3rd Ed.,* p. 1642, McGraw-Hill Book Co.

Two-phase Flow

Two-phase (liquid/vapor) flow is quite complicated and even the long-winded methods do not have high accuracy. You cannot even have complete certainty as to which flow regime exists for a given situation. Volume 2 of Ludwig's design books[1] and the GPSA Data Book[2] give methods for analyzing two-phase behavior.

For our purposes, a rough estimate for general two-phase situations can be achieved with the Lockhart and Martinelli[3] correlation. Perry's[4] has a writeup on this correlation. To apply the method, each phase's pressure drop is calculated as though it alone was in the line. Then the following parameter is calculated:

$$X = [\Delta P_L/\Delta P_G]^{1/2}$$

where: ΔP_L and ΔP_G are the phase pressure drops

The X factor is then related to either Y_L or Y_G. Whichever one is chosen is multiplied by its companion pressure drop to obtain the total pressure drop. The following equation[5] is based on points taken from the Y_L and Y_G curves in Perry's[4] for both phases in turbulent flow (the most common case):

$$Y_L = 4.6X^{-1.78} + 12.5X^{-0.68} + 0.65$$
$$Y_G = X^2 Y_L$$

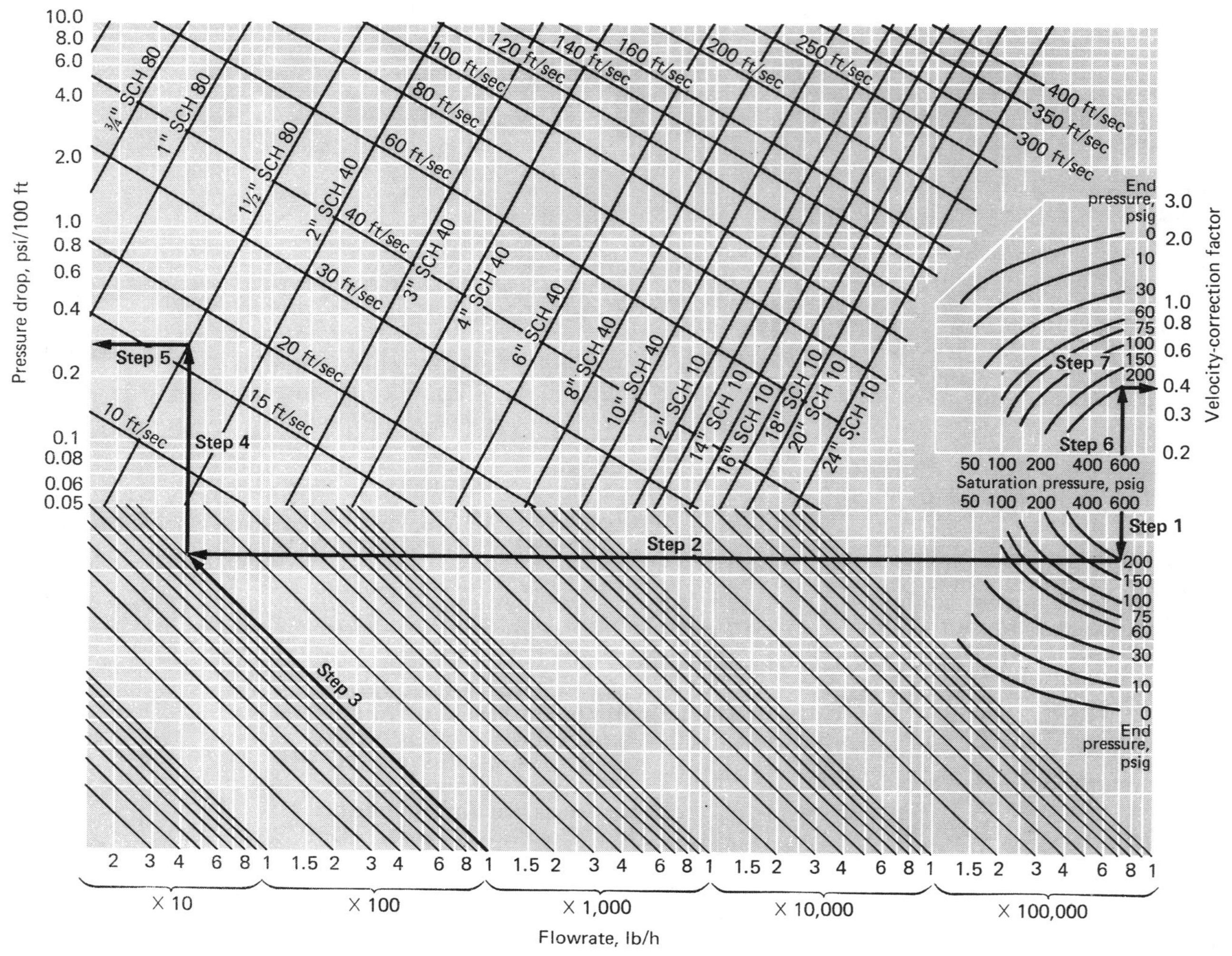

Sizing Lines for Flashing Steam-Condensate

The X range for Lockhart and Martinelli curves is 0.01 to 100.

For fog or spray type flow, Ludwig[1] cites Baker's[6] suggestion of multiplying Lockhart and Martinelli by two.

For the frequent case of flashing steam-condensate lines, Ruskan[7] supplies the handy graph shown above.

This chart provides a rapid estimate of the pressure drop of flashing condensate, along with the fluid velocities. *Example:* If 1,000 lb/h of saturated 600-psig condensate is flashed to 200 psig, what size line will give a pressure drop of 1.0 psi/100 ft or less? Enter at 600 psig below insert on the right, and read down to a 200 psig end pressure. Read left to intersection with 1,000 lb/hr flowrate, then up vertically to select a 1½ in for a 0.28 psi/100 ft pressure drop. Note that the velocity given by this lines up if 16.5 ft/s are used; on the insert at the right read up from 600 psig to 200 psig to find the velocity correction factor 0.41, so that the corrected velocity is 6.8 ft/s.

Sources

1. Ludwig, E. E., *Applied Process Design For Chemical and Petrochemical Plants, Vol. 1,* Gulf Publishing Co.
2. *GPSA Data Book, Vol. II,* Gas Processors Suppliers Association, 10th Ed., 1987.
3. Lockhart, R. W., and Martinelli, R. C., "Proposed Correlation of Data for Isothermal Two-Phase, Two-Component Flow in Pipes," *Chemical Engineering Progress,* 45:39–48, 1949.

4. Perry, R. H., and Green, D., *Perry's Chemical Engineering Handbook, 6th Ed.,* McGraw-Hill Book Co.
5. Branan, C. R., *The Process Engineer's Pocket Handbook, Vol. 2,* Gulf Publishing Co.
6. Baker, O., "Multiphase Flow in Pipe Lines," *Oil and Gas Journal,* November 10, 1958, p. 156.
7. Ruskan, R. P., "Sizing Lines For Flashing Steam-Condensate," *Chemical Engineering,* November 24, 1975, p. 88.

Compressible Flow

For "short" lines, such as in a plant, where $\Delta P > 10\%\ P_1$, either break into sections where $\Delta P < 10\%\ P_1$ or use

$$\Delta P = P_1 - P_2 = \frac{2P_1}{P_1 + P_2}\left[0.323\left(\frac{fL}{d} + \frac{\ln (P_1/P_2)}{24}\right)S_1 U_1^{\,2}\right]$$

from Maxwell[1] which assumes isothermal flow of ideal gas.

where:

ΔP = Line pressure drop, psi
P_1, P_2 = Upstream and downstream pressures in psi ABS
S_1 = Specific gravity of vapor relative to water = $0.00150\ MP_1/T$
d = Pipe diameter in inches
U_1 = Upstream velocity, ft/sec
f = Friction factor (assume .005 for approximate work)
L = Length of pipe, feet
ΔP = Pressure drop in psi (rather than psi per standard length as before)
M = Mol. wt.

For "long" pipelines, use the following from McAllister[2]:

Equations Commonly Used for Calculating Hydraulic Data for Gas Pipe Lines

Panhandle A.

$$Q_b = 435.87 \times (T_b / P_b)^{1.0778} \times D^{2.6182} \times E \times \left[\frac{P_1^{\,2} - P_2^{\,2} - \dfrac{0.0375 \times G \times (h_2 - h_1) \times P_{avg}^{\,2}}{T_{avg} \times Z_{avg}}}{G^{0.8539} \times L \times T_{avg} \times Z_{avg}}\right]^{0.5394}$$

Panhandle B.

$$Q_b = 737 \times (T_b / P_b)^{1.020} \times D^{2.53} \times E \times \left[\frac{P_1^{\,2} - P_2^{\,2} - \dfrac{0.0375 \times G \times (h_2 - h_1) \times P_{avg}^{\,2}}{T_{avg} \times Z_{avg}}}{G^{0.961} \times L \times T_{avg} \times Z_{avg}}\right]^{0.51}$$

Weymouth.

$$Q = 433.5 \times (T_b / P_b) \times \left[\frac{P_1^{\,2} - P_2^{\,2}}{GLTZ}\right]^{0.5} \times D^{2.667} \times E$$

$P_{avg} = 2/3\ [P_1 + P_2 - (P_1 \times P_2)/P_1 + P_2]$

P_{avg} is used to calculate gas compressibility factor Z

Nomenclature for Panhandle Equations

Q_b = flow rate, SCFD
P_b = base pressure, psia
T_b = base temperature, °R
T_{avg} = average gas temperature, °R
P_1 = inlet pressure, psia
P_2 = outlet pressure, psia
G = gas specific gravity (air = 1.0)
L = line length, miles
Z = average gas compressibility
D = pipe inside diameter, in.
h_2 = elevation at terminus of line, ft
h_1 = elevation at origin of line, ft
P_{avg} = average line pressure, psia
E = efficiency factor
E = 1 for new pipe with no bends, fittings, or pipe diameter changes
E = 0.95 for very good operating conditions, typically through first 12–18 months
E = 0.92 for average operating conditions
E = 0.85 for unfavorable operating conditions

Nomenclature for Weymouth Equation

Q = flow rate, MCFD
T_b = base temperature, °R
P_b = base pressure, psia
G = gas specific gravity (air = 1)
L = line length, miles
T = gas temperature, °R
Z = gas compressibility factor
D = pipe inside diameter, in.
E = efficiency factor. (See Panhandle nomenclature for suggested efficiency factors)

Sample Calculations

Q = ?
G = 0.6
T = 100°F
L = 20 miles
P_1 = 2,000 psia
P_2 = 1,500 psia
Elev diff. = 100 ft
D = 4.026-in.
T_b = 60°F
P_b = 14.7 psia
E = 1.0
P_{avg} = 2/3(2,000 + 1,500 − (2,000 × 1,500/2,000 + 1,500))
= 1,762 psia

Z at 1,762 psia and 100°F = 0.835.

Panhandle A.

$$Q_b = 435.87 \times (520/14.7)^{1.0788} \times (4.026)^{2.6182} \times 1 \times \left[\frac{(2,000)^2 - (1,500)^2 - \dfrac{0.0375 \times 0.6 \times 100 \times (1,762)^2}{560 \times 0.835}}{(0.6)^{.8539} \times 20 \times 560 \times .835}\right]^{0.5394}$$

$Q_b = 16,577$ MCFD

Panhandle B.

$$Q_b = 737 \times (520/14.7)^{1.020} \times (4.026)^{2.53} \times 1 \times \left[\frac{(2,000)^2 - (1,500)^2 - \dfrac{0.0375 \times 0.6 \times 100 \times (1,762)^2}{560 \times 0.835}}{(0.6)^{.961} \times 20 \times 560 \times .835}\right]^{0.51}$$

$Q_b = 17,498$ MCFD

Weymouth.

$$Q = 0.433 \times (520/14.7) \times [(2,000)^2 - (1,500)^2/(0.6 \times 20 \times 560 \times 0.835)]^{1/2} \times (4.026)^{2.667}$$

Q = 11,101 MCFD

Source

Pipecalc 2.0, Gulf Publishing Company, Houston, Texas. Note: Pipecalc 2.0 will calculate the compressibility factor, minimum pipe ID, upstream pressure, downstream pressure, and flow rate for Panhandle A, Panhandle B, Weymouth, AGA, and Colebrook-White equations. The flow rates calculated in the above sample calculations will differ slightly from those calculated with Pipecalc 2.0 since the viscosity used in the examples was extracted from Figure 5, p. 147. Pipecalc uses the Dranchuk et. al. method for calculating gas compressibility.

Equivalent Lengths for Multiple Lines Based on Panhandle A

Condition I.

A single pipe line which consists of two or more different diameter lines.

Let L_E = equivalent length
$L_1, L_2, \ldots L_n$ = length of each diameter
$D_1, D_2, \ldots D_n$ = internal diameter of each separate line corresponding to $L_1, L_2, \ldots L_n$
D_E = equivalent internal diameter

$$L_e = L_1\left[\frac{D_E}{D_1}\right]^{4.8539} + L_2\left[\frac{D_E}{D_2}\right]^{4.8539} + \cdots L_n\left[\frac{D_E}{D_n}\right]^{4.8539}$$

Example. A single pipe line, 100 miles in length consists of 10 miles 10¾-in. OD; 40 miles 12¾-in OD and 50 miles of 22-in. OD lines.

Find equivalent length (L_E) in terms of 22-in. OD pipe.

$$L_E = 50 + 40\left[\frac{21.5}{12.25}\right]^{4.8539} + 10\left[\frac{21.5}{10.25}\right]^{4.8539}$$
$$= 50 + 614 + 364$$
$$= 1{,}028 \text{ miles equivalent length of 22-in. OD}$$

Condition II.

A multiple pipe line system consisting of two or more parallel lines of different diameters and different lengths.

Let L_E = equivalent length
$L_1, L_2, L_3, \ldots L_n$ = length of various looped sections
$d_1, d_2, d_3, \ldots d_n$ = internal diameter of the individual line corresponding to length L_1, L_2, L_3 & L_n

$$L_E = L_1\left[\frac{d_E^{2.6182}}{d_1^{2.6182} + d_2^{2.6182} + d_3^{2.6182} + \ldots d_n^{2.6182}}\right]^{1.8539} + \ldots$$
$$L_n\left[\frac{d_E^{2.6182}}{d_1^{2.6182} + d_2^{2.6182} + d_3^{2.6182} + \ldots d_n^{2.6182}}\right]^{1.8539}$$

Let L_E = equivalent length
L_1, L_2, L_3 & L_n = length of various looped sections
d_1, d_2, d_3 & d_n = internal diameter of individual line corresponding to lengths L_1, L_2, L_3 & L_n

$$L_E = L_1\left[\frac{d_E^{2.6182}}{d_1^{2.6182} + d_2^{2.6182} + d_3^{2.6182} + \ldots d_n^{2.6182}}\right]^{1.8539} + \ldots$$
$$L_n\left[\frac{d_E^{2.6182}}{d_1^{2.6182} + d_2^{2.6182} + d_3^{2.6182} + \ldots d_n^{2.6182}}\right]^{1.8539}$$

when L_1 = length of unlooped section
L_2 = length of single looped section
L_3 = length of double looped section
$d_E = d_1 = d_2$

then:

$$L_E = L_1 + 0.27664\, L_2 + L_3\left[\frac{d_1^{2.6182}}{2d_1^{2.6182} + d_3^{2.6182}}\right]^{1.8539}$$

when $d_E = d_1 = d_2 = d_3$

then $L_E = L_1 + 0.27664\, L_2 + 0.1305\, L_3$

Example. A multiple system consisting of a 15 mile section of 3–8⅝-in. OD lines and 1–10¾-in. OD line, and a 30 mile section of 2–8⅝-in. lines and 1–10¾-in. OD line.

Find the equivalent length in terms of single 12-in. ID line.

$$L_E = 15\left[\frac{12^{2.6182}}{3(7.981)^{2.6182} + 10.02^{2.6182}}\right]^{1.8539}$$
$$+ 30\left[\frac{12^{2.6182}}{2(7.981)^{2.6182} + 10.02^{2.6182}}\right]^{1.8539}$$
$$= 5.9 + 18.1$$
$$= 24.0 \text{ miles equivalent of 12-in. ID pipe}$$

Example. A multiple system consisting of a single 12-in. ID line 5 miles in length and a 30 mile section of 3–12-in. ID lines.

Find equivalent length in terms of a single 12-in. ID line.

$$L_E = 5 + 0.1305 \times 30$$
$$= 8.92 \text{ miles equivalent of single 12-in. ID line}$$

References

1. Maxwell, J. B., *Data Book on Hydrocarbons,* Van Nostrand, 1965.
2. McAllister, E. W., *Pipe Line Rules of Thumb Handbook, 3rd Ed.,* Gulf Publishing Co., p. 247–248.
3. Branan, C. R., *The Process Engineer's Pocket Handbook, Vol. 1,* Gulf Publishing Co., p. 4.

Sonic Velocity

To determine sonic velocity, use

$$V_s = \sqrt{KgRT}$$

where

V_s = Sonic velocity, ft/sec
K = C_p/C_v the ratio of specific heats at constant pressure to constant volume. This ratio is 1.4 for most diatomic gases.
g = 32.2 ft./sec^2
R = 1544/mol. wt.
T = Absolute temperature in °R

To determine the critical pressure ratio for gas sonic velocity across a nozzle or orifice use

$$\text{critical pressure ratio} = [2/(K + 1)]^{k/(k-1)}$$

If pressure drop is high enough to exceed the critical ratio, sonic velocity will be reached. When K = 1.4, ratio = 0.53.

Source

Branan, C. R., *The Process Engineer's Pocket Handbook, Vol. 1,* Gulf Publishing Co.

Metering

Orifice

$$(U_o^2 - U_p^2)^{1/2} = C_o(2g\Delta h)^{1/2}$$

Permanent head loss % of Δh

D_o/D_p	Permanent Loss
0.2	95
0.4	82
0.6	63
0.8	40

One designer uses permanent loss = $\Delta h\ (1 - C_o)$

where

U_o = Velocity through orifice, ft/sec
U_p = Velocity through pipe, ft/sec
2g = 64.4 ft/sec^2
Δh = Orifice pressure drop, ft of fluid
D = Diameter
C_o = Coefficient. (Use 0.60 for typical application where D_o/D_p is between 0.2 and 0.8 and Re at vena contracta is above 15,000.)

Venturi

Same equation as for orifice:

$C_o = 0.98$

Permanent head loss approximately 3–4% Δh.

Rectangular Weir

$$F_v = 3.33\ (L—0.2H)H^{3/2}$$

where

F_v = Flow in ft^3/sec
L = Width of weir, ft
H = Height of liquid over weir, ft

Pitot Tube

$$\Delta h = u^2/2g$$

Source

Branan, C. R., *The Process Engineer's Pocket Handbook Vol. 1,* Gulf Publishing Co.

Control Valves

Notes:

1. References 1 and 2 were used extensively for this section. The sizing procedure is generally that of Fisher Controls Company.
2. Use manufacturers' data where available. This handbook will provide approximate parameters applicable to a wide range of manufacturers.
3. For any control valve design be sure to use one of the modern methods, such as that given here, that takes into account such things as control valve pressure recovery factors and gas transition to incompressible flow at critical pressure drop.

Liquid Flow

Across a control valve the fluid is accelerated to some maximum velocity. At this point the pressure reduces to its lowest value. If this pressure is lower than the liquid's vapor pressure, flashing will produce bubbles or cavities of vapor. The pressure will rise or "recover" downstream of the lowest pressure point. If the pressure rises to above the vapor pressure, the bubbles or cavities collapse. This causes noise, vibration, and physical damage.

When there is a choice, design for no flashing. When there is no choice, locate the valve to flash into a vessel if possible. If flashing or cavitation cannot be avoided, select hardware that can withstand these severe conditions. The downstream line will have to be sized for two phase flow. It is suggested to use a long conical adaptor from the control valve to the downstream line.

When sizing liquid control valves first use

$$\Delta P_{allow} = K_m(P_1 - r_c P_v)$$

where

ΔP_{allow} = Maximum allowable differential pressure for sizing purposes, psi
K_m = Valve recovery coefficient (see Table 3)
r_c = Critical pressure ratio (see Figures 1 and 2)
P_1 = Body inlet pressure, psia
P_v = Vapor pressure of liquid at body inlet temperature, psia

This gives the maximum ΔP that is effective in producing flow. Above this ΔP no additional flow will be produced since flow will be restricted by flashing. Do not use a number higher than ΔP_{allow} in the liquid sizing formula. Some designers use as the minimum pressure for flash check the upstream absolute pressure minus two times control valve pressure drop.

Table 1 gives critical pressures for miscellaneous fluids. Table 2 gives relative flow capacities of various types of control valves. This is a rough guide to use in lieu of manufacturer's data.

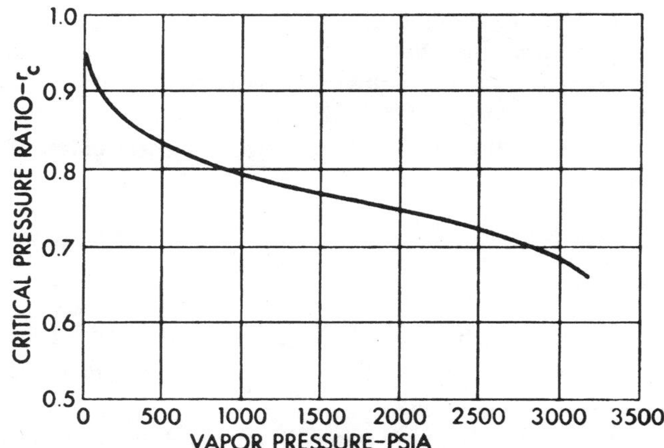

Figure 1. Enter on the abscissa at the water vapor pressure at the valve inlet. Proceed vertically to intersect the curve. Move horizontally to the left to read r_c on the ordinate (Reference 1).

The liquid sizing formula is

$$C_v = Q\sqrt{\frac{G}{\Delta P}}$$

where

C_v = Liquid sizing coefficient
Q = Flow rate in GPM
ΔP = Body differential pressure, psi
G = Specific gravity (water at 60°F = 1.0)

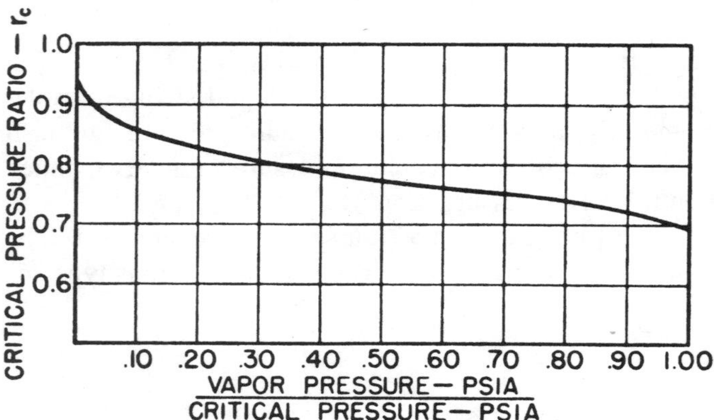

Figure 2. Determine the vapor pressure/critical pressure ratio by dividing the liquid vapor pressure at the valve inlet by the critical pressure of the liquid. Enter on the abscissa at the ratio just calculated and proceed vertically to intersect the curve. Move horizontally to the left and read r_c on the ordinate (Reference 1).

Two liquid control valve sizing rules of thumb are

1. No viscosity correction necessary if viscosity $\leqq$ 20 centistokes.
2. For sizing a flashing control valve add the C_v's of the liquid and the vapor.

Gas and Steam Flow

The gas and steam sizing formulas are

Gas

$$C_g = \frac{Q}{\sqrt{\frac{520}{GT}}\, P_1 \sin\left[\frac{3417}{C_1}\sqrt{\frac{\Delta P}{P_1}}\right]_{deg.}}$$

Steam (under 1000 psig)

$$C_s = \frac{Q_s(1 + 0.00065\, T_{sh})}{P_1 \sin\left[\frac{3417}{C_1}\sqrt{\frac{\Delta P}{P_1}}\right]_{deg.}}$$

Table 1
Critical Pressure of Various Fluids, Psia*

Ammonia	1636	Isobutane	529.2
Argon	705.6	Isobutylene	580
Butane	550.4	Methane	673.3
Carbon Dioxide	1071.6	Nitrogen	492.4
Carbon Monoxide	507.5	Nitrous Oxide	1047.6
Chlorine	1118.7	Oxygen	736.5
Dowtherm A	465	Phosgene	823.2
Ethane	708	Propane	617.4
Ethylene	735	Propylene	670.3
Fluorine	808.5	Refrigerant 11	635
Helium	33.2	Refrigerant 12	596.9
Hydrogen	188.2	Refrigerant 22	716
Hydrogen Chloride	1198	Water	3206.2

**For values not listed, consult an appropriate reference book.*

Steam and Vapors (all vapors, including steam under any pressure conditions)

$$C_s = \frac{Q_s}{1.06\sqrt{d_1 P_1}\, \sin\left[\frac{3417}{C_1}\sqrt{\frac{\Delta P}{P_1}}\right]_{deg.}}$$

When the bracketed quantity in the equations equals or exceeds 90 degrees, critical flow is indicated. The quantity must be limited to 90 degrees. This then becomes unity since sin 90° = 1.

Table 2
Relative Flow Capacities of Control Valves (Reference 2)

Valve Type	C_d*	C_dF_p†	C_dF_L**
Double-seat globe	12	11	11
Single-seat top-guided globe	11.5	10.8	10
Single-seat split body	12	11.3	10
Sliding gate	6–12	6–11	na
Single-seat top-entry cage	13.5	12.5	11.5
Eccentric rotating plug (Camflex)	14	13	12
60° open butterfly	18	15.5	12
Single-seat Y valve (300 & 600 lb)	19	16.5	14
Saunders type (unlined)	20	17	na
Saunders type (lined)	15	13.5	na
Throttling (characterized) ball	25	20	15
Single-seat streamlined angle (flow-to-close)	26	20	13
90° open butterfly (average)	32	21.5	18

Note: This table may serve as a rough guide only since actual flow capacities differ between manufacturer's products and individual valve sizes. (Source: ISA "Handbook of Control Valves" Page 17)

*Valve flow coefficient $C_v = C_d \times d^2$ (d = valve dia., in.)

†C_v/d^2 of valve when installed between pipe reducers (pipe dia. 2 × valve dia.)

**C_v/d^2 of valve when undergoing critical (choked) flow conditions.

Explanation of terms:

$C_l = C_g/C_v$ (some sizing methods use C_f or Y in place of C_l)
C_g = Gas sizing coefficient
C_s = Steam sizing coefficient
C_v = Liquid sizing coefficient
d_l = Density of steam or vapor at inlet, lbs/ft^3
G = Gas specific gravity = mol. wt./29
P_l = Valve inlet pressure, psia
ΔP = Pressure drop across valve, psi
Q = Gas flow rate, SCFH
Q_s = Steam or vapor flow rate, lb/hr
T = Absolute temperature of gas at inlet, °R
T_{sh} = Degrees of superheat, °F

Table 3
Average Valve-Recovery Coefficients, K_m and C_l*
(Reference 2)

Type of Valve	K_m	C_l
Cage-trim globes:		
Unbalanced	0.8	33
Balanced	0.70	33
Butterfly:		
Fishtail	0.43	16
Conventional	0.55	24.7
Ball:		
Vee-ball, modified-ball, etc.	0.40	22
Full-area ball	0.30	
Conventional globe:		
Single and double port (full port)	0.75	35
Single and double port (reduced port)	0.65	35
Three way	0.75	
Angle:		
Flow tends to open (standard body)	0.85	
Flow tends to close (standard body)	0.50	
Flow tends to close (venturi outlet)	0.20	
Camflex:		
Flow tends to close	0.72	24.9
Flow tends to open	0.46	31.1
Split body	0.80	35

**For use only if not available from manufacturer.*

Table 4
Correlations of Control Valve Coefficients (Reference 2)

$C_l = 36.59\ C_f$ $\qquad K_m = C_f^2 = F_L^2$
$C_l = 36.59\ \sqrt{K_m}$ $\qquad C_s = 1.83\ C_fC_v$
$C_g = C_lC_v$ $\qquad C_v = 19.99\ C_s/C_l$
$C_g = 19.99\ C_s$

Values of K_m calculated from C_f agree within 10% of published data of K_m.
Values of C_l calculated from K_m are within 21% of published data of C_l.

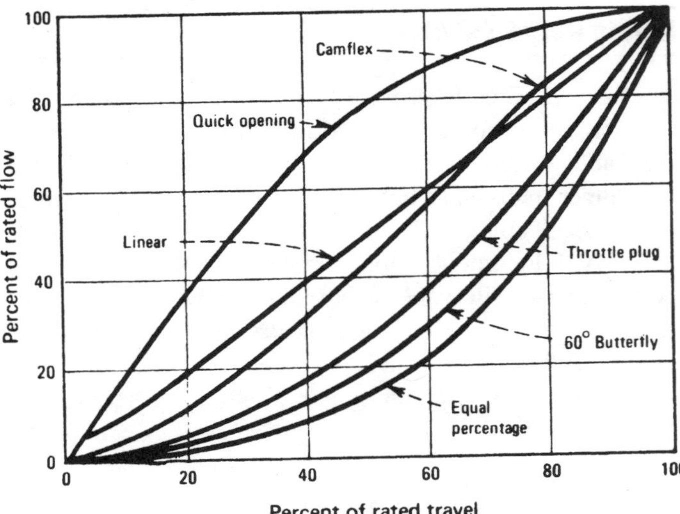

Figure 3. These are characteristic curves of common valves (Reference 2).

The control valve coefficients in Table 4 are for full open conditions. The control valve must be designed to operate at partial open conditions for good control. Figure 3 shows partial open performance for a number of trim types.

General Control Valve Rules of Thumb

1. Design tolerance. Many use the greater of the following:
   $Q_{sizing} = 1.3\ Q_{normal}$
   $Q_{sizing} = 1.1\ Q_{maximum}$
2. Type of trim. Use equal percentage whenever there is a large design uncertainty or wide rangeability is desired. Use linear for small uncertainty cases.

   Limit max/min flow to about 10 for equal percentage trim and 5 for linear. Equal percentage trim usually requires one larger nominal body size than linear.
3. For good control where possible, make the control valve take 50%–60% of the system flowing head loss.
4. For saturated steam keep control valve outlet velocity below 0.25 mach.
5. Keep valve inlet velocity below 300 ft/sec for 2″ and smaller, and 200 ft/sec for larger sizes.

References

1. Fisher Controls Company, Sizing and Selection Data, Catalog 10.
2. Chalfin, Fluor Corp., "Specifying Control Valves," *Chemical Engineering,* October 14, 1974.

Safety Relief Valves

The ASME code provides the basic requirements for over-pressure protection. Section I, *Power Boilers,* covers fired and unfired steam boilers. All other vessels including exchanger shells and similar pressure containing equipment fall under Section VIII, *Pressure Vessels.* API RP 520 and lesser API documents supplement the ASME code. These codes specify allowable accumulation, which is the difference between relieving pressure at which the valve reaches full rated flow and set pressure at which the valve starts to open. Accumulation is expressed as percentage of set pressure in Table 1.

Table 1
Accumulation Expressed as Percentage of Set Pressure

	ASME Section I Power Boilers	ASME Section VIII Pressure Vessels	Typical Design for Compressors Pumps and Piping
LIQUIDS			
thermal expansion	—	10	25
fire	—	20	20
STEAM			
over-pressure	3	10	10
fire	—	20	20
GAS OR VAPOR			
over-pressure	—	10	10
fire	—	20	20

Full liquid containers require protection from thermal expansion. Such relief valves are generally quite small. Two examples are

1. Cooling water that can be blocked in with hot fluid still flowing on the other side of an exchanger.
2. Long lines to tank farms that can lie stagnant and exposed to the sun.

Sizing

Use manufacturer's sizing charts and data where available. In lieu of manufacturer's data use the formula

$$u = 0.4\sqrt{2g\Delta h}$$

where

Δh = Head loss in feet of flowing fluid
u = Velocity in ft/sec
g = 32.2 ft/sec^2

This will give a conservative relief valve area. For compressible fluids use Δh corresponding to ½ P_l if head difference is greater than that corresponding to ½ P_l (since sonic velocity occurs). If head difference is below that corresponding to ½ P_l use actual Δh.

For vessels filled with only gas or vapor and exposed to fire use

$$A = \frac{0.042\,A_s}{\sqrt{P_l}} \qquad \text{(API RP 520, Reference 3)}$$

A = Calculated nozzle area, in.^{2}
P_l = Set pressure (psig) × (1 + fraction accumulation) + atmospheric pressure, psia. For example, if accumulation = 10%, then (1 + fraction accumulation) = 1.10
A_s = Exposed surface of vessel, ft^2

This will also give conservative results. For heat input from fire to liquid containing vessels see "Determination of Rates of Discharge."

The set pressure of a conventional valve is affected by back pressure. The spring setting can be adjusted to compensate for constant back pressure. For a variable back pressure of greater than 10% of the set pressure, it is customary to go to the balanced bellows type which can generally tolerate variable back pressure of up to 40% of set pressure. Table 2 gives standard orifice sizes.

Determination of Rates of Discharge

The more common causes of overpressure are

1. External fire
2. Heat Exchanger Tube Failure
3. Liquid Expansion
4. Cooling Water Failure
5. Electricity Failure
6. Blocked Outlet

Table 2
Relief Valve Designations

Standard Orifice Designation	Orifice Area (in.^{2})	1 × 2	1.5 × 2	1.5 × 2.5	1.5 × 3	2 × 3	2.5 × 4	3 × 4	4 × 6	6 × 8	6 × 10	8 × 10
D	0.110	•	•	•								
E	0.196	•	•	•								
F	0.307	•	•	•								
G	0.503			•	•	•						
H	0.785				•	•						
J	1.287					•	•	•				
K	1.838							•				
L	2.853							•	•			
M	3.60								•			
N	4.34								•			
P	6.38								•			
Q	11.05									•		
R	16.0									•	•	
T	26.0											•

Valve Body Size (Inlet Diameter × Outlet Diameter), in.

7. Failure of Automatic Controls
8. Loss of Reflux
9. Chemical Reaction (this heat can sometimes exceed the heat of an external fire)

Plants, situations, and causes of overpressure tend to be dissimilar enough to discourage preparation of generalized calculation procedures for the rate of discharge. In lieu of a set procedure most of these problems can be solved satisfactorily by conservative simplification and analysis. It should be noted also that, by general assumption, two unrelated emergency conditions will not occur simultaneously.

The first three causes of overpressure on our list are more amenable to generalization than the others and will be discussed.

Fire

The heat input from fire is discussed in API RP 520 (Reference 3). One form of their equation for liquid containing vessels is

$$Q = 21{,}000\ FA_w^{0.82}$$

where

Q = Heat absorption, Btu/hr
A_w = Total wetted surface, ft^2
F = Environment factor

The environmental factors represented by F are

Bare vessel = 1.0
Insulated = 0.3/insulation thickness, in.
Underground storage = 0.0
Earth covered above grade = 0.03

The height above grade for calculating wetted surface should be

1. For vertical vessels—at least 25 feet above grade or other level at which a fire could be sustained.
2. For horizontal vessels—at least equal to the maximum diameter.
3. For spheres or spheroids—whichever is greater, the equator or 25 feet.

Heat Exchanger Tube Failure

1. Use the fluid entering from twice the cross section of one tube as stated in API RP 520 (Reference 3) (one tube cut in half exposes two cross sections at the cut).[3]
2. Use $\Delta h = u^2/2g$ to calculate leakage. Since this acts similar to an orifice, we need a coefficient; use 0.7. So,

$$u = 0.7\ \sqrt{2g\Delta h}$$

For compressible fluids, if the downstream head is less than ½ the upstream head, use ½ the upstream head as Δh. Otherwise use the actual Δh.

Liquid Expansion

The following equation can be used for sizing relief valves for liquid expansion.

$$Q = \frac{BH}{500\ GC} \qquad \text{(API RP 520, Reference 3)}$$

where

Q = Required capacity, gpm
H = Heat input, Btu/hr
B = Coefficient of volumetric expansion per °F:
 = 0.0001 for water
 = 0.0010 for light hydrocarbons
 = 0.0008 for gasoline
 = 0.0006 for distillates
 = 0.0004 for residual fuel oil
G = Specific gravity
C = Specific heat, Btu/lb °F

Rules of Thumb for Safety Relief Valves

1. Check metallurgy for light hydrocarbons flashing during relief. Very low temperatures can be produced.
2. Always check for reaction force from the tailpipe.
3. Hand jacks are a big help on large relief valves for several reasons. One is to give the operator a chance to reseat a leaking relief valve.
4. Flat seated valves have an advantage over bevel seated valves if the plant forces have to reface the surfaces (usually happens at midnight).
5. The maximum pressure from an explosion of a hydrocarbon and air is 7 × initial pressure, unless it occurs in a long pipe where a standing wave can be set up. It may be cheaper to design some small vessels to withstand an explosion than to provide a safety relief system. It is typical to specify ¼″ as minimum plate thickness (for carbon steel only).

Sources

1. Rearick, "How to Design Pressure Relief Systems," Parts I and II, *Hydrocarbon Processing,* August/September 1969.
2. ASME Boiler and Pressure Vessel Code, Sections I and VIII.
3. Recommended Practice for the Design and Installation of Pressure Relieving Systems in Refineries, Part I—"Design," latest edition, Part II—"Installation," latest edition RP 520 American Petroleum Institute.
4. Isaacs, Marx, "Pressure Relief Systems," *Chemical Engineering,* February 22, 1971.

2
Heat Exchangers

TEMA

Nomenclature

Shell and tube heat exchangers are designated by front head type, shell type, and rear head type as shown in Figures 1-4 and Table 1 from the Standards of Tubular Exchanger Manufacturers Association (TEMA).

(text continued on page 23)

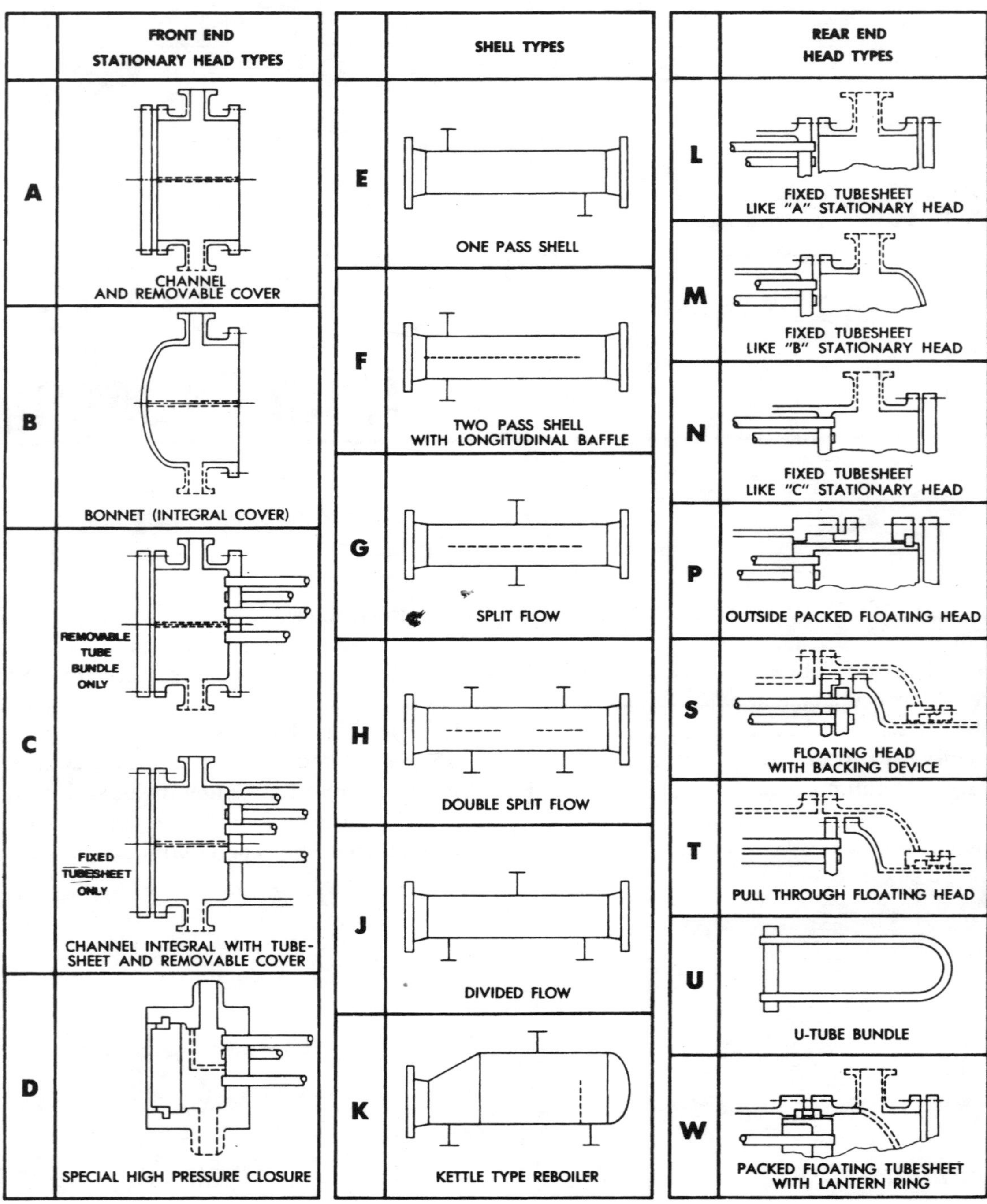

Figure 1. Heat exchangers.

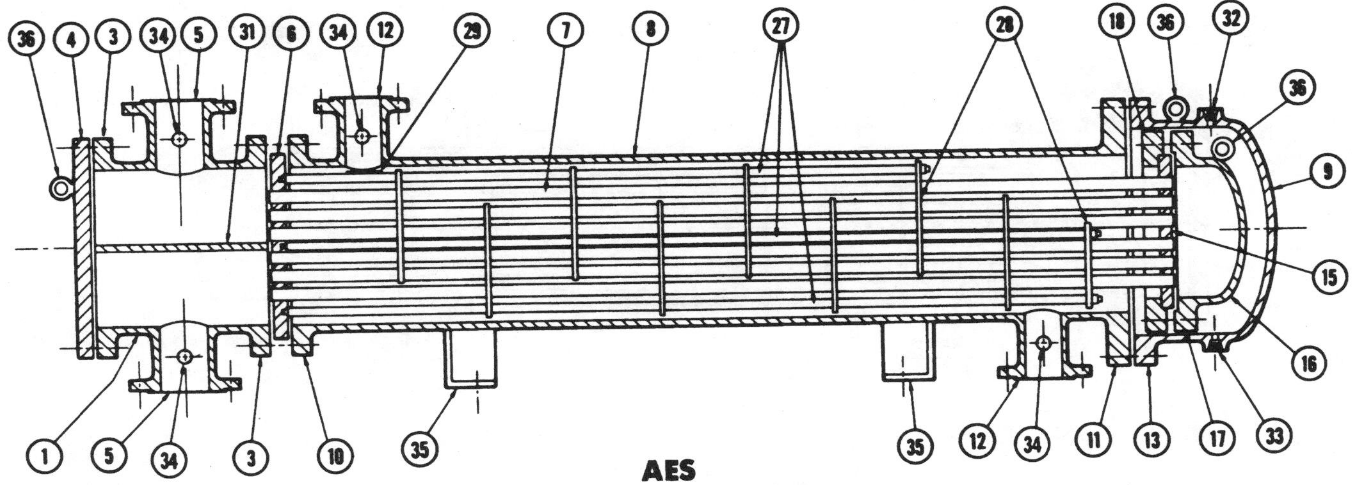

Figure 2. Type AES.

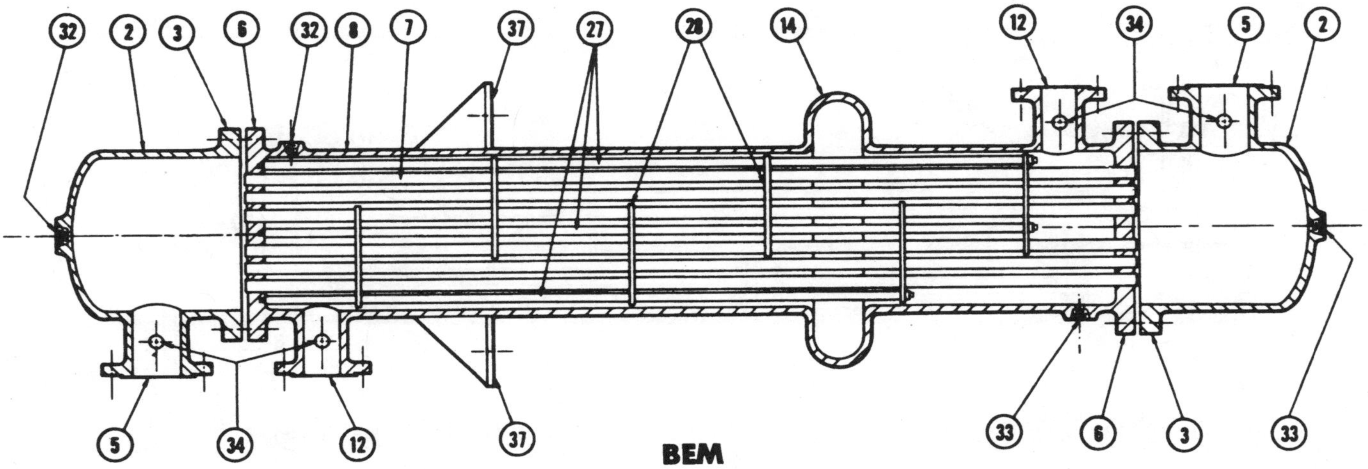

BEM

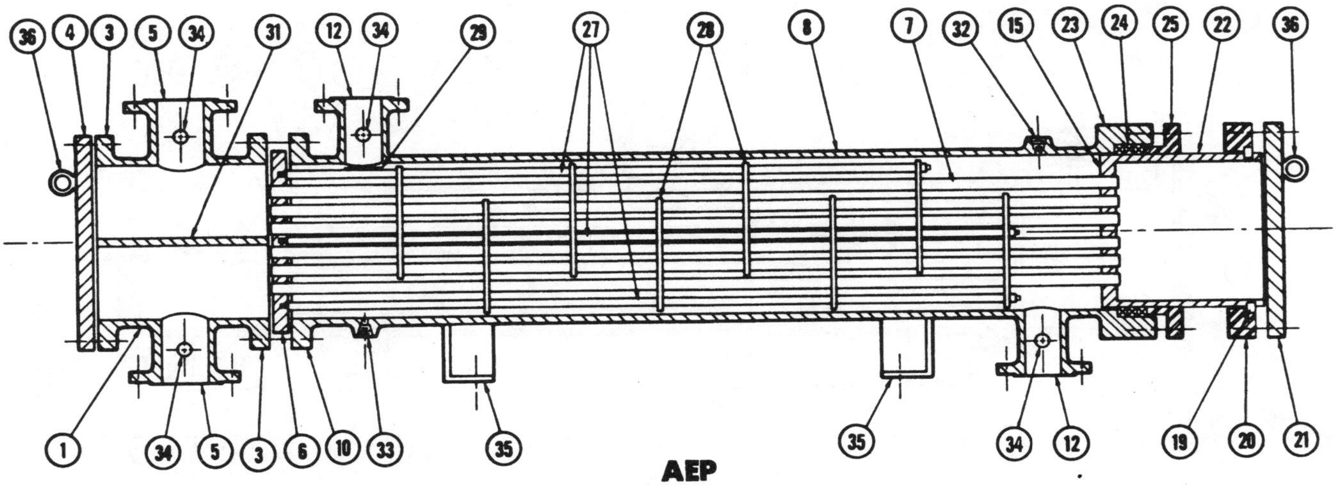

AEP

Figure 3. Types BEM, AEP, CFU.

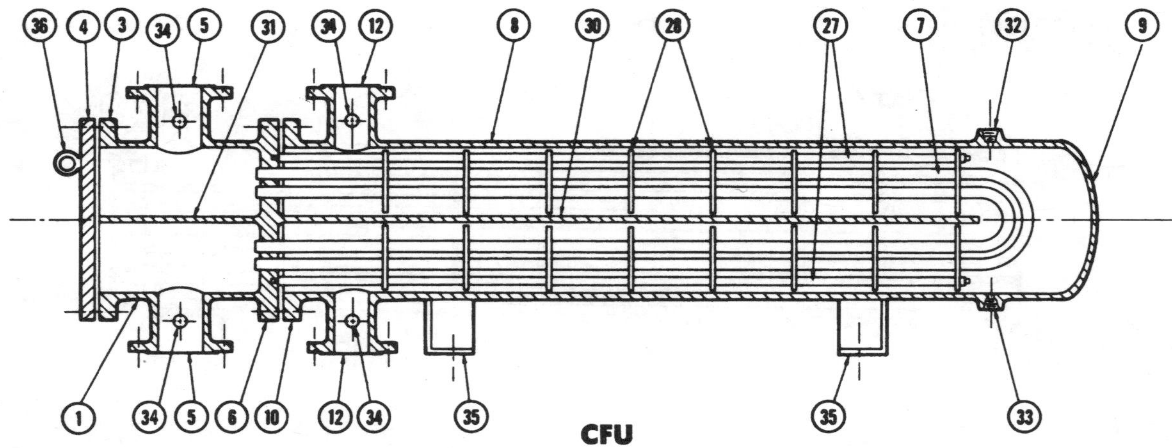

Figure 3. Continued.

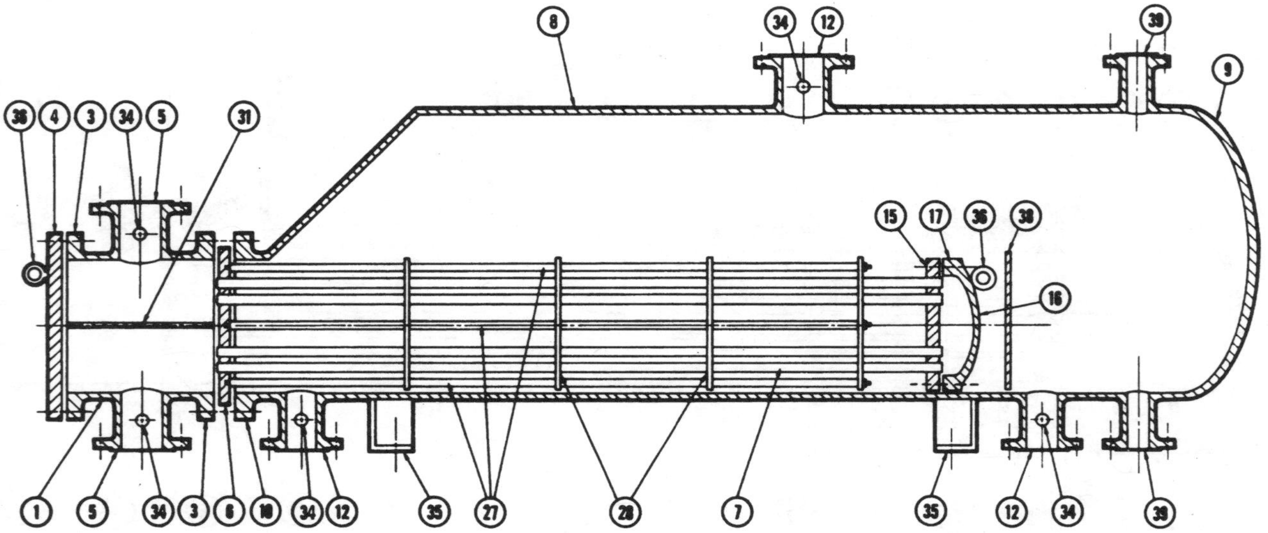

AKT

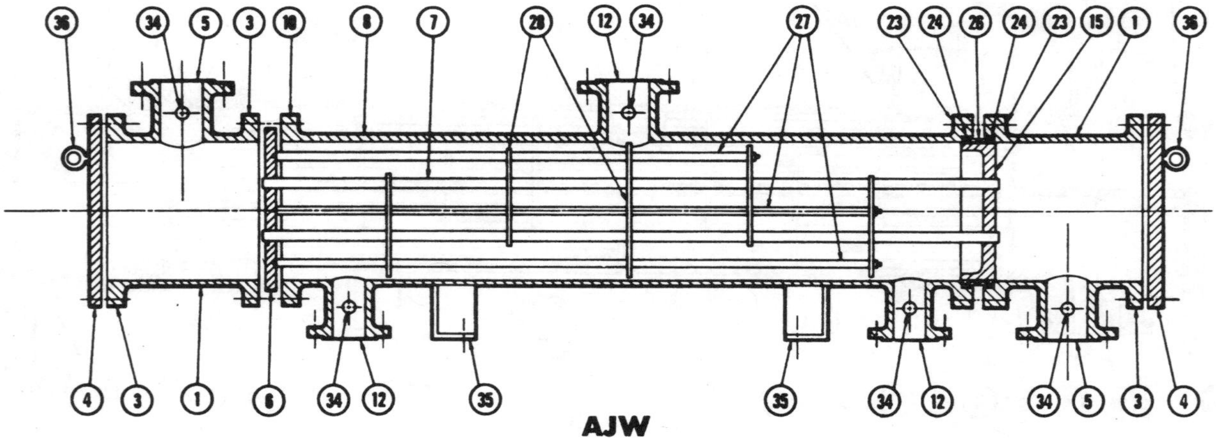

AJW

Figure 4. Types AKT and AJW.

Table 1
Typical Heat Exchanger Parts and Connections

1. Stationary Head—Channel
2. Stationary Head—Bonnet
3. Stationary Head Flange—Channel or Bonnet
4. Channel Cover
5. Stationary Head Nozzle
6. Stationary Tubesheet
7. Tubes
8. Shell
9. Shell Cover
10. Shell Flange—Stationary Head End
11. Shell Flange—Rear Head End
12. Shell Nozzle
13. Shell Cover Flange
14. Expansion Joint
15. Floating Tubesheet
16. Floating Head Cover
17. Floating Head Flange
18. Floating Head Backing Device
19. Split Shear Ring
20. Slip-on Backing Flange
21. Floating Head Cover—External
22. Floating Tubesheet Skirt
23. Packing Box Flange
24. Packing
25. Packing Follower Ring
26. Lantern Ring
27. Tie Rods and Spacers
28. Transverse Baffles or Support Plates
29. Impingement Baffle
30. Longitudinal Baffle
31. Pass Partition
32. Vent Connection
33. Drain Connection
34. Instrument Connection
35. Support Saddle
36. Lifting Lug
37. Support Bracket
38. Weir
39. Liquid Level Connection

(text continued from page 20)

Classes

Table 2 compares TEMA classes R, C, and B.

Table 2*
TEMA Standards—1978
Comparison of Classes R, C, & B

Paragraph	Topic	R	C	B
1.12	Definition	for the generally severe requirements of petroleum and related processing applications.	for the generally moderate requirements of commercial and general process applications.	for general process service.
1.51	Corrosion allowance on carbon steel	⅛ inch	1⁄16 inch	1⁄16 inch
2.2	Tube diameters	¾, 1, 1¼, 1½, and 2 inch od	R + ¼, ⅜, ½, and ⅝	R + ⅝
2.5	Tube pitch and minimum cleaning lane	1.25 × tube od. ¼ inch lane.	R + ⅝ tubes may be located 1.2 × tube od	R + lane may be 3⁄16 inch in 12 inch and smaller shells for ⅝ and ¾ tubes.
3.3	Minimum shell diameter	8 inch tabulated	6 inch tabulated	6 inch tabulated.
4.42	Longitudinal baffle thickness	¼ inch minimum	⅛ inch alloy, ¼ inch CS	⅛ inch alloy, ¼ inch carbon steel
4.71	Minimum tie rod diameter	⅜ inch	¼ inch in 6–15 inch shells	¼ inch 6–15 inch shells.
5.11	Floating head cover cross-over area	1.3 times tube flow area	Same as tube flow area	Same as tube flow area
5.31	Lantern ring construction	375°F maximum. 300 psi up to 24 inch diam shell 150 psi for 25–42 inch shells 75 psi for 43–60 inch shells	600 psi maximum.	(same as TEMA R)

** By permission. Rubin, F. L.*

Table 2* Continued
TEMA Standards—1978
Comparison of Classes R, C, & B

Paragraph	Topic	R	C	B
6.2	Gasket materials	Metal jacketed or solid metal for (a) internal floating head cover. (b) 300 psi and up. (c) all hydrocarbons.	Metal jacketed or solid metal (a) internal floating head. (b) 300 psi and up. Asbestos permitted for 300 psi and lower pressures.	(same as TEMA C)
6.32	Peripheral gasket contact surface	Flatness tolerance specified.	No tolerance specified.	No tolerance specified.
7.131	Minimum tubesheet thickness with expanded tube joints	Outside diameter of the tube.	0.75 × tube od for 1 inch and smaller. ⅞ inch for 1¼ od 1 inch for 1½ od 1.25 inch for 2 od	(same as TEMA C)
7.44	Tube Hole Grooving	Two grooves	Above 300 psi design pressure: above 350°F design temp.-2 grooves	(Same as TEMA R)
7.51	Length of expansion	Smaller of 2 inch or tubesheet thickness	Smaller of 2 × tube od or 2″	(same as TEMA R)
7.7	Tubesheet pass partition grooves	3/16 inch deep grooves required	Over 300 psi 3/16 inch deep grooves required or other suitable means for retaining gaskets in place	(same as TEMA C)
9.3	Pipe Tap Connections	6000 psi coupling with bar stock plug	3000 psi coupling	3000 psi coupling with bar stock plug
9.32	Pressure Gage Connections	required in nozzles 2 inch & up.	(shall be specified by purchaser)	(same as TEMA R)
9.33	Thermometer Connections	required in nozzles 4 inch & up.	(shall be specified by purchaser)	(same as TEMA R)
9.1	Nozzle construction	no reference to flanges	same as TEMA R	All nozzles larger than one inch must be flanged.
10.1	Minimum bolt size	¾ inch	½ inch recommended. smaller bolting may be used	⅝ inch

** By permission. Rubin, F. L.*

Sources

1. Standards of Tubular Exchanger Manufacturers Association (TEMA), 7th Edition.
2. Rubin, F. L. "What's the Difference Between TEMA Exchanger Classes," *Hydrocarbon Processing,* 59, June 1980, p. 92.
3. Ludwig, E. E., *Applied Process Design For Chemical and Petrochemical Plants, 2nd Ed., Vol. 3,* Gulf Publishing Co.

Selection Guides

Here are two handy shell and tube heat exchanger selection guides from Ludwig[1] and GPSA[2].

Table 1
Selection Guide Heat Exchanger Types

Type Designation	Significant Feature	Applications Best Suited	Limitations	Relative Cost in Carbon Steel Construction
Fixed Tube Sheet	Both tube sheets fixed to shell	Condensers; liquid-liquid; gas-gas; gas-liquid; cooling and heating, horizontal or vertical, reboiling	Temperature difference at extremes of about 200°F. Due to differential expansion	1.0
Floating Head or Tube Sheet (Removable and non-removable bundles)	One tube sheet "floats" in shell or with shell, tube bundle may or may not be removable from shell, but back cover can be removed to expose tube ends.	High temperature differentials, above about 200°F. extremes; dirty fluids requiring cleaning of inside as well as outside of shell, horizontal or vertical.	Internal gaskets offer danger of leaking. Corrosiveness of fluids on shell side floating parts. Usually confined to horizontal units.	1.28
U-Tube; U-Bundle	Only one tube sheet required. Tubes bent in U-shape. Bundle is removable.	High temperature differentials which might require provision for expansion in fixed tube units. Clean service or easily cleaned conditions on both tube side and shell side. Horizontal or vertical.	Bends must be carefully made or mechanical damage and danger of rupture can result. Tube side velocities can cause erosion of inside of bends. Fluid should be free of suspended particles.	1.08
Kettle	Tube bundle removable as U-type or floating head. Shell enlarged to allow boiling and vapor disengaging.	Boiling fluid on shell side, as refrigerant, or process fluid being vaporized. Chilling or cooling of tube side fluid in refrigerant evaporation on shell side.	For horizontal installation. Physically large for other applications.	1.2–1.4
Double Pipe	Each tube has own shell forming annular space for shell side fluid. Usually use externally finned tube.	Relatively small transfer area service, or in banks for larger applications. Especially suited for high pressures in tube above 400 psig.	Services suitable for finned tube. Piping-up a large number often requires cost and space.	0.8–1.4
Pipe Coil	Pipe coil for submersion in coil-box of water or sprayed with water is simplest type of exchanger.	Condensing, or relatively low heat loads on sensible transfer.	Transfer coefficient is low, requires relatively large space if heat load is high.	0.5–0.7
Open Tube Sections (Water cooled)	Tubes require no shell, only end headers, usually long, water sprays over surface, sheds scales on outside tubes by expansion and contraction. Can also be used in water box.	Condensing, relatively low heat loads on sensible transfer.	Transfer coefficient is low, takes up less space than pipe coil.	0.8–1.1
Open Tube Sections (Air Cooled) Plain or finned tubes	No shell required, only end heaters similar to water units.	Condensing, high level heat transfer.	Transfer coefficient is low, if natural convection circulation, but is improved with forced air flow across tubes.	0.8–1.8

Table 1 Continued
Selection Guide Heat Exchanger Types

Type Designation	Significant Feature	Applications Best Suited	Limitations	Relative Cost in Carbon Steel Construction
Plate and Frame	Composed of metal-formed thin plates separated by gaskets. Compact, easy to clean.	Viscous fluids, corrosive fluids slurries, High heat transfer.	Not well suited for boiling or condensing; limit 350-500°F by gaskets. Used for Liquid-Liquid only; not gas-gas.	0.8–1.5
Spiral	Compact, concentric plates; no bypassing, high turbulence.	Cross-flow, condensing, heating.	Process corrosion, suspended materials.	0.8–1.5
Small-tube Teflon	Chemical resistance of tubes; no tube fouling.	Clean fluids, condensing, cross-exchange.	Low heat transfer coefficient.	2.0–4.0

Table 2
Shell and Tube Exchanger Selection Guide
(Cost Increases from Left to Right)

Type of Design	"U" Tube	Fixed Tubesheet	Floating Head Outside Packed	Floating Head Split Backing Ring	Floating Head Pull-Through Bundle
Provision for differential expansion	Individual tubes free to expand	expansion joint in shell	floating head	floating head	floating head
Removable bundle	yes	no	yes	yes	yes
Replacement bundle possible	yes	not practical	yes	yes	yes
Individual tubes replaceable	only those in outside row	yes	yes	yes	yes
Tube interiors cleanable	difficult to do mechanically can do chemically	yes, mechanically or chemically	yes, mechanically or chemically	yes, mechanically or chemically	yes, mechanically or chemically
Tube exteriors with triangular pitch cleanable	chemically only	chemically only	chemically only	chemically only	chemically only
Tube exteriors with square pitch cleanable	yes, mechanically or chemically	chemically only	yes, mechanically or chemically	yes, mechanically or chemically	yes, mechanically or chemically
Number of tube passes	any practical even number possible	normally no limitations	normally no limitations	normally no limitations	normally no limitations
Internal gaskets eliminated	yes	yes	yes	no	no

Sources

1. Ludwig, E. E., *Applied Process Design for Chemical and Petrochemical Plants, 2nd Ed., Vol. 3,* Gulf Publishing Co.

2. *GPSA Engineering Data Book,* Gas Processors Suppliers Association, 10th Ed. 1987.

Pressure Drop Shell and Tube

Tubeside Pressure Drop

This pressure drop is composed of several parts which are calculated as shown in Tables 1 and 2.

Table 1
Calculation of Tubeside Pressure Drop in Shell and Tube Exchangers

Part	Pressure Drop in Number of Velocity Heads	Equation
Entering plus exiting the exchanger	1.6	$\Delta h = 1.6 \frac{U_p^2}{2g}$ (This term is small and often neglected)
Entering plus exiting the tubes	1.5	$\Delta h = 1.5 \frac{U_T^2}{2g} N$
End losses in tubeside bonnets and channels	1.0	$\Delta h = 1.0 \frac{U_T^2}{2g} N$
Straight tube loss	See Chapter 1, Fluid Flow, Piping Pressure Drop	

Δh = Head loss in feet of flowing fluid
U_p = Velocity in the pipe leading to and from the exchanger, ft/sec
U_T = Velocity in the tubes
N = Number of tube passes

Table 2
Calculation of Tubeside Pressure Drop in Air-Cooled Exchangers

Part	Pressure Drop In Approximate Number of Velocity Heads	Equation
All losses except for straight tube	2.9	$\Delta h = 2.9 \frac{U_T^2}{2g} N$
Straight tube loss	See Chapter 1, Fluid Flow, Piping Pressure Drop	

Table 3
Tube Pattern Relationships

	Triangular	Square Inline	Square Staggered
a	a = c	a = c	a = 1.414c
b	b = 0.866c	b = c	b = 0.707c

Shellside Pressure Drop

Tube Patterns

With segmental baffles, where the shellside fluid flows across the tube bundle between baffles, the following tube patterns are usual:

1. Triangular—Joining the centers of 3 adjacent tubes forms an equilateral triangle. Any side of this triangle is the tube pitch c.
2. Square inline—Shellside fluid has straight lanes between tube layers, unlike triangular where alternate tube layers are offset. This pattern makes for easy cleaning since a lance can be run completely through the bundle without interference. This pattern has less pressure drop than triangular but shell requirements are larger and there is a lower heat transfer coefficient for a given velocity at many velocity levels. Joining the centers of 4 adjacent tubes forms a square. Any side of this square is the tube pitch c.
3. Square staggered, often referred to as square rotated—Rotating the square inline pitch 45° no longer gives the shellside fluid clear lanes through the bundle. Tube pitch c is defined as for square inline.

Two other terms need definition: transverse pitch a and longitudinal pitch b. For a drawing of these dimensions see the source article. For our purposes appropriate lengths are shown in Table 3.

Turbulent Flow

For turbulent flow across tube banks, a modified Fanning equation and modified Reynold's number are given.

$$\Delta P_f = \frac{4f'' N_R N_{sp} \rho U^2_{max}}{2g}$$

$$Re' = \frac{D_o U_{max} \rho}{\mu}$$

where

ΔP_f = Friction loss in lb/ft^3
f'' = Modified friction factor

N_R = Rows of tubes per shell pass (N_R is always equal to the number of minimum clearances through which the fluid flows in series. For square staggered pitch the maximum velocity, U_{max}, which is required for evaluating Re′ may occur in the transverse clearances a or the diagonal clearances c. In the latter case N_R is one less than the number of tube rows.
N_{sp} = Number of shell passes
ρ = Density, lb/ft^3

The modified friction factor can be determined by using Tables 4 and 5.

Table 4
Determination of f″ for 5 Tube Rows or More

C/D$_o$ Both in Same Length Units		Triangular		
Re′ × 10^{-3}→	2	8	20	40
1.25 (min)	0.210	0.155	0.130	0.107
1.50	.145	.112	.090	.074
2.00	.118	.096	.081	.066
3.00	.089	.076	.063	.052
		Square Inline		
Re′ × 10^{-3}→	2	8	20	40
1.25 (min)	0.139	0.135	0.116	0.099
1.50	.081	.079	.080	.071
2.00	.056	.057	.055	.053
3.00	.052	.050	.045	.038
		Square Staggered		
Re′ × 10^{-3}→	2	8	20	40
1.25 (min)	0.130	0.106	0.088	0.063
1.50	.125	.103	.079	.061
2.00	.108	.090	.071	.058
3.00	—	—	—	—

Table 5
f″ Correction Factor For Less Than 5 Tube Rows

Number of Rows	1	2	3	4
Correction Factor	1.30	1.30	1.15	1.07

U_{max} = Maximum linear velocity (through minimum cross-sectional area), ft/sec
g = 32.2 ft/sec^2
Re′ = Modified Reynold's number
D_o = Outside tube diameter, ft
μ = Viscosity, lb/ft sec; centipoises × 0.000672

An equation has been developed for five tube rows or more. For each C/D$_o$, the approximate general relationship is as follows:

$$f'' = Y/Re^{0.21}$$

The value of Y is tabulated as follows:

C/D$_o$	Type of Tube Pitch: Triangular	Square Inline	Square Rotated
1.25 (min.)	1.0225	0.8555	0.6571
1.50	0.7150	0.5547	0.6234
2.00	0.6188	0.3948	0.5580
3.00	0.4816	0.3246	

Laminar Flow

Below $\frac{D_c U_{max} \rho}{\mu} = 40$ where D_c is the tube clearance in feet, the flow is laminar. For this region use

$$\Delta P_f = \frac{1.68 \mu U_{max} L}{D_e^2}$$

where

L = Length of flow path, ft
D_e = Equivalent diameter, ft; 4 times hydraulic radius

$$D_e = 4 \frac{\text{(cross-sectional flow area)}}{\text{(wetted perimeter)}} = D_o\left(\frac{4ab}{\pi D_o^2} - 1\right)$$

Pressure Drop for Baffles

Previous equations determine the pressure drop across the tube bundle. For the additional drop for flow through the free area above, below, or around the segmental baffles use

$$\Delta P_f = \frac{W^2 N_B N_{sp}}{\rho S_B^2 g}$$

where

W = Flow in lb/sec
N_B = Number of baffles in series per shell pass
S_B = Cross-sectional area for flow around segmental baffle, ft^2

Flow Parallel to Tubes

For flow parallel to tubes or in an annular space, e.g., a double-pipe heat exchanger, use

$$\Delta P = \frac{2f\rho U^2 L}{gD_e}$$

where

ΔP = Pressure drop, lb/ft^2
f = Friction factor

Source

Scovill Heat Exchanger Tube Manual, Scovill Manufacturing Company, Copyright 1957.

Temperature Difference

Only countercurrent flow will be considered here. It is well known that the log mean temperature is the correct temperature difference to be used in the expression:

$$q = UA\Delta T_M$$

where

q = Heat duty in Btu/hr
U = Overall heat transfer coefficient in Btu/hr ft^2 °F
A = Tube surface area in ft^2
ΔT_M = Mean temperature difference in °F. For our case it is the log mean temperature difference.

$$\Delta T_M = \frac{GTD - LTD}{\ln(GTD/LTD)}$$

where

GTD = Greater temperature difference
LTD = Lesser temperature difference

When GTD/LTD < 2 the arithmetic mean is within about 2% of the log mean.

These refer to hot and cold fluid terminal temperatures, inlet of one fluid versus outlet of the other. For a cross exchanger with no phase change, the ΔT_M gives exact results for true countercurrent flow. Most heat exchangers, however, deviate from true countercurrent so a correction factor, F, is needed.

These correction factors are given in various heat transfer texts. In lieu of correction factor curves use the following procedure to derive the factor:

1. Assume shellside temperature varies linearly with length.
2. For first trial on tubeside assume equal heat is transferred in each pass with constant fluid heat capacity.
3. Using the end temperatures of each shell and tube pass calculate ΔT_M for each tube pass. From this the fraction of total duty for each tube pass is determined.
4. For the new end temperatures calculate the new ΔT_M for each tube pass.
5. The arithmetic average of the tube pass ΔT_M's is the ΔT_M corrected for number of passes. F = ΔT_M corrected/ΔT_M uncorrected.

The above procedure will quickly give numbers very close to the curves.

One thing to be careful of in cross exchangers is a design having a so-called "temperature cross." An example is shown in Figure 1.

In Figure 1, the colder fluid being heated emerges hotter than the outlet temperature of the other fluid. For actual heat exchangers that deviate from true countercurrent flow the following things can happen under temperature cross conditions:

1. The design can prove to be impossible in a single shell.
2. The correction factor can be quite low requiring an uneconomically large area.
3. The unit can prove to be unsatisfactory in the field if conditions change slightly.

For Figure 1, assuming one shell pass and two or more tube passes, the correction factor is roughly 0.7. This shows the undesirability of a temperature cross in a single shell pass.

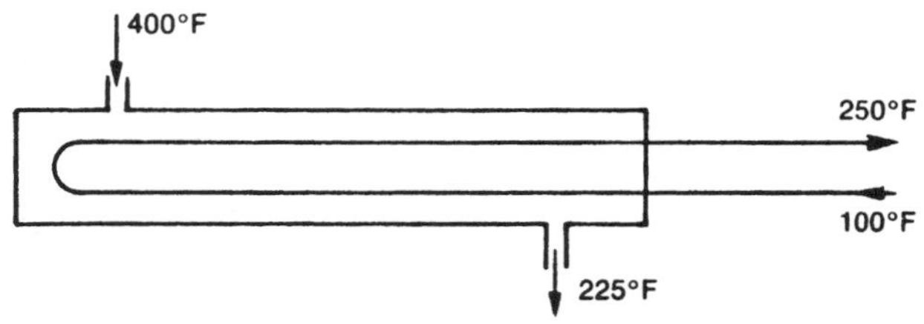

Figure 1. Shown here is an example of a temperature cross.

The calculation procedure for temperature correction factors won't work for a temperature cross in a single shell pass, but this is an undesirable situation anyway.

Some conditions require breaking up the exchanger into multiple parts for the calculations rather than simply using corrected terminal temperatures. For such cases one should always draw the q versus temperature plot to be sure no undesirable pinch points or even intermediate crossovers occur.

An example of a multisection calculation would be a propane condenser. The first section could be a desuperheating area, where q versus T would be a steeply sloped straight line followed by a condensing section with a straight line parallel to the q axis (condensing with no change in temperature). Finally, there could be a subcooling section with another sloped line. One can calculate this unit as three separate heat exchangers.

Source

Branan, C. R., *The Process Engineer's Pocket Handbook, Vol. 1,* Gulf Publishing Co., p. 55.

Shell Diameter

Determination of Shell and Tube Heat Exchanger Shell Diameter

For triangular pitch proceed as follows:

1. Draw the equilateral triangle connecting three adjacent tube centers. Any side of the triangle is the tube pitch (recall 1.25 D_o is minimum).
2. Triangle area is ½ bh where b is the base and h is the height.
3. This area contains ½ tube.
4. Calculate area occupied by all the tubes.
5. Calculate shell diameter to contain this area.
6. Add one tube diameter all the way around (two tube diameters added to the diameter calculated above).
7. The resulting is minimum shell diameter. There is no firm standard for shell diameter increments. Use 2-inch increments for initial planning.

For square pitch proceed similarly.

Source

Branan, C. R., *The Process Engineer's Pocket Handbook, Vol. 1,* Gulf Publishing Co., p. 54.

Shellside Velocity Maximum

This graph shows maximum shellside velocities; these are rule-of-thumb maximums for reasonable operation.

Source

Ludwig, E. E., *Applied Process Design For Chemical and Petrochemical Plants, 2nd Ed.,* Gulf Publishing Co.

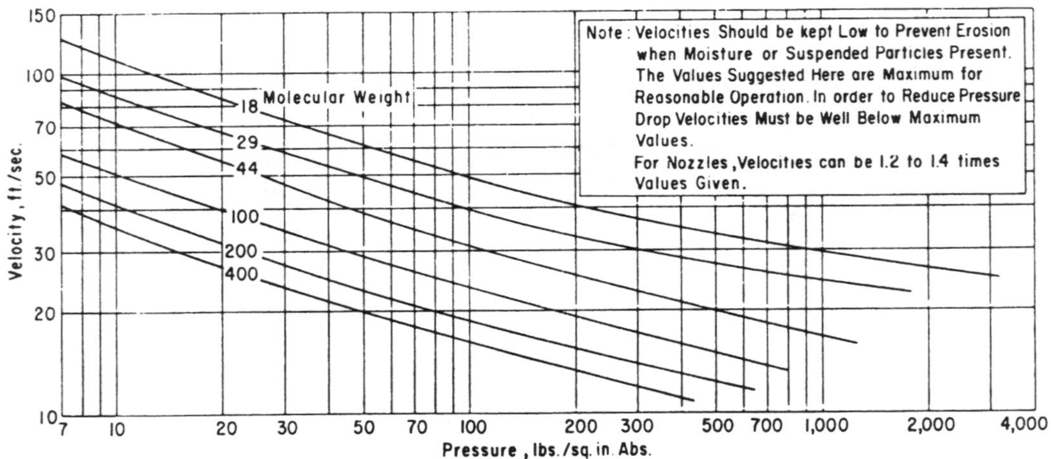

Figure 1. Maximum velocity for gases and vapors through heat exchangers on shell-side.

Nozzle Velocity Maximum

The following table gives suggested maximum velocities in exchanger nozzles. The pressure drop through nozzles should be checked, especially where pressure losses are a problem such as in low pressure systems.

Maximum Recommended VelocitiesThrough Nozzle Connections, Piping, etc. Associated With Shell and/or Tube Sides of Heat Exchanger

Viscosity in centipoise	Maximum velocity, ft/sec	Remarks
Liquids:		
Over 1500	2	Very heavy oils
1000–500	2.5	Heavy oils
500–100	2.5	Medium oils
100– 35	5	Light oils
35– 1	6	Light oils
Below 1	8	. . .
Vapors and Gases:		
Use 1.2 to 1.4 of value shown on Figure 1 in the previous section, shellside velocity maximum, for velocity through exchangers.		

Source

Ludwig, E. E., *Applied Process Design For Chemical and Petrochemical Plants, 2nd Ed.*, Gulf Publishing Co.

Heat Transfer Coefficients

Determining Heat Exchanger Heat Transfer Coefficient

To do this one must sum all the resistances to heat transfer. The reciprocal of this sum is the heat transfer coefficient. For a heat exchanger the resistances are

Tubeside fouling	R_{FT}
Shellside fouling	R_{FS}
Tube metal wall	R_{MW}
Tubeside film resistance	R_T
Shellside film resistance	R_S

For overall tubeside plus shellside fouling use experience factors or 0.002 for most services and 0.004 for extremely fouling materials. Neglect metal wall resistance for overall heat transfer coefficient less than 200 or heat flux less than 20,000. These will suffice for ballpark work.

For film coefficients many situations exist. Table 1 gives ballpark estimates of film resistance at reasonable design velocities.

For liquid boiling the designer is limited by a maximum flux q/A. This handbook cannot treat this subject in detail. For most applications assuming a limiting flux of 10,000 will give a ballpark estimate.

The literature has many tabulations of typical coefficients for commercial heat transfer services. A number of these follow.

(text continued on page 37)

Table 1
Film Resistances

Liquids	R
Water	0.0013
Gasoline	.0067
Gas Oils	.0115
Viscous Oils Heating	.0210
Viscous Oils Cooling	.0333
Organic Solvents	.0036
Gases	
Hydrocarbons	
Low Pressure	.0364
High Pressure	.0200
Air	
Low Pressure	.0500
High Pressure	.0250
Vapors Condensing	
Steam	
No Air	0.0006
10% Air by Vol.	.0010
20% Air by Vol.	.0040
Gasoline	
Dry	.0067
With Steam	.0044
Propanes, Butanes, Pentanes	
Pure	.0033
Mixed	.0067
Gas Oils	
Dry	.0133
With Steam	.0090
Organic Solvents	.0030
Light Oils	.0033
Heavy Oils (vacuum)	.0285
Ammonia	.0133
Evaporation	
Water	.0007
Organic Solvents	.0050
Ammonia	.0033
Light Oils	.0044
Heavy Oils	.0333

Table 2
Overall Coefficients in Typical Petrochemical Applications U, BTU/hr (sq.ft.) (°F.)

In Tubes	Outside Tubes	Type Equipment	Velocities Ft./Sec.		Overall Coef-ficient	Temp. Range, °F	Estimated Fouling		
			Tube	Shell			Tube	Shell	Overall
A. Heating-Cooling									
Butadiene mix. (Super-heating)	Steam	H	25–35	....	12	400–100	....	....	0.04
Solvent	Solvent	H	....	1.0–1.8	35–40	110–30	....	....	0.0065
Solvent	Propylene (vaporization)	K	1–2	....	30–40	40–0	....	....	0.006
C_4 Unsaturates	Propylene (vaporization)	K	20–40	....	13–18	100–35	....	....	0.005
Solvent	Chilled Water	H	....	....	35–75	115–40	0.003	0.001	....
Oil	Oil	H	....	....	60–85	150–100	0.0015	0.0015	....
Ethylene-vapor	Condensate & Vapor	K	....	....	90–125	600–200	0.002	0.001	....
Ethylene vapor	Chilled water	H	....	....	50–80	270–100	0.001	0.001	....
Condensate	Propylene (refrigerant)	K-U	....	....	60–135	60–30	0.001	0.001	....
Chilled water	Transformer oil	H	....	....	40–75	75–50	0.001	0.001	....
Calcium Brine-25%	Chlorinated C_1	H	1–2	0.5–1.0	40–60	–20–+10	0.002	0.005	....
Ethylene liquid	Ethylene vapor	K-U	....	....	10–20	–170–(–100)	....	....	0.002
Propane vapor	Propane liquid	H	....	....	6–15	–25–100	....	....	0.002
Lights & chlor. HC	Steam	U	....	....	12–30	–30–260	0.001	0.001	....
Unsat. light HC, CO, CO_2, H_2	Steam	H	....	....	10–2	400–100	....	....	0.3
Ethanolamine	Steam	H	....	....	15–25	400–40	0.001	0.001	....
Steam	Air mixture	U	....	....	10–20	–30–220	0.0005	0.0015	....
Steam	Styrene & Tars	U (in tank)	....	....	50–60	190–230	0.001	0.002	....
Chilled Water	Freon-12	H	4–7	....	100–130	90–25	0.001	0.001	....
Water*	Lean Copper Solvent	H	4–5	....	100–120	180–90	....	....	0.004
Water	Treated water	H	3–5	1–2	100–125	90–110	....	....	0.005
Water	C_2-chlor. HC, lights	H	2–3	....	6–10	360–100	0.002	0.001	....
Water	Hydrogen chloride	H	....	....	7–15	230–90	0.002	0.001	....
Water	Heavy C_2-Chlor.	H	....	....	45–30	300–90	0.001	0.001	....
Water	Perchlorethylene	H	....	....	55–35	150–90	0.001	0.001	....
Water	Air & Water Vapor	H	....	....	20–35	370–90	0.0015	0.0015	....
Water	Engine Jacket Water	H	....	....	230–160	175–90	0.0015	0.001	....
Water	Absorption Oil	H	....	....	80–115	130–90	0.0015	0.001	....
Water	Air-Chlorine	U	4–7	....	8–18	250–90	....	....	0.005
Water	Treated Water	H	5–7	....	170–225	200–90	0.001	0.001	....
B. Condensing									
C_4 Unsat	Propylene refrig.	K	v	....	58–68	60–35	....	....	0.005
HC Unsat. lights	Propylene refrig.	K	v	....	50–60	45–3	....	....	0.0055
Butadiene	Propylene refrig.	K	v	....	65–80	20–35	....	....	0.004
Hydrogen Chloride	Propylene refrig.	H	....	....	110–60	0–15	0.012	0.001	....
Lights & Chloro-ethanes	Propylene refrig.	KU	....	....	15–25	130–(–20)	0.002	0.001	....
Ethylene	Propylene refrig.	KU	....	....	60–90	120–(–10)	0.001	0.001	....
Unsat. Chloro HC	Water	H	7–8	....	90–120	145–90	0.002	0.001	....
Unsat. Chloro HC	Water	H	3–8	....	180–140	110–90	0.001	0.001	....
Unsat. Chloro HC	Water	H	6	....	15–25	130–(–20)	0.002	0.001	....
Chloro-HC	Water	KU	....	....	20–30	110–(–10)	0.001	0.001	....
Solvent & Non Cond	Water	H	....	....	25–15	260–90	0.0015	0.004	....
Water	Propylene Vapor	H	2–3	....	130–150	200–90	....	....	0.003
Water	Propylene	H	....	....	60–100	130–90	0.0015	0.001	....
Water	Steam	H	....	....	225–110	300–90	0.002	0.0001	....

**Unless specified, all water is untreated, brackish, bay, or sea.*

Notes: H = Horizontal, Fixed or Floating Tube Sheet.
U = U—Tube Horizontal Bundle.
K = Kettle Type.
V = Vertical.
R = Reboiler.
T = Thermosiphon.
v = Variable.
HC = Hydrocarbon.
(C) = Cooling range Δt.
(Co) = Condensing range Δt.

(table continued on next page)

Table 2 Continued
Overall Coefficients in Typical Petrochemical Applications U, BTU/hr (sq.ft.) (°F.)

In Tubes	Outside Tubes	Type Equipment	Velocities Ft./Sec.		Overall Coef- ficient	Temp. Range, °F	Estimated Fouling		
			Tube	Shell			Tube	Shell	Overall
Water	Steam	H	....	....	190–235	230–130	0.0015	0.001	....
Treated Water	Steam (Exhaust)	H	....	....	20–30	220–130	0.0001	0.0001	....
Oil	Steam	H	....	....	70–110	375–130	0.003	0.001	....
Water	Propylene Cooling & Cond.	H	....	....	25–50 110–150	30–45 (C) 15–20 (Co)	0.0015	0.001	....
Chilled Water	Air-Chlorine (Part. Cond)	U	....	....	8–15 20–30	8–15 (C) 10–15 (C0)	0.0015	0.005	....
Water	Light HC, Cool & Cond.	H	....	....	35–90	270–90	0.0015	0.003	....
Water	Ammonia	H	....	....	140–165	120–90	0.001	0.001	....
Water	Ammonia	U	....	....	280–300	110–90	0.001	0.001	....
Air-Water Vapor	Freon	KU	....	....	10–50 10–20	60–10	....	....	0.01
C. Reboiling									
Solvent, Copper-NH_3	Steam	H	7–8	....	130–150	180–160	....	....	0.005
C_4 Unsat	Steam	H	....	....	95–115	95–150	....	....	0.0065
Chloro. HC	Steam	VT	....	....	35–25	300–350	0.001	0.001	....
Chloro. Unsat. HC	Steam	VT	....	....	100–140	230–130	0.001	0.001	....
Chloro. ethane	Steam	VT	....	....	90–135	300–350	0.001	0.001	....
Chloro. ethane	Steam	U	....	....	50–70	30–190	0.002	0.001	....
Solvent (heavy)	Steam	H	....	....	70–115	375–300	0.004	0.0005	....
Mono-di-ethanolamines	Steam	VT	....	....	210–155	450–350	0.002	0.001	....
Organics, acid, water	Steam	VT	....	....	60–100	450–300	0.003	0.0005	....
Amines and water	Steam	VT	....	....	120–140	360–250	0.002	0.0015	....
Steam	Naphtha frac.	Annulus, Long, F.N.	....	....	15–20	270–220	0.0035	0.0005	....
Propylene	C_2, $C_2^{=}$	KU	....	....	120–140	150–40	0.001	0.001	....
Propylene-Butadiene	Butadiene, Unsat.	H	....	25–35	15–18	400–100	....	....	0.02

**Unless specified, all water is untreated, brackish, bay, or sea.*

Notes: *H* = *Horizontal, Fixed or Floating Tube Sheet.*
U = *U—Tube Horizontal Bundle.*
K = *Kettle Type.*
V = *Vertical.*
R = *Reboiler.*
T = *Thermosiphon.*
v = *Variable*
HC = *Hydrocarbon.*
(C) = *Cooling range* Δt.
(Co) = *Condensing range* Δt.

Table 3
Approximate Overall Heat Transfer Coefficient, U*

Use as guide as to order of magnitude and not as limits to any value. Coefficients of actual equipment may be smaller or larger than the values listed.

Hot Fluid	Cold Fluid	U, BTU/Hr. (Sq. Ft.) (°F.)
Condensing		
Steam (pressure)	Water	350–750
Steam (vacuum)	Water	300–600
Saturated organic solvents near atmospheric	Water	100–200
Saturated organic solvents, vacuum with some non-cond	Water, brine	50–120
Organic solvents, atmospheric and high non-condensable	Water, brine	20–80
Aromatic vapors, atmospheric with non-condensables	Water	5–30
Organic solvents, vacuum and high non-condensables	Water, brine	10–50
Low boiling atmospheric	Water	80–200
High boiling hydrocarbon, vacuum	Water	10–30
Heaters		
Steam	Water	250–750
Steam	Light oils	50–150
Steam	Heavy oils	10–80
Steam	Organic solvents	100–200
Steam	Gases	5–50
Dowtherm	Gases	4–40
Dowtherm	Heavy oils	8–60
Flue gas	Aromatic HC and Steam	5–15
Evaporators		
Steam	Water	350–750
Steam	Organic solvents	100–200
Steam	Light oils	80–180
Steam	Heavy oils (vacuum)	25–75
Water	Refrigerants	75–150
Organic solvents	Refrigerants	30–100
Heat Exchangers (no change of phase)		
Water	Water	150–300
Organic solvents	Water	50–150
Gases	Water	3–50
Light oils	Water	60–160
Heavy oils	Water	10–50
Organic solvents	Light oil	20–70
Water	Brine	100–200
Organic solvents	Brine	30–90
Gases	Brine	3–50
Organic solvents	Organic solvents	20–60
Heavy oils	Heavy oils	8–50

**By permission, The Pfaudler Co., Rochester, N.Y., Bulletin 949.*

Table 4
Approximate Overall Heat Transfer Coefficient, U

Condensation		
Process Side (Hot)	**Condensing Fluid (Cold)**	
Hydrocarbons (light)	Water	100–160
Hydrocarbons w/inerts (traces)	Water	30–75
Organic Vapors	Water	70–160
Water Vapor	Water	150–340
Water Vapor	Hydrocarbons	60–150
Exhaust Steam	Water	280–450
Hydrocarbons (light)	Refrigerant	45–110
Organics (light)	Cooling brine	50–120
Gasoline	Water	65–130
Ammonia	Water	135–260
Hydrocarbons (heavy)	Water	40–75
Dowtherm Vapor	Liquid Organic	75–115
Vaporization		
Process side (Vaporized)	**Heating Fluid**	
Hydrocarbons, light	Steam	90–210
Hydrocarbons, C4-C8	Steam	75–150
Hydrocarbons, C3-C4 (vac)	Steam	45–175
Chlorinated HC	Steam	90–210
HCl Solution (18–22%)	Steam	120–240
Chlorine	Steam	130–220
Coils in Tank		
Coil Fluid	**Tank Fluid**	
Steam	Aqueous Sol'n (agitated)	140–210
Steam	Aqueous Sol'n (no agitation)	60–100
Steam	Oil-heavy (no agitation)	10–25
Steam	Oil-heavy (with agitation)	25–55
Steam	Organics (agitated)	90–140
Hot Water	Water (no agitation)	35–65
Hot Water	Water (agitation)	90–150
Hot Water	Oil-heavy (no agitation)	6–25
Heat Transfer Oil	Organics (agitation)	25–50
Salt Brine	Water (agitation)	50–110
Water (cooling)	Glycerine (agitation)	50–75

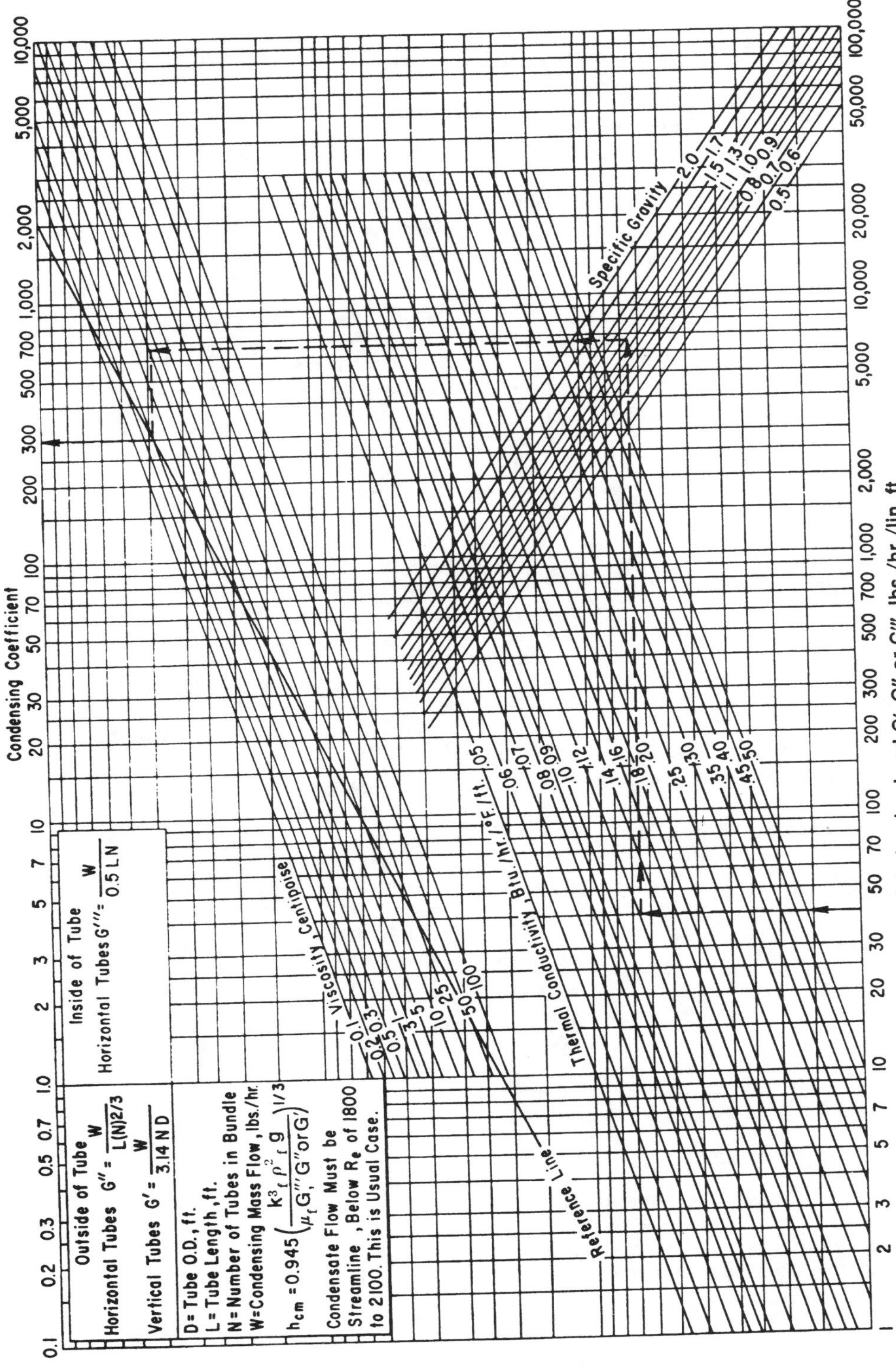

Figure 1. Condensing film coefficients. (By permission, D. Q. Kern, *Process Heat Transfer,* first edition, McGraw-Hill Book Co., 1950.)

Table 5
Typical Overall Heat-Transfer Coefficients for Air Coolers

Service		1 in. Fintube	
		½ in. by 9	⅝ in. by 10
1. Water & water solutions		U_b U_x	U_b U_x
Engine jacket water		(See Note Below)	
(r_d = 0.001)............................		110–7.5	130–6.1
Process water			
(r_d = 0.002)............................		95–6.5	110–5.2
50–50 ethylene glycol-water			
(r_d = 0.001)............................		90–6.2	105–4.9
50–50 ethylene glycol-water			
(r_d = 0.002)............................		80–5.5	95–4.4
2. Hydrocarbon liquid coolers			
	Viscosity, cp, at avg. temp.	U_b U_x	U_b U_x
	0.2	85–5.9	100–4.7
	0.5	75–5.2	90–4.2
	1.0	65–4.5	75–3.5
	2.5	45–3.1	55–2.6
	4.0	30–2.1	35–1.6
	6.0	20–1.4	25–1.2
	10.0	10–0.7	13–0.6
3. Hydrocarbon gas coolers			
	Pressure, psig	U_b U_x	U_b U_x
	50	30–2.1	35–1.6
	100	35–2.4	40–1.9
	300	45–3.1	55–2.6
	500	55–3.8	65–3.0
	750	65–4.5	75–3.5
	1000	75–5.2	90–4.2
4. Air and flue-gas coolers			
Use one-half of value given for hydrocarbon gas coolers.			
5. Steam condensers			
(Atmospheric pressure & above)		U_b U_x	U_b U_x
Pure steam (r_d = 0.0005)		125–8.6	145–6.8
Steam with non-condensibles		60–4.1	70–3.3
6. HC condensers	Condensing* Range, °F	U_b U_x	U_b U_x
	0° range	85–5.9	100–4.7
	10° range	80–5.5	95–4.4
	25° range	75–5.2	90–4.2
	60° range	65–4.5	75–3.5
	100° & over range	60–4.1	70–3.3
7. Other condensers		U_b U_x	U_b U_x
	Ammonia	110–7.6	130–6.1
	Freon 12	65–4.5	75–3.5

Notes: U_b is overall rate based on bare tube area, and U_x is overall rate based on extended surface.
Based on approximate air face mass velocities between 2600 and 2800 lb/(hr • sq ft of face area).
**Condensing range = hydrocarbon inlet temperature to condensing zone minus hydrocarbon outlet temperature from condensing zone.*

Table 6
Typical Transfer Coefficients for Air-Cooled Heat Exchangers

Condensing service	U Btu/hr, ft², °F
Amine reactivator	90–100
Ammonia	100–120
Freon 12	60–80
Heavy naphtha	60–70
Light gasoline	80
Light hydrocarbons	80–95
Light naphtha	70–80
Reactor effluent—Platformers, Rexformers, Hydroformers	60–80
Steam (0–20 psig)	130–140
Still overhead—light naphthas, steam and non-condensible gas	60–70
Gas cooling service	
Air or flue gas @ 50 psig. (ΔP = 1 psi.)	10
Air or flue gas @ 100 psig. (ΔP = 2 psi.)	20
Air or flue gas @ 100 psig. (ΔP = 5 psi.)	30
Ammonia reactor stream	80–90
Hydrocarbon gases @ 15–50 psig. (ΔP = 1 psi.)	30–40
Hydrocarbon gases @ 50–250 psig. (ΔP = 3 psi.)	50–60
Hydrocarbon gases @ 250–1,500 psig. (ΔP = 5 psi.)	70–90
Liquid cooling service	
Engine jacket water	120–130
Fuel oil	20–30
Hydroformer and Platformer liquids	70
Light gas oil	60–70
Light hydrocarbons	75–95
Light naphtha	70
Process water	105–120
Residuum	10–20
Tar	5–10

Coefficients are based on outside bare tube surface for 1-in. O.D. tubes with 8 extruded Al fins/in., ⅝ in. high, 16.9 surface ratio.

Table 7
Typical Transfer Coefficients For Air-Cooled Exchangers Based on Outside Bare Tube Surface

Condensing Service	U, BTU/hr. (Sq. Ft.) (°F.)
Amine Reactivator	90–100
Ammonia	100–120
Acetic acid at 14.7 psia	90*
Freon 12	60–80
Heavy Naphtha	60–70
Light Gasoline	80
Light Hydrocarbons, thru C4's	80–95
Light Naphtha	70–80
Reactor effluent—Platformers, Rexformers, Hydroformers	60–80
Steam (0–20 psig)	130–140, 145*
Still overhead—light naphthas, Steam and non-condensable gas	60–70
Organics, Low Boiling, @ 14.7 psia	105*
Methanol @ 15 psig	90*
Methanol @ 0 psig	95*
Gas Cooling Service	
Air or flue gas @ 50 psig (ΔP = 1 psi)	10
Air or flue gas @ 100 psig (ΔP = 2 psi)	20
Air or flue gas @ 100 psig (ΔP = 5 psi)	30
Ammonia reactor stream	80–90
Hydrocarbon gases @ 15–50 psig (ΔP = 1 psi)	30–40
Hydrocarbon gases @ 50–250 psig (ΔP = 3 psi)	50–60
Hydrocarbon gases @ 250–1,500 psig (ΔP = 5 psi)	70–90
Liquid Cooling Service	
Engine Jacket Water	120–134
Fuel oil	120–30
Hydroformer and Platformer liquids	70
Light gas oil	60–70
Light hydrocarbons, C3–C6	75–95
Light Naphtha	70
Process water	105–120
Residuum	10–20
Tar	5–10

**Based on inside of tube surface.*

Table 8
Overall Transfer Rates For Air-Cooled Heat Exchanger

Service	* Stab Transfer Rate	**Suggested No. of Tube Layers
Cooling Service		
Engine Jacket Water	6–7	4
Light Hydrocarbons	4–5	4 or 6
Light Gas Oil	3–4	4 or 6
Heavy Gas Oil	2.5–3	4 or 6
Lube Oil	1–2	4 or 6
Bottoms	0.75–1.5	6 or more
Flue Gas @ 100 psig & 5 psi ΔP	2–2.5	4
Condensing Service		
Steam	7–8	4
Light Hydrocarbon	4–5	4 or 6
Reactor Effluent	3–4	6
Still Overhead	2.75–3.5	4 or 6

**Transfer rate, BTU/(hr.) (sq. ft.) (°F.), based on outside fin tube surface for 1″ O.D. tubes with ⅝″ high aluminum fins spaced 11 per inch.*

***The suggested number of tube layers cannot be accurately predicted for all services. In general, coolers having a cooling range up to 80°F. and condensers having a condensing range up to 50°F. are selected with 4 tube layers. Cooling and condensing services with ranges exceeding these values are generally figured with 6 tube layers.*

Courtesy Griscom-Russell Co.

(text continued from page 31)

Sources

1. Branan, C. R., *The Process Engineer's Pocket Handbook, Vol. 1,* Gulf Publishing Co. (Table 1).
2. Ludwig, E. E., *Applied Process Design For Chemical and Petrochemical Plants, 2nd Ed., Vol. 3,* Gulf Publishing Co. (Tables 2, 4 and 7).
3. Bulletin 949, The Pfaudler Co., Rochester, N. Y., (Table 3).
4. Kern, D. Q., *Process Heat Transfer, 1st Ed.,* McGraw-Hill Book Co. 1950. (Figure 1).
5. *GPSA Engineering Data Book,* Gas Processors Suppliers Association, 10th Ed., 1987. (Table 5).
6. Smith, Ennis C., Cooling With Air-Technical Data Relevant to Direct Use of Air for Process Cooling, Hudson Products Corporation. (Table 6).
7. Griscom-Russell Co. data. (Table 8).

Fouling Resistances

Tables 1 and 2 have fouling resistances suggested by TEMA[1]. In Table 3 they are from Ludwig[2].

Table 1
Fouling Resistances for Water

Temperature of Heating Medium	Up to 240° F.		240°-400° F.*	
Temperature of Water	125° F. or Less		Over 125° F.	
Types of Water	Water Velocity Ft./Sec.		Water Velocity Ft./Sec.	
	3 and Less	Over 3	3 and Less	Over 3
Sea Water	.0005	.0005	.001	.001
Brackish Water	.002	.001	.003	.002
Cooling Tower and Artificial Spray Pond:				
Treated Makeup	.001	.001	.002	.002
Untreated	.003	.003	.005	.004
City or Well Water (Such as Great Lakes)	.001	.001	.002	.002
River Water:				
Minimum	.002	.001	.003	.002
Average	.003	.002	.004	.003
Muddy or Silty	.003	.002	.004	.003
Hard (Over 15 grains/gal.)	.003	.003	.005	.005
Engine Jacket	.001	.001	.001	.001
Distilled or Closed Cycle Condensate	.0005	.0005	.0005	.0005
Treated Boiler Feedwater	.001	.0005	.001	.001
Boiler Blowdown	.002	.002	.002	.002

**Ratings in columns 3 and 4 are based on a temperature of the heating medium of 240°–400° F. If the heating medium temperature is over 400° F. and the cooling medium is known to scale, these ratings should be modified accordingly.*

Fouling Resistances for Industrial Fluids

Oils	
Fuel oil	.005
Transformer oil	.001
Engine lube oil	.001
Quench oil	.004
Gases and Vapors	
Manufactured gas	.01
Engine exhaust gas	.01
Steam (non-oil bearing)	.0005
Exhaust steam (oil bearing)	.001
Refrigerant vapors (oil bearing)	.002
Compressed air	.002
Industrial organic heat transfer media	.001
Liquids	
Refrigerant liquids	.001
Hydraulic fluid	.001
Industrial organic heat transfer media	.001
Molten heat transfer salts	.0005

Fouling Resistances for Chemical Processing Streams

Gases and Vapors	
Acid gas	.001
Solvent vapors	.001
Stable overhead products	.001
Liquids	
MEA & DEA solutions	.002
DEG & TEG solutions	.002
Stable side draw and bottom product	.001
Caustic solutions	.002
Vegetable oils	.003

Fouling Resistances for Natural Gas-Gasoline Processing Streams

Gases and Vapors	
Natural gas	.001
Overhead products	.001
Liquids	
Lean oil	.002
Rich oil	.001
Natural gasoline & liquefied petroleum gases	.001

Fouling Resistances for Oil Refinery Streams

Crude and Vacuum Unit Gases and Vapors	
Atmospheric tower overhead vapors	.001
Light naphthas	.001
Vacuum overhead vapors	.002

Table 2
Crude and Vacuum Liquids

Crude Oil	0–199° F. Velocity Ft./Sec. Under 2	2–4	4 and Over	200°–299° F. Velocity Ft./Sec. Under 2	2–4	4 and Over
Dry	.003	.002	.002	.003	.002	.002
Salt†	.003	.002	.002	.005	.004	.004

	300°–499° F. Velocity Ft./Sec. Under 2	2–4	4 and Over	500°F and over Velocity Ft./Sec. Under 2	2–4	4 and Over
Dry	.004	.003	.002	.005	.004	.003
Salt†	.006	.005	.004	.007	.006	.005

†Normally desalted below this temperature range. († to apply to 200–299° F., 300–499° F., 500° F. and over.)

Gasoline	.001
Naphtha & light distillates	.001
Kerosene	.001
Light gas oil	.002
Heavy gas oil	.003
Heavy fuel oils	.005
Asphalt & residuum	.010

Cracking and Coking Unit Streams

Overhead vapors	.002
Light cycle oil	.002
Heavy cycle oil	.003
Light coker gas oil	.003
Heavy coker gas oil	.004
Bottoms slurry oil (4½ ft./sec. minimum)	.003
Light liquid products	.002

Catalytic Reforming, Hydrocracking, and Hydrodesulfurization Streams

Reformer charge	.002
Reformer effluent	.001
Hydrocracker charge & effluent**	.002
Recycle gas	.001
Hydrodesulfurization charge & effluent**	.002
Overhead vapors	.001
Liquid product over 50° A.P.I.	.001
Liquid product 30°–50° A.P.I.	.002

*** Depending on charge characteristics and storage history, charge resistance may be many times this value.*

Light Ends Processing Streams

Overhead vapors & gases	.001
Liquid products	.001
Absorption oils	.002
Alkylation trace acid streams	.002
Reboiler streams	.003

(table continued)

Table 2 Continued
Crude and Vacuum Liquids

Lube Oil Processing Streams

Feed stock	.002
Solvent feed mix	.002
Solvent	.001
Extract†	.003
Raffinate	.001
Asphalt	.005
Wax slurries†	.003
Refined lube oil	.001

†Precautions must be taken to prevent wax deposition on cold tube walls.

Table 3
Suggested Fouling Factors in Petrochemical Processes
r = 1/BTU/Hr. (sq. ft.) (°F.)

Fluid	Velocity, Ft./Sec.	Temperature Range <100° F.	>100° F.
Waters:			
Sea (limited to 125° F. max.)	<4	0.002	0.003
	>7	0.0015	0.002
River (settled)	<2	0.002	0.002–0.003
	>4	0.0005–0.0015	0.001–0.0025
River (treated and settled)	<2	0.0015	0.002
	>4	0.001	0.0015
4 mils baked phenolic coating[65]		0.0005	
15 mils vinyl-aluminum coating		0.001	
Condensate (100°–300° F)	<2	0.001	0.002–0.004
	>4	0.0005	0.001
Steam (saturated) oil free			0.0005–0.0015
With traces oil			0.001–0.002
Light Hydrocarbon Liquids (methane, ethane, propane, ethylene, propylene, butane-clean)	}	0.001	
Light Hydrocarbon Vapors: (clean)	}		
Chlorinated Hydrocarbons (carbon tetrachloride, chloroform, ethylene dichloride, etc.)			
Liquid		0.001	0.002
Condensing		0.001	0.0015
Boiling		0.002	0.002
Refrigerants (vapor condensing and liquid cooling)			

(table continued)

Table 3 Continued
Suggested Fouling Factors in Petrochemical Processes
r = 1/BTU/Hr. (sq. ft.) (°F.)

Fluid	Velocity, Ft./Sec.	Temperature Range <100° F.	>100° F.
Refrigerants (continued)			
Ammonia		0.001	
Propylene		0.001	
Chloro-fluoro-refrigerants		0.001	
Caustic liquid, Salt-free			
20% (steel tube)	3–8	0.0005	
50% (nickel tube)	6–9	0.001	
73% (nickel tube)	6–9	0.001	
Gases (Industrially clean)			
Air (atmos.)		0.0005–0.001	
Air (compressed)		0.001	
Flue Gases		0.001–0.003	
Nitrogen		0.0005	
Hydrogen		0.0005	
Hydrogen (saturated with water)		0.002	

(table continued)

Table 3 Continued
Suggested Fouling Factors in Petrochemical Processes
r = 1/BTU/Hr. (sq. ft.) (°F.)

Fluid	Velocity, Ft./Sec.	Temperature Range <100° F.	>100° F.
Polymerizable vapors with inhibitor		0.003–0.03	
High temperature cracking or coking, polymer buildup		0.02–0.06	
Salt Brines (125° F. max.)	<2	0.003	0.004
	>4	0.002	0.003
Carbon Dioxide[93] (Sublimed at low temp.)		0.2–0.3	

Sources

1. *Standards of Tubular Exchanger Manufacturers Association, Inc.*, (TEMA) *7th Ed.*
2. Ludwig, E. E., *Applied Process Design for Chemical and Petrochemical Plants, 2nd Ed., Vol. 3,* Gulf Publishing Co.

Metal Resistances

Here is an extremely useful chart for tube and pipe resistances for various metals.

Sources

1. Griscom-Russell Co. data
2. Ludwig, E. E., *Applied Process Design for Chemical and Petrochemical Plants, 2nd Ed.,* Gulf Publishing Co.

Metal Resistance of Tubes and Pipes

Thermal Conductivity, K			Values of r_w: 25	63	89	21	17	14.6	238	34.4	85	10	57.7
OD Tube	Gauge	Factor	Carbon Steel	Admiralty	Red Brass, .45% Ars. Copper	4-6% Chrome ½% Moly Steel 80-20 CU-NI	70-30 CU-NI	Monel	Copper 99.9+% CU	Nickel	Aluminum	Stainless AISI Type 302 & 304	Yorkalbro, Alum. Brass
⅝″	18	.00443	.000177	.000070	.000050	.000211	.000261	.000303	.000019	.000129	.000052	.000443	.000077
	16	.00605	.000242	.000096	.000068	.000288	.000356	.000414	.000025	.000176	.000071	.000605	.000105
	14	.00798	.000319	.000127	.000090	.000380	.000469	.000547	.000034	.000232	.000094	.000798	.000138
	18	.00437	.000175	.000069	.000049	.000208	.000257	.000299	.000018	.000127	.000051	.000437	.000076
	16	.00593	.000237	.000094	.000067	.000282	.000349	.000406	.000025	.000172	.000070	.000593	.000103
¾″	14	.00778	.000311	.000123	.000087	.000370	.000458	.000533	.000033	.000226	.000092	.000778	.000135
	13	.00907	.000363	.000144	.000102	.000432	.000534	.000621	.000038	.000264	.000107	.000907	.000157
	12	.01063	.000425	.000169	.000119	.000506	.000625	.000728	.000045	.000309	.000125	.001063	.000184

(table continued)

Metal Resistance of Tubes and Pipes Continued

Thermal Conductivity, K			Values of r_w										
			25	63	89	21	17	14.6	238	34.4	85	10	57.7
OD Tube	Gauge	Factor	Carbon Steel	Admiralty	Red Brass, .45% Ars. Copper	4-6% Chrome ½% Moly Steel 80-20 CU-NI	70-30 CU-NI	Monel	Copper 99.9+% CU	Nickel	Aluminum	Stainless AISI Type 302 & 304	Yorkalbro, Alum. Brass
	18	.00429	.000172	.000068	.000048	.000204	.000252	.000294	.000018	.000125	.000050	.000429	.000074
	16	.00579	.000232	.000092	.000065	.000276	.000341	.000397	.000024	.000168	.000068	.000579	.000100
1"	14	.00754	.000302	.000120	.000085	.000359	.000444	.000516	.000032	.000219	.000089	.000754	.000131
	13	.00875	.000350	.000139	.000098	.000417	.000515	.000599	.000037	.000254	.000103	.000875	.000152
	12	.01019	.000408	.000162	.000114	.000485	.000599	.000698	.000043	.000296	.000120	.001019	.000177
	10	.01289	.000516	.000205	.000145	.000614	.000758	.000883	.000054	.000375	.000152	.001289	.000223
	8	.01647	.000659	.000261	.000185	.000784	.000967	.001128	.000069	.000479	.000194	.001647	.000285
	18	.00425	.000170	.000067	.000048	.000202	.000250	.000291	.000018	.000124	.000050	.000425	.000074
	16	.00571	.000228	.000091	.000064	.000272	.000336	.000391	.000024	.000166	.000067	.000571	.000099
1¼"	14	.00741	.000296	.000118	.000083	.000353	.000436	.000508	.000031	.000215	.000087	.000741	.000128
	13	.00857	.000343	.000136	.000096	.000408	.000504	.000587	.000036	.000249	.000101	.000857	.000149
	12	.00995	.000398	.000158	.000112	.000474	.000585	.000682	.000042	.000289	.000117	.000995	.000172
	10	.01251	.000500	.000199	.000141	.000596	.000736	.000857	.000053	.000364	.000147	.001251	.000217
	8	.01584	.000634	.000251	.000178	.000754	.000932	.001085	.000067	.000460	.000186	.001584	.000275
	18	.00422	.000169	.000067	.000047	.000201	.000248	.000289	.000018	.000123	.000050	.000422	.000073
	16	.00566	.000226	.000090	.000064	.000270	.000333	.000388	.000024	.000165	.000067	.000566	.000098
1½"	14	.00732	.000293	.000116	.000082	.000349	.000431	.000501	.000031	.000213	.000086	.000732	.000127
	13	.00845	.000338	.000134	.000095	.000402	.000497	.000579	.000036	.000246	.000099	.000845	.000146
	12	.00979	.000392	.000155	.000110	.000466	.000576	.000671	.000041	.000285	.000115	.000979	.000170
	10	.01226	.000490	.000195	.000138	.000584	.000721	.000840	.000052	.000356	.000144	.001226	.000212
	8	.01545	.000618	.000245	.000174	.000736	.000909	.001058	.000065	.000449	.000182	.001545	.000268
	18	.00419	.000168	.000067	.000047	.000200	.000246	.000287	.000018	.000122	.000049	.000419	.000073
	16	.00560	.000224	.000089	.000063	.000267	.000329	.000384	.000024	.000163	.000066	.000560	.000097
2"	14	.00722	.000289	.000115	.000081	.000344	.000425	.000495	.000030	.000210	.000085	.000722	.000125
	13	.00831	.000332	.000132	.000093	.000396	.000489	.000569	.000035	.000242	.000098	.000831	.000144
	12	.00961	.000384	.000153	.000108	.000458	.000565	.000658	.000040	.000279	.000113	.000961	.000167
	10	.01197	.000479	.000190	.000134	.000570	.000704	.000820	.000050	.000348	.000141	.001197	.000207
	8	.01499	.000600	.000238	.000168	.000714	.000882	.001027	.000063	.000436	.000176	.001499	.000260
Pipe ¾" IPS	St'd	.01055	.000422	.000167	.000119	.000502	.000621	.000723	.000044	.000307	.000124	.001055	.000183
	X Hvy	.01504	.000602	.000239	.000169	.000716	.000885	.001030	.000063	.000437	.000177	.001504	.000261
	Sch. 160	.0229	.000916	.000363	.000257	.001090	.001347	.00157	.000096	.000666	.000269	.00229	.000397
	XX Hvy.	.0363	.00145	.000576	.000408	.00173	.00214	.00249	.000153	.001055	.000427	.00363	.000629
Pipe 1½" IPS	St'd	.01308	.000523	.000208	.000147	.000623	.000769	.000896	.000055	.000380	.000154	.001308	.000227
	X Hvy	.01863	.000745	.000296	.000209	.000887	.001096	.001276	.000078	.000542	.000219	.001863	.000323
	Sch. 160	.0275	.00110	.000437	.000309	.00131	.00162	.00188	.000116	.000799	.000324	.00275	.000477
	XX Hvy	.0422	.00169	.000670	.000474	.00201	.00248	.00289	.000177	.001227	.000496	.00422	.000731

r_w for 1″ OD 16 BWG steel tube with 18 BWG admiralty liner = .00031

For other materials divide factor number by the thermal conductivities of various materials:

60-40 Brass 55	Chrome Vanadium Steel SAE 6120 ....23.2	Tin 35	Everdur #50.................19
Zinc 64		Lead...............20	Wrought Iron...............40
			Alcunic.....................56.5

L = Length of flow path, ft

D_e = Equivalent diameter, ft; 4 times hydraulic radius

The Griscom-Russell Co., by permission

Vacuum Condensers

Outlet Temperature and Pressure. It is important to have proper subcooling in the vent end of the unit to prevent large amounts of process vapors from going to the vacuum system along with the inerts.

Control. It is necessary to have some over-surface and to have a proper baffling to allow for pressure control during process swings, variable leakage of inerts, etc. One designer adds 50% to the calculated length for the over-surface. The condenser must be considered part of the control system (similar to extra trays in a fractionator to allow for process swings not controlled by conventional instrumentation).

The inerts will "blanket" a portion of the tubes. The blanketed portion has very poor heat transfer. The column pressure is controlled by varying the percentage of the tube surface blanketed. When the desired pressure is exceeded, the vacuum system will suck out more inerts, and lower the percentage of surface blanketed. This will increase cooling and bring the pressure back down to the desired level. The reverse happens if the pressure falls below that desired. This is simply a matter of adjusting the heat transfer coefficient to heat balance the system.

Figure 1 shows typical baffling. The inerts move through the first part of the condenser as directed by the baffles. The inerts then pile up at the outlet end lowering heat transfer as required by the controller. A relatively large section must be covered by more or less stagnant inerts which are sub-cooled before being pulled out as needed. Without proper baffles, the inerts build up in the condensing section and decrease heat transfer until the pressure gets too high. Then the vacuum valve opens wider, pulling process vapor and inerts into the vacuum system. Under these conditions pressure control will be very poor.

Pressure Drop. Baffling must be designed to keep the pressure drop as low as possible. The higher the pressure drop the higher the energy consumption and the harder the job of attaining proper vent end sub-cooling. Pressure drop is lower at the outlet end because of smaller mass flow.

Bypassing. Baffles should prevent bypass of inlet vapor into the vent. This is very important.

Typical Condenser. Figure 1 illustrates an inlet "bathtub" used for low vacuums to limit pressure drop at entrance to exchanger and across first rows of tubes. Note staggered baffle spacing with large spacing at inlet, and the side to side (40% cut) baffles. Enough baffles must be used in the inlet end for minimum tube support. In the last 25% of the outlet end a spacing of 1/10 of a diameter is recommended.

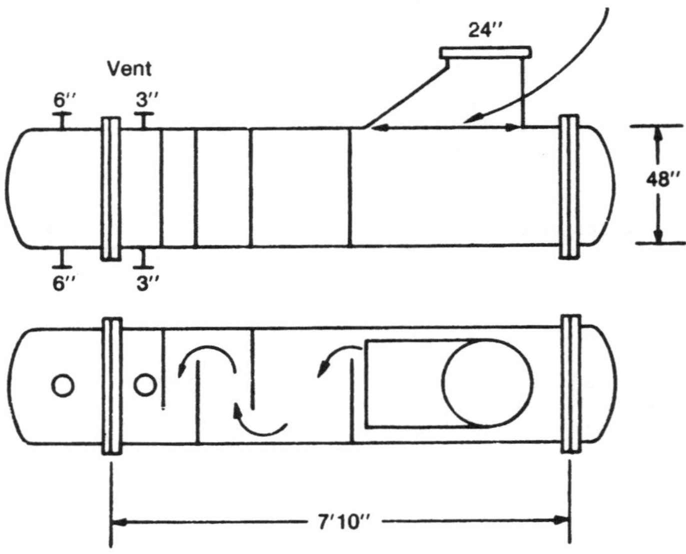

Figure 1. Baffling and inlet "bathtub" are shown in this typical vacuum condenser design.

Source

Based on notes provided by Jack Hailer and consultation by Guy Z. Moore while employed at El Paso Products Co.

Air-cooled Heat Exchangers: Forced vs Induced Draft

Advantages and disadvantages of forced and induced draft air-cooled heat exchangers are shown here to aid in selection.

Advantages of Induced Draft

- Better distribution of air across the section.
- Less possibility of the hot effluent air recirculating around to the intake of the sections. The hot air is dis-

charged upward at approximately 2½ times the velocity of intake, or about 1500 ft/min.

- Less effect of sun, rain, and hail, since 60% of the surface area of the sections is covered.
- Increased capacity in the event of fan failure, since the natural draft stack effect is much greater with induced draft.

Disadvantages of Induced Draft

- Higher horsepower since the fan is located in the hot air.
- Effluent air temperature should be limited to 200° F, to prevent potential damage to fan blades, bearings, V-belts, or other mechanical components in the hot air stream.
- The fan drive components are less accessible for maintenance, which may have to be done in the hot air generated by natural convection.
- For inlet process fluids above 350° F, forced draft design should be used; otherwise, fan failure could subject the fan blades and bearings to excessive temperatures.

Advantages of Forced Draft

- Slightly lower horsepower since the fan is in cold air. (Horsepower varies directly as the absolute temperature.)
- Better accessibility of mechanical components for maintenance.
- Easily adaptable for warm air recirculation for cold climates.

Disadvantages of Forced Draft

- Poor distribution of air over the section.
- Greatly increased possibility of hot air recirculation, due to low discharge velocity from the sections and absence of stack.
- Low natural draft capability on fan failure due to small stack effect.
- Total exposure of tubes to sun, rain, and hail.

Source

GPSA Engineering Data Book, Gas Processors Suppliers Association, 10th Ed.

Air-cooled Heat Exchangers: Pressure Drop Air Side

This method will approximate required fan horsepower.

$$\Delta P_a = \frac{F_p N}{D_R}$$

where

ΔP_a = Static pressure drop, inches of water
F_p = Static pressure drop factor, see Table 1
N = Number of tube rows

$$D_R = \frac{\text{Actual density at average air temperature}}{\text{Air density at sea level and 70°F}}$$

Use perfect gas law to calculate D_R

$$\text{Altitude (above sea level), ft} = 25{,}000 \ln\left(\frac{14.7}{P}\right)$$

where

P = Atmospheric pressure in psia

Once ΔP_a is obtained, the pressure that the fan has to provide is then calculated.

$$P_F = \Delta P_a + \left[\frac{\text{ACFM Per Fan}}{3{,}140\ D^2}\right]^2 D_R$$

where

P_F = Total pressure that fan has to provide, inches of water
ACFM = Actual cubic feet per minute
D = Fan diameter, feet

$$D_R = \frac{\text{Actual density at fan temperature}}{\text{Air density at sea level and 70°F}}$$

$$HP = \frac{\text{ACFM Per Fan } (P_F)}{4{,}460}$$

where

HP = Approximate horsepower per fan

An equation has been developed for the F_p data in Table 1 as follows:

$F_p = 6 \times 10^{-8}$ (face velocity)$^{1.825}$
(Face velocity range is 1,400–3,600, lb/hr/ft^2 face area)

Source

GPSA Engineering Data Book, Gas Processors Suppliers Association, 10th Ed. 1987.

Table 1
Determination of F_p

Air Face Mass Velocity lb/Hr/ft^2 Face Area	F_p
1,400	0.033
1,600	.0425
1,800	.0525
2,000	.0625
2,200	.075
2,400	.0875
2,600	.100
2,800	.115
3,000	.130
3,200	.150
3,400	.165
3,600	.185

Air-cooled Heat Exchangers: Rough Rating

A suggested procedure is as follows (References 1 and 2):

1. Calculate exchanger duty (MMBtu/hr).
2. Select an overall U_x from Tables 5 and 6 in the Heat Transfer Coefficients section (based on finned area). Arbitrarily use ½″-fins, 9 to the inch for determining U_x.
3. Calculate approximate air temperature rise from

$$\Delta T_a = \left(\frac{U_x + 1}{10}\right)\left(\frac{T_1 + T_2}{2} - t_1\right)$$

where

ΔT_a = Air side temperature rise, °F
U_x = Overall coefficient based on finned area, Btu/hr ft^2 °F
T_1, T_2 = Process side inlet and outlet temperatures, °F
t_1 = Air inlet temperature, °F

4. Calculate ΔT_M and apply appropriate correction factor F.
5. Calculate exchanger extended area from

$$A_x = \frac{q}{U_x \Delta T_M}$$

where

A_x = Extended (finned) surface, ft^2
q = Duty, Btu/hr
ΔT_M = Log mean temperature difference, °F

6. Estimate number of tube rows from Hudson Company optimum bundle depth curve, Figure 1. Use 4 to 6 tube rows if curve comes close to that number.
7. Arbitrarily choose 1″ OD tubes, ½″-fins, 9 to the inch at 2″ Δ pitch. This will give "middle of the road" face area.
8. Calculate face area (F_a), ft^2

$$F_a = \frac{A_x}{APSF}$$

9. Pick a desirable combination of tube length and section width to achieve the approximate face area.

Table 1
Design Values for Rough Rating

Typical Section Width, ft	Typical Tube Length, ft
6	6, 10, 15, 20, 24, 30
8	10, 15, 20, 24, 30
12	12, 16, 24, 32, 40
16	16, 24, 32, 40

APF = Extended area ft^2/ft of tube = 3.80
AR = Extended area/bare area = 14.5
APSF = Extended area/bundle face area = No. tube rows (22.8)

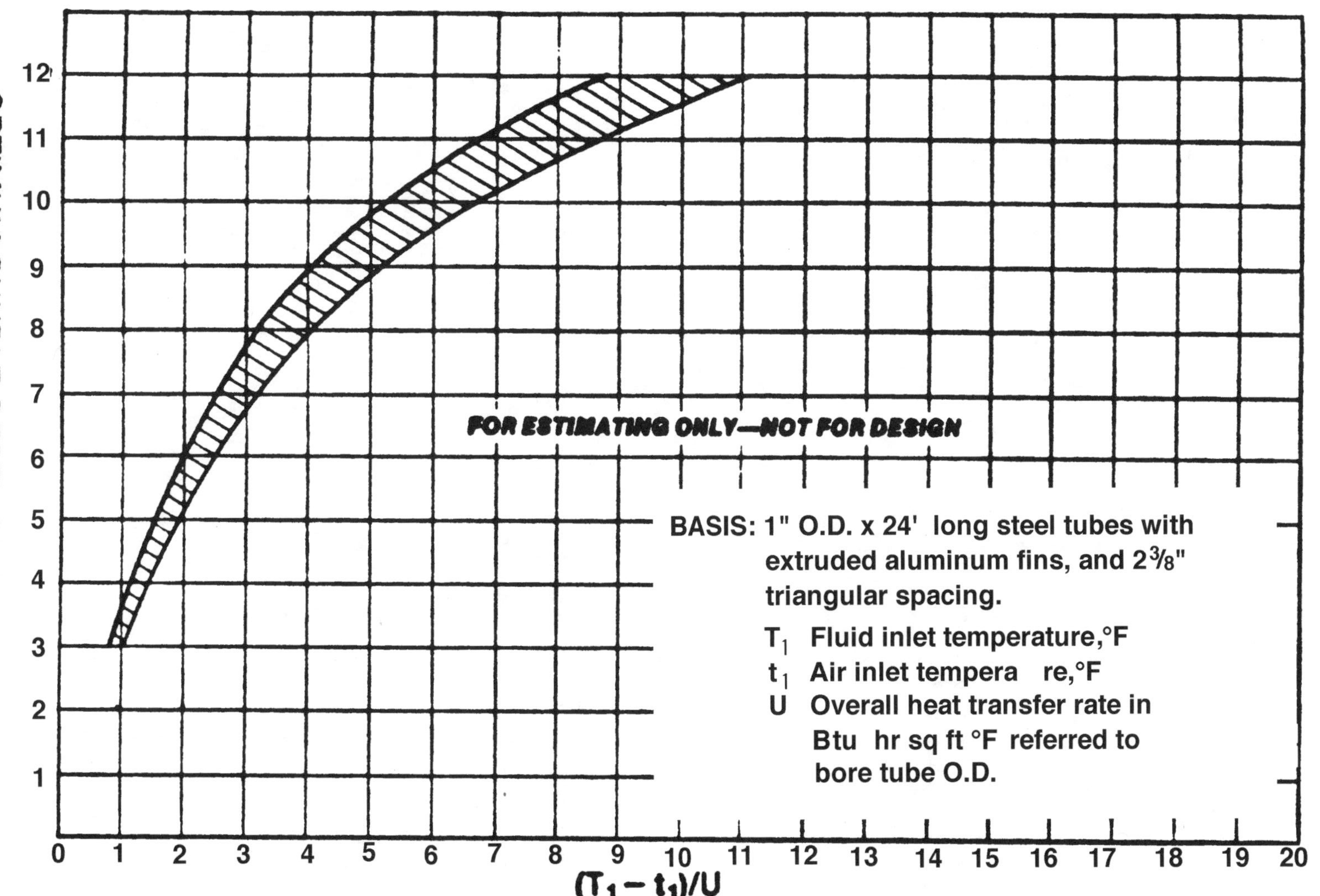

Figure 1. This is the effect of temperature level and overall transfer rate upon optimum bundle depth (Reference 3).

10. Estimate number of fans. Use the fact that fanned section length divided by bay width seldom exceeds 1.8. A 16-foot wide bay with 24-foot tubes would have one fan (ratio = 1.5). The same 16-foot wide bay with 32-foot tubes would have 2 fans (ratio would be 2.0 for 1 fan).
11. Estimate minimum fan area by

$$FAPF = \text{Fan Area Per Fan} = \frac{0.40\,(F_a)}{\text{Number of Fans}}$$

12. From above calculate fan diameter rounded up to the next even feet.
13. Calculate horsepower per fan from the previous section, Air-cooled Heat Exchangers, Pressure Drop, Air Side.
14. Number of tubes (N_T) can be obtained from

$$N_T = \frac{A_x}{APF\,(\text{Tube length})}$$

Sources

1. Lerner, J. E., Flour Corporation, "Simplified Air Cooler Estimating," *Hydrocarbon Processing,* February 1972.
2. *GPSA Engineering Data Book,* Gas Processors Suppliers Association, 10th Ed., 1987.
3. Hudson Products Corporation, *The Basics of Air Cooled Heat Exchangers,* Copyright © 1986.
4. Branan, C. R., *The Process Engineer's Pocket Handbook, Vol. 1,* Gulf Publishing Co.

Air-cooled Heat Exchangers: Temperature Control

Here are three primary means of controlling temperature for air-cooled heat exchangers.

Adjustable Shutters

Shutters mounted above the cooling sections serve to protect them from overhead wind, snow, ice, and hail. In addition they are also used to regulate, either manually or automatically, the flow of air across the finned tubes and thus control the process fluid outlet temperature.

Controllable Pitch Fan

The controllable pitch fan provides an infinitely variable air delivery across the K-Fin Sections through automatic changes in the fan blade angle. Temperature can be closely controlled to meet the varying demands of operating conditions and fluctuating atmospheric temperatures with appreciable power savings under low load conditions. For certain control applications, the ability of this fan to pump air backwards with negative blade angles is used.

Combination Controls

The combination of adjustable shutters with variable speed drive or with controllable pitch fan is frequently used for close fluid temperature control. This system is particularly useful during start-up and shut-down procedures where fluids are subject to freezing in cold ambient temperatures. It is also well adapted to fluid temperature control while operating under high wind and freezing conditions. This diagram shows a two-speed electric motor with automatically adjustable shutters. The arrangement lends itself to many cooling services while effecting a horse power saving through the use of the two-speed motor.

Source

Griscom-Russell Co. material.

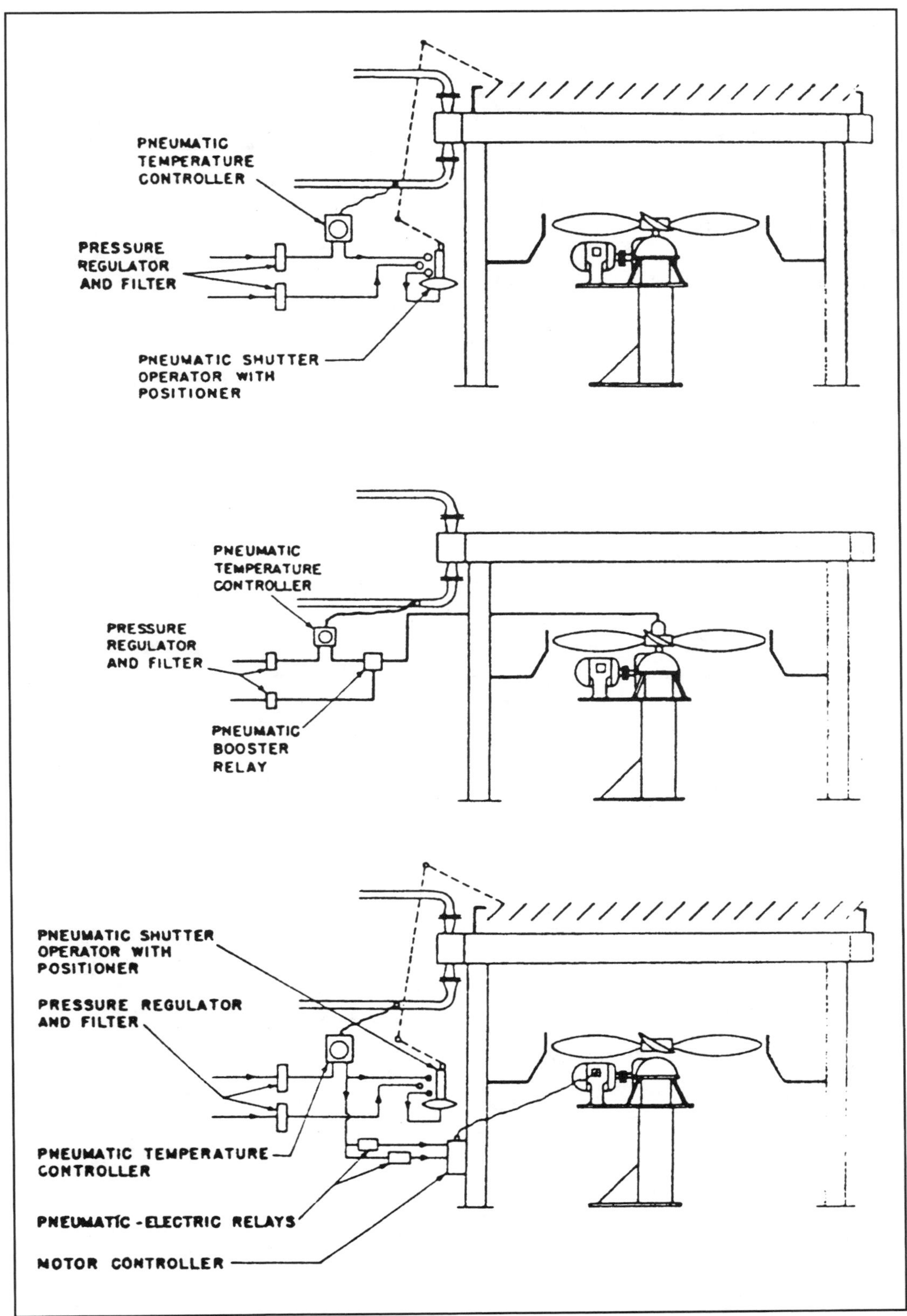

Figure 1. These are schemes for the temperature control of air coolers.

Miscellaneous Rules of Thumb

1. For fixed tubesheet design of shell and tube heat exchangers don't allow too high a temperature difference between tubeside and shellside without providing a shellside expansion joint. The author has seen 70° F (one company) and 100° F (another company) used as this limit. An easy way to calculate the maximum stress is as follows:

a. Assume the tubes are at tubeside bulk temperature and the shell is at shellside bulk temperature.

b. Calculate the elongation of tubes, if unhampered, and shell, if unhampered, starting at 70° F and using the respective coefficients of expansion.

c. If the tubes would have grown more or less than the shell, the difference will set up stress in both members, one in tension and the other in compression.

d. Assume the deformation (strain) in each member is inversely proportional to its cross-sectional area. In other words, the fraction of the total strain in inches/inch of length will be proportionally more for the member (tubes or shell) having the smallest cross section.

e. For each member, multiply its strain by Young's modulus (Modulus of Elasticity) in consistent units to get stress. Strain is dimensionless so Young's modulus in lb/in^2 will yield stress in lb/in^2.

f. Compare this with the maximum allowable stress for the material used.

g. The tensile and compressive module for steel are essentially the same.

2. Typical handling of design parameters.

DP = MOP + 10% but not less than MOP + 30 psi

DP for vacuum use 15 psi external pressure for cases having atmospheric pressure on the other side.

DT: Below 32° F DT = minimum operating temperature
Above 32° F DT = MOT + 25° F but not less than 150° F

where

DP = Design pressure
DT = Design temperature
MOP = Max. operating pressure
MOT = Max. operating temperature

3. 40% baffle cut = 40% open area is a good rule of thumb maximum for shell and tube heat exchangers.

Source

Branan, C. R., *The Process Engineer's Pocket Handbook, Vol. 1,* Gulf Publishing Co.

3
Fractionators

Introduction

More shortcut design methods and rules of thumb have been developed for fractionation than probably any other unit operation. For example the paper[1] reprinted in Appendix 5 on development of shortcut equipment design methods contains 18 references for fractionation shortcut methods out of 37 total. Both the process and mechanical aspects of fractionation design have useful rules of thumb. Many of the mechanical design rules of thumb become included in checklists of do's and don'ts.

For troubleshooting fractionators see the troubleshooting section in this book. For a general understanding of how trayed and packed fractionating columns work, illustrated by actual field cases, I recommend Lieberman's design book[2].

Sources

1. Branan, C. R., Development of Short-cut Equipment Design Methods, ASEE Annual Conference Proceedings, Computer Aided Engineering, American Society for Engineering Education, 1985.
2. Lieberman, N. P., *Process Design For Reliable Operations, 2nd Ed.,* Gulf Publishing Co.

Relative Volatility

For quick estimates, a relative volatility can be estimated as follows:

The equilibrium vaporization constant K is defined for a compound by

$$K_i = \frac{Y_i}{X_i}$$

where

Y_i = Mole fraction of component i in the vapor phase
X_i = Mole fraction of component i in the liquid phase

To calculate a distillation, the relative volatility α is needed. It is defined as

$$\alpha = \frac{K_i}{K_j}$$

where i and j represent two components to be separated.

Raoult's Law for ideal systems is

$$p_i = P_i X_i$$

where

p_i = Partial pressure of i
P_i = Vapor pressure of pure component i

By definition

$$p_i = \Pi Y_i$$

where

Π = Total pressure of the system

so

$$P_i X_i = \Pi Y_i$$

and

$$\frac{P_i}{\Pi} = \frac{Y_i}{X_i} = K_i$$

Therefore for systems obeying Raoult's Law

$$\alpha = \frac{P_i}{P_j}$$

Source

Branan, C. R., *The Process Engineer's Pocket Handbook, Vol. 1,* Gulf Publishing Co., 1976.

Minimum Reflux

For shortcut design of a distillation column, the minimum reflux calculation should be made first.

Minimum Reflux Introduction by Keith Claiborne[6]

A quick overview of minimum reflux requirements as well as improved understanding of the meaning of the term can be had by considering separation of a binary feed.

For a binary, let's denote as V the fractional molar split of the feed into overhead product and as L the fractional split into bottom product. Calculate compositions of the flash separation of feed into vapor v and liquid l to give v/l = V/L. The resulting vapor can be regarded as being composed of a portion d of the overhead product composition and a portion r of the flash liquid composition.

It is then easy to calculate what the proportions must be so that d × overhead composition + r × flash liquid composition = flash vapor composition. Thus:

$$r / d = \frac{a - b}{b - c}$$

where

a = fraction high volatile component in overhead specification
b = fraction high volatile component in flash vapor
c = fraction high volatile component in flash liquid.

The reason for this simple relationship is that the concept of minimum reflux implies an infinite number of stages and thus no change in composition from stage to stage for an infinite number of stages each way from the "pinch point" (the point where the McCabe-Thiele operating lines intersect at the vapor curve; for a well behaved system, this is the feed zone). The liquid refluxed to the feed tray from the tray above is thus the same composition as the flash liquid.

The total feed tray vapor can be regarded as comprising some material that passes unchanged through the upper section into product. The unchanged material can then be regarded as being conveyed by other material (at the composition of the reflux to the feed tray) endlessly recirculating as vapor to the tray above and, reliquified, refluxing onto the feed tray.

Binary minimum reflux so calculated implies feed enthalpy just equal to the above started vapor V and liquid L. Any increase or decrease in that enthalpy must be matched by increase or decrease in total heat content of overhead reflux. Note that the Underwood[1] binary reflux equation essentially computes the flash versus specification composition relationship along with enthalpy correction.

For more than two components, calculation is not so easy. Bounds can, however, often be perceived. If in the feed there is only an insignificant amount of materials of volatility intermediate between the light and heavy keys, the following applies:

Minimum total reflux (lbs or mols/hr) corresponding to given total feed will be greater than if only the actual total mols of heavy and light key components were present. Reflux need will be less than if the actual total mols of feed were present, but composed only of light and heavy keys. The more closely non-keyed components are clustered to volatilities of the keys, the nearer are reflux needs to that calculated for the binary and total feed volume.

Minimum Reflux Multicomponent

The Underwood Method[1] will provide a quick estimate of minimum reflux requirements. It is a good method to use when distillate and bottoms compositions are specified. Although the Underwood Method will be detailed here, other good methods exist such as the Brown-Martin[2] and Colburn[3] methods. These and other methods are discussed and compared in Van Winkle's book[4]. A method to use for column analysis when distillate and bottoms compositions are not specified is discussed by Smith[5].

The Underwood Method involves finding a value for a constant, Θ, that satisfies the equation

$$\sum_{1}^{n} \frac{X_{iF}\alpha_i}{\alpha_i - \Theta} = 1 - q = \frac{X_{1F}\alpha_1}{\alpha_1 - \Theta} + \frac{X_{2F}\alpha_2}{\alpha_2 - \Theta} + \ldots.$$

The value of Θ will lie between the relative volatilities of the light and heavy key components, which must be adjacent.

After finding Θ, the minimum reflux ratio is determined from

$$R_m + 1 = \sum_1^n \frac{\alpha_i X_{iD}}{\alpha_i - \Theta}$$

Nomenclature

α_i = Relative volatility of component i verses the heavy key component
Θ = Underwood minimum reflux constant
X_{iF} = Mol fraction of component i in the feed
X_{iD} = Mol fraction of component i in the distillate
q = Thermal condition of the feed
  Bubble point liquid q = 1.0
  Dew point vapor q = 0
  General feed $q = (L_S - L_R)/F$
  where:
   L_S = Liquid molar rate in the stripping section
   L_R = Liquid molar rate in the rectification section
   F = Feed molar rate
R_m = Minimum reflux ratio

Sources

1. Underwood, A. J. V., "Fractional Distillation of Multicomponent Mixtures," *Chemical Engineering Progress,* 44, 603 (1948).
2. Martin, H. Z. and Brown, G. G., *Trans. A.I.Ch.E.,* 35:679 (1939).
3. Colburn, A. P., *Trans. A.I.Ch.E.,* 37:805 (1941).
4. Van Winkle, Matthew, *Distillation,* McGraw-Hill, 1967.
5. Smith, B. D., *Design of Equilibrium Stage Processes,* McGraw-Hill, 1963.
6. Branan, C. R., *The Fractionator Analysis Pocket Handbook,* Gulf Publishing Co., 1978.

Minimum Stages

The Fenske Method gives a quick estimate for the minimum theoretical stages at total reflux.

$$N_m + 1 = \frac{\ln[(X_{LK}/X_{HK})_D (X_{HK}/X_{LK})_B]}{\ln(\alpha_{LK/HK})_{AVG}}$$

Nomenclature

B = subscript for bottoms
D = subscript for distillate
HK = subscript for heavy key
LK = subscript for light key
N_m = minimum theoretical stages at total reflux
X_{HK} = mol fraction of heavy key component
X_{LK} = mol fraction of the light key component
$\alpha_{LK/HK}$ = relative volatility of component vs the heavy key component

Source

Fenske, M., *Ind. Eng. Chem. 24,* 482, (1932).

Actual Reflux and Actual Theoretical Stages

The recommended[1] method to use to determine the actual theoretical stages at an actual reflux ratio is the Erbar/Maddox[2] relationship. In the graph, N is the theoretical stages and R is the actual reflux ratio L/D, where L/D is the molar ratio of reflux to distillate. N_m is the minimum theoretical stages and R_m is the minimum reflux ratio.

The actual reflux ratio that one picks should be optimized from economics data. For a ballpark estimate use 1.1–1.2 R_m for a refrigerated system, and 1.2–1.35 R_m for a hot system.

Another useful molar ratio is reflux/feed, L/F. In binary systems, L/F for all practical purposes is unchanging for wide differences in feed composition, so long as the following hold:

1. The distillate and bottoms compositions, but not necessarily the quantities, are held constant.

2. The feed tray is kept matched in composition to the feed (which means the feed tray moves with feed composition changes).

The reader can verify the above using the Underwood[3] equations and the tower material balance. The author once calculated a case where a large feed change would change L/D by 46%, whereas L/F changed only 1%. The stability of L/F is well proven in the field and is a good factor to use in predicting the effect of feed changes for design and in an operating plant.

The author has curve-fit the Erbar/Maddox curves since readability for the graph is limited. For simplicity, let:

$x = N_m/N$
$y = R/(R + 1)$
$z = R_m/(R_m + 1)$

For $y = A + Bx + Cx^2 + Dx^3 + Ex^4 + Fx^5$, the following table is presented:

Z	A	B	C	D	E	F
0	.00035	.16287	−.23193	5.09032	−8.50815	4.48718
.1	.09881	.32725	−2.57575	10.20104	−12.82050	5.76923
.2	.19970	.14236	−.58646	2.60561	−3.12499	1.76282
.3	.29984	.09393	−.23913	1.49008	−2.43880	1.79486
.4	.40026	.12494	−.49585	2.15836	−3.27068	2.08333
.5	.50049	−.03058	.81585	−2.61655	3.61305	−1.28205
.6	.60063	−.00792	.60063	−2.06912	3.39816	−1.52243
.7	.70023	−.01109	.45388	−1.25263	1.94348	−.83334
.8	.80013	−.01248	.76154	−2.72399	3.85707	−1.68269
.9	.89947	.00420	.38713	−1.14962	1.40297	−.54487
1.0	1.0	-0-	-0-	-0-	-0-	-0-

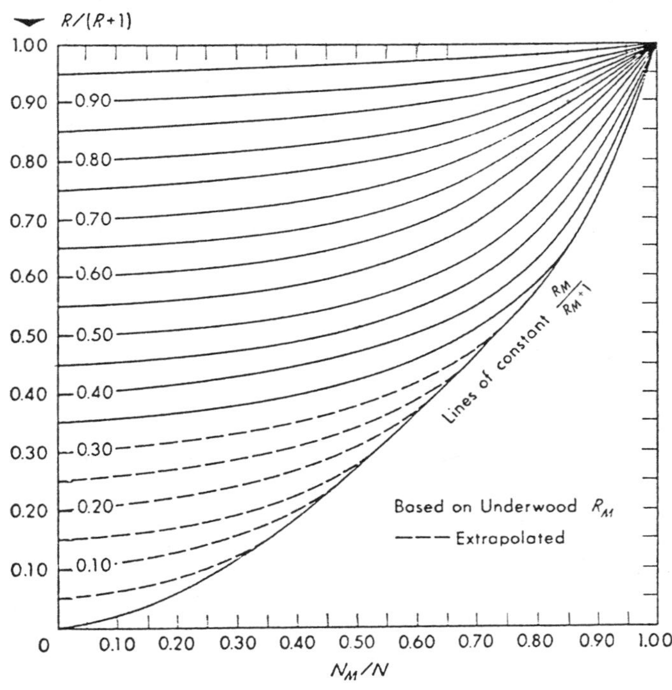

Figure 1. Plates-reflux correlation of Erbar and Maddox.

Sources

1. Fair, J. R. and Bolles, W. L., "Modern Design of Distillation Columns," *Chemical Engineering,* April 22, 1968.
2. Erbar, J. H. and Maddox, R. N., "Latest Score: Reflux vs Trays," *Petroleum Refiner,* May 1961, p. 185.
3. Equations were generated using FLEXCURV, V.2.0, Gulf Publishing Co.

Actual Trays

After actual theoretical trays are determined (see Actual reflux and theoretical stages) one needs to estimate the actual physical number of trays required in the distillation column. This is usually done by dividing the actual theoretical trays by the overall average fractional tray efficiency. Then a few extra trays are normally added for offload conditions, such as a change in feed composition. Experience for a given service is the best guide for extra trays.

Source

Branan, C. R., *The Process Engineer's Pocket Handbook, Vol. 2,* Gulf Publishing Co., 1983, p. 4.

Tray Efficiency

Actual stages depend upon the tray efficiency, which will probably be the weakest number in the design. Using operating data from a similar system is certainly best where possible. Tables 1 and 2 give some shortcut correlations.

Ludwig[1] discusses new work by the A.I.Ch.E. which has produced a method more detailed than the previous shortcut methods. He states that some of the shortcut methods can be off by 15–50% as indicated by the A.I.Ch.E. work. The spread of the Drickamer and Bradford correlation shown in the Ludwig plot is about 10 points of efficiency or ±5 efficiency points around the curve. Ludwig states that comparisons between shortcut and A.I.Ch.E. values indicate that deviations for the shortcut methods are usually on the safe or low side.

Maxwell's[3] correlation was generated from hydrocarbon data only. Ludwig states that the Drickamer and Bradford correlation is good for hydrocarbons, chlorinated hydrocarbons, glycols, glycerine and related compounds, and some rich hydrocarbon absorbers and strippers.

Table 1
Fractionator Overall Tray Efficiency, %

Viscosity Centipoises	Gunness and Other Data Plotted Versus Reciprocal Viscosity in Maxwell. (Average viscosity of liquid on the plates)	Drickamer and Bradford Correlation Plotted in Ludwig. (Molal average viscosity of the feed)
0.05*	. . .	98
0.10	104**	79
0.15	86	70
0.20	76	60
0.30	63	50
0.40	56	42
0.50	50	36
0.60	46	31
0.70	43	27
0.80	40	23
0.90	38	19
1.00	36	17
1.50	30	7
1.70	28	5

**Extrapolated slightly*
***Maxwell explains how efficiencies above 100% are quite possible.*

Ludwig also presents correlations of O'Connell[4]. He warns that O'Connell may give high results. Ludwig suggests using the O'Connell absorber correlation only in areas where it gives a lower efficiency than the fractionator correlation of Drickamer and Bradford or O'Connell.

For high values of α, low tray efficiency usually results.

Table 2
Overall Tray Efficiency, % (O'Connell)

Correlating Factor	Fractionators*	Absorbers $\left(\frac{HP}{\mu} = \frac{\rho}{KM\mu}\right)^{**}$
0.01	. .	8
0.05	. .	17
0.10	87	22
0.15	80	23
0.20	74	26
0.30	67	29
0.40	62	32
0.50	57	33
0.60	55	35
0.70	52	36
0.80	51	37
0.90	49	38
1	48	39
1.5	44	42
2	41	45
3	37	48
4	35	52
5	33	53
6	32	56
7	32	57
8	31	58
9	. .	60
10	. .	61

**Fractionators = (Relative volatility of key components) (Viscosity of feed in centipoises). Relative volatility and viscosity are taken at average tower conditions between top and bottom.*
***H = Henry's law constant, mols/ft³ (ATM)*
P = Pressure, ATM
μ = Viscosity, centipoises
ρ = Density, lbs/ft³
K = Equilibrium K of key components
M = Mol. wt. of liquid

Sources

1. Ludwig, E. E., *Applied Process Design For Chemical and Petrochemical Plants, Vol. 2,* Gulf Publishing Co.
2. Drickamer, H. G. and Bradford, J. B., "Overall Plate Efficiency of Commercial Hydrocarbon Fractionating Columns," *Trans. A.I.Ch.E.* 39, 319, (1943).

3. Maxwell, J. B., *Data Book on Hydrocarbons,* Van Nostrand, 1965.

4. O'Connell, H. E., "Plate Efficiency of Fractionating Columns and Absorbers," *Trans. A.I.Ch.E.* 42, 741, (1946).

Diameter of Bubble Cap Trays

Here are two quick approximation methods for bubble cap tray column diameter.

Standard Velocity

The standard velocity based on total cross section is defined as:

$$U_{STD} = 0.227\ (\rho_L/\rho_V - 1)^{0.5}$$

The standard velocity is multiplied by a factor depending upon the service as follows:

Service	%U_{STD}
Vacuum towers	110*
Refinery group**	100*
Debutanizers or other 100–250 PSIG towers	80
Depropanizers or other high pressure towers	60

*Applies for 24″ tray spacing and above

**"Refinery group" = low pressure naphtha fractionators, gasoline splitters, crude flash towers, etc.

Souders—Brown[1]

The maximum allowable mass velocity for the total column cross section is calculated as follows:

$$W = C[\rho_V(\rho_L - \rho_V)]^{1/2}$$

The value of W is intended for general application and to be multiplied by factors for specific applications as follows:

Absorbers: 0.55
Fractionating section of absorber oil stripper: 0.80
Petroleum column: 0.95
Stabilizer or stripper: 1.15

The developed equation for the Souders-Brown C factor is

$$C = (36.71 + 5.456T - 0.08486T^2) \ln S - 312.9 + 37.62T - 0.5269T^2$$

Correlation ranges are:

$C = 0$ to 700
$S = 0.1$ to 100
$T = 18$ to 36

The top, bottom, and feed sections of the column can be checked to see which gives the maximum diameter.

The Souders-Brown correlation considers entrainment as the controlling factor. For high liquid loading situations and final design, complete tray hydraulic calculations are required.

Ludwig[2] states that the Souders-Brown method appears to be conservative for a pressure range of 5 to 250 psig allowing W to be multiplied by 1.05 to 1.15 if judgment and caution are exercised.

The Souders-Brown correlation is shown in Figure 1.

Nomenclature

C = Souders-Brown maximum allowable mass velocity factor
S_{STD} = Standard maximum allowable velocity for tower cross section, ft/sec
W = Souders-Brown maximum allowable mass velocity, lbs/(ft^2 cross section) (hr)
ρ_L = Liquid density, lbs/ft^3
ρ_V = Vapor density, lbs/ft^3
T = Tray spacing, in.
S = Surface tension, dynes/cm

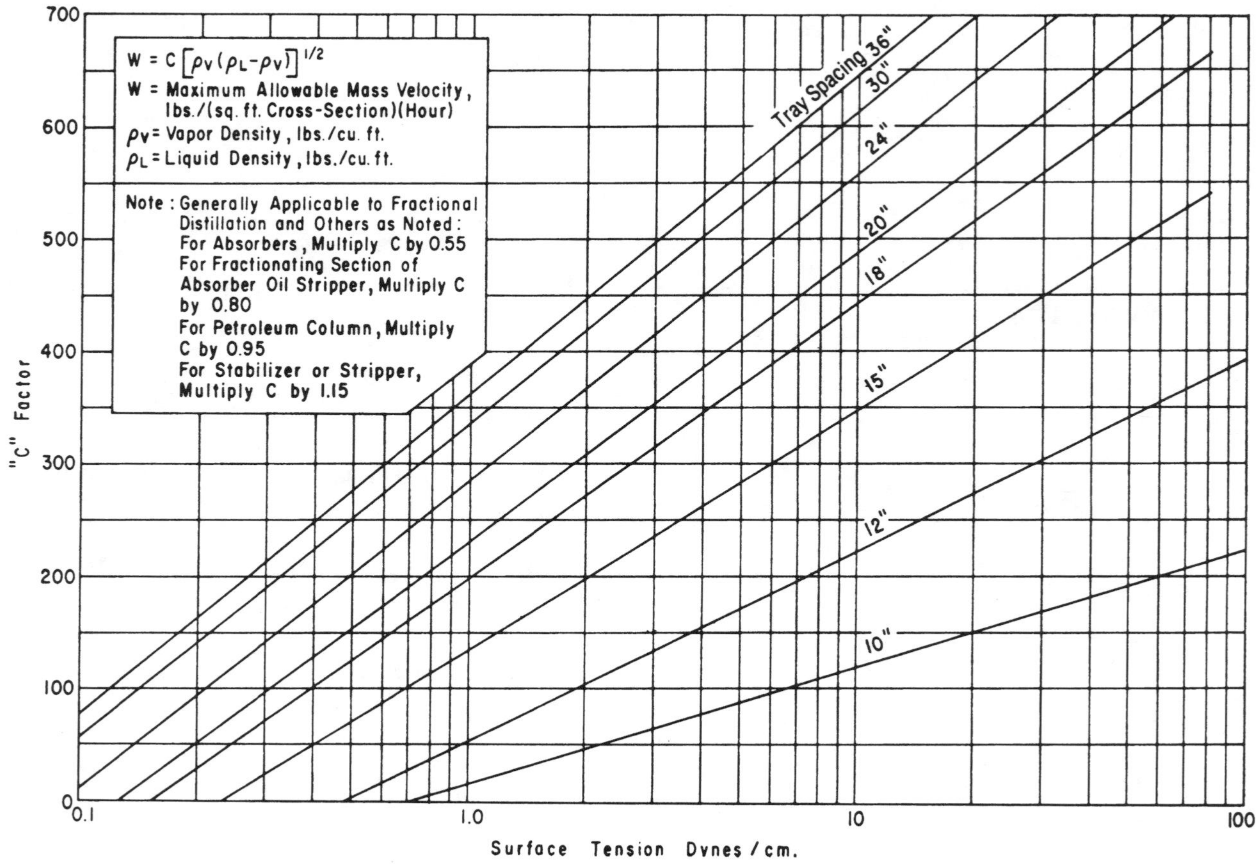

Figure 1. Souders-Brown correlation.

Sources

1. Souders, M. and Brown, G., "Design of Fractionating Columns—I Entrainment and Capacity," *Ind. Eng. Chem.,* 26, 98, (1934).
2. Ludwig, E. E., *Applied Process Design for Chemical and Petrochemical Plants, Vol. 2,* Gulf Publishing Co.

Diameter of Sieve/Valve Trays (F Factor)

The F factor is used in the expression $U = F/(\rho_V)^{0.5}$ to obtain the allowable superficial vapor velocity based on free column cross-sectional area (total column area minus the downcomer area). For foaming systems, the F factor should be multiplied by 0.75.

The F factor correlation from Frank[1] is shown in Figure 1:

The author has developed an equation for the F factor as follows:

$$F = (547 - 173.2T + 2.3194T^2)\ 10^{-6}P + 0.32 + 0.0847T - 0.000787T^2$$

Correlation ranges are:

F = 0.8 to 2.4
P = 0 to 220
T = 18 to 36

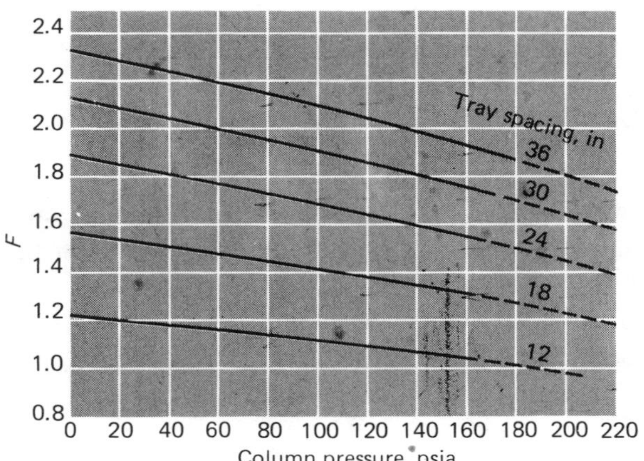

Figure 1. F Factor as a function of column pressure drop and tray spacing.

For estimating downcomer area, Frank gives the following correlation:

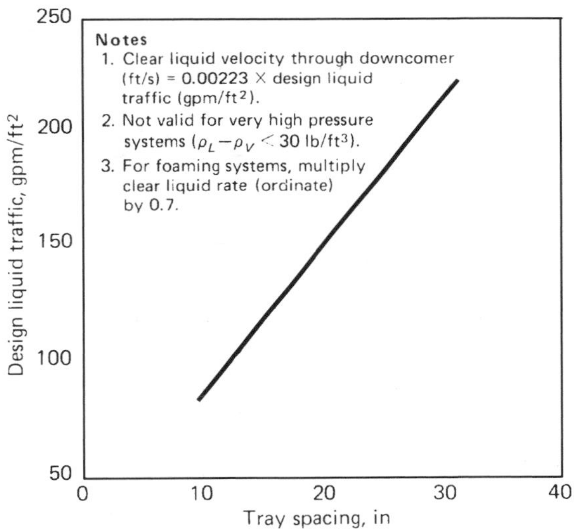

Figure 2. Estimation of downcomer area for a tray-type distillation column.

The author has developed the following equation to fit the correlation:

$$DL = 6.667T + 16.665$$

Clear liquid velocity (ft/sec) through the downcomer is then found by multiplying DL by 0.00223. The correlation is not valid if $\rho_L - \rho_V$ is less than 30 lb/ft^3 (very high pressure systems). For foaming systems, DL should be multiplied by 0.7. Frank recommends segmental downcomers of at least 5% of total column cross-sectional area, regardless of the area obtained by this correlation.

For final design, complete tray hydraulic calculations are required.

For even faster estimates, the following rough F factor guidelines have been proposed:

Situation	F factor
Fractionating column total cross section vapor velocity	1.0–1.5
Sieve tray hole velocity to avoid weeping	>12
Disengaging equipment for separating liquid droplets from vapor	<6

Nomenclature

- DL = Design downcomer liquid traffic, gpm/ft^2
- F = Factor for fractionation allowable velocity
- P = Column pressure, psia
- T = Tray spacing, in.
- ρ_L = Liquid density, lb/ft^3
- ρ_V = Vapor density, lb/ft^3

Source

Frank, O., "Shortcuts for Distillation Design," *Chemical Engineering,* March 14, 1977.

Diameter of Sieve/Valve Trays (Smith)

Smith uses settling height as a correlating factor which is intended for use with various tray types. The correlation is shown in Figure 1. Curves are drawn for a range of settling heights from 2 to 30 inches. Here, U is the vapor velocity above the tray not occupied by downcomers.

The developed equations for the curves are (Y subscript is settling height in inches):

$$Y = A + BX + CX^2 + DX^3$$

	A	B	C	D
Y_{30}	−1.68197	−.67671	−.129274	−.0046903
Y_{24}	−1.77525	−.56550	−.083071	+.0005644
Y_{22}	−1.89712	−.59868	−.080237	+.0025895
Y_{20}	−1.96316	−.55711	−.071129	+.0024613
Y_{18}	−2.02348	−.54666	−.067666	+.0032962
Y_{16}	−2.19189	−.51473	−.045937	+.0070182
Y_{14}	−2.32803	−.44885	−.014551	+.0113270
Y_{12}	−2.47561	−.48791	−.041355	+.0067033
Y_{10}	−2.66470	−.48409	−.040218	+.0064914
Y_8	−2.78979	−.43728	−.030204	+.0071053
Y_6	−2.96224	−.42211	−.030618	+.0056176
Y_4	−3.08589	−.38911	+.003062	+.0122267
Y_2	−3.22975	−.37070	−.000118	+.0110772

Smith recommends obtaining the settling height (tray spacing minus clear liquid depth) by applying the familiar Francis Weir formula. For our purposes of rapidly checking column diameter, a faster approach is needed.

For applications having 24-in. tray spacing, the author has observed that use of a settling height of 18 inches is good enough for rough checking. The calculation yields a superficial vapor velocity that applies to the tower cross section not occupied by downcomers. A downcomer area of 10% of column area is minimum except for special cases of low liquid loading.

For high liquid loading situations and final design, complete tray hydraulic calculations are required.

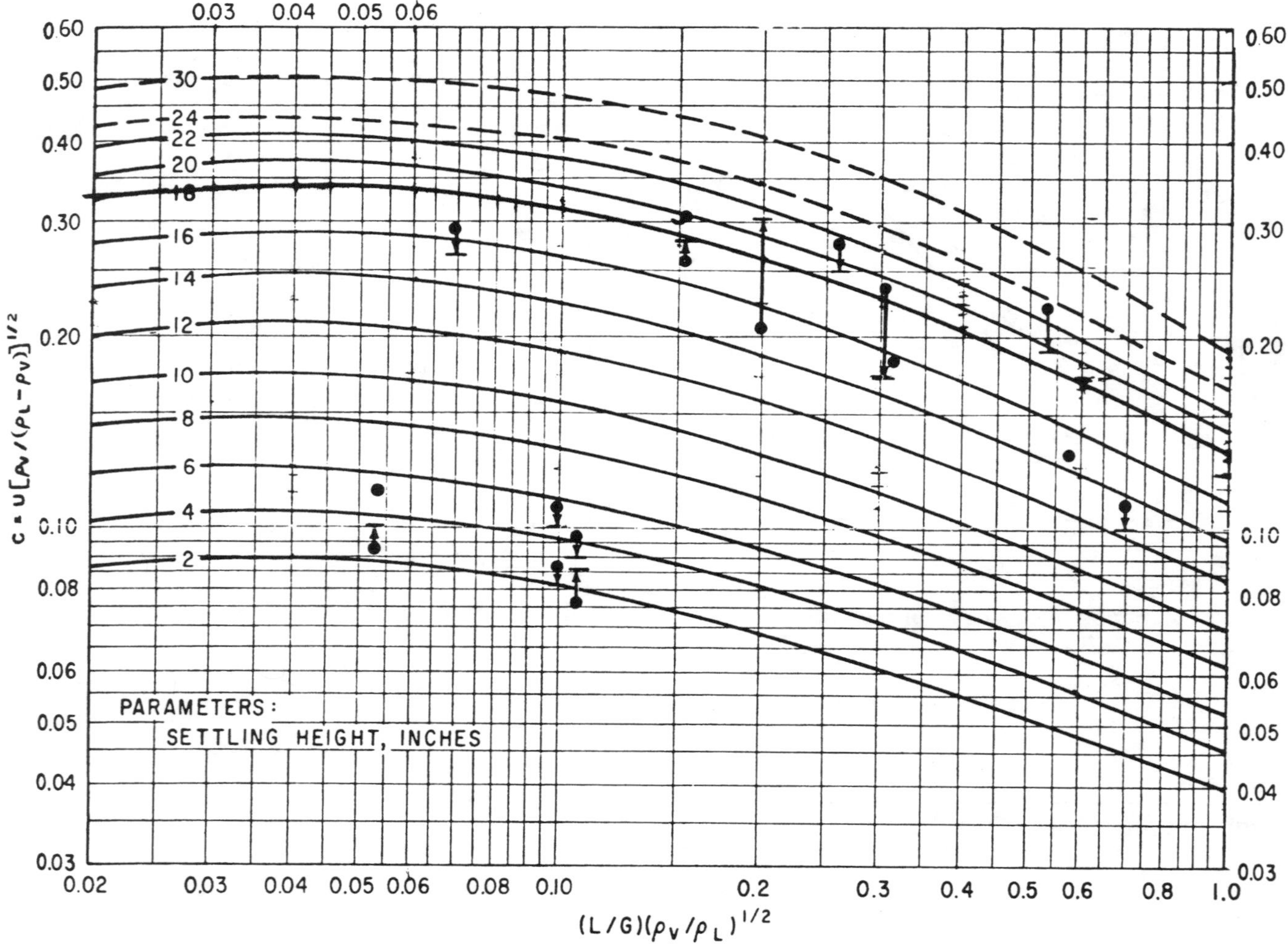

Figure 1. The vapor velocity function is plotted against a capacity factor with parameters of settling heights.

Nomenclature

G = Vapor rate, lb/hr
L = Liquid rate, lb/hr
U = Superficial vapor velocity above the tray not occupied by downcomers, ft/sec
ρ_L = Liquid density, lb/ft^3
ρ_V = Vapor density, lb/ft^3

Sources

1. Smith, R., Dresser, T., and Ohlswager, S., "Tower Capacity Rating Ignores Trays," *Hydrocarbon Processing and Petroleum Refiner,* May 1963.

2. The equations were generated using FLEXCURV V.2, Gulf Publishing Co.

Diameter of Sieve/Valve Trays (Lieberman)

Lieberman gives two rules of thumb for troubleshooting fractionators that could also be used as checks on a design. First, the pressure drops across a section of trays must not exceed 22% of the space between the tray decks, to avoid incipient flood. Mathematically, hold

$P/(SG)(T_n)(T_s) < 22\%$

Where

P = Pressure drop in inches of water
SG = Specific gravity of the liquid on the tray at the appropriate temperature
T_n = Number of trays
T_s = Tray spacing, in.

For sieve trays, a spray height of 15 inches is obtained when the jetting factor is 6–7.

jetting factor = $U^2 \rho_v/\rho_L$

Where:

U = Hole vapor velocity, ft/sec
ρ_L = Liquid density
ρ_V = Vapor density

For a 15-inch spray height, a tray spacing of at least 21 inches is recommended.

Source

Lieberman, N. P., *Process Design For Reliable Operations, 2nd Ed.,* Gulf Publishing Co.

Diameter of Ballast Trays

Calculation/Procedure for Ballast Tray Minimum Tower Diameter

The following method will give quick approximate results, but for complete detailed rating, use the Glitsch Manual[1].

1. Determine vapor capacity factor CAF.

   $CAF = CAF_o \times \text{System Factor}$

   The proper system factor can be found in Table 1.

2. For $\rho_V < 0.17$ use

   $CAF_o = (TS)^{0.65} \times (\rho_V)^{1/6}/12$

   For $\rho_V > 0.17$ use

   12–30″ TS: $CAF_o = (0.58 - 3.29/(TS) - [(TS \times \rho_V - TS)/1560]$

Table 1
System Factors[1]

Service	System Factor
Non-foaming, regular systems	1.00
Fluorine systems, e.g., BF_3., Freon	.90
Moderate foaming, e.g., oil absorbers, amine and glycol regenerators	.85
Heavy foaming, e.g., amine and glycol absorbers	.73
Severe foaming, e.g., MEK units	.60
Foam-stable systems, e.g., caustic regenerators	.30–.60

36 & 48″ TS: $CAF_o = (0.578 - 3.29/TS) - [(0.415TS \times \rho_V - TS)/1560] - 0.0095\rho_V$

3. The capacity factor CAF_o increases with increasing tray spacing up to a limiting value. Limits reached

below 48″ spacing are not experienced until the vapor density exceeds 4 lbs/ft^2. The author developed an equation to determine the limiting CAF_o value for vapor densities above 4 lbs/ft^2. It should be used for situations having 18″ tray spacing and above.

$$CAF_o\ (limiting) = 0.648 - 0.0492\rho_V$$

Absorbers and strippers frequently operate with a liquid having essentially the same physical characteristics regardless of the pressure. An example of this is a gas absorber. The same lean oil is used if the tower is operating at 100 or 1000 psi. This type of system is excluded from the CAF_o *limiting value.*

4. Calculate V_{load}

$$V_{load} = CFS\sqrt{\rho_v/(\rho_L - \rho_v)}$$

5. Using V_{load} and GPM (column liquid loading in gallons per minute), obtain approximate tower diameter for calculating flow path length. Use

$$(DTA)^2 = 3.025A + 0.0012AB + 7.647 \times 10^{-7} AB^2 + 2.81 \times 10^{-4} B^{1.719}$$

where

DTA = Approximate tower diameter, ft
A = V_{load}
B = GPM

This is the author's correlation of a nomograph in the Glitsch Manual. It gives results within 5% of the nomograph for diameter 4 feet or greater and within 15% for smaller diameters. This is adequate for this first approximation of tower diameter. It applies for

a. Single pass trays
b. 24″ tray spacing at 80% flood
c. DT = 2′–10′
d. V_{load} = 0–30
e. GPM = 0–1500

6. For 2-pass trays
   a. Divide V_{load} by 2
   b. Divide GPM by 2
   c. Obtain diameter from single pass equation
   d. Multiply diameter by $\sqrt{2}$

For 4-pass trays, replace the 2's by 4's.

7. Calculate flow path length.

$$FPL = 9 \times DTA/NP$$

8. Calculate minimum active area.

$$AAM = \frac{V_{load} + (GPM \times FPL/13000)}{CAF \times FF}$$

A flood factor of .65 to .75 should be used for column diameters under 36″ to compensate for wall effects. Larger columns are typically designed for about 80% of flood.

9. Obtain downcomer design velocity, VD_{dsg}. Use the smallest value from these three equations

$$VD_{dsg} = 250 \times \text{system factor}$$
$$VD_{dsg} = 41\sqrt{\rho_L - \rho_v} \times \text{system factor}$$
$$VD_{dsg} = 7.5\sqrt{TS} \times \sqrt{\rho_L - \rho_v} \times \text{system factor}$$

where

VD_{dsg} = Design velocity, GPM/ft^2

For the system factor, use values out of Table 1 except for the last item in that table (foam-stable systems) use 0.30.

10. Calculate downcomer minimum area.

$$ADM = GPM/(VD_{dsg} \times FF)$$

where

ADM = Minimum downcomer area, ft^2

If the downcomer area calculated above is less than 11% of the active area, use the smaller of

ADM = Double that of the equation
ADM = 11% of the active area

11. Calculate minimum column cross-sectional area. Use the larger of

$$ATM = AAM + 2 \times ADM$$

$$ATM = \frac{V_{load}}{0.78 \times CAF \times FF}$$

where

ATM = Minimum column cross-sectional area, ft^2. Further detailed design calculations may result in a change in tower diameter.

12. Calculate column minimum diameter.

$$DT = \sqrt{ATM / 0.7854}$$

Percentage of Flood for Existing Ballast Tray Columns

$$\% \text{ Flood} = \frac{V_{load} + (GPM \times FPL / 13000)}{AA \times CAF}$$

Minimum diameter for multipass trays is given in Table 2.

Holding GPM/WFP below about 8 is preferred, although liquid rates as high as 20 GPM/WFP can and have been used. WFP is the width of the flow path in inches. Some companies like to use trays having no more than two passes.

Table 2
Minimum Practical Diameter for Multipass Ballast Trays

Multipass Tray Number of Passes	Minimum Diameter (ft)	Preferred Diameter (ft)
2	5	6
3	8	9
4	10	12
5	13	15

Nomenclature

AA = Active area, ft^2
AAM = Minimum active area, ft^2
AD = Downcomer area, ft^2
ADM = Minimum downcomer area, ft^2
ATM = Minimum column cross-sectional area, ft^2
CAF = Vapor capacity factor
CAF_o = Flood capacity factor at zero liquid load
CFS = Vapor rate, actual ft^3/sec
DT = Tower diameter, ft
DTA = Approximate tower diameter, ft
FF = Flood factor or design percent of flood, fractional
FPL = Tray flow path length, in.
GPM = Column liquid loading, gal/min
NP = Tray number of flow paths or passes
TS = Tray spacing, in.
VD_{dsg} = Downcomer design velocity, GPM/ft^2
V_{load} = Column vapor load factor
WFP = Width of tray flow path, in.
ρ_L = Liquid density, lbs/ft^3
ρ_V = Vapor density, lbs/ft^3

Sources

1. *Glitsch Ballast Tray Design Manual, 5th Ed.* Bulletin No. 4900, Copyright 1974, Glitsch Inc.
2. Branan, C. R., *The Fractionator Analysis Pocket Handbook,* Gulf Publishing Co., 1978.

Diameter of Fractionators, General

For checking designs one can roughly relate tower diameter to reboiler duty as follows:

Situation	Reboiler Duty, MM/BTU/hr
Pressure distillation	$0.5\ D^2$
Atmospheric pressure distillation	$0.3\ D^2$
Vacuum distillation	$0.15\ D^2$

Where

D = Tower diameter, ft

Source

Branan, C. R., *The Process Engineer's Pocket Handbook, Vol. 1,* Gulf Publishing Co., 1976.

Control Schemes

A good understanding of control schemes is essential for understanding fractionation systems. It is not possible to discuss all possible control hookups. The following discussions will therefore include many of the major means of control and attempt to provide enough control philosophy to allow custom design or analysis for individual circumstances. The discussions will be limited to conventional instrumentation since computer control is beyond the scope of this handbook.

Column Control Introduction by Keith Claiborne[5]

For orientation in trying to control fractionators, it is well to emphasize the territory within which fractionators, and therefore control, move.

Fractionators produce two results only: (1) stream splitting, with so many pounds going out one "end' and all other feed pounds going out the other and (2) component segregation toward one or the other of the product streams, characterized by the Fenske ratio:

$$\frac{\text{mols lt in O.H.}}{\text{mols lt in Btm}} \times \frac{\text{mols hvy in Btm}}{\text{mols hvy in O.H.}}$$

Result (1) is achieved by control of product flow from one end of the fractionator. Result (2) is achieved by control of the heat load on the fractionator.

By a combination of the two control handles, one can affect component specification in the product streams, but there is always some maximum possible extent of separation (value of Fenske ratio) in a given system. Nearly all control measures are designed to permit control of the two handles. It should be realized that column operation needs to be kept reasonably smooth, otherwise separation already achieved may in the following five minutes (or six hours) partially be undone by surging in the system.

As total and component rates vary in system feed, it becomes necessary to vary controlled parameters to maintain desired properties in product streams. There may also be subsidiary constraints. If column pressure gets too high, the feed system may be unable to input feed or allowable design pressures of equipment may be exceeded. With too low a pressure, product may not flow from the system. Too high a bottom temperature may cause excesses over equipment design or product degradation. Subsidiary constraints can produce "rapid" variations in flows or pressures. Even when these variations are not excessive for a given fractionator, they may impose impossible control problems on equipment connected in a larger system with that fractionator—upstream, downstream, or laterally.

In the long term, the best understanding and optimization of a system will usually require attention to attaining good heat and component balances in operation. Sometimes control performance, otherwise acceptable, may interact with measurements to inhibit accurate accounting.

All results besides (1) and (2) are, however, usually of more distant concern. Primarily emphasis on understanding and operating controls should be placed on those two items. It will sometimes be found that no pressure or temperature controls are needed if good, quick sensors are available to show what overall split is being made and what separation is occurring.

Pressure Control

Pressure has been classically controlled at a fixed value in fractionating columns. Shinskey[1] and Smith/Brodmann[2] have discussed variable pressure control. The author has also been told of unpublished successful variable pressure applications. Only the classical pressure control will be discussed here.

For partial condenser systems, the pressure can be controlled by manipulating vapor product or a noncondensible vent stream. This gives excellent pressure control. To have a constant top vapor product composition, the condenser outlet temperature also needs to be controlled. For a total condenser system, a butterfly valve in the column overhead vapor line to the condenser has been used. Varying the condenser cooling by various means such as manipulation of coolant flow is also common.

Pressure can also be controlled by variable heat transfer coefficient in the condenser. In this type of control, the condenser must have excess surface. This excess surface becomes part of the control system. One example of this is a total condenser with the accumulator running full and the level up in the condenser. If the pressure is too high, the level is lowered to provide additional cooling, and vice versa. This works on the principle of a slow moving liquid film having poorer heat transfer than a condensing vapor film. Sometimes it is necessary to put a partially flooded condenser at a steep angle rather than horizontal for proper control response.

Another example of pressure control by variable heat transfer coefficient is a vacuum condenser. The vacuum system pulls the inerts out through a vent. The control valve between the condenser and vacuum system varies the amount of inerts leaving the condenser. If the pressure gets too high, the control valve opens to pull out more inerts and produce a smaller tube area blanketed by inerts. Since relatively stagnant inerts have poorer heat transfer than condensing vapors, additional inerts withdrawal will increase cooling and lower pressure. The reverse happens if the pressure gets too low.

For vacuum applications having column temperature control above the feed point, put the measuring elements for the temperature and pressure controllers close together. This will make for good composition control at varying column loads (varying column differential pressure). In vacuum systems, a slight pressure change will produce large equilibrium temperature changes.

Another pressure control method is a hot gas bypass. The author has seen a retrofitted hot gas bypass system solve an unsatisfactory pressure control scheme. In the faulty system, a pressure control valve was placed between the overhead condenser and the lower reflux drum. This produced erratic reflux drum pressures. A pressure safety valve (psv) vented the reflux drum to the flare to protect against overpressure. A "psv saver" control valve, set slightly below the psv relieving pressure, kept the psv from having to operate very often. The "psv saver" is a good idea since psv's can develop leaks if forced to open and reseat frequently.

The fix for the erratic reflux drum pressure problem was to provide for separate pressure control of the fractionator column and the reflux drum. A new pressure control valve was installed upstream of the condenser and the old condenser outlet control valve was removed. A hot gas bypass, designed for 20% vapor flow, was installed around the pressure control valve and condenser. A control valve was installed in the hot gas bypass line. The column pressure was then maintained by throttling the new control valve upstream of the condenser. The reflux drum pressure was controlled by the hot gas bypass control valve and the "psv saver" working in split range. The new system is shown in the figure below.

Chin[3] states that a scheme having only a control valve in the hot gas bypass line manipulated by the column pressure controller will not work for the case of zero net vapor

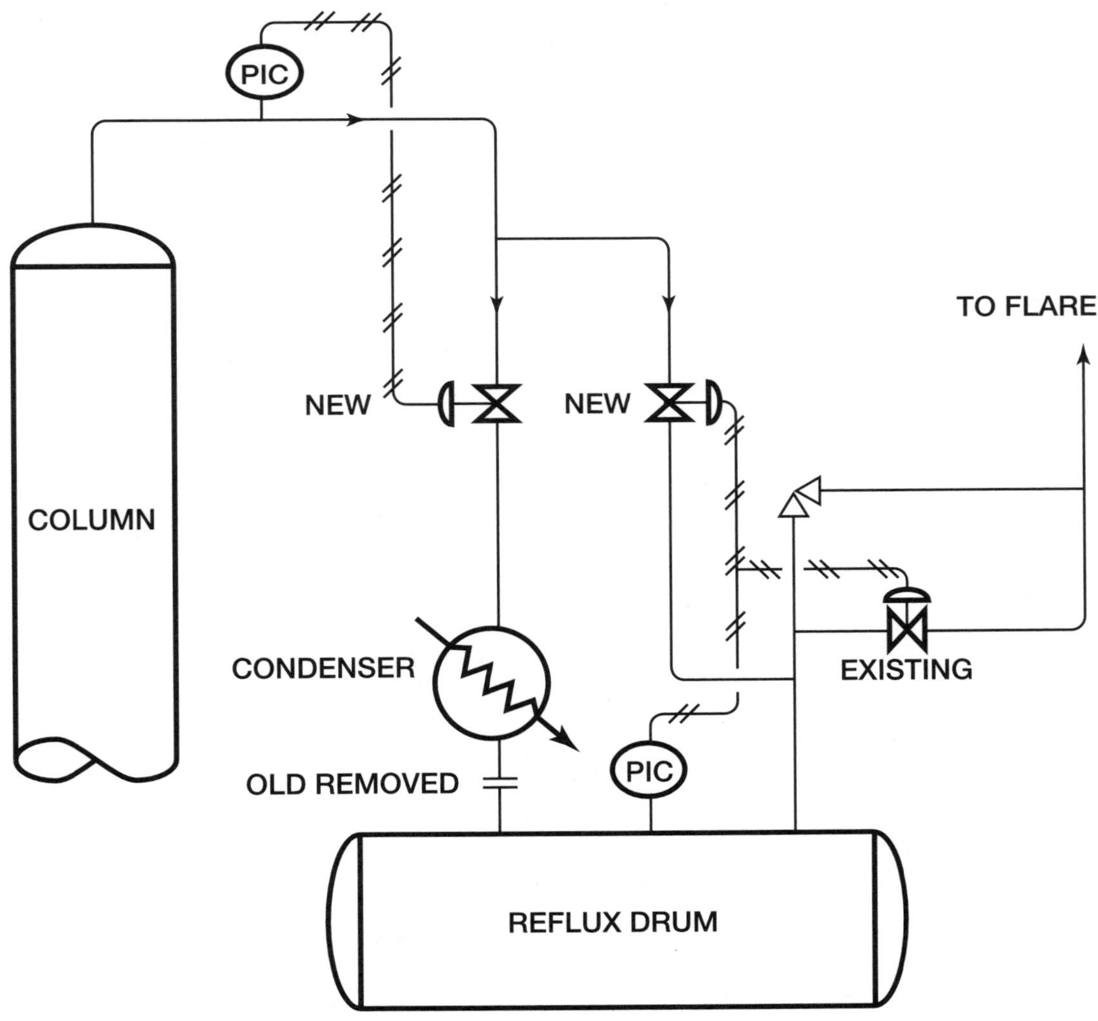

product. He cites an actual plant case and offers his analysis of the reasons. Chin's article shows 21 different distillation control methods with advantages and disadvantages for each.

Temperature Control

General. With simple instrumentation discussed here, it is not possible to satisfactorily control the temperature at both ends of a fractionation column. Therefore, the temperature is controlled either in the top or bottom section, depending upon which product specification is the most important. For refinery or gas plant distillation where extremely sharp cut points are probably not required, the temperature on the top of the column or the bottom is often controlled. For high purity operation, the temperature will possibly be controlled at an intermediate point in the column. The point where $\Delta T/\Delta C$ is maximum is generally the best place to control temperature. Here, $\Delta T/\Delta C$ is the rate of change of temperature with concentration of a key component. Control of temperature or vapor pressure is essentially the same. Manual set point adjustments are then made to hold the product at the other end of the column within a desired purity range. The technology does exist, however, to automatically control the purity of both products.

Temperature is the hardest parameter to control in a fractionation system. It exhibits high process and measurement lag. Temperature can also be ambivalent as a measure of composition. Pressure changes are reflected quickly up and down the column. Temperature changes are not. It is typical to provide three-mode controllers for all temperature applications.

Column Top Temperature. This temperature can be controlled automatically or manually by manipulating reflux, depending upon whether the automatic control point is at the top or bottom of the column. For partial condenser applications, it is typical to control condenser outlet temperature instead of column top temperature as discussed previously. In this case, a variety of methods are available, such as manipulating cooling water flow or bypass, or varying louvers or fan pitch on an air-cooled condenser. Manipulation of cooling water flow has the drawbacks of heat transfer being relatively insensitive to this variable and possible fouling if flow is not held at maximum.

Column Bottom Temperature. The bottom temperature is often controlled on the reboiler outlet line with a control valve in the heating medium line. The control point can also be on a bottom section tray. Care must be exercised in location of the temperature control point. It is recommended, especially for large columns, that a cascade arrangement be used. The recommended scheme has a complete flow recorder/controller (FRC) in the heating medium line including orifice and control valve. The set point of this FRC is manipulated by the temperature recorder/controller (TRC). This eliminates the TRC from manipulating the control valve directly (recall that temperature is the most difficult parameter to control). This makes for smoother control for normal operations. Also, it is handy for startup to be able to uncouple the TRC and run the reboiler on FRC for a period.

For a fired reboiler, a pump-around system is used with an FRC to maintain constant flow. There will be a low flow alarm plus fuel shutoff. There will also be a high flow alarm plus fuel shutoff, since a tube rupture would reflect itself in a high flow.

Feed Temperature. This is often allowed to vary. Often a feed-to-bottoms cross exchanger is used for preheat. Often the amount of heat available in the bottoms is close to the optimum feed preheat. Care must be taken not to heat the feed too hot and put inordinate vapor into the top section of the column. This will be detrimental to rectification. Feed preheat can be controlled by manipulating heating medium or by bypass for a cross exchanger. In deciding whether or not to control feed temperature, the designer sometimes has to choose between column stability versus energy recovery similarly to the case of constant versus variable pressure control. Feed temperature control becomes more critical as its amount of liquid and vapor increases relative to that of the traffic produced by the reflux and reboiler. For column analysis, it is good to check on how constantly the feed temperature is running.

Level Control

For a total condenser, accumulator level is typically set by varying distillate draw. For a partial condenser, it can be controlled with a condenser hot gas bypass.

The column bottom level is sometimes controlled by bottoms draw. Varying reboiler heating medium is another possibility. For some cases, bottoms draw level control works better and for others, heating medium level control. Bojnowski[4] gives a case where heating medium level control was desired for two columns in a plant. One column had to be put on bottoms draw level control, however, when improper reboiler thermosyphoning caused sluggish response with heating medium level control. In the case of variable heat content reboiler steam supply, use heating medium level control with bottoms draw rate setting composition. This provides the best bottoms composition control.

Flow Control

The feed flow is often not controlled but is rather on level control from another column or vessel. The liquid product flows (distillate and bottoms) are often on level rather than flow control. Top vapor product is, however, usually on pressure control. The reflux is frequently on FRC, but also may be on column TRC or accumulator level.

Differential Pressure Control

For column analysis and troubleshooting it is important to have pressure drop measured with a DP cell. The differential pressure can also be used to control column traffic. A good way to do this would be to let the differential pressure control the heating medium to the reboiler. The largest application for differential pressure control is with packed columns where it is desirable to run at 80 to 100% of flood for best efficiency.

Differential Temperature Control

Temperature sensing points at various points along the tower shell are often a useful troubleshooting tool. Many common operating problems produce a decrease in differential temperature across a section of the column. A differential temperature as well as a differential pressure recorder across a troublesome tower section would provide invaluable operator assistance.

Differential temperature as well as differential pressure can be used as a primary control variable. In one instance, it was hard to meet purity on a product in a column having close boiling components. The differential temperature across several bottom section trays was found to be the key to maintaining purity control. So a column side draw flow higher in the column was put on control by the critical temperature differential. This controlled the liquid "reflux" running down to the critical zone by varying the liquid drawn off at the side draw. This novel scheme solved the control problem.

Sources

1. Shinskey, F. G., "Energy-Conserving Control Systems for Distillation Units," *Chemical Engineering Progress,* May 1976.
2. Smith, C. L., and Brodmann, M. T., "Process Control: Trends for the Future," *Chemical Engineering,* June 21, 1976.
3. Chin, T. G., "Guide to Distillation Pressure Control Methods," *Hydrocarbon Processing,* October 1979, p. 145.
4. Bojnowski, J. J., Groghan, R. M., Jr., and Hoffman, R. M., "Direct and Indirect Material Balance Control," *Chemical Engineering Progress,* September 1976.
5. Branan, C. R., *The Fractionator Analysis Pocket Handbook,* Gulf Publishing Co., 1978.

Optimization Techniques

This section deals with optimization of an existing column. Higher throughput or improved product quality is balanced against higher operating costs such as labor, energy, or maintenance. If the product purity and production are fixed, then the optimum is the minimum energy to do the set job. Optimization would involve holding the reflux at the minimum to deliver the distillate purity and boilup at the minimum to deliver the bottoms purity.

Often, there is latitude in production rate and purity specifications, requiring optimization calculations to determine the best set of column operating conditions such as reflux, feed temperature, and reboiler heat input. The first step will be to determine by plant testing, and calculations to extrapolate the plant data, the ultimate limiting capacities of all pieces of equipment in the system such as column shell, reboiler, condenser, and pumps.

The search for the optimum can involve many case studies. Often, if time is limited, a rigorous canned computer program is utilized. However, this is expensive. If time is not as much of a factor, the method presented here will allow the calculations to be handled conveniently by hand or computer having limited core.

Smith-Brinkley Method

The procedure proposed for the optimization work is the Smith-Brinkley Method[1]. It is especially good for the uses described in this section. The more accurately known the operating parameters, such as tray temperatures and internal traffic, the more advantageous the Smith-Brinkley Method becomes.

The Smith-Brinkley Method uses two sets of separation factors for the top and bottom parts of the column, in contrast to a single relative volatility for the Underwood Method[2]. The Underwood Method requires knowing the distillate and bottoms compositions to determine the required reflux. The Smith-Brinkley Method starts with the column parameters and calculates the product compositions. This is a great advantage in building a model for hand or small computer calculations. Starting with a base case, the Smith-Brinkley Method can be used to calculate the effect of parameter changes on the product compositions.

Distillation

Smith[1] fully explains the Smith-Brinkley Method and presents a general equation from which a specialized equation for distillation, absorption, or extraction can be obtained. The method for distillation columns is discussed here.

For distillation component i,

$$f = \frac{(1 - S_n^{N-M}) + R(1 - S_n)}{(1 - S_n^{N-M}) + R(1 - S_n) + hS_n^{N-M}(1 - S_m^{M+1})}$$

where

$f_i = (BX_B/FX_F)_i$

$S_{ni} = K_iV/L$

$S_{mi} = K'_iV'/L'$

h = Correlating factor; if the feed is mostly liquid, use Equation 1 and if mostly vapor, Equation 2

$$h_i = \frac{K'_i}{K_i}\left(\frac{L}{L'}\right)\left(\frac{1 - S_n}{1 - S_m}\right)_i \quad (1)$$

$$h_i = \frac{L}{L'}\left(\frac{1 - S_n}{1 - S_m}\right)_i \quad (2)$$

The effective top and bottom section temperatures are used to determine K_i and K'_i. These are used along with the effective top and bottom section molar liquid and vapor rates to determine S_n and S_m.

Building a Distillation Base Case

If data is meager, section temperatures can be simply the average of the top and feed temperatures for the top section and bottom and feed temperatures for the bottom section. Also, the molar flows can be obtained, assuming equal molar overflow. However, usually a rigorous computer run is available for a plant case or a design case. Having this, the effective values can be determined by averaging values for all stages in a section. Some columns can be calculated adequately with the simple averages first discussed, but in others the stage-by-stage averaged values from a computer run are necessary for satisfactory results.

The Smith-Brinkley Method can therefore be used to generate a hand base case beginning with either a heat and material balanced plant case, a rigorous computer solution of a plant case, or computer solution of a design case. Once the hand base case is established, alternate cases can be done by hand (or small computer having limited core) using the Smith-Brinkley Method.

Alternate Distillation Cases

Changing feed temperature can be approximated by adding one half of the change in feed temperature to the top and bottom effective temperatures. For a large feed temperature change, the feed flash and the column heat balance will have to be redone.

Small changes in reflux can be approximated by assuming no change in effective temperatures, thus avoiding trial and error calculation. The extra liquid and vapor are added to S_n and S_m. For larger reflux changes, the extra trial and error of determining the new end temperatures by bubble point and dew point calculations can be included. The effective section temperatures change only one half as fast as end temperatures, since the feed temperature remains unchanged.

The preceding discussion on reflux assumes that the condenser is not limiting when the reflux is raised. For a severely limited condenser, an evaluation must first be made of the condenser heat transfer before analyzing the effect of a reflux increase with Smith-Brinkley. Likewise, a limiting reboiler or trays close to flood would have to be evaluated prior to Smith-Brinkley calculations.

A small change in reboiler duty can be approximated by adding the increased traffic to S_m and S_n. If the top temperature is automatically controlled, extra reflux would also result.

In summary, for each alternate operation, an approximate net effect or set of effects can be quickly built into the Smith-Brinkley input to determine new end compositions. The instrumentation scheme will affect the result produced by an operating change.

Collecting Field Data

When collecting meaningful field fractionating column data, the column must not only have constant flow rates, but the flows must give a good material balance (no accumulation). In addition, a steady state condition must exist for the given flow rates.

To determine if steady state conditions exist, the temperatures and pressures in the column can be tabulated to assure that they are reasonably unchanging. Laboratory analyses are usually too slow and expensive for checking lined out conditions. Monitoring reflux accumulator boiloff is often an effective way of noting concentration changes. Simply let a sample of the accumulator liquid boil at atmospheric pressure in a bottle with a thermometer inserted. This method is limited to light hydrocarbons and is not accurate enough for precision fractionation.

The accumulator holdup will often be the limiting item setting time required for samples to be representative of the column operating conditions. One "changeout" or "turnover" of the accumulator may be insufficient time, if test conditions are much different from conditions prior to the test. The following example will serve to show this.

Assume an accumulator having a liquid holdup of 2,000 gal. Assume that the liquid leaving as reflux plus distillate is 40 GPM. Further assume that the accumulator liquid starts at 2.0 vol % of a key component and that the test conditions produce 0.0 vol. % of the key component. How long will it take for the concentration to reach 0.1 vol. % (the assumed laboratory test accuracy)?

Establish the following symbols:

x = Concentration, vol. fraction
dx = Differential concentration element, vol. fraction
dT = Differential time element, min.

Then:

$$\frac{dx}{dT} = \frac{40x}{2000}; \qquad \int_{0.02}^{0.001} \frac{dx}{x} = 0.02 \int_{o}^{T} dT;$$

$$\ln x\Big]_{0.02}^{0.001} = 0.02T\Big]_{o}^{T}$$

$$-6.908 + 3.912 = 0.02T; T = -150 \text{ or } T = 150 \text{ minutes}$$

whereas: 1 "turnover" = 2000/40
= 50 minutes

Nomenclature

B = Bottoms total molar rate or subscript for bottoms
F = Feed total molar rate or subscript for feed
f_i = Ratio of molar rate of component i in the bottoms to that in the feed
K_i = Equilibrium constant of component i in the top section equals y/x
K'_i = Equilibrium constant of component i in the bottom section equals y/x
L = Effective total molar liquid rate in top section
L′ = Effective total molar liquid rate in bottom section
M = Total equilibrium stages below the feed stage including reboiler
N = Total equilibrium stages in the column including reboiler and partial condenser
R = Actual reflux ratio
S_m = Stripping factor for component i in bottom section
S_n = Stripping factor for component i in top section
V = Effective total molar vapor rate in top section
V′ = Effective total molar vapor rate in bottom section
X = Mol fraction in the liquid
Y = Mol fraction in the vapor

Sources

1. Smith, B. D., *Design of Equilibrium Stage Processes,* McGraw-Hill, 1963.
2. Underwood, A. J. V., "Fractional Distillation of Multicomponent Mixtures," *Chemical Engineering Progress,* 44, 603, (1948).
3. Branan, C. R., *The Fractionator Analysis Pocket Handbook,* Gulf Publishing Co., 1978.

Reboilers

Reboiler Types

The *GPSA Engineering Data Book*[1] has an excellent section on reboilers. The most common types are the following:

- Forced circulation (Figure 1)
- Natural circulation (Figure 2)
  - Once-through
  - Recirculating
- Vertical thermosyphon (Figure 3)
- Horizontal thermosyphon (Figure 4)
- Flooded bundle (Kettle) (Figure 5)
- Recirculating—Baffled bottom (Figure 6)

For thermosyphon reboilers, the hydraulic aspects are as important as the heat transfer aspects. The design of thermosyphon reboiler piping is too broad a subject for this handbook. Some good articles on the subject can be found in References 2–14. Reference 3 is particularly good for horizontal thermosyphon reboilers. Table 1 has typical vertical thermosyphon design standards.

Reboiler Selection

Figure 7 provides an overview of reboiler selection choices. The accompanying notes provide information for a quick or "first cut" estimate of the appropriate type for a given application. Tables 2[12] and 3[13] provide additional, more detailed, selection data. Table 2 gives advantages and disadvantages for all the major reboiler types. Table 3 is limited to thermosyphon types.

For reboilers, especially thermosyphon types, "the devil is in the details." The information presented herein is intended for preliminary work. Final design is performed by experienced engineers using detailed design techniques.

General notes for thermosyphon reboilers:

1. Never use inclined piping for two-phase flow in a process plant. This is particularly true for reboiler return piping. Use only horizontal or vertical runs.
2. If the reboiler heating medium is condensing steam, provide a desuperheater if the superheat is more than 40–60°F.

(*text continued on page 72*)

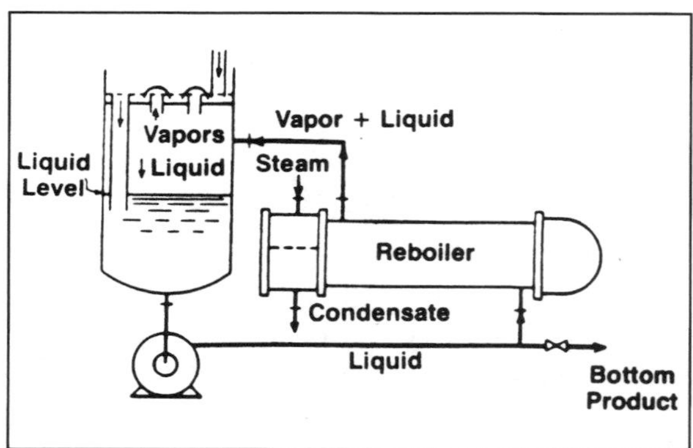

Figure 1. Forced-circulation reboiler arrangement.

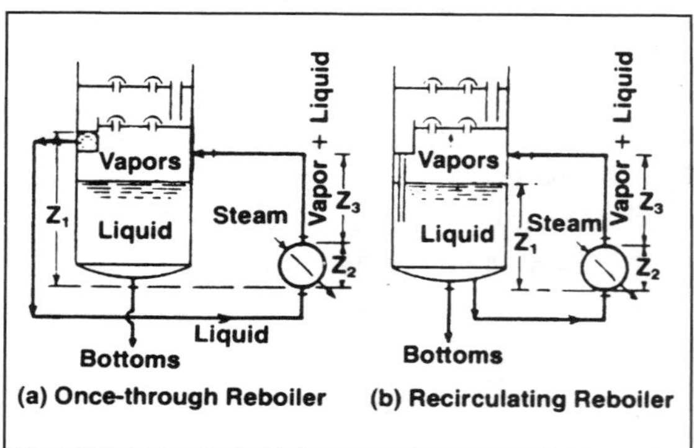

Figure 2. Natural-circulation reboiler arrangements.

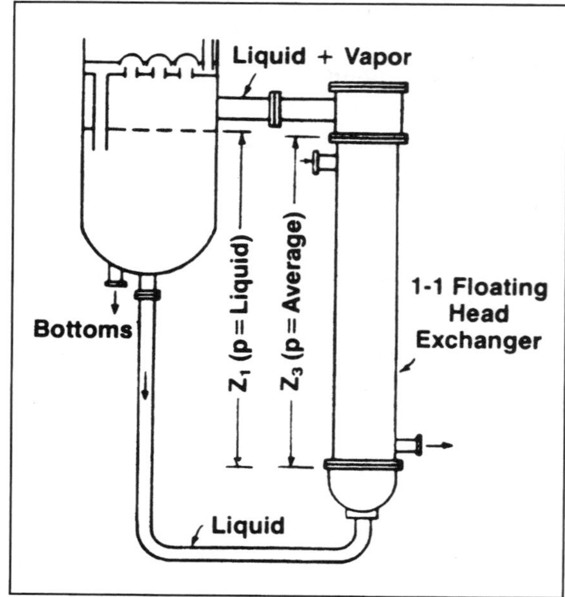

Figure 3. Vertical thermosyphon reboiler connected to tower.

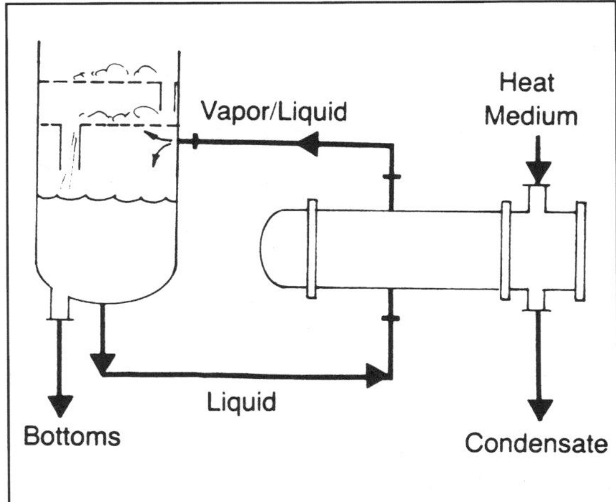

Figure 4. Horizontal thermosyphon reboiler.

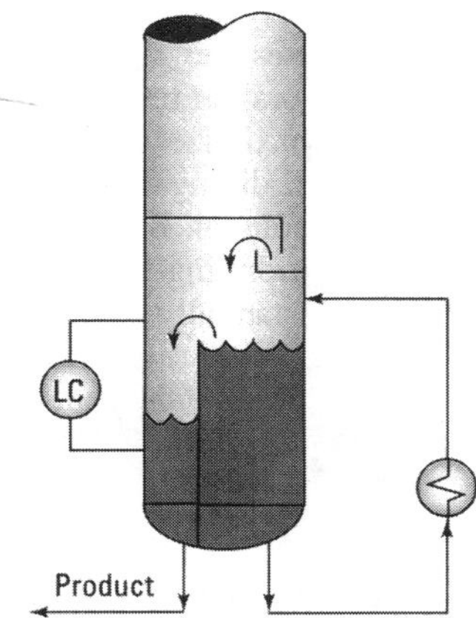

Figure 6. Recirculating—Baffled bottom[13] (Reproduced with permission of the American Institute of Chemical Engineers, copyright 1997 AIChE, all rights reserved.)

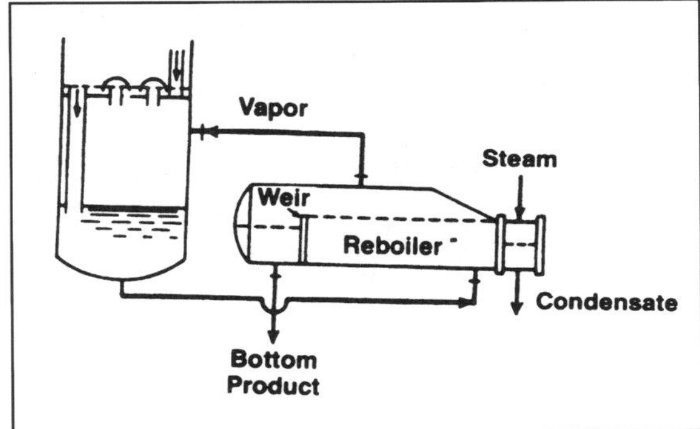

Figure 5. Kettle reboiler arrangement.

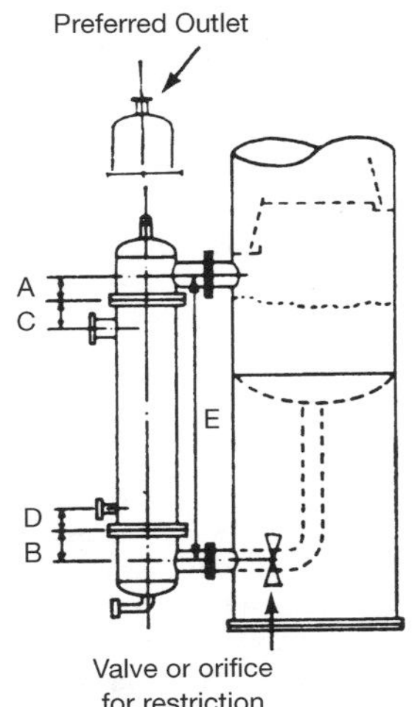

Table 1
Typical Thermosiphon Reboiler Design Standards*

Shell O.D., inches	No. tubes	Tube sheet faces	Approx. area	Nozzle sizes: Vapor, inches	Nozzle sizes: Liquid, inches	Nozzle sizes: Steam, inches	Nozzle sizes: Cond., inches	Dimensions: A inches	Dimensions: B inches	Dimensions: C inches	Dimensions: D inches	Dimensions: E
16	108	4′-11¾″	132□′	6	4	4	1½	7 15/16	5 15/16	8	5¾	6′-1¾″
20	176	4′-11¾″	215□′	8	6	4	2	8 15/16	7 15/16	8	5¾	6′-4¾″
24	272	4′-11¾″	333□′	10	6	6	3	9 15/16	7 15/16	9	7¾	6′-5¾″
30	431	4′-11¾″	527□′	12	6	6	3	11 7/16	7 15/16	9	7¾	6′-7¼″
36	601	4′-11¾″	735□′	16	8	8	4	13 3/16	9 15/16	10¾	8	6′-11″
24	272	6′-7¾″	448□′	10	6	6	3	9 15/16	7 15/16	9	6¾	8′-1¾″
30	431	6′-7¾″	710□′	12	6	6	3	11 7/16	7 15/16	9	6¾	8′-3¼″
36	601	6′-7¾″	990□′	16	8	8	4	13 3/16	9 15/16	10¾	8	8′-7″
42	870	6′-7¾″	1,440□′	16	10	8	4	17 11/16	10 15/16	10 1/16	6 13/16	9′-0½″
30	431	9′-11¾″	1,065□′	12	6	8	3	11 7/16	7 15/16	9	6¾	11′-7¼″
36	601	9′-11¾″	1,520□′	16	8	8	4	13 3/16	9 15/16	10¾	6¾	11′-11″
42	870	9′-11¾″	2,180□′	16	10	8	4	17 11/16	10 15/16	10 1/16	6 13/16	12′-4½″

**By permission, D. C. Lee, J. W. Dorsey, G. Z. Moore and F. D. Mayfield,* Chem. Eng. Prog., *Vol. 52, No. 4., p. 160 (1956).*

***Cross section area of vapor nozzle off to channel must be minimum of 1.25 times total flow area of all tubes.*

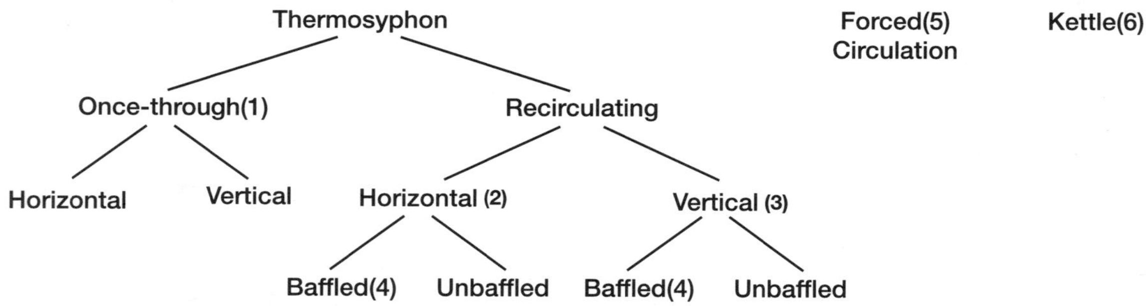

Figure 7. Quick Selection Guide

[1]Preferable to recirculating where acceptable vaporization rates can be maintained (less than 25–30%). This type is chosen when there is a need to minimize exposure of degradable and/or fouling substances to high temperatures.
[2]Used for large duties, dirty process, and frequent cleaning required. The process is usually in the shell side. This type is used in 95% of oil refinery thermosyphon applications.
[3]Used for small duties, clean process, and only infrequent cleaning required. Vaporization is usually less than 30%, but less than 15% if the fractionator pressure is below 50 psig. The viscosity of the reboiler feed should be less than 0.5 cp. Put a butterfly valve in the reboiler inlet piping. This type is used in nearly 100% of chemical plant thermosyphon applications (70% of petrochemical).
[4]Greater stability than unbaffled.
[5]Usually used where piping pressure drop is high and therefore natural circulation is impractical.
[6]Very stable and easy to control. Has no two-phase flow. Allows low tower skirt height. This type is expensive, however.

Table 2
Reboiler Comparison[12]

Type	Advantage	Disadvantage
Kettle	• One theoretical tray • Ease of maintenance • Vapor disengaging • Low skirt height • Handles viscosity greater than 0.5 cp • Ease of control • No limit on vapor load	• Extra piping and space • High cost • Fouls with dirty fluids • High residence time in heat zone for degradation tendency of some fluids • Low residence time surge section of reboiler
Vertical once-through	• One theoretical tray • Simple piping and compact • Not easily fouled • Less cost than kettle	• Difficult maintenance • High skirt height • No control of circulation • Moderate controllability
Vertical natl. circ.	• Good controllability • Simple piping and compact • Less cost than kettle	• No theoretical tray • Accumulation of high boiling point components in feed line, i.e., temperature may be slightly higher than tower bottom • Too high liquid level above design could cause reboiler to have less capacity • Fouls easier • Difficult maintenance • High skirt height
Horizontal once-through	• One theoretical tray • Simple piping and compact	• No control of circulation • Moderate controllability

(*table continued on next page*)

Kettle(6)

Table 2 (Continued)
Reboiler Comparison[12]

Type	Advantage	Disadvantage
	• Not easily fouled • Lower skirt height than vertical • Less pressure drop than vertical • Longer tubes possible • Ease of maintenance • Less cost than kettle	• High skirt height
Horizontal natl. circ.	• Ease of maintenance • Lower skirt height than vertical • Less pressure drop than vertical • Longer tubes possible • Less cost than kettle	• No theoretical tray • Extra space and piping as compared to vertical • Fouls easier as compared to vertical • Accumulation of higher boiling point components in feed line, i.e., temperature may be slightly higher than tower bottom
Forced circ.	• One theoretical tray • Handles high viscous solids-containing liquids • Circulation controlled • Higher transfer coefficient	• Highest cost with additional piping and pumps • Higher operating cost • Requires additional plant area

Table 3
Thermosyphon Selection Criteria[13]

Factor	Vertical	Vertical	Horizontal	Horizontal
Process side	Tube side	Tube side	Shell side	Shell side
Process flow	Once-through	Circulating	Once-through	Circulating
Heat-transfer coefficient	High	High	Moderately high	Moderately high
Residence time in heated zone	Low	Medium	Low	Medium
ΔT required	High	High	Medium	Medium
Design data	Available	Available	Some available	Some available
Capital cost (total)	Low	Low	Medium	Medium
Plot requirements	Small	Small	Large	Large
Piping needed	Low cost	Low cost	High cost	High cost
Size possible*	Small	Small	Large	Large
Shells	3 maximum	3 maximum	As needed	As needed
Skirt height	High	High	Lower	Lower
Distillation stages	One	Less than one	One	Less than one
Maintenance	Difficult	Difficult	Easy	Easy
Circulation control	None	Possible	None	Possible
Controllability	Moderate	Moderate	Moderate	Moderate
Fouling suitability (process)	Good	Moderate	Good	Moderate
Vaporization range, minimum	5%+	10%+	10%+	15%+
Vaporization range†	25%	25%	25%	25%
Vaporization range, maximum‡	35%	35%	35%	35%

*Small = less than 8,000 ft^2/shell, large = more than 8,000 ft^2/shell.
†Normal upper limit of standard design range; design for vaporization above this level should be handled with caution.
‡Maximum if field data are available.

(*text continued from page 68*)

Sources

1. *GPSA Engineering Data Book,* 10th Ed., Gas Processors Suppliers Association.
2. Figure 8, from "Design Data for Thermosiphon Reboilers," by D. C. Lee, J. W. Dorsey, G. Z. Moore & F. Drew Mayfield, *Chemical Engineering Progress,* Vol. 52, No. 4, pp. 160–164 (1956). "Reproduced by permission of the American Institute of Chemical Engineers. © 1956 AIChE."
3. Collins, Gerald K., "Horizontal-Thermosiphon Reboiler Design," *Chemical Engineering,* July 19, 1976.
4. Fair, J. R., "Design Steam Distillation Reboilers," *Hydrocarbon Processing and Petroleum Refiner,* February 1963.
5. Fair, J. R., "Vaporizer and Reboiler Design," parts 1 and 2, *Chemical Engineering,* July 8 and August 5, 1963.
6. Fair, J. R., "What You Need to Design Thermosyphon Reboilers," *Petroleum Refiner,* February 1960.
7. Frank, O. and Prickett, R. D., "Designing Vertical Thermosyphon Reboilers," *Chemical Engineering,* September 3, 1973.
8. Kern, Robert, "How to Design Piping for Reboiler Systems," *Chemical Engineering,* August 4, 1975.
9. Kern, Robert, "Thermosyphon Reboiler Piping Simplified," *Hydrocarbon Processing,* December 1968.
10. Orrell, W. H., "Physical Considerations in Designing Vertical Thermosyphon Reboilers," *Chemical Engineering,* September 17, 1973.
11. Branan, C. R., *The Fractionator Analysis Pocket Handbook,* Gulf Publishing Co., 1978.
12. Love, D. L., "No Hassle Reboiler Selection," *Hydrocarbon Processing,*" October 1992.
13. Sloley, A. W., "Properly Design Thermosyphon Reboilers," *Chemical Engineering Progress,* March 1997.
14. McCarthy, A. J. and Smith, B. R., "Reboiler System Design—The Tricks of the Trade," *Chemical Engineering Progress, May* 1995.

Packed Columns

Packed columns are gaining ground on trayed columns. Lieberman[1] states that based on his design and operating experience, a properly designed packed tower can have 20–40% more capacity than a trayed tower with an equal number of fractionation stages.

Improved packings are being developed as well as internals and techniques for assuring proper operation. Uniform liquid distribution is imperative for maximum performance. Structured packings are being applied in an expanding range of applications such as high-pressure distillation.

Final design of packed columns should be performed by experts, but the layman is often required to provide preliminary designs for studies. This packed column section provides the information necessary for such estimates.

Packed Column Internals

In preliminary design one usually picks a reasonable packing and uses literature data to determine its height in one or more beds. Standard internals are provided to round out the estimate. See Figure 1 and Notes.

Once packing heights are determined in other sections from HETP (distillation) or K_{GA} (absorption), the height allowances for the internals (from Figure 1) can be added to determine the overall column height. Column diameter is determined in sections on capacity and pressure drop for the selected packing (random dumped or structured).

Typical packed column internals are shown in Figures 2–11.

Packed Column Internals—Notes for Figure 1

A. Total Tower Height. This is an assumed maximum.
B. Top Section. This is from the top of the distributor to the top tangent.
C. Liquid Distributor (see Figures 4–9). According to Strigle[3], "This is the most important column internal from a process standpoint. A liquid distributor is required at all locations in the tower where an external liquid stream is introduced. High-performance distributors provide a flow rate variation per irrigation point that is a maximum of ±5% of the average flow. The geometric uniformity of liquid distribution has more effect on pack-

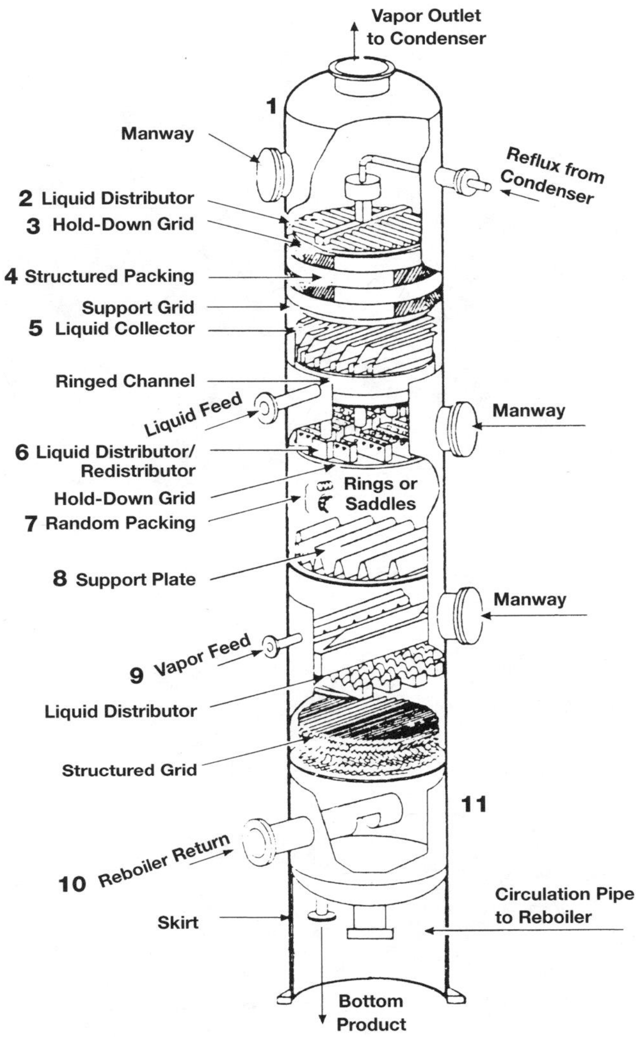

Note Number	Section	Notes	Recommended Height Allowance for Studies	Reference
1.	Maximum Tower Height	A	175 ft	7
1.	Top Section	B	4 ft	2, 7
2.	Liquid Distributor	C	1 ft 8 in.	2, 3, 5, 6
3.	Hold-Down Grid	D	—	3
4.	Structured Packing	E	From Text	3
5.	Liquid Collector	F	3 ft	2, 3
6.	Liquid Redistributor	G	3 ft wo/manway	2, 3
			4 ft w/manway	2, 3
7.	Random Packing	H	From Text	3
8.	Support Plate	I	Part of bed	3
9.	Vapor Feed	J	2 nozzle diameters plus 12 in.	2
10.	Reboiler Return	K	Use discussion K	4
11.	Bottom Section	L	6 ft	2, 7

Figure 1. Packed column internals.[9] (Reprinted by special permission from *Chemical Engineering,* March 5, 1984, copyright 1984 by McGraw-Hill, Inc.)

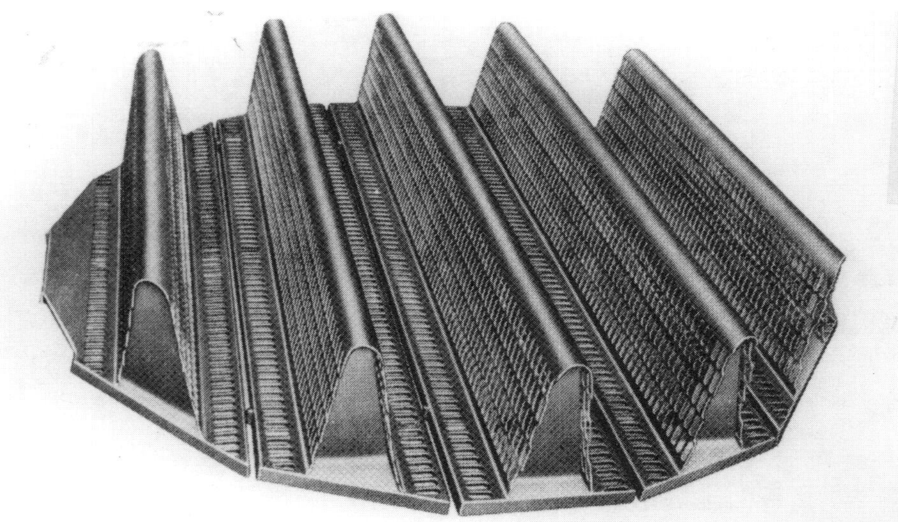

Figure 2. Multibeam packing support plate. (Courtesy of Norton Chemical Process Products Corporation.)

Figure 3. Bed limiter, Model 823. (Courtesy of Norton Chemical Process Products Corporation.)

ing efficiency than the number of distribution points per square foot."

Frank[5] reports "The liquid fall between the distributor and the top of the packing should be no greater than 12 in."

Graf[6] states "For spray type distributors in vacuum towers the main header of the distributor will be 18–36 in. above the top of the packing."

A private contact[2] stated, "There will be 6–8 in. from the bottom of the distributor to the top of the bed. The distributor is 1 ft thick."

D. Hold-Down Grid (see Figure 3). According to Strigle[3], "Normally, the upper surface of the packed bed is at least 6 in. below a liquid distributor or redistributor. The bed limiter or hold-down plate is located on top of the packed bed in this space. It is important to provide such a space to permit gas disengagement from the packed bed. Such a space allows the gas to accelerate to the velocity necessary to pass through the distributor without exceeding the capacity of the packing.

"Bed limiters commonly are used with metal or plastic tower packings. The primary function of these devices

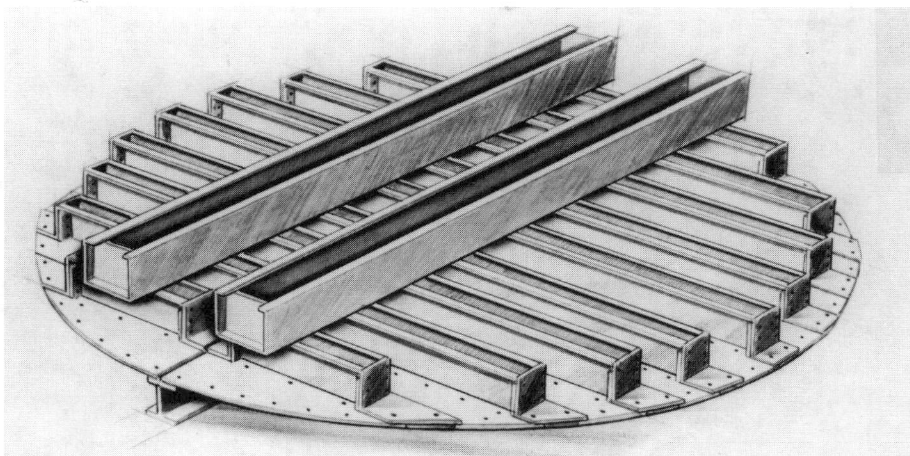

Figure 4. Orifice deck liquid distributor. (Courtesy of Norton Chemical Process Products Corporation.)

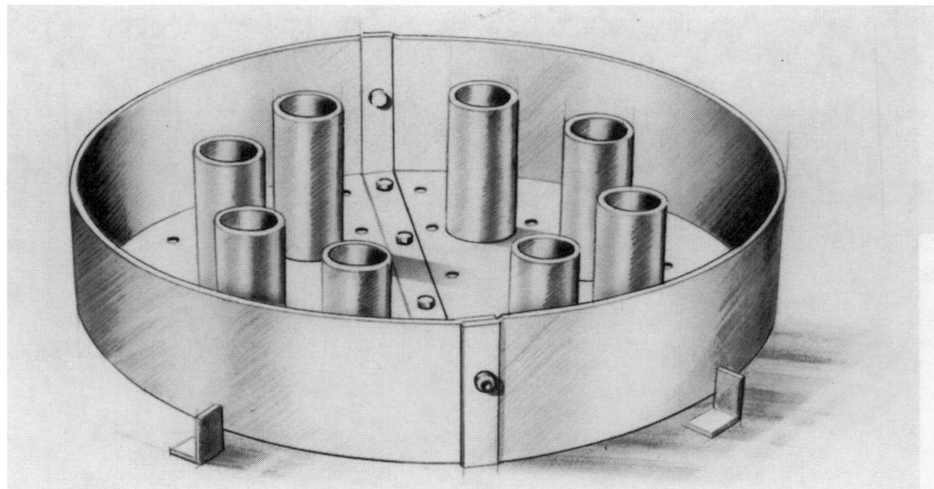

Figure 5. Orifice pan liquid distributor. (Courtesy of Norton Chemical Process Products Corporation.)

is to prevent expansion of the packed bed, as well as to maintain the bed top surface level. In large diameter columns, the packed bed will not fluidize over the entire surface. Vapor surges fluidize random spots on the top of the bed so that after return to normal operation the bed top surface is quite irregular. Thus the liquid distribution can be effected by such an occurrence.

"Bed limiters are fabricated as a light weight metal or plastic structure. Usually a mesh backing is used to prevent passage of individual pieces of tower packing. Because the bed limiter rests directly on the top of the packed bed, structurally it must be only sufficiently rigid to resist any upward forces acting on this packed bed.

"Hold-down plates are weighted plates used with ceramic or carbon tower packings. The hold-down plate must rest freely on the top of the packed bed because beds of ceramic and carbon packings tend to settle during operation. These plates usually act by their own weight to prevent bed expansion: therefore the plates weigh 20 to 30 lbs/ft^2."[3]

E. Structured Packing. This will be discussed in the appropriate section.

(*text continued on page 78*)

Figure 6. Orifice trough distributor with side-wall orifice, Model 137. (Courtesy of Norton Chemical Process Products Corporation.)

Figure 7. Weir trough liquid distributor. (Courtesy of Norton Chemical Process Products Corporation.)

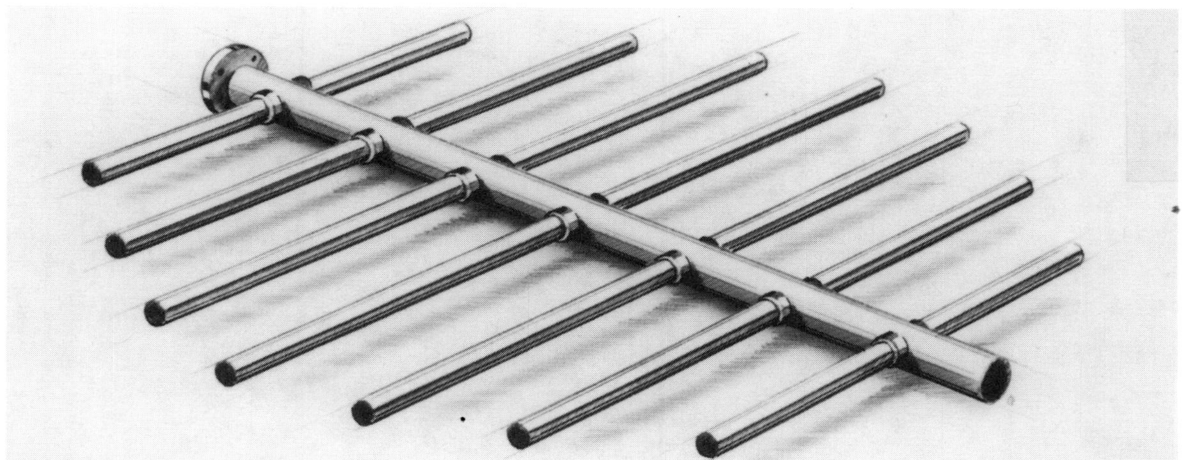

Figure 8. Orifice ladder liquid distributor. (Courtesy of Norton Chemical Process Products Corporation.)

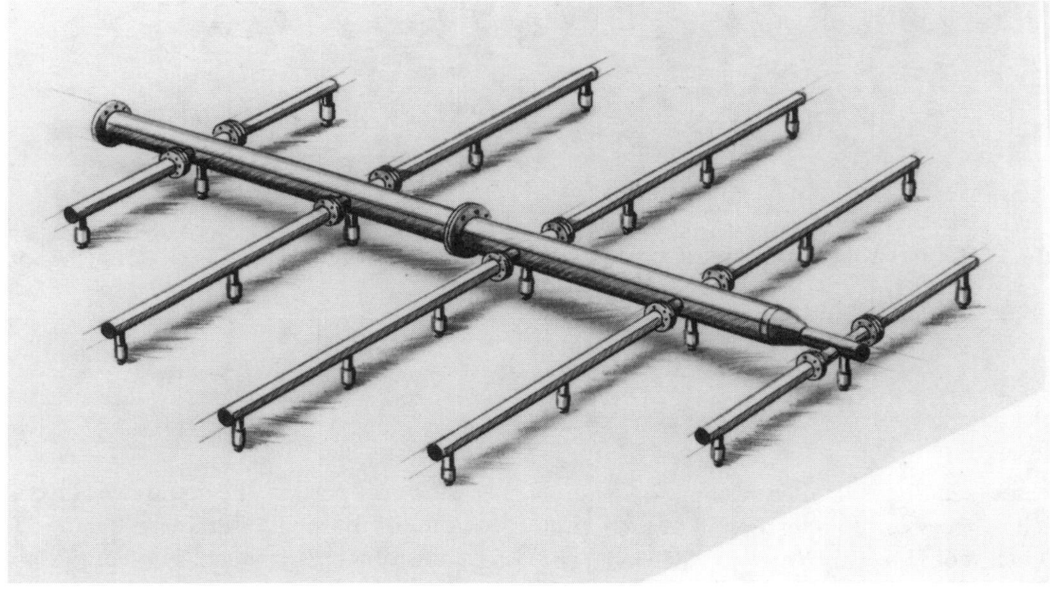

Figure 9. Spray nozzle liquid distributor. (Courtesy of Norton Chemical Process Products Corporation.)

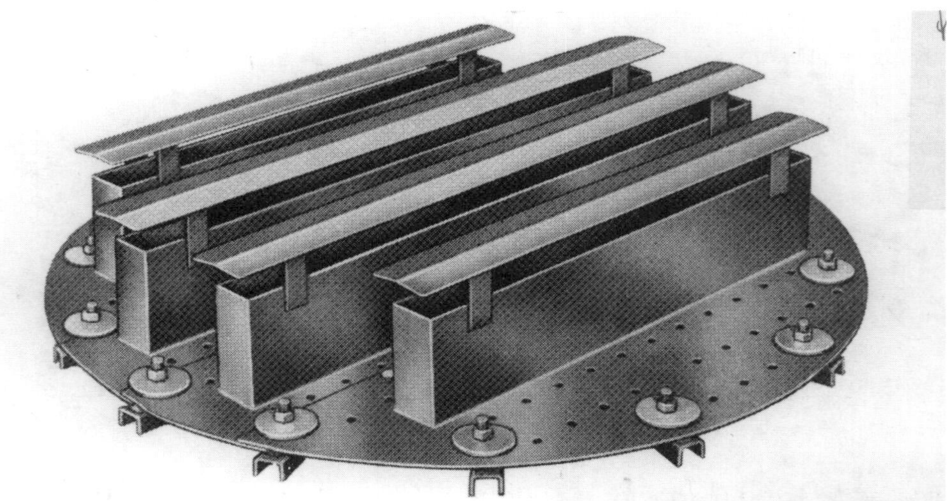

Figure 10. Orifice deck liquid redistributor. (Courtesy of Norton Chemical Process Products Corporation.)

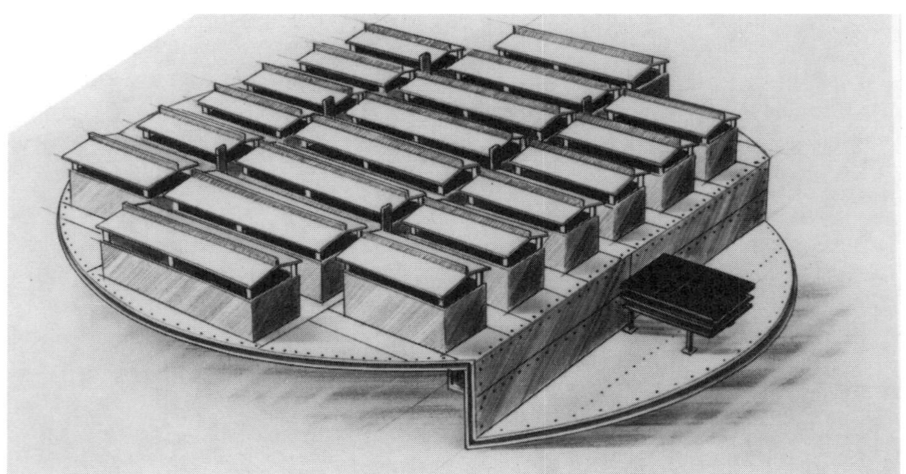

Figure 11. Liquid collector plate. (Courtesy of Norton Chemical Process Products Corporation.)

(*text continued from page 75*)

F. Liquid Collector (see Figure 11). According to Strigle,[3] "Sometimes it is necessary to intercept all of the liquid blowing down the column. This may occur due to an enlargement or contraction of the column diameter when operating at liquid rates in excess of 15 gpm/ft^2. If the lower portion of the column is of a larger diameter than the upper portion, the liquid must be collected at the bottom of the smaller diameter section. It then is fed to a redistributor located at the top of the larger diameter section to irrigate uniformly the lower section. If the lower portion of the column is of a smaller diameter than the upper portion, the liquid must be collected at the bottom of the larger diameter section. It then is fed to a redistributor at the top of the smaller diameter section to prevent excessive wall flow in the lower section.

"In a number of applications liquid pumparound sections are installed. Such sections are used in caustic scrubbers where only a small addition of sodium hydroxide is required to neutralize the absorbed acid gas. However, the liquid irrigation rate of each section must be adequate to ensure good gas scrubbing in the packed bed. In absorbers where the solute has a high heat of vaporization or heat of solution, the downflowing liquid progressively increases in temperature. This heated liquid must be removed from the column and cooled before it is returned to irrigate the next lower packed bed. The absorption otherwise is limited by the high vapor pressure of the solute above the hot rich solution. In some operations, liquid is removed and recirculated back to the top of the same packed bed after cooling. Such a situation is common when the packed bed is used as a total, or partial, condenser.

"In applications that must use a feed location that will not match the internal liquid composition, the feed must be mixed with the downflowing liquid. One method to accomplish this is the use of a collector plate that is installed beneath the rectifying section of the column. The liquid feed then is added to the pool of downflowing liquid on this collector plate. Such an arrangement provides at least partial mixing of the feed with the rectifier effluent in the liquid downcomer to give a more uniform composition of liquid onto the distributor irrigating the stripping section of the column. In other cases, a liquid cross-mixing device is installed in the column at the feed elevation.

"For columns in which there is a substantial flash of the feed liquid, or in which the feed is a vapor of a different composition than the internal vapor, a collector plate can be installed above the feed point. The purpose of this plate is to provide mixing of the vapor phase in the gas risers so that a more uniform vapor composition enters the rectifying section of the column.

"Figure 11 shows a typical liquid collector plate for a column that uses one side downcomer to withdraw the liquid. The maximum diameter for such a design is about 12 ft, which is limited by the hydraulic gradient necessary for such a liquid flow-path length. For larger diameter columns, two opposite side downcomers or a center downcomer normally is used unless the total amount of liquid collected is relatively small.

"A liquid collector plate must be of gasketed construction so that it can be sealed to the supporting ledge and will be liquid-tight. The sections could be seal-welded together; however, future removal would be difficult. There must be sufficient liquid head available to cause the liquid to flow out the side nozzle in the column shell. The use of sumps provides such liquid head without pooling liquid across the entire plate, where leakage and the weight of liquid to be supported would be greater.

"The gas risers must have a sufficient flow area to avoid a high gas-phase pressure drop. In addition, these gas risers must be uniformly positioned to maintain proper gas distribution. The gas risers should be equipped with covers to deflect the liquid raining onto this collector plate and prevent it from entering the gas risers where the high gas velocity could cause entrainment. These gas riser covers must be kept a sufficient distance below the next packed bed to allow the gas phase to come to a uniform flow rate per square foot of column cross-sectional area before entering the next bed.

"Where only about 5% or less of the liquid downflow is to be withdrawn from the column, a special collector box can be installed within the packed bed. This box can remove small quantities of intermediate boiling components that otherwise would accumulate in a sufficient quantity to interfere with the fractionation operation. Such a collector box must be designed very carefully to avoid interference with the vapor distribution above it or reduction in the quality of liquid distribution below it."[3]

The height allowance depends upon the style of collector used. The height can be as little as 2 ft or as much as 5 ft (if the collector is 3 ft thick).[2]

G. Liquid Redistributor (see Figure 10). According to Strigle[3], "A liquid redistributor is required at the top of each packed bed. The liquid flow from a typical support plate is not sufficiently uniform to properly irrigate the next lower packed bed. The multibeam type of packing support plate, which is widely used, tends to segregate the liquid downflow from a packed bed into a pair of parallel rows of liquid streams about 2-in. apart with a 10-in. space between adjacent pairs. Gas-distributing support plates likewise, do not give a sufficiently uniform liquid irrigation pattern, because gas riser locations take precedence in the design of such a plate.

"Liquid redistributors must operate in the same manner as gravity-fed distributors. To intercept all of the liquid downflow, these distributors usually have a deck that is sealed to a supporting ledge, as shown in Figure 10. If liquid-tight pans or boxes are used for a redistributor, gas riser covers and wall wipers must be used, or a collector plate must be installed above it, to intercept all the liquid downflow. Whenever the liquid falls directly onto a deck-type redistributor, the gas risers must be provided with covers to prevent liquid from raining into this area of high vapor velocity. To avoid excessive disturbance of the liquid pool on the deck, the redistributor should be located no more than 2 ft below the support plate for the bed above it.

"In any mass transfer operation, the compositions of the liquid and vapor phases are assumed to follow the relationship illustrated by the column operating line.This line represents the overall calculated profile down the column; however, the composition on each individual square foot of a particular column cross-section may vary from that represented by the operating line. These variations are the result of deviations in the hydraulic flow rates of the vapor and liquid phases, as well as incomplete mixing of the phases across the entire column.

"One of the functions of a liquid redistributor is to remix the liquid phase so as to bring the entire liquid flow onto the next lower bed at a more uniform composition. To perform this function, the liquid redistributor must intercept all the liquid that is flowing down the column, including any liquid on the column walls. If a new feed is to be introduced, a feed sparger or parting box must be used to predistribute this feed onto the redistributor. The redistributor must maintain the uniform vapor distribution that should have been established at the column bottom. To perform these functions, the redistributor must have a large flow area available that is transverse to the gas risers, because only very low gradient heads are available for cross-mixing of the liquid. In addition, the vapor flow area must be sufficient to avoid a high pressure drop in the gas phase.

"There has been considerable speculation regarding the depth of packing that could be installed before liquid redistribution is required. Silvey and Keller found no loss of efficiency in distillation with 18-ft deep beds of 1½-in. ceramic Raschig rings and 35-ft deep beds of 3-in ceramic Raschig rings. On the basis of their reported average HETP values, these beds were developing only 7 and 10 theoretical stages, respectively. Strigle and Fukuyo showed that with high performance redistributors over 20 theoretical stages per packed bed could be obtained repeatedly with 1-in. size packing.

"Packed bed heights typically vary from 20 ft to 30 ft. Many times the location of manholes to provide access to the redistributor will determine the packed depth. Whenever more than 15 theoretical stages are required in one packed bed, good liquid distribution is critical."[3]

Without a manway in the section the height allowance will run 3 ft (2-ft space plus a 1-ft thick redistributor). With a manway the allowance will be 4 ft.[2]

H. Random Packing. This will be discussed in the appropriate section.

I. Support Plate (see Figure 2). According to Strigle[3], "The primary function of the packing support plate is to serve as a physical support for the tower packing plus the weight of the liquid holdup. In the design of the packing support plate, no allowance is made for the buoyancy due to the pressure drop through the packed bed nor for the support offered by the column walls. In addition, the packing support plate must pass both the downwardly flowing liquid phase as well as the upwardly flowing gas phase to the limit of the capacity of the tower packing itself.

"The Figure 2 plate uses the gas injection principle to put the gas phase into the packed bed through openings in the almost vertical sidewalls of each beam. The liquid phase flows down along the sides of the beams to the horizontal bottom legs where it leaves the bed through perforations. Thus, the gas phase flows through one set of openings, while the liquid phase passes through a different set of openings.

"The beam-style design provides a high mechanical strength permitting single spans 12 ft long or greater. The modular design allows installation through standard 18-in. manholes and permits its use in any column 48-in. ID and larger. A modification of this multibeam support plate is available for use in smaller diameter columns. Pressure drop through this packing support plate is very low for the current design. In fact, its pressure drop actually has been included in the packing factor, which is calculated from a measurement of the pressure drop through a 10-ft depth of tower packing resting on such a support plate.

"In extremely corrosive services, ceramic support plates may be required. In larger diameter columns (such as sulfuric acid plant towers), a series of ceramic grids is installed resting on brick arches or piers. Then a layer of cross-partition rings or grid blocks is stacked on these grid bars to support the dumped ceramic packing.

"Plastic packing support plates also are available. Thermoplastic support plates are similar in design to metal plates. Fiberglass-reinforced plastic internals use metal internal designs adapted for hand lay-up. These plates are designed for use with plastic tower packings because their load-bearing capabilities are limited. Generally free-span length does not exceed 4 ft, due to possible high-temperature creep.

"A more simple design of support plates can be used for structured packings, because such packings are installed in discrete modules. Support plates for such packings provide a horizontal contact surface designed to prevent distortion of the packing while possessing sufficient structural strength to support the weight of the packing and liquid holdup over the length of the span. Such support plates have a very high open area for gas and liquid passage and do not add any significant pressure drop."[3]

J. Vapor Feed. The vapor inlet nozzle centerline should be at least one nozzle diameter plus 12 in. below the support plate.[2]

K. Reboiler Return. The top of the reboiler return nozzle should be at least one nozzle diameter plus 12 in. (18 in. minimum) below the bed support. The bottom of the reboiler return nozzle should be 12 in. minimum above the high liquid level and at least one nozzle diameter (18 in. minimum) above the normal liquid level.[4]

L. Bottom Section. This is from the bottom tangent to the bottom bed support.

Typical Applications

This subsection will help you select the type of packing to use for your studies and gives typical HETP and HTU values for ballpark estimates when time is short.

Table 3 compares packings as an aid to initial selection. For simple applications use Pall Rings for studies as a "tried and true" packing. This will give conservative results when compared with more recent random packings.

Table 4 gives typical packing depths for random packing for a variety of applications along with HETP, HTU, and other sizing data.

Random Dumped Packings

For studies using random dumped packings one needs to be able to determine column diameter and height. Column diameter is determined with use of generalized pressure drop correlations. Column height consists of space occupied by internals (discussed earlier under "Packed Column Internals") and the height of the packing. This subsection gives you methods to determine these items.

First let's discuss column diameter. There is a minimum column diameter for a given sized packing. Table 5 shows this relationship.

Table 6 shows maximum liquid loading rates per ft^2 of column diameter. Minimum liquid rate runs 0.5 to 2 gpm/ft^2.

Table 7 shows ranges of pressure drop for design. Pressure drop sets the allowable vapor flow rate. The flood pressure drop, for random or structure packings, is given in Reference 15 as:

$$\Delta P_{flood} = 0.115F^{0.7}$$

For this equation to apply the updated packing factors from Reference 3 or 13 must be used.

The pressure drop for a given proposed set of conditions is determined from Figure 12 or Figure 13, which has a linear Y-axis so is easier to interpolate. Table 8 gives packing factors (F) to be used with Figures 12 and 13.

Next, to determine packed column height use Table 9 for distillation HETP values, leaning towards the high side of the range for studies. For use of K_{GA} values see Section 4—Absorbers. Bed height per packed bed runs up to 20–30 ft for metal or ceramic packings, but plastic packing is usually limited to 24 ft.

At high liquid viscosities (often occurring in vacuum distillation) the HETP will increase. Table 10 shows this relationship.

Pictures of various random packings are shown in Figures 14–19.

(text continued on page 84)

Table 3
Packing Type Application[10]

Packing	Application Features
Raschig Rings	Earliest type, usually cheaper per unit cost, but sometimes less efficient than others. Available in widest variety of materials to fit service. Very sound structurally. Usually packed by dumping wet or dry, with larger 4–6-inch sizes sometimes hand stacked. Wall thickness varies between manufacturers, also some dimensions; available surface changes with wall thickness. Produce considerable side thrust on tower. Usually has more internal liquid channeling, and directs more liquid to walls of tower. Low efficiency.
Berl Saddles	More efficient than Raschig Rings in most applications, but more costly. Packing nests together and creates "tight" spots in bed which promotes channeling but not as much as Raschig rings. Do not produce as much side thrust, has lower HTU and unit pressure drops with higher flooding point than Raschig rings. Easier to break in bed than Raschig rings.
Intalox Saddles[1] and Other Saddle-Designs	One of most efficient packings, but more costly. Very little tendency or ability to nest and block areas of bed. Gives fairly uniform bed. Higher flooding limits and lower pressure drop than Raschig rings or Berl saddles; lower HTU values for most common systems. Easier to break in bed than Raschig rings, as ceramic.
Pall Rings[2]	Lower pressure drop (less than half) than Raschig rings, also lower HTU (in some systems also lower than Berl saddles), higher flooding limit. Good liquid distribution, high capacity. Considerable side thrust on column wall. Available in metal, plastic and ceramic.
Metal Intalox[1] Hy-Pak[1] Chempak[8]	High efficiency, low pressure drop, reportedly good for distillations.
Spiral Rings	Usually installed as stacked, taking advantage of internal whirl of gas-liquid and offering extra contact surface over Raschig ring, Lessing rings or crosspartition rings. Available in single, double and triple internal spiral designs. Higher pressure drop. Wide variety of performance data not available.
Lessing Rings	Not much performance data available, but in general slightly better than Raschig ring, pressure drop slightly higher. High side wall thrust.
Cross-Partition Rings	Usually used stacked, and as first layers on support grids for smaller packing above. Pressure drop relatively low, channeling reduced for comparative stacked packings. No side wall thrust.
Grid Tile	Available with plain side and bottom or serrated sides and drip-point bottom. Used stacked only. Also used as support layer for dumped packings. Self supporting, no side thrust. Pressure drop lower than most dumped packings and some stacked, lower than some ¼-inch × 1-inch and ¼-inch × 2-inch wood grids, but greater than larger wood grids. Some HTU values compare with those using 1-inch Raschig rings.
Teller Rosette (Tellerette)	Available in plastic, lower pressure drop and HTU values, higher flooding limits than Raschig rings or Berl saddles. Very low unit weight, low side thrust.
Spraypak[3]	Compared more with tray type performance than other packing materials. Usually used in large diameter towers, above about 24-inch dia., but smaller to 10-inch dia. available. Metal only.
Panapak[4]	Available in metal only, compared more with tray type performance than other packing materials. About same HETP as Spraypak for available data. Used in towers 24 inches and larger. Shows some performance advantage over bubble cap trays up to 75 psia in fractionation service, but reduced advantages above this pressure or in vacuum service.
Stedman Packing	Available in metal only, usually used in batch and continuous distillation in small diameter columns not exceeding 24-inches dia. High fractionation ability per unit height, best suited for laboratory work. Conical and triangular types available. Not much industrial data available.
Sulzer, Flexipac, and Similar	High efficiency, generally low pressure drop, well suited for distillation of clean systems, very low HETP.
Goodloe Packing[5] and Wire Mesh Packing	Available in metal and plastic, used in large and small towers for distillation, abosrption, scrubbing, liquid extraction. High efficiency, low HETP, low pressure drop. Limited data available.
Cannon Packing	Available in metal only, low pressure drop, low HETP, flooding limit probably higher than Raschig rings. Not much literature data available. Used mostly in small laboratory or semi-plant studies.
Wood Grids	Very low pressure drop, low efficiency of contact, high HETP or HTU, best used in atmospheric towers of square or rectangular shape. Very low cost.
Dowpac FN-90[6]	Plastic packing of very low pressure drop (just greater than wood slats), transfer Coefficients about same as 2-inch Raschig rings. Most useful applications in gas cooling systems or biological trickling filters.
Poly Grid[7]	Plastic packing of very low pressure drop, developed for water-air cooling tower applications.

[1]Trade name, Norton Co.
[2]Introduced by Badische Anilin and Sodafabrik, Ludwigshafen am Rhein
[3]Trade name of Denholme Inc., Licensed by British Government
[4]Trade name of Packed Column Corp.
[5]Trade name Packed Column Corp.
[6]Trade name of The Dow Chemical Co.
[7]Trade name The Fluor Products Co.
[8]Trade name Chem-Pro Equip. Corp.

Table 4
Typical Packing Depths[11]

System	L/G lb(/hr-ft²)	Diam. in.	Packing Type	Packing Size, in.	Bed depth, ft	HETP, ft	HTU, ft	System press., psia	ΔP, in H_2O/ ft pkg	% Overhead
Absorber	8300/11000	36	Pall rings	2	23.0	2.8	-	865	0.55	-
L.O.-Top fractionator	3600/ 4700	36	Pall rings	2	17.0	2.5	-	157	0.12	-
L.O.-Bottom fractionator	10600/ 5600	48	Pall rings	2	17.0	2.8	-	157	0.30	-
Deethanizer top	11000/ 4600	18	Pall rings	1½	20.0	2.9	-	300	0.20	-
Deethanizer bottom	19000/ 4500	30.0	Pall rings	2	18.0	3.3	-	300	0.30	-
Depropanizer top	5700/ 4200	23.4	Pall rings	1½	16.0	3.2	-	270	0.30	-
Depropanizer bottom	5700/ 4200	23.4	Pall rings	1½	24.0	2.4	-	270	0.30	-
Debutanizer top	1900/ 3100	19.5	Pall rings	1½	12.0	2.4	-	90	0.12	-
Debutanizer bottom	1900/ 3100	19.5	Pall rings	1½	18.0	2.0	-	90	0.12	-
Pentane-iso-pentane	2100/ 1900	18.0	Pall rings	1	9.0/7.6	1.5	-	Atmos.	0.40	-
Light and heavy naphtha	660/ 1250	15.0	Pall rings	1	10.0	2.00	2.05	100 mm. Hg	1.10	95.0
	320/ 600	15	Pall rings	1	10.0	3.25	2.50	100 mm. Hg	0.20	95.0
	620/ 1450	15	Pall rings	1	10.0	1.45	1.25	100 mm. Hg	1.75	97.5
	260/ 650	15	Pall rings	1	10.0	1.45	1.30	100 mm. Hg	0.20	97.5
	370/ 850	15	Intalox	1	10.0	2.30	1.90	100 mm. Hg	0.80	93.0
	210/ 500	15	Intalox	1	10.0	2.70	2.10	100 mm. Hg	0.22	99.0
	340/ 800	15	Raschig rings	1	10.0	1.95	1.40	100 mm. Hg	1.11	91.6
	210/ 500	15	Raschig rings	1	10.0	2.70	1.97	100 mm. Hg	0.40	96.5
Iso-octane Toluene	1970/ 2300	15	Pall rings	1	10.0	1.34	1.35	Atmos.	0.70	82.0
	950/ 2100	15	Pall rings	1	10.0	1.90	2.17	Atmos.	0.10	76.0
	2100/ 2660	15	Pall rings	1	10.0	0.80	1.02	Atmos.	1.70	84.0
	960/ 1200	15	Pall rings	1	10.0	1.53	1.42	Atmos.	0.15	74.0
	1110/ 1300	15	Pall rings	1	10.0	1.34	1.29	100 mm. Hg	1.08	92.5
	510/ 600	15	Pall rings	1	10.0	1.88	1.81	100 mm. Hg	0.20	87.0
	1020/ 1300	15	Pall rings	1	10.0	1.67	1.60	100 mm. Hg	1.14	92.0
	470/ 600	15	Pall rings	1	10.0	2.07	2.00	100 mm. Hg	0.20	89.0
Gas plant absorber	4700/ 6000	48	Pall rings	2	23.0	2.90	-	900	0.11	92.0 propane absorbed

Table 5
Maximum Packing Size[3]

Nominal Packing Size (in.)	Minimum Column ID (in.)
1	12
1½	18
2	24
3½	42

Table 6
Maximum Recommended Liquid Loading for Random Packings[3]

Packing Size (in.)	Liquid Rate (gpm/ft²)
¾	25
1	40
1½	55
2	70
3½	125

Table 7
Design Pressure Drop[12]

Service	Pressure Drop (inches water/ ft packed depth)
Absorbers and Regenerators (Non-foaming Systems)	0.25 to 0.40
Absorbers and Regenerators (Moderate Foaming Systems)	0.15 to 0.25
Fume Scrubbers (Water Absorbent)	0.40 to 0.60
Fume Scrubbers (Chemical Absorbent)	0.25 to 0.40
Atmospheric or Pressure Fractionators	0.40 to 0.80
Vacuum Fractionators	0.15 to 0.40

Table 8
Packing Factors (F)—Random Dumped Packings[3]

	Nominal Packing Size (in.)							
	½	⅝	¾	1	1¼	1½	2	3 or 3½
IMTP® Packing (Metal)		51		41		24	18	12
Hy-Pak® Packing (Metal)				45		32	26	16
Super Intalox® Saddles (Ceramic)				60			30	
Super Intalox® Saddles (Plastic)				40			28	18
Intalox Snowflake® (Plastic)							13	
Pall Rings (Plastic)		95		55		40	26	17
Pall Rings (Metal)		81		56		40	27	18
Intalox® Saddles (Ceramic)	200		145	92		52	40	22
Raschig Rings (Ceramic)	580	380	255	179	125	93	65	37
Raschig Rings (1/32″ Metal)	300	170	155	115				
Raschig Rings (1/16″ Metal)	410	300	220	144	110	83	57	32
Berl Saddles (Ceramic)	240		170	110		65	45	
Tellerettes (Plastic)				35			24	17

Table 9
Separation Efficiency In Standard Distillation Systems[12]

Nominal Packing Size (in.)	HETP (in.)
¾	11 to 16
1	14 to 20
1½	18 to 27
2	22 to 34
3 or 3½	31 to 45

Table 10
Effect of Liquid Viscosity on Packing Efficiency[3]

Liquid Viscosity (cps)	Relative HETP (%)
0.22	100
0.35	110
0.75	130
1.5	150
3.0	175

(text continued from page 81)

Structured Packings

Generalizations are not as easy for structured packings as for random packings. See the following subsection, "Comparison To Trays," for study type structured packing data.

Reference 13 is recommended for its large database of structured packing information. One parameter for structured packing within the narrative of Reference 13 is that the minimum wetting rate is 0.1 to 0.2 gpm/ft^2 compared to 0.5 to 2 gpm/ft^2 for random packings.

The flood pressure drop for structured packing, already shown in the subsection on random dumped packings, is repeated here[15]:

$$\Delta P_{flood} = 0.115F^{0.7}$$

For this equation to apply, the updated packing factors from Reference 3 or 13 must be used.

In addition, K_{GA} data are given for some structured packing in Chapter 4 Absorbers—Inorganic Type.

Comparison To Trays

We opened the "Packed Columns" subsection with the statement by Lieberman[1] that, based on his design and operating experience, a properly designed packed tower can have 20–40% more capacity than a trayed tower with equal number of fractionation stages. Kister[14] states that packing's pressure drop is typically three to five times lower than that of trays.

Kister[14] points out that, in the final analysis, design comparisons of packing versus trays must be evaluated

(text continued on page 88)

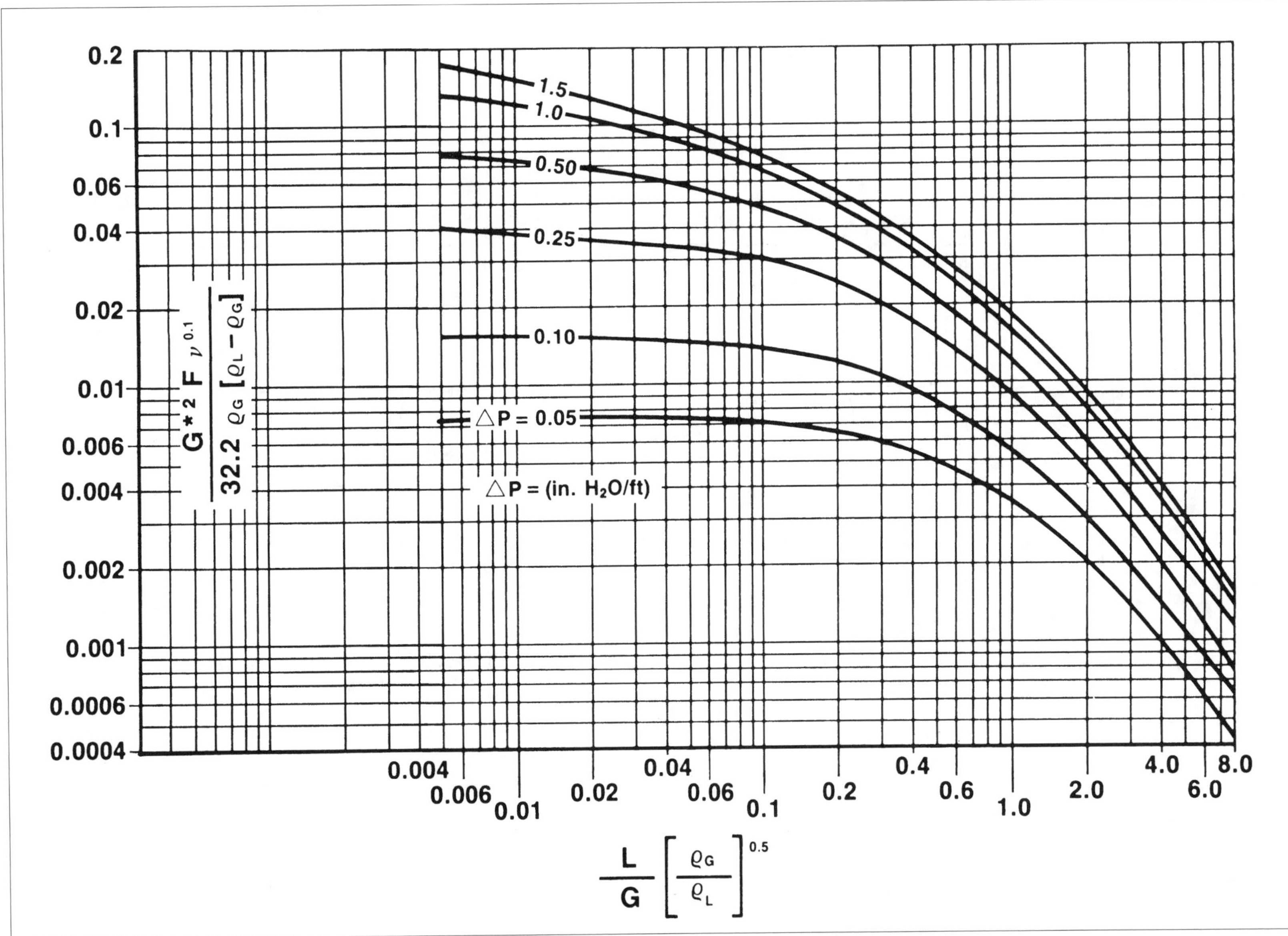

Figure 12. Generalized pressure drop correlation.[3]

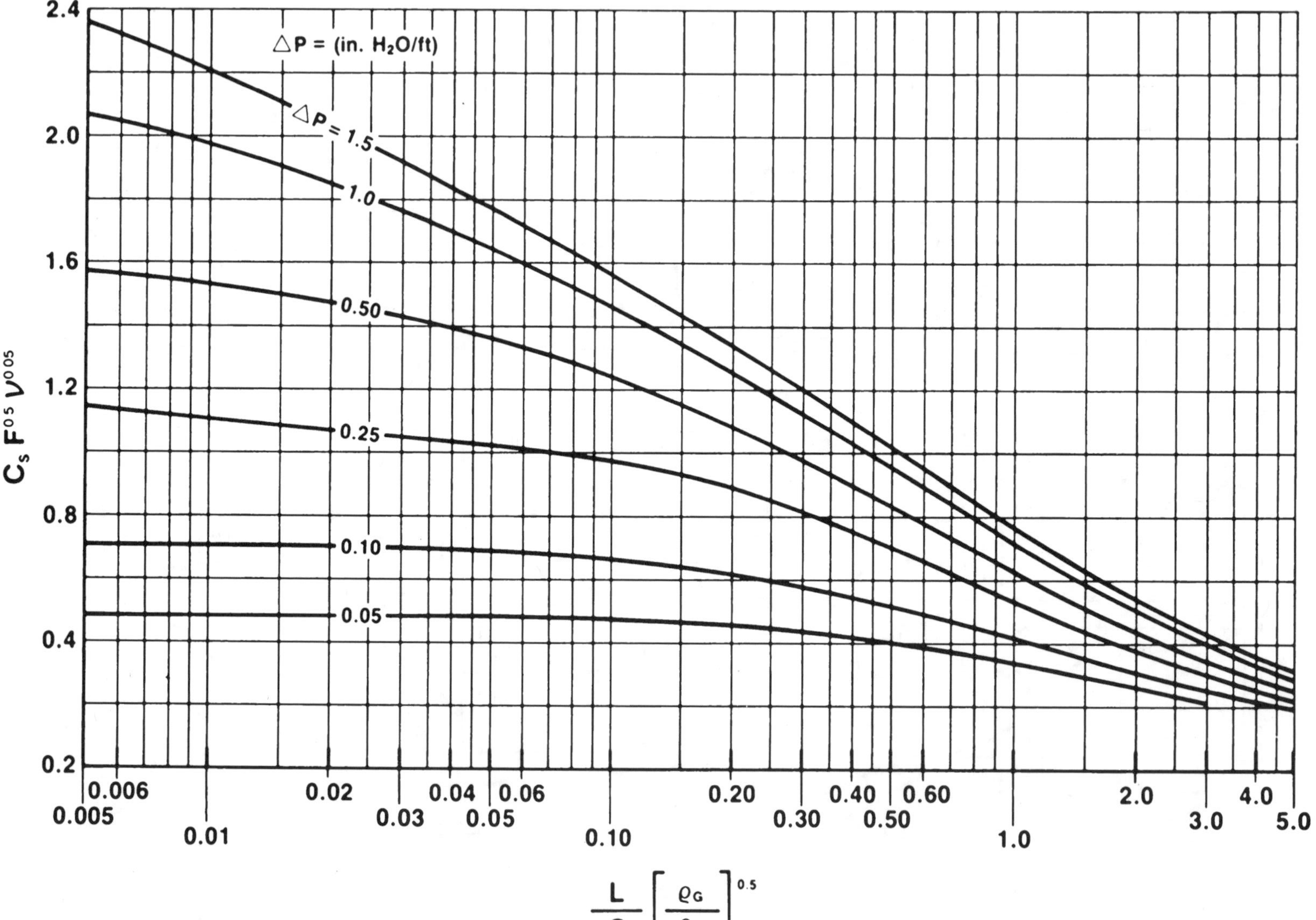

Figure 13. Alternate generalized pressure drop correlation.[3]

Figure 14. Koch Plastic Flexiring. (Courtesy of Koch Engineering Co., Inc., Bul. PFR-1.)

Figure 16. Plastic Pall Ring. Note: Ballast Ring® of Glitsch, Inc. is quite similar. (By permission Norton Chemical Process Products Corporation.)

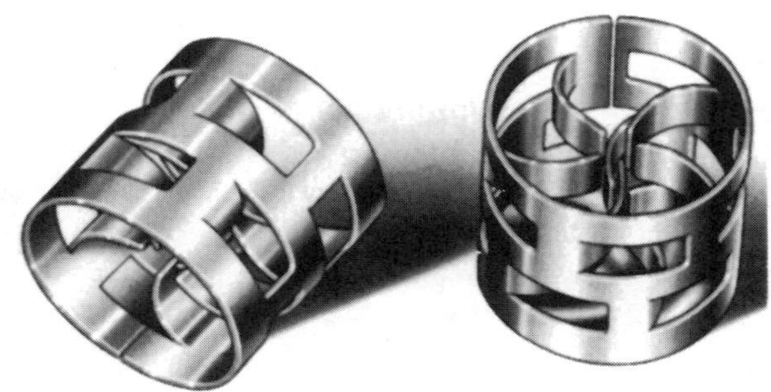

Figure 17. Metal Pall Ring. Note: Ballast Ring® of Glitsch, Inc. is quite similar, also Koch Engineering Flexiring® (By permission, Norton Chemical Process Products Corporation.)

Figure 15. Metal Hy-Pak®. (By permission, Norton Chemical Process Products Corporation.)

Figure 18. Intalox® Metal Tower Packing. (By permission, Norton Chemical Process Products Corporation.)

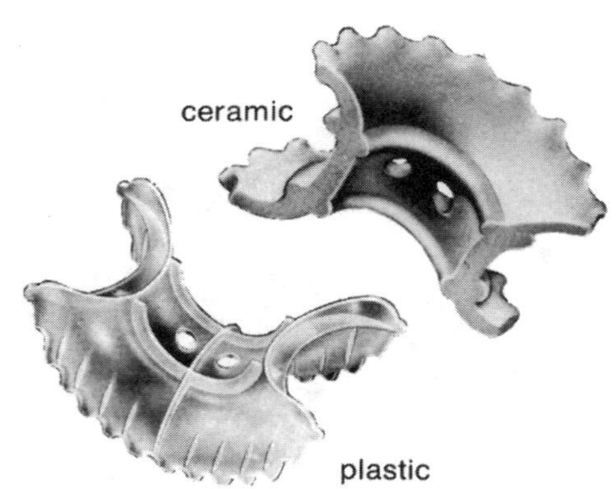

Figure 19. Super Intalox® saddles. Note Ballast Saddle® of Glitsch Inc. is similar, also Koch Engineering Flexisaddle®. (By permission, Norton Chemical Process Products Corporation.)

(text continued from page 84)

on a case-by-case basis. For example, the lower pressure drop for packing can be used for gains in relative volatility and/or capacity. These gains, however, are more important at vacuum than in atmospheric or pressure towers.[14] It is also noted that comparisons between packing and trays must be made on an "apples-to-apples" basis with each optimized for the test conditions.

Kister[14] discusses his comparison tests in great detail and presents Figures 20 and 21 for comparison of capacity and efficiency. We quote from his summary as follows (Reproduced with permission of the American Institute of Chemical Engineers, copyright 1994 AIChE, all rights reserved):

"At FPs of ≈ 0.02 – ≈ 0.1:

- The trays and the random packing have much the same efficiency and capacity.
- The structured packing efficiency is about 50% higher than either the trays or the random packing.
- As FP increases from 0.02 to 0.1, the capacity advantage of the structured packing (over the trays or over the random packing) declines to 0 from 30–40%.

"At FPs of 0.1–0.3:

- The trays and the random packing have much the same efficiency and capacity.
- The structured packing has much the same capacity as the trays and the random packing.
- As FP increases from 0.1 to 0.3, the efficiency advantage of the structured packing over the random pack-

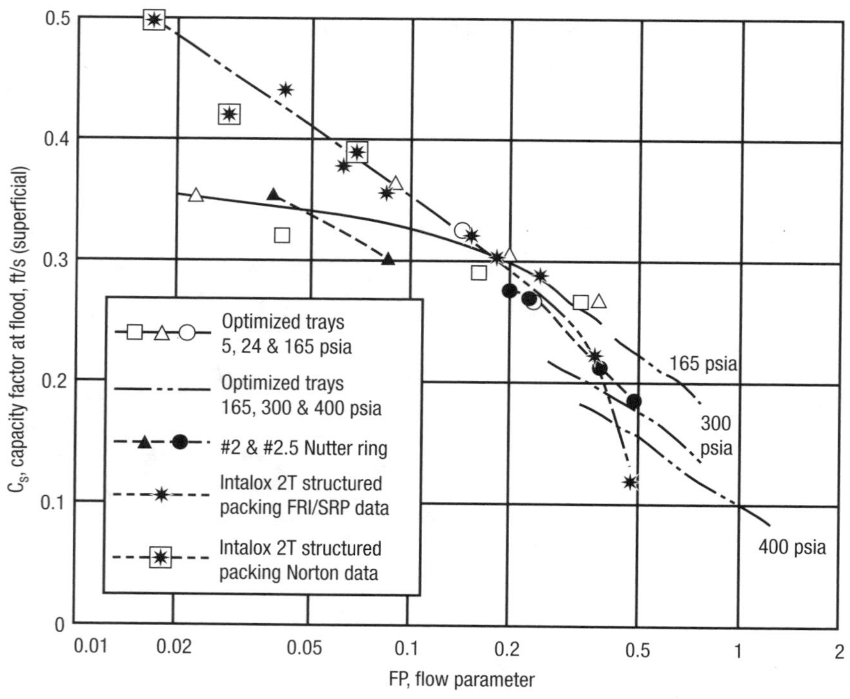

Figure 20. Overall comparison of capacity.[14] (Reproduced with permission of the American Institute of Chemical Engineers. Copyright © 1994 AIChE. All rights reserved.)

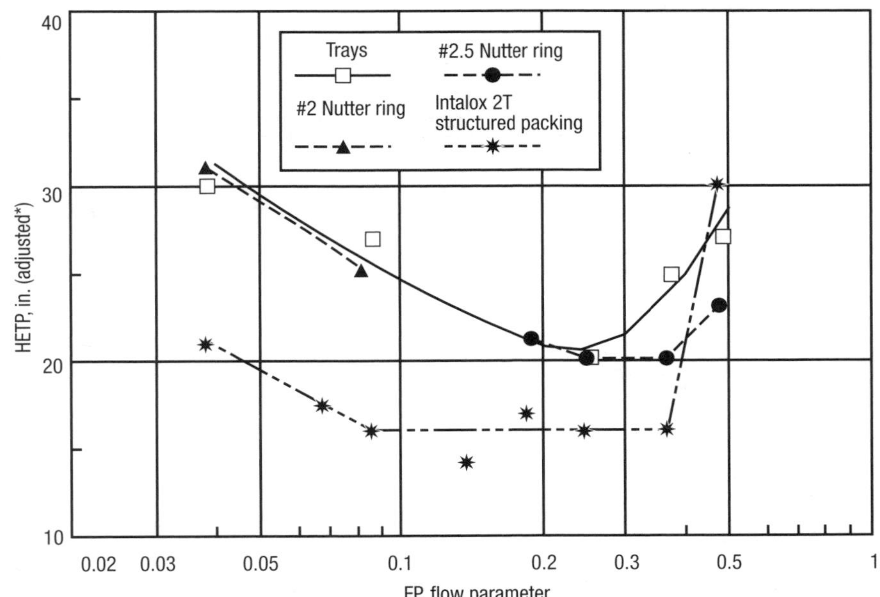

Figure 21. Overall comparison of efficiency.[14] (Reproduced with permission of the American Institute of Chemical Engineers. Copyright © 1994 AIChE. All rights reserved.)

ing and over the trays declines to about 20% from about 50%.

"At FPs of 0.3–0.5:

- Efficiency and capacity for the trays, the random packing, and the structured packing decline with a rise in flow parameter.
- The capacity and efficiency decline is steepest in the structured packing, shallowest in the random packing.
- At an FP of 0.5 and 400 psia, the random packing appears to have the highest capacity and efficiency, and the structured packing the least.

"The above results are based on data obtained for optimized designs and under ideal test conditions. To translate our findings to the real world, one must factor in liquid and vapor maldistribution, which is far more detrimental to the efficiency of packings than trays. In addition, one also must account for poor optimization or restrictive internals, which are far more detrimental to the capacity of trays than packings. We also have cited several other factors that need to be considered when translating the findings of our analysis to real-world towers."

In addition, Chen[9] gives typical performance data for trays, random packing, and structured packing in Table 11.

We conclude this subsection with a bravo! to Bravo[16] who provides useful rules-of-thumb for tower retrofits. These are intended for engineers comparing structured packing to high-performance trays for replacing older conventional trays.

Figure 22 gives his backup capacity data for his selection graph (Figure 23). It is easy to see why he preselects packing over trays for a flow parameter of less than 0.1 for example.

Finally, Table 12 gives cost rules-of-thumb for completing the comparison economics.

Table 11
Typical Performance of Tower Packings[9]

	Type of Packing		
Characteristic	**Trays**	**Random packing**	**Structured packing**
Capacity			
F-factor, $(ft/s)(lb/ft^3)^{1/2}$	0.25–2.0	0.25–2.4	0.1–3.6
C-factor, ft/s	0.03–0.25	0.03–0.3	0.01–0.45
Pressure drop, mm Hg/ theoretical stage	3–8	0.9–1.8	0.01–0.8
Mass-transfer efficiency, HETP, in.	24–48	18–60	4–30

Reproduced with special permission from Chemical Engineering, *March 5, 1984, copyright 1984 by McGraw-Hill Inc., New York.*

(*text continued on page 92*)

Table 12
Cost-estimation Tools for Internals in Distillation-Tower Retrofits

(Basis: Type-304 stainless-steel construction, 1997 installation)

Hardware cost for structured packings including distributors: (for packings in the range of 100–250 m^2/m^3 of surface and using high performance distributors)

$$\$ = D^2 [79S_H + 160 (2N_b - 1)]$$

where D = column diameter, ft
S_H = summation of all bed heights, ft
N_b = number of beds
\$ = cost of hardware in U.S. dollars

Hardware cost for high performance trays including distributors:

$$\$ = D^2 (52N_t + 160)$$

where D = column diameter, ft
N_t = number of trays
\$ = cost of hardware in U.S. dollars

Hardware installation and removal: (including support installation but no removal costs)
Installation factor for structured packings:

0.5–0.8 times cost of hardware

Installation factor for high performance trays:

1.0–1.4 times cost of hardware at same tray spacing
1.2–1.5 times cost of hardware at different tray spacing
1.3–1.6 times cost of hardware when replacing packing

Removal factor for removing trays to install new trays:

0.1 times cost of new hardware at same tray spacing
0.2 times cost of new hardware at different tray spacing

Factor to remove trays to install packing: 0.1 times cost of new hardware
Factor to remove packing to install trays: 0.07 times cost of new hardware

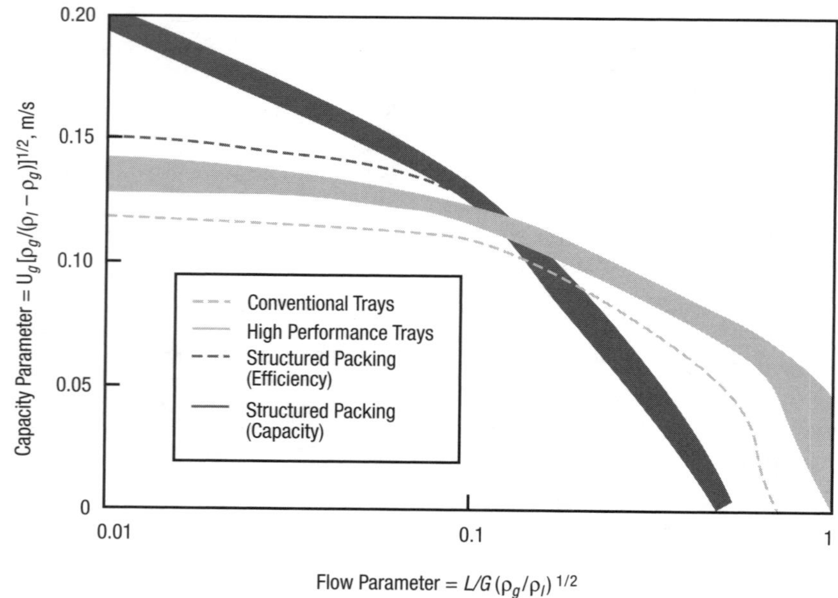

Figure 22. Maximum useful capacity of column internals.[16] (Reproduced with permission of the American Institute of Chemical Engineers. Copyright © 1997 AIChE. All rights reserved.)

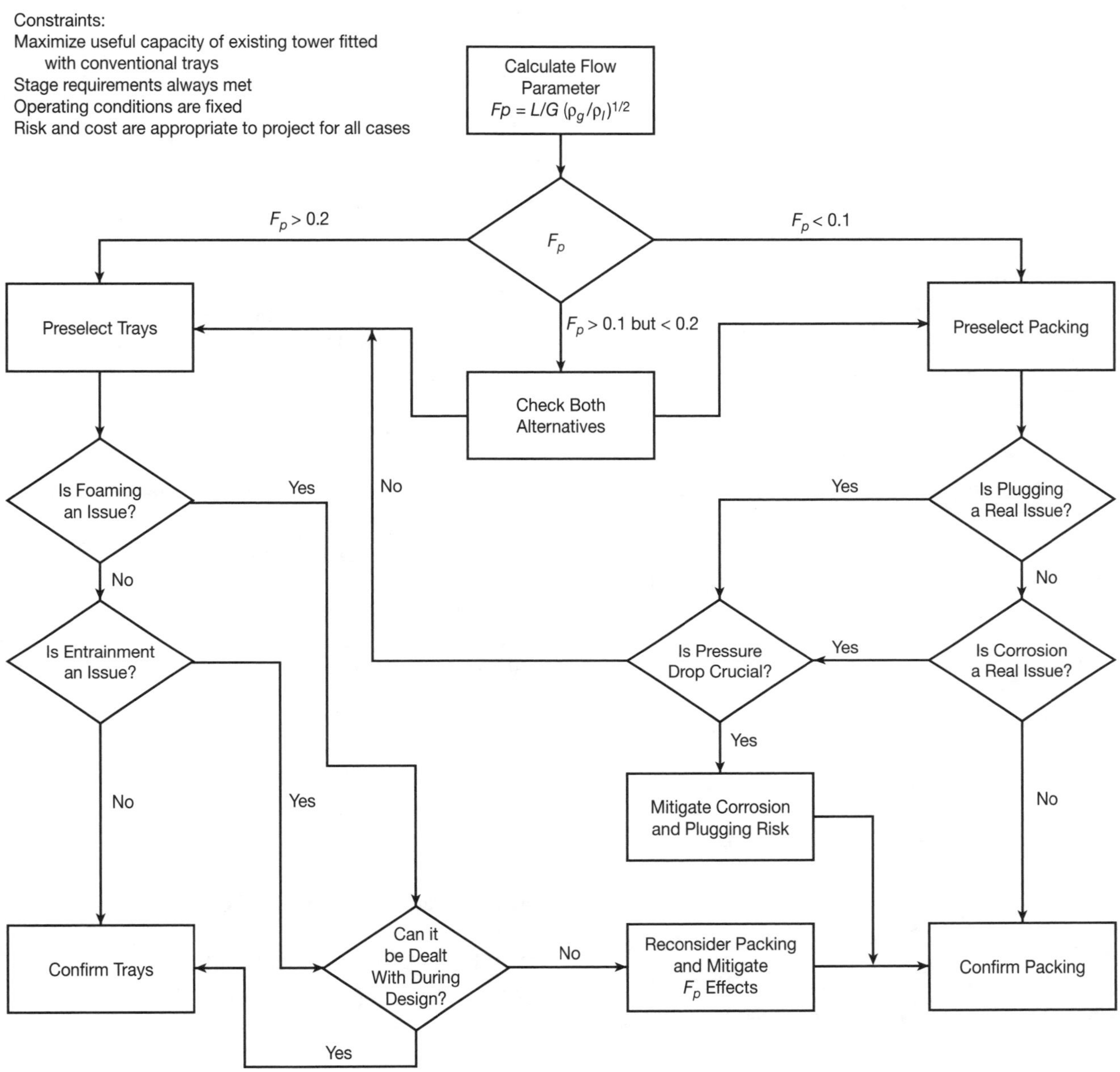

Figure 23. Choosing between structured packing or high performance trays for distillation retrofits.[16] (Reproduced with permission of the American Institute of Chemical Engineers. Copyright © 1997 AIChE. All rights reserved.)

(*text continued from page 89*)

Nomenclature

C_s or C_p = Capacity factor or capacity parameter, ft/s or m/s
= $U_G\,[\rho_G/(\rho_L - \rho_G)]^{0.5}$
= C factor
= $G^*/[\rho_G(\rho_L - \rho_G)]^{0.5}$
$C_S F^{0.5}\,\nu^{0.05}$ = Capacity parameter

(This is a different capacity parameter from C_P)

F = Packing factor, dimensionless
FP or F_P = Flow parameter = $L/G(\rho_G/\rho_L)^{0.5}$
G = Gas mass velocity, lb/ft^2 hr
G* = Gas mass velocity, lb/ft^2 sec
HETP = Height equivalent to a theoretical stage, in. or ft
HTU = Height of a transfer unit, ft
K_{GA} = Overall gas mass-transfer coefficient, lb moles/(hr) (ft^2) (atm)
L = Liquid mass velocity, lb/ft^{2}hr
U_G = Gas velocity, ft/sec
$U_G(\rho_G)^{0.5}$ = F-factor
ΔP = Bed pressure drop, inches of water per foot of packing
ρ_G = Gas density, lb/ft^3
ρ_L = Liquid density, lb/ft^3
ν = Kinematic liquid viscosity, centistokes (centistokes = centipoises/sp.gr.)

Sources

1. Lieberman, N. P., *Process Design for Reliable Operations,* 2nd Ed., Gulf Publishing Co., Houston, Texas, 1988.
2. Private correspondence, June 2, 1997.
3. Strigle, R. F., *Packed Tower Design and Applications,* 2nd Ed., Gulf Publishing Co., Houston, Texas, 1994.
4. Kitterman, Layton, "Tower Internals and Accessories," paper presented at Congresso Brasileiro de Petro Quimiea, Rio de Janeiro, November 8–12, 1976.
5. Frank, Otto, "Shortcuts for Distillation Design," *Chemical Engineering,* March 14, 1977.
6. Graf, Kenneth, "Correlations for Design Evaluation of Packed Vacuum Towers," *Oil and Gas Journal,* May 20, 1985.
7. Dr. Richard Long, Dept. of Chemical Engineering, New Mexico State University, "Guide to Design for Chemical Engineers," handout to students for plant designs, 1994.
8. *GPSA Engineering Data Book,* Gas Processors Suppliers Association, Vol. II, 10th Ed.
9. Chen, G. K., "Packed Column Internals," *Chemical Engineering,* March 5, 1984.
10. Ludwig, E. E., *Applied Process Design For Chemical and Petrochemical Plants,* Vol. 2, 3rd Ed. Gulf Publishing Co., Houston, Texas, 1997.
11. Eckert, J. S., "Tower Packings—Comparative Performance," *Chemical Engineering Progress,* 59 (5), 76–82, 1963.
12. Branan, C. R., *The Process Engineer's Pocket Handbook,* Vol. 2, Gulf Publishing Co., Houston, Texas, 1983.
13. Kister, H. Z., *Distillation Design,* McGraw-Hill, New York, 1992.
14. Kister, H. Z., K. F. Larson and T. Yanagi, "How Do Trays and Packing Stack Up?" *Chemical Engineering Progress,* February 1994.
15. Kister, H. Z., "Troubleshoot Distillation Simulations," *Chemical Engineering Progress,* June 1995.
16. Bravo, J. L., "Select Structured Packings or Trays?" *Chemical Engineering Progress,* July 1997.

4
Absorbers

Introduction

A general study of absorption can be confusing since the calculation methods for the two major types are quite different. First, there is hydrocarbon absorption using a lean oil having hydrocarbon components much heavier than the component absorbed from the gas stream. These absorbers may or may not be reboiled. For designing these, one uses equilibrium vaporization constants (K values) similarly to distillation. Another similarity to distillation is the frequent use of fractionating trays instead of packing. Canned computer distillation programs usually include hydrocarbon absorber options.

The other major type is gas absorption of inorganic components in aqueous solutions. For this type design one uses mass transfer coefficients. Packed towers are used so often for this type that its discussion is often included under sections on packed towers. However, in this book it is included here.

Source

Branan, C. R., *The Process Engineer's Pocket Handbook, Vol. 1,* Gulf Publishing Co., 1976.

Hydrocarbon Absorber Design

Because of its similarity to distillation, many parts of this subject have already been covered, such as

1. Tray Efficiency
2. Tower Diameter Calculations
3. K Values

As for distillation, shortcut hand calculation methods exist for hydrocarbon absorption. In distillation, relative volatility (α) values are generated from the K values. For hydrocarbon absorption the K values are used to generate absorption and stripping factors. The 1947 Edmister[7] method using effective overall absorption and stripping factors and the well known Edmister graphs are very popular for hand calculations. An excellent write-up on this and the Kremser-Brown-Sherwood methods are on pages 48–61 of Ludwig[1].

Edmister Method (1947). Briefly, the Edmister absorption method (1947) with a known rich gas going to a fixed tower is as follows:

1. Assume theoretical stages and operating temperature and pressure.
2. Knowing required key component recovery E_a, read A_e from Figure 1 at known theoretical trays n.

$$E_a = \frac{A_e^{\,n+1} - A_e}{A_e^{\,n+1} - 1} \tag{1}$$

$$E_s = \frac{S_e^{\,m+1} - S_e}{S_e^{\,m+1} - 1} \tag{2}$$

where

n = Number of theoretical stages in absorber
m = Number of theoretical stages in stripper
E_a = Fraction absorbed
E_s = Fraction stripped
A_e = Effective absorption factor
S_e = Effective stripping factor

3. Assume
   a. Total mols absorbed
   b. Temperature rise of lean oil (normally 20–40 °F)
   c. Lean oil rate, mols/hr
4. Use Horton and Franklin's[9] relationship for tower balance in mols/hr. This is shown in Table 1.
5. Calculate L_1/V_1 and L_n/V_n.
6. Use Horton/Franklin method to estimate tower temperatures. This is shown in Table 2.
7. Obtain top and bottom K values.

Table 1
Tower Balance

Quantity	Symbol	Equation
Rich gas entering at bottom	V_{n+1}	Known
Gas Absorbed	ΔV	Assumed
Lean gas leaving absorber	V_1	$V_1 = V_{n+1} - \Delta V$
Gas leaving bottom tray	V_n	$V_n = V_{n+1}(V_1/V_{n+1})^{1/N}$
Gas leaving tray 2 from top	V_2	$V_2 = V_1/(V_1/V_{n+1})^{1/N}$
Lean oil	L_0	Known
Liquid leaving top tray	L_1	$L_1 = L_0 + V_2 - V_1$
Liquid leaving bottom tray	L_n	$L_n = L_o + \Delta V$

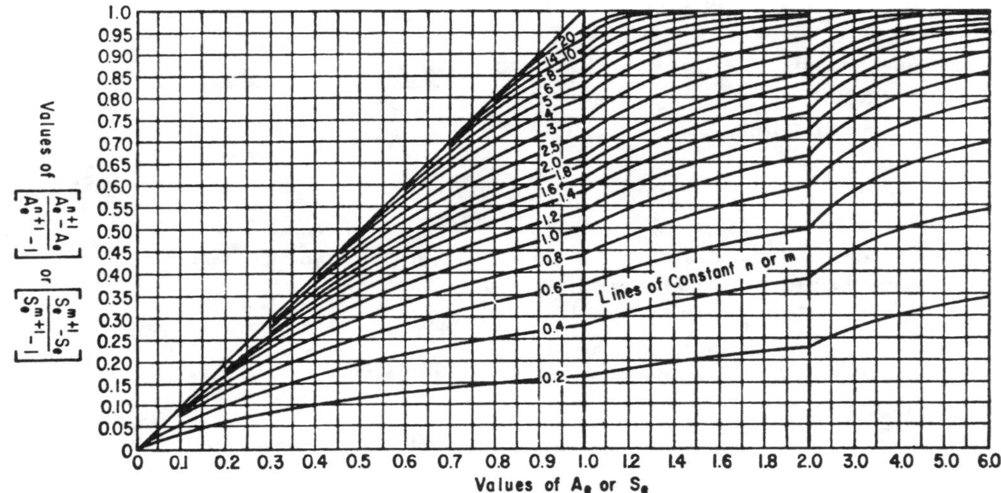

Figure 1. This graph shows the abosrption and stripping factors, E_a and E_s, versus effective values, A_e and S_e (efficiency functions). (By permission, W. C. Edmister, *Petroleum Engineer,* September, 1947 Series to January, 1948.)

Table 2
Tower Temperatures

Temperature	Symbol	Equation
Rich gas inlet	T_{n+1}	Known
Lean oil	T_0	Known
Temperature rise	Δ_T	Assumed
Bottom Tray	T_n	$T_n = T_{n+1} + \Delta T$
Top Tray	T_1	$T_1 = T_n - \Delta T \dfrac{(V_{n+1} - V_2)}{(V_{n+1} - V_1)}$

8. Calculate absorption factors for each component i at the top and bottom

$$A_{Ti} = L_l/K_{li}\, V_l \quad (3)$$

$$A_{Bi} = L_n/K_{ni}\, V_n \quad (4)$$

For stripping factors

$$S_{Ti} = V_l\, K_{li}/L_l \quad (5)$$

$$S_{Bi} = V_n\, K_{ni}/L_n \quad (6)$$

9. Obtain A_{ei} from

$$A_e = [A_B(A_T + 1) + 0.25]^{1/2} - 0.5 \quad (7)$$

Similarly

$$S_e = [S_T(S_B + 1) + 0.25]^{1/2} - 0.5 \quad (8)$$

10. Read E_{ai} values from Figure 1.
11. Calculate mols of each component absorbed.
12. Compare to assumed total mols absorbed and reassume lean oil rate if necessary.

Edmister Method (1957). Edmister has developed an improved procedure[8] that features equations combining absorption and stripping functions as follows:

$$V_1 = \phi_a\, V_{n+1} + (1 - \phi_s)\, L_o \quad \text{(Absorption Section)} \quad (9)$$

$$L_1 = \phi_s\, L_{m+1} + (1 - \phi_a)\, V_o \quad \text{(Stripping Section)} \quad (10)$$

where

L_1 = Liquid from bottom stripping tray
L_{m+1} = Liquid to top stripping tray
$\phi_a = 1 - E_a$, fraction not absorbed
$\phi_s = 1 - E_s$, fraction not stripped
V_o = Vapor to bottom stripping tray

Other symbols are defined in Tables 1 and 2. Figure 1 and Equations 3-8 are used as before.

V_1 and L_1 are found from Equations 9 and 10. The improved procedure is better for rigorous solution of complicated absorber designs.

Lean Oil. The selection of lean oil for an absorber is an economic study. A light lean oil sustains relatively high lean oil loss, but has the advantage of high mols/gal compared to a heavier lean oil. The availability of a suitable material has a large influence on the choice. A lean oil 3 carbon numbers heavier than the lightest component absorbed is close to optimum for some applications. In natural gas plant operations, however, the author generally sees a lean oil heavier by about 10–14 carbon numbers.

Presaturators. A presaturator to provide lean oil/gas contact prior to feeding the lean oil into the tower can be a good way of getting more out of an older tower. Absorber tray efficiencies run notoriously low. A presaturator that achieves equilibrium can provide the equivalent of a theoretical tray. This can easily equal 3–4 actual trays. Some modern canned computer distillation/absorption programs provide a presaturator option.

Sources

1. Ludwig, *Applied Process Design For Chemical and Petrochemical Plants,* 2nd Ed., Vol. 2, Gulf Publishing Company, 1979.
2. Fair, James R., "Sorption Processes for Gas Separation," *Chemical Engineering,* July 14, 1969.
3. Zenz, F. A., "Designing Gas-Absorption Towers," *Chemical Engineering,* November 13, 1972.
4. *NGPSA Engineering Data Book,* Natural Gas Processors Suppliers Association, 9th ed., 1972.
5. Norton, *Chemical Process Products,* Norton Company, Chemical Process Products Division.
6. Treybal, R. E., *Mass Transfer Operations,* McGraw-Hill Book Co., Inc., New York, 1955.
7. Edmister, W. C., *Petroleum Engineer,* September, 1947 Series to January, 1948.
8. Edmister, W. C., "Absorption and Stripping-factor Functions for Distillation Calculation by Manual- and Digital-computer Methods," *A.I.Ch.E. Journal,* June 1957.
9. Horton, G. W. and Franklin, B., "Calculation of Absorber Performance and Design," *Ind. Eng. Chem.* 32, 1384, 1940.
10. Rousseau, R. W., and Staton, S. J., "Analyzing Chemical Absorbers and Strippers," *Chemical Engineering,* July 18, 1988.
11. Diab, S. and Maddox, R. N., "Absorption," *Chemical Engineering,* December 27, 1982.
12. Branan, C. R., *The Process Engineer's Pocket Handbook, Vol. 1,* Gulf Publishing Co., 1976.

Hydrocarbon Absorbers, Optimization

This section is a companion to the section titled Fractionators-Optimization Techniques. In that section the Smith-Brinkley[1] method is recommended for optimization calculations and its use is detailed. This section gives similar equations for simple and reboiled absorbers.

For a simple absorber the Smith-Brinkley equation is for component i:

$$f = \frac{(1 - S^N) + q_s(S^N - S)}{1 - S^{N+1}}$$

where

$f_i = (BX_B/FX_F)_i$
$S_{mi} = K'_i V'/L'$
$S_{ni} = K_i V/L'$

For a reboiled absorber:

$$f = \frac{(1 - S_n^{N-M}) + q_s(S_n^{N-M} - S_n)}{(1 - S_n^{N-M}) + hS_n^{N-M}(1 - S_m^{M+1})}$$

where

h_i = correlating factor; if the feed is mostly liquid, use Equation 1 and if mostly vapor, Equation 2.

$$h_i = (K'_i/K_i)(L/L')[(1 - S_n)/(1 - S_m)]_i \quad (1)$$

$$h_i = (L/L')[(1 - S_n)/(1 - S_m)]_i \quad (2)$$

Nomenclature

B = Bottoms total molar rate, or subscript for bottoms
F = Feed total molar rate, or subscript for feed
f_i = Ratio of the molar rate of component i in the bottoms to that in the feed
K_i = Equilibrium constant of component i in top section = y/x
K'_i = Equilibrium constant of component i in bottom section = y/x
L = Effective total molar liquid rate in top section
L′ = Effective total molar liquid rate in bottom section

M = Total equilibrium stages below feed stage including reboiler
N = Total equilibrium stages including reboiler and partial condenser
q_s = Fraction of the component in the lean oil of a simple or reboiled absorber
S = Overall stripping factor for component i
S_m = Stripping factor for component i in bottom section
S_n = Stripping factor for component i in top section
V = Effective total molar vapor rate in top section
V′ = Effective total molar vapor rate in bottom section
X = Mol fraction in the liquid
Y = Mol fraction in the vapor

Sources

1. Smith, B. E., *Design of Equilibrium Stage Processes,* McGraw-Hill, 1963.
2. Branan, C. R., *The Process Engineer's Pocket Handbook, Vol. 1,* Gulf Publishing Co., 1976.

Inorganic Type

Design of inorganic absorbers quite often involves a system whose major parameters are well defined such as system film control, mass transfer coefficient equations, etc. Ludwig[1] gives design data for certain well known systems such as NH_3-Air-H_2O, Cl_2-H_2O, CO_2 in alkaline solutions, etc. Likewise, data for commercially available packings is well documented such as packing factors, HETP, HTU, etc.

Film Control. The designer needs to know whether his system is gas or liquid film controlling. For commercial processes this is known.

In general, an absorption is gas film controlling if essentially all resistance to mass transfer is in the gas film. This happens when the gas is quite soluble in, or reactive with, the liquid. Ludwig[1] gives a listing of film control for a number of commercial systems. If a system is essentially all gas or liquid film controlling, it is common practice to calculate only the controlling mass transfer coefficient. Norton[2] states that for gas absorption, the gas mass transfer coefficient is usually used, and for stripping the liquid mass transfer coefficient is usually used.

Mass Transfer Coefficients. General equations for mass transfer coefficients are given in various references if specific system values are not available. These must, however, be used in conjunction with such things as packing effective interfacial areas and void fractions under operating conditions for the particular packing selected. It is usually easier to find K_{GA} for the packing used with a specific system than effective interfacial area and operation void fraction. Packing manufacturers' data or references, such as Ludwig[1] can provide specific system K_{GA} or K_{LA} data. Tables 1–6 show typical K_{GA} data for various systems and tower packings.

If K_{GA} values are available for a known system, those of an unknown system can be approximated by

$$K_{GA}\,(\text{unknown}) = K_{GA}\,(\text{known})\left(\frac{D_v\ \text{unknown}}{D_v\,\text{known}}\right)^{0.56}$$

where

K_{GA} = Gas film overall mass transfer coefficient, lb mols/hr (ft^3) (ATM)
D_v = Diffusivity of solute in gas, ft^2/hr

Diffusivities. The simplest gas diffusivity relationship is the Gilliland relationship.

$$D_v = 0.0069\,\frac{T^{3/2}(1/M_A + 1/M_B)^{1/2}}{P(V_A^{\ 1/3} + V_B^{\ 1/3})^2}$$

where

T = Absolute temperature, °R
M_A, M_B = Molecular weights of the two gases, A and B
P = Total pressure, ATM
V_A, V_B = Molecular volumes of gases, cc/gm mol

Height of Overall Transfer Unit. Transfer unit heights are found as follows:

$$H_{OG} = \frac{G_m}{K_{GA}\,P_{AVR}}$$

$$H_{OL} = \frac{L_m}{K_{LA}\,\rho_L}$$

Table 1
K_{GA} For Various Systems[4]

Solute Gas	Absorbent Liquid	K_{GA} Lb mols/(hr • ft³ • atm)
Br_2	5% NaOH	5.0
Cl_2	Water	4.6
Cl_2	8% NaOH	14.0
CO_2	4% NaOH	2.0
HBr	Water	5.9
HCHO	Water	5.9
HCl	Water	19.0
HCN	Water	5.9
HF	Water	8.0
H_2S	4% NaOH	5.9
NH_3	Water	17.0
SO_2	Water	3.0
SO_2	11% Na_2CO_3	12.0

Table 2
Relative K_{GA} For Various Packings[4]

Type of Packing	Material	Relative K_{GA}
Super Intalox®	Plastic	1.00
Intalox® Saddles	Ceramic	0.94
Hy-Pak®	Metal	1.11
Pall Rings	Metal	1.06
Pall Rings	Plastic	0.97
Maspac	Plastic	1.00
Tellerettes	Plastic	1.19
Raschig Rings	Ceramic	0.78

Table 3
Overall Mass Transfer Coefficient CO_2/NaOH System[5]
Metal Tower Packings

Packing	K_{GA} (lb-mol/h • ft³ • atm)
#25 IMTP® Packing	3.42
#40 IMTP® Packing	2.86
#50 IMTP® Packing	2.44
#70 IMTP® Packing	1.74
1 in. Pall Rings	3.10
1½ in. Pall Rings	2.58
2 in. Pall Rings	2.18
3½ in. Pall Rings	1.28
#1 Hy-Pak® Packing	2.89
#1½ Hy-Pak® Packing	2.42
#2 Hy-Pak® Packing	2.06
#3 Hy-Pak® Packing	1.45

Table 4
Overall Mass Transfer Coefficient CO_2/NaOH System[5]
Plastic Tower Packings

Packing	K_{GA} (lb-mol/h • ft³ • atm)
#1 Super Intalox® Packing	2.80
#2 Super Intalox® Packing	1.92
#3 super Intalox® Packing	1.23
Intalox® Snowflake® Packing	2.37
1 in. Pall Rings	2.64
1½ in. Pall Rings	2.25
2 in. Pall Rings	2.09
3½ in. Pall Rings	1.23

Table 5
Overall Mass Transfer Coefficient CO_2/NaOH System[5]
Ceramic Tower Packings

Packing	K_{GA} (lb-mol/h • ft³ • atm)
1 in. Intalox® Saddles	2.82
1½ in. Intalox® Saddles	2.27
2 in. Intalox® Saddles	1.88
3 in. Intalox® Saddles	1.11
1 in. Raschig Rings	2.31
1½ in. Raschig Rings	1.92
2 in. Raschig Rings	1.63
3 in. Raschig Rings	1.02

Table 6
Overall Mass Transfer Coefficient CO_2/NaOH System[5]
Structured Tower Packings

Packing	K_{GA} (lb-mol/h • ft³ • atm)
Intalox® Structured Packing 1T	4.52
Intalox® Structured Packing 2T	3.80
Intalox® Structured Packing 3T	2.76

where

H_{OG}, H_{OL} = Height of transfer unit based on overall gas or liquid film coefficients, ft
G_m, L_m = Gas or liquid mass velocity, lb mols/(hr)(ft^2)
K_{GA}, K_{LA} = Gas or liquid mass transfer coefficients, consistent units
P_{AVR} = Average total pressure in tower, ATM
ρ_L = Liquid density, lb/ft³

Number of Transfer Units. For dilute solutions the number of transfer units N_{OG} is obtained by

$$N_{OG} = \frac{\dfrac{Y_1 - Y_2}{(Y - Y^*)_1 - (Y - Y^*)_2}}{\ln \dfrac{(Y - Y^*)_1}{(Y - Y^*)_2}}$$

where

$(Y - Y^*)$ = Driving force, expressed as mol fractions
Y = Mol fraction of one component (solute) at any point in the gas phase
Y^* = Mol fraction gas phase composition in equilibrium with a liquid composition, X
X = Mol fraction in the liquid at the same corresponding point in the system as Y
1, 2 = Inlet and outlet of the system, respectively

Sources

1. Ludwig, E. E., *Process Design For Chemical and Petrochemical Plants, Vol. 2,* Gulf Publishing Co., 1965.
2. *Norton Chemical Process Products,* Norton Company, Chemical Process Products Division.
3. Branan, C. R., *The Process Engineer's Pocket Handbook, Vol. 1,* Gulf Publishing Co., 1976.
4. Branan, C. R., *The Process Engineer's Pocket Handbook,* Vol. 2, Gulf Publishing Co., Houston, Texas, 1983.
5. Strigle, R. F., *Packed Tower Design and Applications,* 2nd Ed., Gulf Publishing Co., Houston, Texas, 1994.

5

Pumps

Affinity Laws

Dynamic type pumps obey the affinity laws:

1. Capacity varies directly with impeller diameter and speed.
2. Head varies directly with the square of impeller diameter and speed.
3. Horsepower varies directly with the cube of impeller diameter and speed.

Horsepower

The handiest pump horsepower formula for a process engineer is:

$$HP = GPM(\Delta P)/1715(Eff)$$

where:

HP = Pump horsepower
GPM = Gallons per minute
ΔP = Delivered pressure (discharge minus suction), psi
Eff = Pump efficiency, fraction

Efficiency

An equation was developed by the author from the pump efficiency curves in the eighth edition of The GPSA Engineering Data Book[1], provided by the M. W. Kellogg Co. The curves were found to check vendor data well. The equation admittedly appears bulky, but is easy to use.

$$\text{Eff.} = 80 - 0.2855F + 3.78 \times 10^{-4}\,FG - 2.38 \times 10^{-7}\,FG^2 + 5.39 \times 10^{-4}\,F^2 - 6.39 \times 10^{-7}\,F^2G + 4 \times 10^{-10}\,F^2G^2$$

where

Eff. = Pump percentage efficiency
F = Developed head, ft
G = Flow, GPM

Ranges of applicability:

F = 50–300 ft
G = 100–1000 GPM

The equation gives results within about 7% of the aforementioned pump curves. This means within 7% of the curve valve, not 7% absolute, i.e., if the curve valve is 50%, the equation will be within the range 50 ± 3.5%.

For flows in the range 25–99 GPM a rough efficiency can be obtained by using the equation for 100 GPM and then subtracting 0.35%/GPM times the difference between 100 GPM and the low flow GPM. For flows at the bottom of the range (25–30 GPM), this will give results within about 15% for the middle of the head range and 25% at the extremes. This is adequate for ballpark estimates at these low flows. The horsepower at the 25–30 GPM level is generally below 10.

Sources

1. *GPSA Engineering Data Book,* Natural Gas Processors Suppliers Association, 8th Ed., 1966 and 9th Ed., 1972.
2. Branan, C. R., *The Process Engineer's Pocket Handbook, Vol. 1,* Gulf Publishing Co., 1976.

Minimum Flow

Most pumps need minimum flow protection to protect them against shutoff. At shutoff, practically all of a pump's horsepower turns into heat, which can vaporize the liquid and damage the pump. Such minimum flow protection is particularly important for boiler feedwater pumps that handle water near its boiling point and are multistaged for high head. The minimum flow is a relatively constant flow going from discharge to suction. In the case of boiler feedwater pumps, the minimum flow is preferably piped back to the deaerator.

The process engineer must plan for minimum flow provisions when making design calculations. For preliminary work, approximate the required minimum flow by assuming all the horsepower at blocked-in conditions turns into heat. Then, provide enough minimum flow to carry away this heat at a 15°F rise in the minimum flow stream's temperature.

This approach will provide a number accurate enough for initial planning. For detailed design, the process engineer should work closely with the mechanical engineer and/or vendor representative involved to set exact requirements, including orifice type and size for the minimum flow line. Also, a cooler may be required in the minimum flow line or it may need to be routed to a vessel. For boiler feedwater pumps, a special stepped type orifice is often used to control flashing.

Source

Branan, C. R., *The Process Engineer's Pocket Handbook, Vol. 2,* Gulf Publishing Co., 1983.

General Suction System

This is an important part of the pump system and should be thought of as a very specialized piping design. Considerable attention must be directed to the pump suction piping to ensure satisfactory pump operation.

A pump is designed to handle liquid, not vapor. Unfortunately, for many situations, it is easy to get vapor into the pump if the design is not carefully done. Vapor forms if the pressure in the pump falls below the liquid's vapor pressure. The lowest pressure occurs right at the impeller inlet where a sharp pressure dip occurs. The impeller rapidly builds up the pressure, which collapses vapor bubbles, causing cavitation and damage. This must be avoided by maintaining sufficient net positive suction head (NPSH) as specified by the manufacturer.

Vapor must also be avoided in the suction piping to the pump. It is possible to have intermediate spots in the system where the pressure falls below the liquid's vapor pressure if careful design is not done.

Therefore, the suction system must perform two major jobs: maintain sufficient NPSH; and maintain the pressure above the vapor pressure at all points.

NPSH is the pressure available at the pump suction nozzle after vapor pressure is subtracted. It is expressed in terms of liquid head. It thus reflects the amount of head loss that the pump can sustain internally before the vapor pressure is reached. The manufacturer will specify the NPSH that his pump requires for the operating range of flows when handling water. This same NPSH is normally used for other liquids.

For design work, the known pressure is that in the vessel from which the pump is drawing. Therefore, the pressure and NPSH available at the pump suction flange must be calculated. The vessel pressure and static head pressure are added. From this must be subtracted vapor pressure and any pressure losses in the entire suction system such as:

1. Friction losses in straight pipe, valves and fittings
2. Loss from vessel to suction line
3. Loss through equipment in the suction line (such as a heat exchanger)

The NPSH requirement must be met for all anticipated flows. Maximum flow will usually have a higher NPSH than normal flow. For some pumps, extremely low flows can also require higher NPSH.

It is usually necessary for the process engineer to have an idea of NPSH requirements early in the design phase of a project. The NPSH sets vessel heights and influences other design aspects. The choice of pumps is an economic balance involving NPSH requirements and pump speed. The lower speed pump will usually have lower NPSH requirements and allow lower vessel heights. A low-speed pump may also have a better maintenance record. However, the higher-speed pump will usually deliver the required head in a cheaper package.

The suction system piping should be kept as simple as reasonably possible and adequately sized. Usually the suction pipe should be larger than the pump suction nozzle.

The second major job of the suction system is to maintain the pressure above the vapor pressure at all points. Usually, possible points of intermediate low pressure occur in the area of the vessel outlet (drawoff) nozzle. Kern[1] gives some good rules of thumb on this point:

1. The minimum liquid head above the drawoff nozzle must be greater than the nozzle exit resistance. Based on a safety factor of 4 and a velocity head "K" factor of 0.5:

   $h_L = 2u^2/2g$

   where:

   h_L = Liquid level above nozzle, ft
   u = Nozzle velocity, ft/sec
   g = 32.2 ft/sec^2

2. For a saturated (bubble point) liquid, pipe vertically downward from the drawoff nozzle as close to the nozzle as possible. This gives maximum static head above any horizontal sections or piping networks ahead of the pump.

A vortex breaker should be provided for the vessel drawoff nozzle. Kern[1] shows some types.

Some suction line rules of thumb are:

1. Keep it short and simple.
2. Avoid loops or pockets that could collect vapor or dirt.
3. Use an eccentric reducer with the flat side up (to prevent trapping vapor) as the transition from the larger suction line to the pump suction nozzle.
4. Typical suction line pressure drops:
   Saturated liquids = 0.05 to 0.5 psi/100 ft
   Subcooled liquids = 0.5 to 1.0 psi/100 ft

Sources

1. Kern, R., "How to Design Piping for Pump-Suction Conditions," *Chemical Engineering,* April 28, 1975.
2. Branan, C. R., *The Process Engineer's Pocket Handbook, Vol. 2,* Gulf Publishing Co., 1983.

Suction System NPSH Available

Typical NPSH calculations keep the pump's lowest pressure below the liquid's vapor pressure as illustrated by the following three examples:

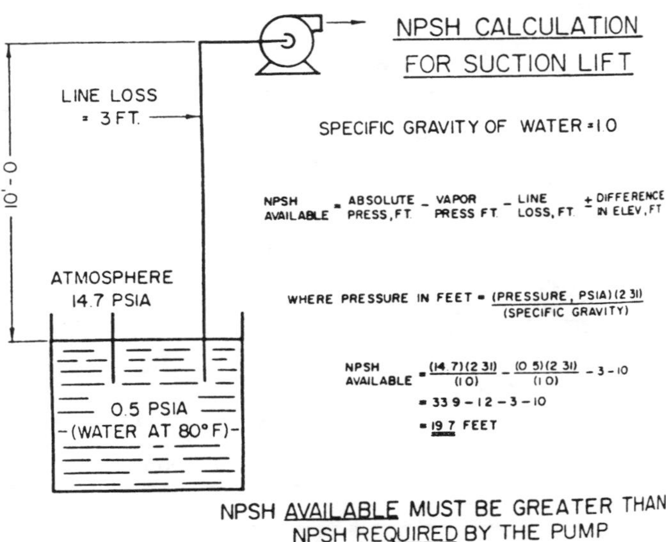

Figure 1. NPSH calculation for suction lift.

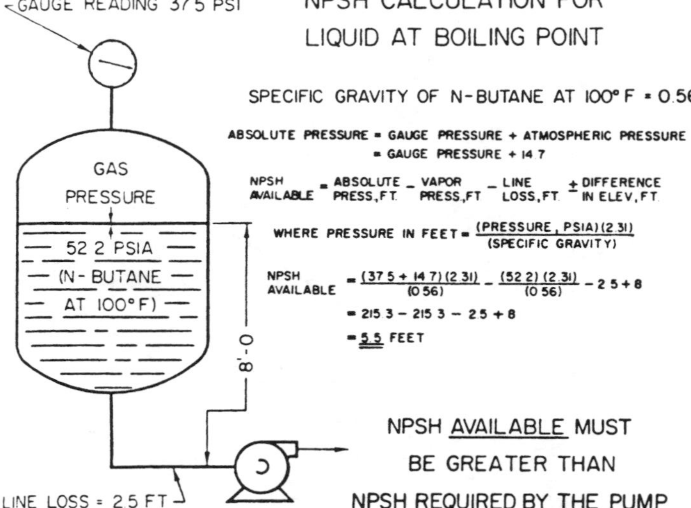

Figure 2. NPSH calculation for liquid at boiling point.

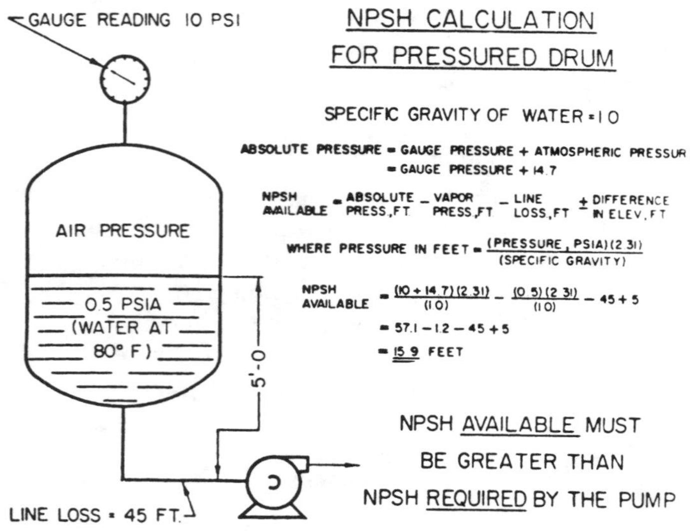

Figure 3. NPSH calculation for pressured drum.

Sources

1. McAllister, E. W., *Pipe Line Rules of Thumb Handbook, 3rd Ed.,* Gulf Publishing Co., 1993.

Suction System NPSH for Studies

For studies or initial design it is good to have quick estimates of pump NPSH. Evans discusses the general formula

$$n(Q)^{0.5}/(NPSH)^{3/4} = C$$

where

n = Speed, rpm
Q = Capacity, gpm
C = A constant between 7,000 and 10,000

Evans plotted the relationship with C = 9,000 resulting in the following graph for initial estimates of minimum NPSH required.

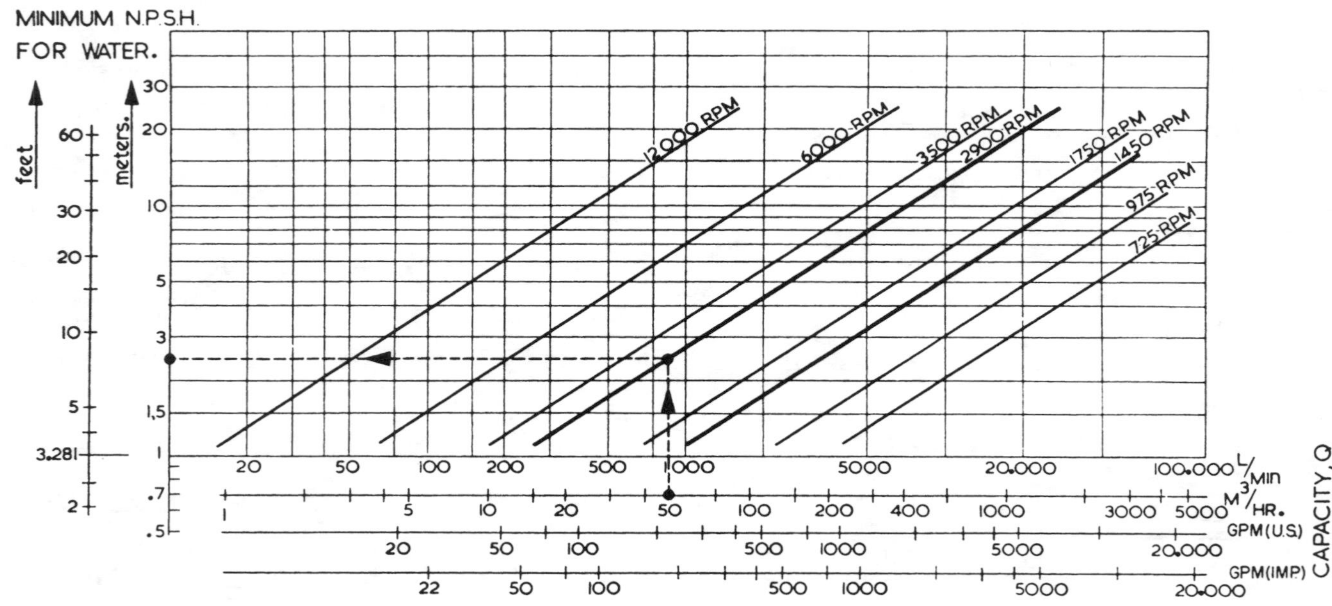

Figure 1. Use these curves to find minimum NPSH required.

Source

Evans, F. L., *Equipment Design Handbook for Refineries and Chemical Plants, Vol. 1, 2nd Ed.,* Gulf Publishing Co., 1979.

Suction System NPSH with Dissolved Gas

NPSH calculations might have to be modified if there are significant amounts of dissolved gas in the pump suction liquid. See "Suction System NPSH Available" in this handbook for calculations when dissolved gas does not need to be considered. In that case the suction liquid's vapor pressure is a term in the equation. With dissolved gases, the gas saturation pressure is often much higher than the liquid's vapor pressure.

Penney discusses contemporary methods of addressing dissolved gases and makes recommendations of his own with plant examples. Basically his recommendations are as follows. When mechanical damage is of prime concern, use the gas saturation pressure in place of the liquid vapor pressure in the calculations. This is the surest, but most expensive, method. Otherwise, use a lower pressure that will allow no more than 3% by volume of flashed gas in the pump suction. Process engineers are best equipped to perform such phase equilibrium calculations.

Source

Penney, R. W., "Inert Gas in Liquid Mars Pump Performance," *Chemical Engineering,* July 3, 1978, p. 63.

Larger Impeller

When considering a larger impeller to expand a pump's capacity, try to stay within the capacity of the motor. Changing impellers is relatively cheap, but replacing the motor and its associated electrical equipment is expensive. The effects of a larger diameter (d) are approximately as follows:

$Q_2 = Q_1(d_2/d_1)$	flow rate
$h_2 = h_1(d_2/d_1)^2$	head delivered
$A_2 = A_1(d_2/d_1)^3$	amperage drawn by motor

Lieberman suggests measuring the motor amperage in the field with the pump discharge control valve wide open. Knowing the motor's rated amperage, including service factor (usually 10–15%), the maximum size impeller that can be used with the existing motor can be found.

Source

Lieberman, N. P., *Process Design For Reliable Operations, 2nd Ed.,* Gulf Publishing Co., 1988.

Construction Materials

Table 1[1] shows materials of construction for general planning when specific service experience is lacking.

Table 1
Pump Materials of Construction
Table materials are for general use, specific service experience is preferred when available

Liquid	Casing & Wear Rings	Impeller & Wear Rings	Shaft	Shaft Sleeves	Type of Seal	Seal Cage	Gland	Remarks
Ammonia, Anhydrous & Aqua	Cast Iron	Cast Iron	Carbon Steel	Carbon Steel	Mechanical	....	Mall. Iron	NOTE: Materials of Construction shown will be revised for some jobs.
Benzene	Cast Iron	Cast Iron	Carbon Steel	Nickel Moly. Steel	Ring Packing	Cast Iron	Mall. Iron	
Brine (Sodium Chloride)	Ni-Resist*	Ni-Resist*	K Monel	K Monel	Ring Packing	Ni-Resist**	Ni-Resist**	*Cast Iron acceptable. ** Malleable Iron acceptable.
Butadiene	Casing: C. Steel—Rings: C.I.	Impeller: C.I.—Rings: C. Steel	Carbon Steel	13% Chrome Steel	Mechanical	....	Carbon Steel	
Carbon Tetrachloride	Cast Iron	Cast Iron	Carbon Steel	Carbon Steel	Mechanical	....	Mall. Iron	
Caustic, 50% (Max. Temp. 200° F.)	Misco C	Misco C	18-8 Stainless Steel	Misco C	Ring Packing	Misco C	Carbon Steel	Misco C manufactured by Michigan Steel Casting Company. 29 Cr-9 Ni Stainless Steel, or equal.
Caustic, 50% (Over 200° F) & 73%	Nickel	Nickel	Nickel or 18-8 Stainless Steel	Nickel	Ring Packing	Nickel	Nickel	
Caustic, 10% (with some sodium chloride)	Cast Iron	23% Cr. 52% Ni Stainless Steel	23% Cr. 52% Ni Stainless Steel	23% Cr. 52% Ni Stainless Steel	Ring Packing	Cast Iron	....	Specifications for 50% Caustic (Maximum Temperature 200° F) also used.
Ethylene	Cast Steel	Carbon Steel	Carbon Steel	Carbon Steel	Mechanical	Cast Iron	Mall. Iron	
Ethylene Dichloride	Cast Iron	Cast Iron	Steel	K Monel	Mechanical	....	K Monel	
Ethylene Glycol	Bronze	Bronze	18-8 Stainless Steel	18-8 Stainless Steel	Ring Packing	....	Bronze	
Hydrochloric Acid, 32%	Impregnated Carbon	Impregnated Carbon	18-8 Stainless Steel	Impregnated Carbon	Mechanical	....	Impregnated Carbon	
Hydrochloric Acid, 32% (Alternate)	Rubber Lined C. Iron	Hard Rubber	Carbon Steel	Rubber or Plastic	Ring Packing	Rubber	Rubber	
Methyl Chloride	Cast Iron	Cast Iron	18-8 Stainless Steel	18-8 Stainless Steel	Mechanical	....	Mall. Iron	
Propylene	Casing: C. Steel–Rings: C.I.	Imp.: C.I.–Rings: C. Stl	Carbon Steel	Carbon Steel	Mechanical	Cast Iron	Mall. Iron	

(table continued on next page)

Table 1 Continued
Pump Materials of Construction
Table materials are for general use, specific service experience is preferred when available

Liquid	Casing & Wear Rings	Impeller & Wear Rings	Shaft	Shaft Sleeves	Type of Seal	Seal Cage	Gland	Remarks
Sulfuric Acid, Below 55%	Hard Rubber Lined C.I.	Special Rubber	Carbon Steel	Hastelloy C	Ring Packing	Special Rubber	Special Rubber	
Sulfuric Acid, 55 to 95%	Cast Si-Iron	Si-Iron	Type 316 Stn. Stl.	Si-Iron	Ring Packing	Teflon	Si-Iron	
Sulfuric Acid, Above 95%	Cast Iron	Cast Iron	Carbon Steel	13% Cr	Mechanical	....	Mall. Iron	
Styrene	Cast Iron	Cast Iron	Carbon Steel	13% Chrome Steel	Ring Packing	Cast Iron	Mall. Iron	
Water, River	Cast Iron	Bronze	18-8 Stainless Steel	Bronze	Mechanical	Cast Iron	Mall. Iron	
Water, Sea	Casing, 1–2% Ni, Cr 3-0.5% Cast Iron Rings: Ni-Resist, 2B	Impeller: Monel Rings: S-Monel	K Monel (Aged)	K Monel or Alloy 20 SS	Ring Packing	Monel or Alloy 20 SS	Monel or Alloy 20 SS	

Source

1. Ludwig, E. E., *Applied Process Design for Chemical and Petrochemical Plants, Vol. 1,* Gulf Publishing Co.

6
Compressors

Ranges of Application

The following graph from the Compressor Handbook is a good quick guide for comparing ranges of application for centrifugal, reciprocating and axial flow compressors.

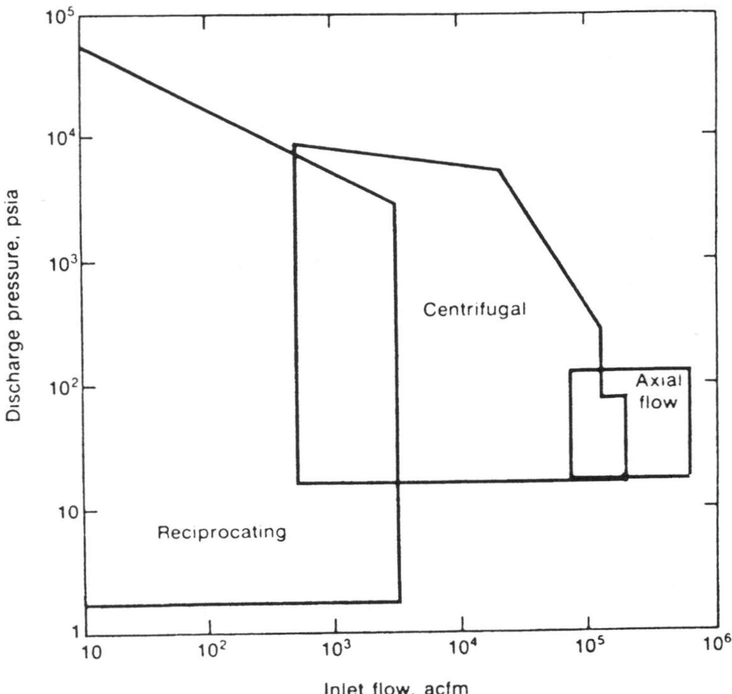

Figure 1. Approximate ranges of application for reciprocating, centrifugal and axial-flow compressors.

Source

Dimoplon, William, "What Process Engineers Need To Know About Compressors," *Compressor Handbook for the Hydrocarbon Industries,* Gulf Publishing Co., 1979.

Generalized Z

A quick estimate of the compressibility factor Z can be made from the nomograph[1,2] shown as Figure 1.

where

T = Temperature in consistent absolute units
T_c = Critical temperature in consistent absolute units
$T_R = T/T_c$
P = Pressure in consistent absolute units

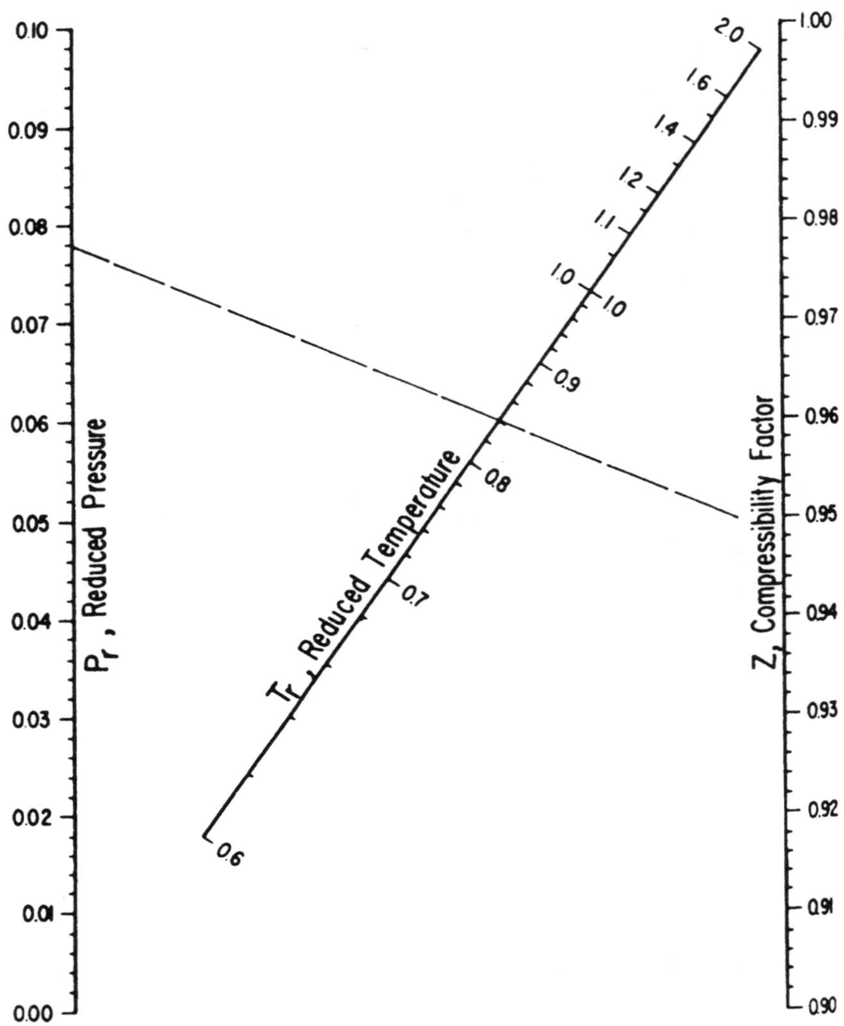

Figure 1. Generalized compressibility factor. (Reproduced by permission *Petroleum Refiner,* Vol. 37, No. 11, copyright 1961, Gulf Publishing Co., Houston.)

P_c = Critical pressure in consistent absolute units
$P_R = P/P_c$

An accuracy of one percent is claimed, but the following gases are excluded: helium, hydrogen, water, and ammonia.

Sources

1. McAllister, E. W., *Pipe Line Rules of Thumb Handbook, 3rd Ed.,* Gulf Publishing Co., 1993.
2. Davis, D. S., *Petroleum Refiner, 37,* No. 11, 1961.

Generalized K

The following handy graph from the *GPSA Engineering Data Book* allows quick estimation of a gas's heat-capacity ratio ($K = C_p/C_v$) knowing only the gas's molecular weight.

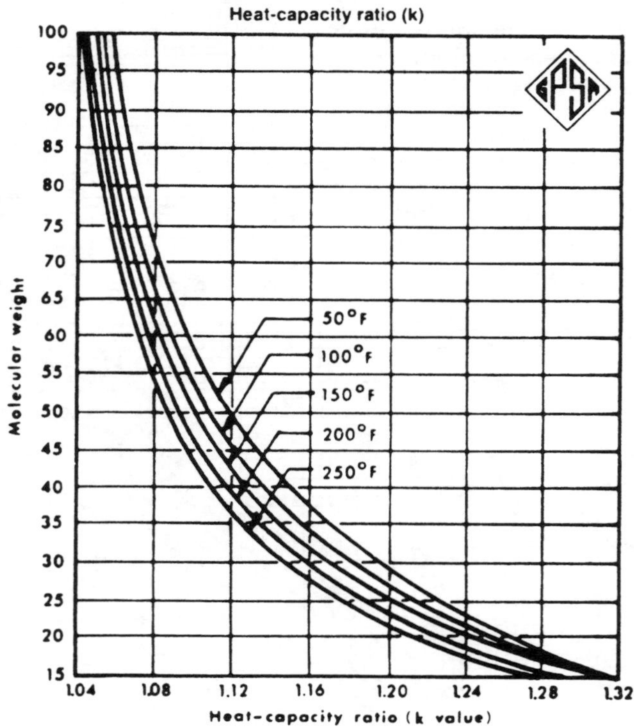

Figure 1. Approximate heat-capacity ratios of hydrocarbon gases.

where

C_p = Heat capacity at constant pressure, Btu/lb°F
C_v = Heat capacity at constant volume, Btu/lb°F

Source

GPSA Engineering Data Book, Gas Processors Suppliers Association, Vol. 1, 10th Ed., 1987.

Horsepower Calculation

For centrifugal compressors the following method is normally used.

First, the required head is calculated. Either the polytropic or adiabatic head can be used to calculate horsepower so long as the polytropic or adiabatic efficiency is used with the companion head.

Polytropic Head

$$H_{poly} = \frac{ZRT_1}{(N-1)/N}\left[\left(\frac{P_2}{P_1}\right)^{(N-1)/N} - 1\right]$$

Adiabatic Head

$$H_{AD} = \frac{ZRT_1}{(K-1)/K}\left[\left(\frac{P_2}{P_1}\right)^{(K-1)/K} - 1\right]$$

where

Z = Average compressibility factor; using 1.0 will yield conservative results
R = 1,544/mol. wt.
T_1 = Suction temperature, °R
P_1, P_2 = Suction, discharge pressures, psia
K = Adiabatic exponent, C_p/C_v
N = Polytropic exponent, $\frac{N-1}{N} = \frac{K-1}{KE_p}$
E_p = Polytropic efficiency
E_A = Adiabatic efficiency

The polytropic and adiabatic efficiencies are related as follows:

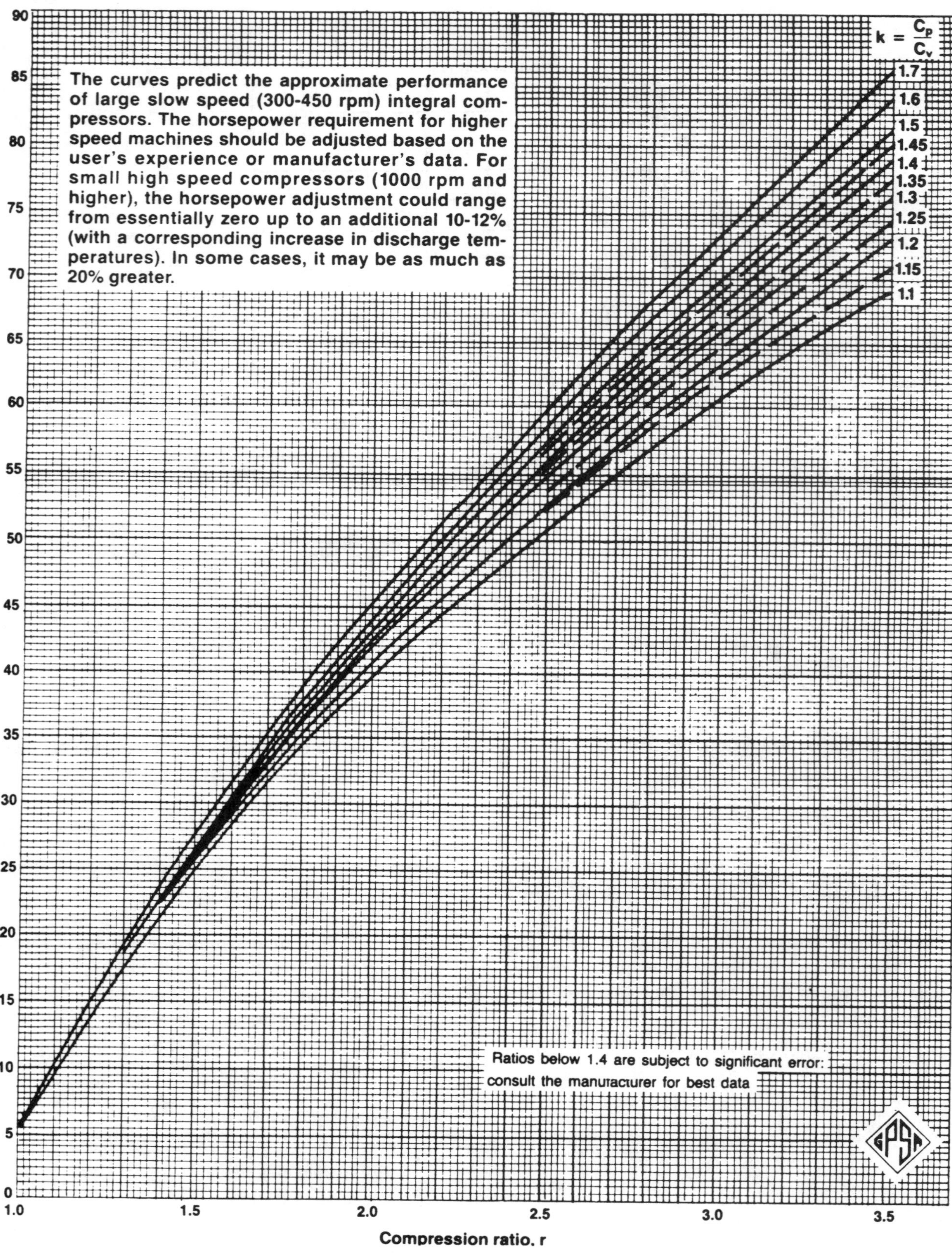
The curves predict the approximate performance of large slow speed (300-450 rpm) integral compressors. The horsepower requirement for higher speed machines should be adjusted based on the user's experience or manufacturer's data. For small high speed compressors (1000 rpm and higher), the horsepower adjustment could range from essentially zero up to an additional 10-12% (with a corresponding increase in discharge temperatures). In some cases, it may be as much as 20% greater.
k = Cp/Cv
1.7
1.6
1.5
1.45
1.4
1.35
1.3
1.25
1.2
1.15
1.1
Ratios below 1.4 are subject to significant error: consult the manufacturer for best data
GPSA
Compression ratio, r
1.0
1.5
2.0
2.5
3.0
3.5
0
5
10
15
20
25
30
35
40
45
50
55
60
65
70
75
80
85
90

Figure 1. Bhp per million curve mechanical efficiency-95% gas velocity through valve-3000 ft/min (API equation).

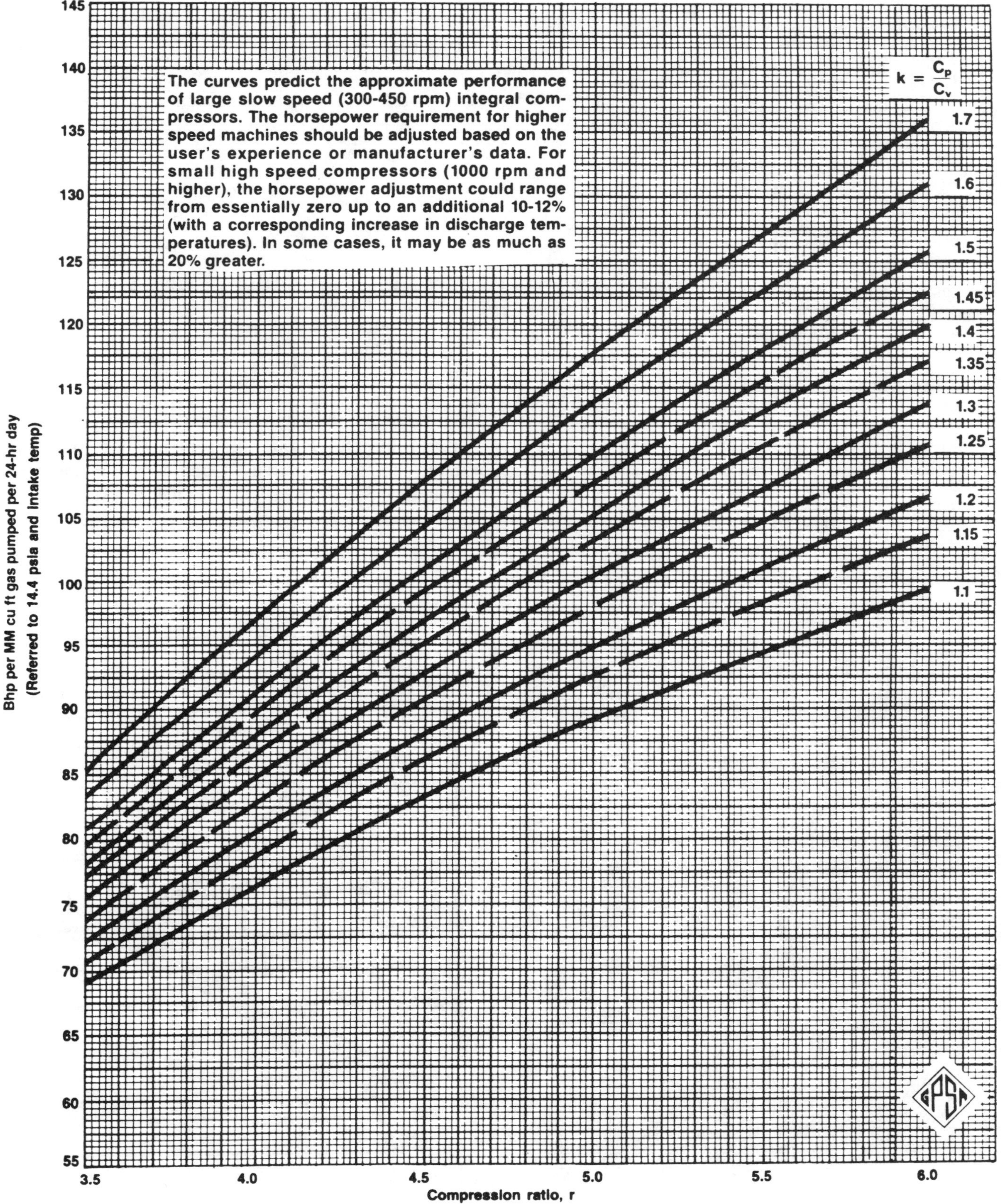
The curves predict the approximate performance of large slow speed (300-450 rpm) integral compressors. The horsepower requirement for higher speed machines should be adjusted based on the user's experience or manufacturer's data. For small high speed compressors (1000 rpm and higher), the horsepower adjustment could range from essentially zero up to an additional 10-12% (with a corresponding increase in discharge temperatures). In some cases, it may be as much as 20% greater.
$k = \frac{C_p}{C_v}$
1.7
1.6
1.5
1.45
1.4
1.35
1.3
1.25
1.2
1.15
1.1
Bhp per MM cu ft gas pumped per 24-hr day
(Referred to 14.4 psia and intake temp)
145
140
135
130
125
120
115
110
105
100
95
90
85
80
75
70
65
60
55
3.5
4.0
4.5
5.0
5.5
6.0
Compression ratio, r
GPSA

Figure 2. Bhp per million curve mechanical efficiency-95% gas velocity through valve-3000 ft/min (API equation).

$$E_A = \left[\frac{\left(\frac{P_2}{P_1}\right)^{(K-1)/K} - 1}{\left(\frac{P_2}{P_1}\right)^{(N-1)/N} - 1} \right] = \left[\frac{\left(\frac{P_2}{P_1}\right)^{(K-1)/K} - 1}{\left(\frac{P_2}{P_1}\right)^{(K-1)/KE_p} - 1} \right]$$

The gas horsepower is calculated using the companion head and efficiency.

From Polytropic Head.

$$HP = \frac{W\ H_{poly}}{E_p\ 33{,}000}$$

From Adiabatic Head.

$$HP = \frac{W\ H_{AD}}{E_A\ 33{,}000}$$

where

HP = Gas horsepower
W = Flow, lb/min

While the foregoing equations will work equally well for reciprocating compressors the famous "horsepower per million" curves[1] are often used, particularly in the natural gas industry. See Figures 1, 2, 3, and 4.

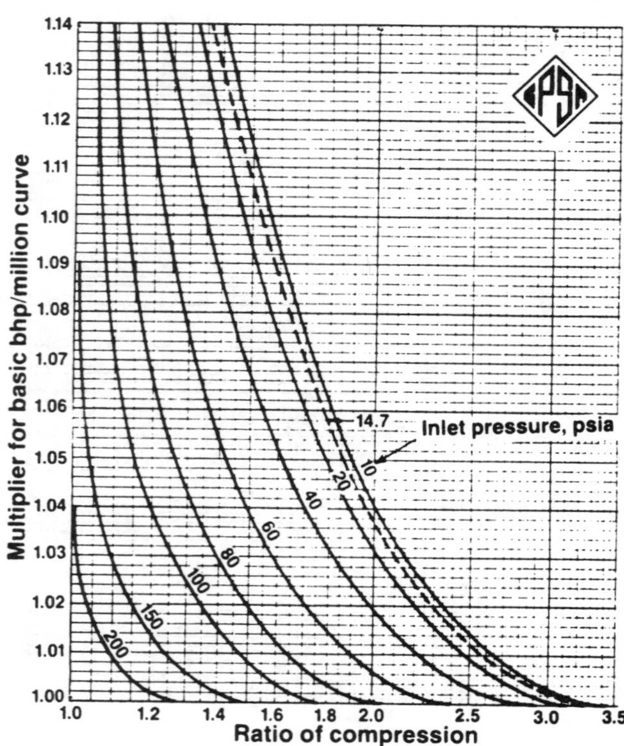

Figure 3. Correction factor for low intake pressure.

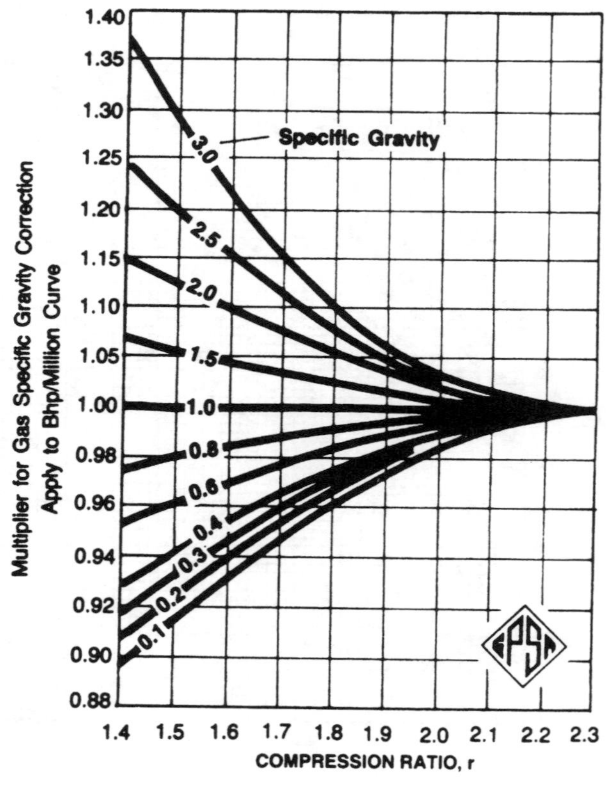

Figure 4. Correction factor for specific gravity.

To develop an equation that will approximate the "horsepower per million" curves, substituting $Q_1 = WZRT_1/144p_1$ into the foregoing horsepower equations will yield:

$$HP = (144/33{,}000)[K/(K-1)]\ (P_1Q_1)[r^{(k-1)/k} - 1]/E_0$$

where

Q_1 = Actual capacity, cfm—measured at inlet conditions
P_1 = Suction pressure, psia
r = Compression ratio = P_2/P_1
E_o = Overall efficiency, fraction

The bases for the "horsepower per million" curves are:

$Q_1 = 694.44$ cfm $= 10^6$ ft^3/day
$P_1 = 14.4$ psia
Mechanical Efficiency = 95%
(We will use 80% overall efficiency)

$$HP = (144/(33{,}000 \times .8))[K/(K-1)](10{,}000)[r^{(k-1)/k} - 1]$$

or

$$HP = 54[K/(K-1)][r^{(k-1)/k} - 1]$$

Sources

1. *GPSA Engineering Data Book,* Gas Processors Suppliers Association, Vol. 1, 10th Ed., 1987.
2. Branan, C. R., *The Process Engineer's Pocket Handbook, Vol. 1,* Gulf Publishing Co., 1976.

Efficiency

Here are some approximate efficiencies to use for centrifugal, axial flow, and reciprocating compressors.

Centrifugal and Axial Flow

Figure 1 gives polytropic efficiencies.

Figure 2 gives the relationship between polytropic and adiabatic efficiencies.

Reciprocating

Figure 3 gives reciprocating compressor efficiencies (see Tables 1 and 2 for correction factors).

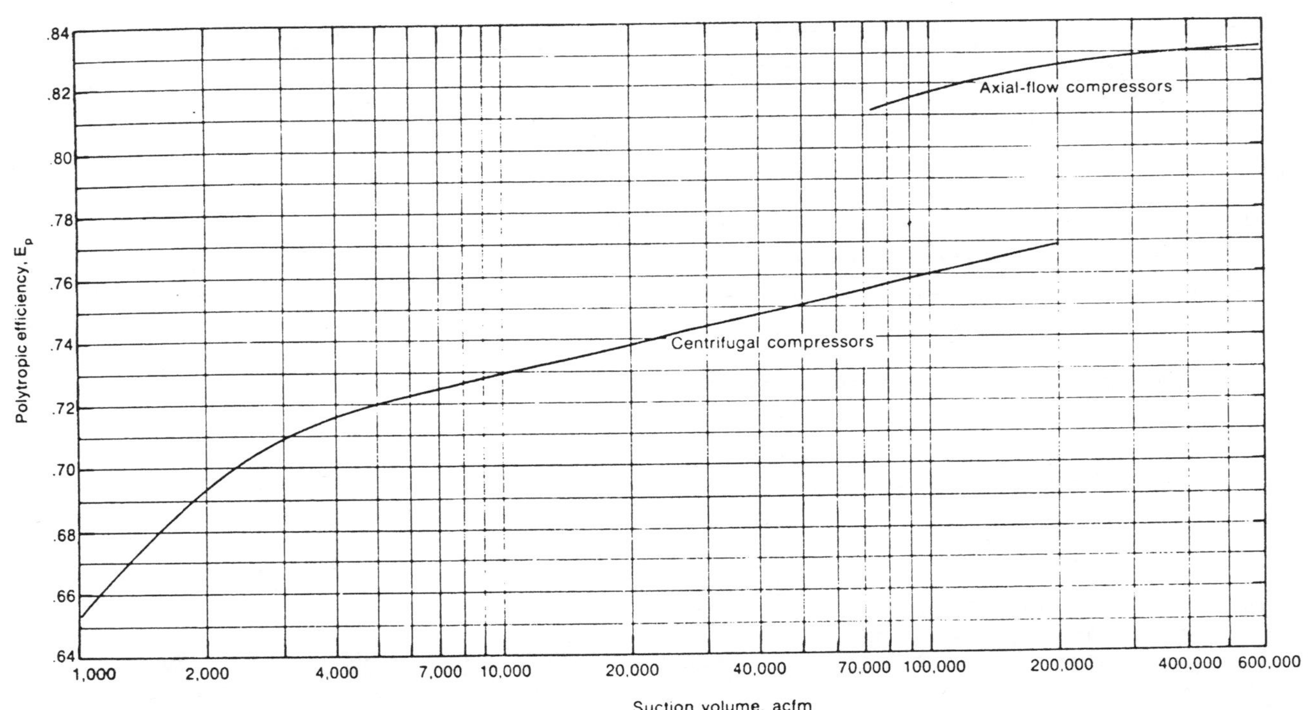

Figure 1. Approximate polytropic efficiencies for centrifugal and axial flow compressors.

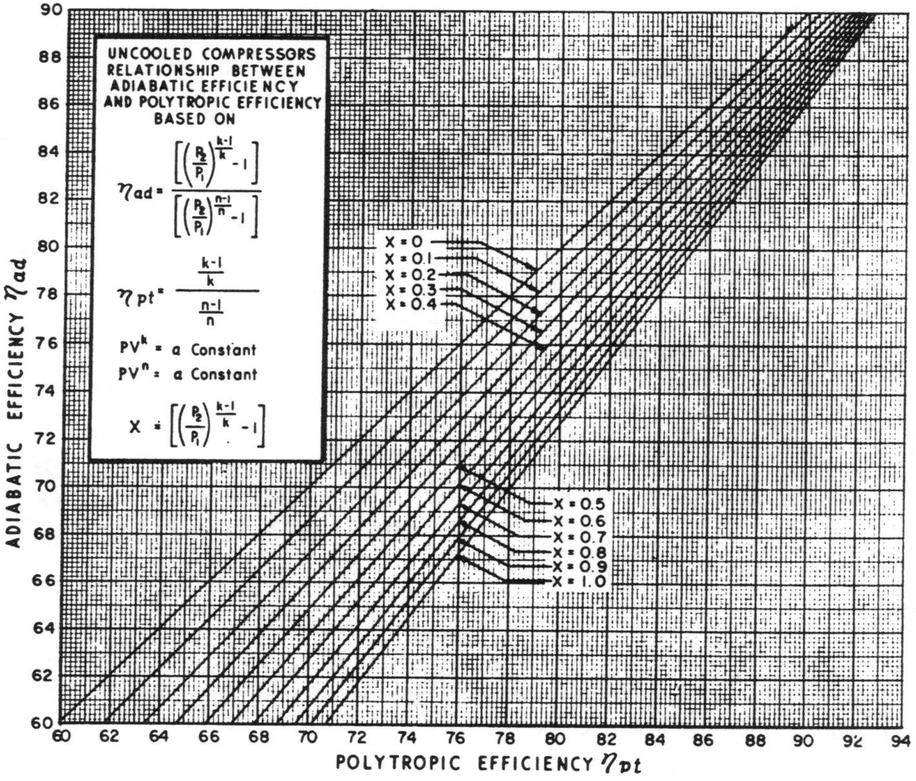

Figure 2. Uncooled compressor relationship between adiabatic efficiency and polytropic efficiency.

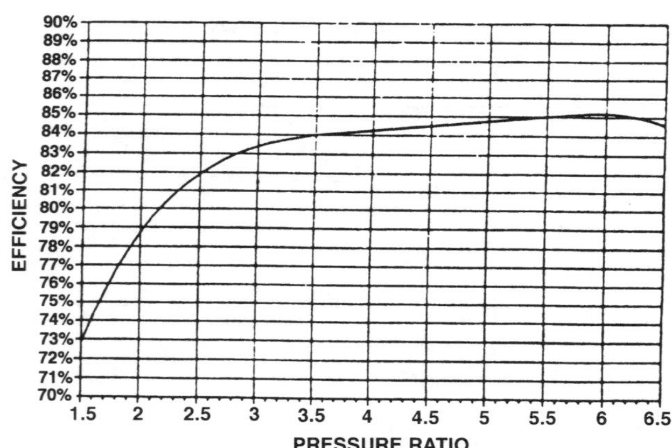

Figure 3. Reciprocating compressor efficiencies.

Table 1
Efficiency Multiplier for Low Pressure

	Pressure Psia							
r_p	10	14.7	20	40	60	80	100	150
3.0	.990	1.00	1.00	1.00	1.00	1.00	1.00	1.00
2.5	.980	.985	.990	.995	1.00	1.00	1.00	1.00
2.0	.960	.965	.970	.980	.990	1.00	1.00	1.00
1.5	.890	.900	.920	.940	.960	.980	.990	1.00

Source: Modified courtesy of the Gas Processors Suppliers Association and Ingersoll-Rand.

Table 2
Efficiency Multiplier for Specific Gravity

	SG				
r_p	1.5	1.3	1.0	0.8	0.6
2.0	0.99	1.0	1.0	1.0	1.01
1.75	0.97	0.99	1.0	1.01	1.02
1.5	0.94	0.97	1.0	1.02	1.04

Source: Modified courtesy of the Gas Processors Suppliers Association.

Sources

1. *Compressor Handbook For The Hydrocarbon Processing Industries,* Gulf Publishing Co., 1979.
2. Evans, F. L., *Equipment Design Handbook For Refineries and Chemical Plants, Vol. 1, 2nd Ed.,* Gulf Publishing Co., 1979.

Temperature Rise

The temperature ratio across a compression stage is

$$T_2/T_1 = (P_2/P_1)^{(K-1)/K} \quad \text{Adiabatic}$$
$$T_2/T_1 = (P_2/P_1)^{(N-1)/N} \quad \text{Polytropic}$$

where

K = Adiabatic exponent, C_p/C_v
N = Polytropic exponent, $(N - 1)/N = (K - 1)/KE_p$
P_1, P_2 = Suction, discharge pressures, psia
T_1, T_2 = Suction, discharge temperatures, °R
E_p = Polytropic efficiency, fraction

Ludwig[1] states that the usual centrifugal compressor is uncooled internally and thus follows a polytropic path.

Often the temperature of the gas must be limited. Sometimes temperature is limited to protect against polymerization as in olefin or butadiene plants. At temperatures greater than 450–500°F, the approximate mechanical limit, problems of sealing and casing growth start to occur. High temperature requires a special and more costly machine. Most multistage applications are designed to stay below 250–300°F.

Intercooling can be used to hold desired temperatures for high overall compression ratio applications. This can be done between stages in a single compressor frame or between series frames. Sometimes economics rather than a temperature limit dictate intercooling.

Sometimes for high compression ratio applications the job cannot be done in a single compressor frame. Usually a frame will not contain more than about eight (8) stages (wheels). There is a maximum head that one stage can handle. This depends upon the gas properties and inlet temperature. Usually this will run 7,000 to 11,000 feet for a single stage. In lieu of manufacturer's data use eight (8) maximum stages per frame. Then subtract one stage for every side nozzle such as to and from an intercooler, side gas injection, etc. For many applications the compression ratio across a frame will run 2.5–4.0.

Sources

1. Ludwig, E. E., *Applied Process Design for Chemical and Petrochemical Plants, Vol. 3,* Gulf Publishing Co.
2. Branan, C. R., *The Process Engineer's Pocket Handbook, Vol. 1,* Gulf Publishing Co., 1976.

Surge Controls

A compressor surges at certain conditions of low flow, and the "compressor map," a plot of head versus flow, has a surge line defining the limits. Surge controls help the machine avoid surge by increasing flow. For an air compressor, a simple spill to the atmosphere is sufficient. For a hydrocarbon compressor, recirculation from discharge to suction is used.

There are several types of surge controls and the process engineer need not be familiar with them all. The main thing to avoid is a low-budget system that has a narrow effective range, especially for large compressors. The author prefers the flow/ΔP type, because the flow and differential pressure relate to the compressor map and are thus easy to understand and work with. This type of surge control system works well over a broad range of operations, and can be switched from manual to automatic control early in a startup.

The correct flow to use is the compressor suction. However, a flow element such as an orifice in the compressor suction can rob inordinate horsepower. Therefore, sometimes the discharge flow is measured and the suction flow computed within the controller by using pressure measurements. Other times the compressor intake nozzle is calibrated and used as a flow element. The correct ΔP to use is the discharge minus the suction pressure.

Source

Branan, C. R., *The Process Engineer's Pocket Handbook, Vol. 2,* Gulf Publishing Co., 1978.

7 Drivers

Motors: Efficiency

Table 1 from the *GPSA Engineering Data Book*[1] compares standard and high efficiency motors. Table 2 from GPSA compares synchronous and induction motors. Table 3 from Evans[2] shows the effect of a large range of speeds on efficiency.

Table 1
Energy Evaluation Chart
NEMA Frame Size Motors, Induction

HP	Approx. Full Load RPM	Amperes Based on 460V: Standard Efficiency	Amperes Based on 460V: High Efficiency	Efficiency in Percentage at Full Load: Standard Efficiency	Efficiency in Percentage at Full Load: High Efficiency
1	1,800	1.9	1.5	72.0	84.0
	1,200	2.0	2.0	68.0	78.5
1½	1,800	2.5	2.2	75.5	84.0
	1,200	2.8	2.6	72.0	84.0
2	1,800	2.9	3.0	75.5	84.0
	1,200	3.5	3.2	75.5	84.0
3	1,800	4.7	3.9	75.5	87.5
	1,200	5.1	4.8	75.5	86.5
5	1,800	7.1	6.3	78.5	89.5
	1,200	7.6	7.4	78.5	87.5
7½	1,800	9.7	9.4	84.0	90.2
	1,200	10.5	9.9	81.5	89.5
10	1,800	12.7	12.4	86.5	91.0
	1,200	13.4	13.9	84.0	89.5
15	1,800	18.8	18.6	86.5	91.0
	1,200	19.7	19.0	84.0	89.5
20	1,800	24.4	25.0	86.5	91.0
	1,200	25.0	24.9	86.5	90.2
25	1,800	31.2	29.5	88.5	91.7
	1,200	29.2	29.1	88.5	91.0
30	1,800	36.2	35.9	88.5	93.0
	1,200	34.8	34.5	88.5	91.0
40	1,800	48.9	47.8	88.5	93.0
	1,200	46.0	46.2	90.2	92.4
50	1,800	59.3	57.7	90.2	93.6
	1,200	58.1	58.0	90.2	91.7
60	1,800	71.6	68.8	90.2	93.6
	1,200	68.5	69.6	90.2	93.0
75	1,800	92.5	85.3	90.2	93.6
	1,200	86.0	86.5	90.2	93.0
100	1,800	112.0	109.0	91.7	94.5
	1,200	114.0	115.0	91.7	93.6
125	1,800	139.0	136.0	91.7	94.1
	1,200	142.0	144.0	91.7	93.6
150	1,800	167.0	164.0	91.7	95.0
	1,200	168.0	174.0	91.7	94.1
200	1,800	217.0	214.0	93.0	94.1
	1,200	222.0	214.0	93.0	95.0

Table 2
Synchronous vs. Induction 3 Phase, 60 Hertz, 2300 or 4000 Volts

HP	Speed RPM	Synch. Motor Efficiency Full Load 1.0 PF	Induction Motor: Efficiency Full Load	Induction Motor: Power Factor
3,000	1,800	96.6	95.4	89.0
	1,200	96.7	95.2	87.0
3,500	1,800	96.6	95.5	89.0
	1,200	96.8	95.4	88.0
4,000	1,800	96.7	95.5	90.0
	1,200	96.8	95.4	88.0
4,500	1,800	96.8	95.5	89.0
	1,200	97.0	95.4	88.0
5,000	1,800	96.8	95.6	89.0
	1,200	97.0	95.4	88.0
5,500	1,800	96.8	95.6	89.0
	1,200	97.0	95.5	89.0
6,000	1,800	96.9	95.6	89.0
	1,200	97.1	95.5	87.0
7,000	1,800	96.9	95.6	89.0
	1,200	97.2	95.6	88.0
8,000	1,800	97.0	95.7	89.0
	1,200	97.3	95.6	89.0
9,000	1,800	97.0	95.7	89.0
	1,200	97.3	95.8	88.0

Table 3
Full Load Efficiencies

hp	3,600 rpm	1,200 rpm	600 rpm	300 rpm
5	80.0	82.5	—	—
	—	—	—	—
20	86.0	86.5	—	—
	—			82.7*
100	91.0	91.0	93.0	—
	—		91.4*	90.3*
250	91.5	92.0	91.0	—
	—	93.9*	93.4*	92.8*
1,000	94.2	93.7	93.5	92.3
	—	95.5*	95.5*	95.5*
5,000	96.0	95.2	—	—
	—	—	97.2*	97.3*

**Synchronous motors, 1.0 PF*

Sources

1. *GPSA Engineering Data Book,* Gas Processors Suppliers Association, Vol. I, 10th Ed.
2. Evans, F. L., *Equipment Design Handbook for Refineries and Chemical Plants, Vol. 1, 2nd Ed.,* Gulf Publishing Co., 1979.

Motors: Starter Sizes

Here are motor starter (controller) sizes.

Polyphase Motors

NEMA Size	Maximum Horsepower Full Voltage Starting 230 Volts	460–575 Volts
00	1.5	2
0	3	5
1	7.5	10
2	15	25
3	30	50
4	50	100
5	100	200
6	200	400
7	300	600

Single Phase Motors

NEMA Size	Maximum Horsepower Full Voltage Starting (Two Pole Contactor) 115 Volts	230 Volts
00	1.3	1
0	1	2
1	2	3
2	3	7.5
3	7.5	15

Source

McAllister, E. W., *Pipe Line Rules of Thumb Handbook,* 3rd Ed., Gulf Publishing Co., 1993.

Motors: Service Factor

Over the years, oldtimers came to expect a 10–15% service factor for motors. Things are changing, as shown in the following section from Evans[1].

For many years it was common practice to give standard open motors a 115% service factor rating; that is, the motor would operate at a safe temperature at 15% overload. This has changed for large motors, which are closely tailored to specific applications. Large motors, as used here, include synchronous motors and all induction motors with 16 poles or more (450 rpm at 60 Hz).

New catalogs for large induction motors are based on standard motors with Class B insulation of 80°C rise by resistance, 1.0 service factor. Previously, they were 60°C rise by thermometer, 1.15 service factor.

Service factor is mentioned nowhere in the NEMA standards for large machines; there is no definition of it. There is no standard for temperature rise or other characteristics at the service factor overload. In fact, the standards are being changed to state that the temperature rise tables are for motors with 1.0 service factors. Neither standard synchronous nor enclosed induction motors have included service factor for several years.

Today, almost all large motors are designed specifically for a particular application and for a specific driven machine. In sizing the motor for the load, the hp is usually selected so that additional overload capacity is not required. Therefore, customers should not be required to pay for capability they do not require. With the elimination of service factor, standard motor base prices have been reduced 4–5% to reflect the savings.

Users should specify standard hp ratings, without service factor for these reasons:

1. All of the larger standard hp are within or close to 15% steps.
2. As stated in NEMA, using the next larger hp avoids exceeding standard temperature rise.
3. The larger hp ratings provide increased pull-out torque, starting torque, and pull-up torque.
4. The practice of using 1.0 service factor induction motors would be consistent with that generally followed in selecting hp requirements of synchronous motors.
5. For loads requiring an occasional overload, such as startup of pumps with cold water followed by continuous operation with hot water at lower hp loads, using a motor with a short time overload rating will probably be appropriate.

Induction motors with a 15% service factor are still available. Large open motors (except splash-proof) are available for an addition of 5% to the base price, with a specified temperature rise of 90° C for Class B insulation by resistance at the overload horsepower. This means the net price will be approximately the same. At nameplate hp the service factor rated motor will usually have less than 80° C rise by resistance.

Motors with a higher service factor rating such as 125% are also still available, but not normally justifiable. Most smaller open induction motors (i.e., 200 hp and below, 514 rpm and above) still have the 115% service factor rating. Motors in this size range with 115% service factor are standard, general purpose, continuous-rated, 60 Hz, design A or B, drip-proof machines. Motors in this size range which normally have a 100% service factor are totally enclosed motors, intermittent rated motors, high slip design D motors, most multispeed motors, encapsulated motors, and motors other than 60 Hz.

Source

Evans, F. L., *Equipment Design Handbook for Refineries and Chemical Plants, Vol. 1, 2nd Ed.,* Gulf Publishing Co., 1979.

Motors: Useful Equations

The following equations are useful in determining the current, voltage, horsepower, torque, and power factor for AC motors:

Full Load I = [hp(0.746)]/[1.73 E (eff.) PF (three phase)
= [hp(0.746)]/[E (eff.) PF] (single phase)
kVA input = IE (1.73)/1,000 (three phase)
= IE/1,000 (single phase)
kW input = kVA input (PF)
hp output = kW input (eff.)/0.746
= Torque (rpm)/5,250
Full Load Torque = hp (5,250 lb.-ft.)/rpm
Power Factor = kW input/kVA input

where

E = Volts (line-to-line)
I = Current (amps)
PF = Power factor (per unit = percent PF/100)
eff = Efficiency (Per unit = percent eff./100)
hp = Horsepower
kW = Kilowatts
kVA = Kilovoltamperes

Source

Evans, F. L., *Equipment Design Handbook for Refineries and Chemical Plants, Vol. 1, 2nd Ed.,* Gulf Publishing Co., 1979.

Motors: Relative Costs

Evans gives handy relative cost tables for motors based on voltages (Table 1), speeds (Table 2), and enclosures (Table 3).

Table 1
Relative Cost at Three Voltage Levels of Drip-Proof 1,200-rpm Motors

	2,300-Volts	4,160-Volts	13,200-Volts
1,500-hp	100%	114%	174%
3,000-hp	100	108	155
5,000-hp	100	104	145
7,000-hp	100	100	133
9,000-hp	100	100	129
10,000-hp	100	100	129

Table 2
Relative Cost at Three Speeds of Drip-Proof 2,300-Volt Motors

	3,600-Rpm	1,800-Rpm	1,200-Rpm
1,500-hp	124%	94%	100%
3,000-hp	132	100	100
5,000-hp	134	107	100
7,000-hp	136	113	100
9,000-hp	136	117	100
10,000-hp	136	120	100

Table 3
Relative Cost of Three Enclosure Types 2,300-volt, 1,200-rpm Motors

	Drip-proof	Force Ventilated*	Totally-Enclosed Inert Gas or Air Filled**
1,500-hp	100%	115%	183%
3,000-hp	100	113	152
5,000-hp	100	112	136
7,000-hp	100	111	134
9,000-hp	100	111	132
10,000-hp	100	110	125

**Does not include blower and duct for external air supply.*
***With double tube gas to water heat exchanger. Cooling water within manufacturer's standard conditions of temperature and pressure.*

Source

Evans, F. L., *Equipment Design Handbook for Refineries and Chemical Plants, Vol. 1, 2nd Ed.,* Gulf Publishing Co., 1979.

Motors: Overloading

When a pump has a motor drive, the process engineer must verify that the motor will not overload from extreme process changes. The horsepower for a centrifugal pump increases with flow. If the control valve in the discharge line fully opens or an operator opens the control valve bypass, the pump will tend to "run out on its curve," giving more flow and requiring more horsepower. The motor must have the capacity to handle this.

Source

Branan, C. R., *The Process Engineer's Pocket Handbook, Vol. 2,* Gulf Publishing Co., 1978.

Steam Turbines: Steam Rate

The theoretical steam rate (sometimes referred to as the water rate) for steam turbines can be determined from Keenan and Keyes[1] or Mollier charts following a constant entropy path. The theoretical steam rate[1] is given as lb/hr/kw which is easily converted to lb/hr/hp. One word of caution—in using Keenan and Keyes, steam pressures are given in *PSIG.* Sea level is the basis. For low steam pressures at high altitudes appropriate corrections must be made. See the section on Pressure Drop Air-Cooled Air Side Heat Exchangers, in this handbook, for the equation to correct atmospheric pressure for altitude.

The theoretical steam rate must then be divided by the efficiency to obtain the actual steam rate. See the section on Steam Turbines: Efficiency.

Sources

1. Keenan, J. H., and Keyes, F. G., "Theoretical Steam Rate Tables," *Trans. A.S.M.E.* (1938).
2. Branan, C. R., *The Process Engineer's Pocket Handbook, Vol. 1,* Gulf Publishing Co., 1976.

Steam Turbines: Efficiency

Evans[1] provides the following graph of steam turbine efficiencies.

Smaller turbines can vary widely in efficiency depending greatly on speed, horsepower, and pressure conditions.

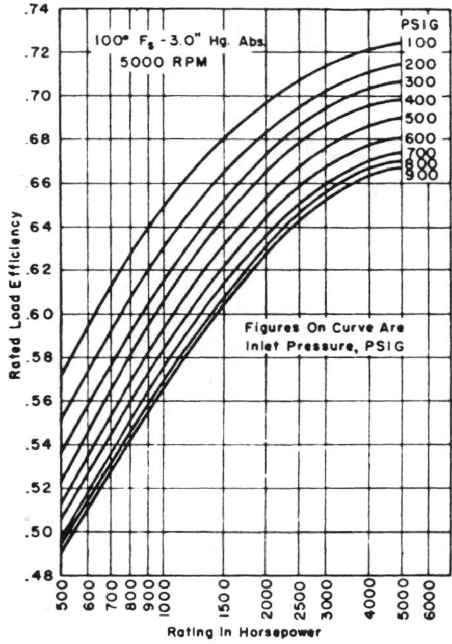

Figure 1. Typical efficiencies for mechanical drive turbines.

Very rough efficiencies to use for initial planning below 500 horsepower at 3500 rpm are

Horsepower	Efficiency, %
1–10	15
10–50	20
50–300	25
300–350	30
350–500	40

Some designers limit the speed of the cheaper small steam turbines to 3600 rpm.

Sources

1. Evans, F. L., *Equipment Design Handbook for Refineries and Chemical Plants, Vol. 1, 2nd Ed.,* Gulf Publishing Co., 1979.
2. Branan, C. R., *The Process Engineer's Pocket Handbook, Vol. 1,* Gulf Publishing Co., 1976.

Gas Turbines: Fuel Rates

Gas turbine fuel rates (heat rates) vary considerably, however Evans[1] provides the following fuel rate graph for initial estimating. It is based on gaseous fuels.

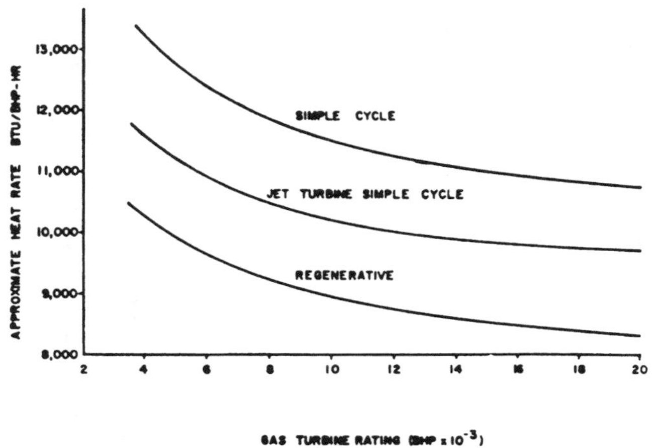

Figure 1. Approximate gas turbine fuel rates.

The *GPSA Engineering Data Book*[2] provides the following four graphs (Figures 2–5) showing the effect of altitude, inlet pressure loss, exhaust pressure loss, and ambient temperature on power and heat rate.

GPSA[2] also provides a table showing 1982 Performance Specifications for a worldwide list of gas turbines, in their Section 15.

Sources

1. Evans, F. L., *Equipment Design Handbook for Refineries and Chemical Plants, Vol. 1, 2nd Ed.,* Gulf Publishing Co., 1979.
2. *GPSA Engineering Data Book,* Gas Processors Suppliers Association, Vol. I, 10th Ed.

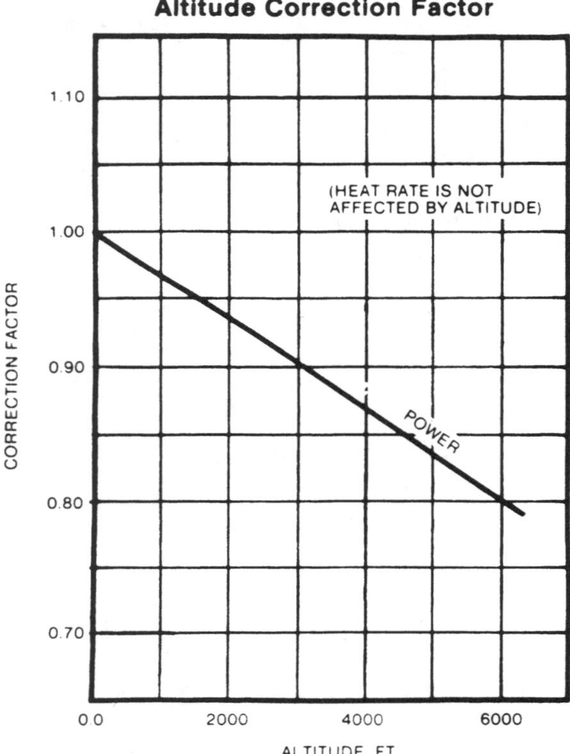

Figure 2. Altitude Correction Factor.

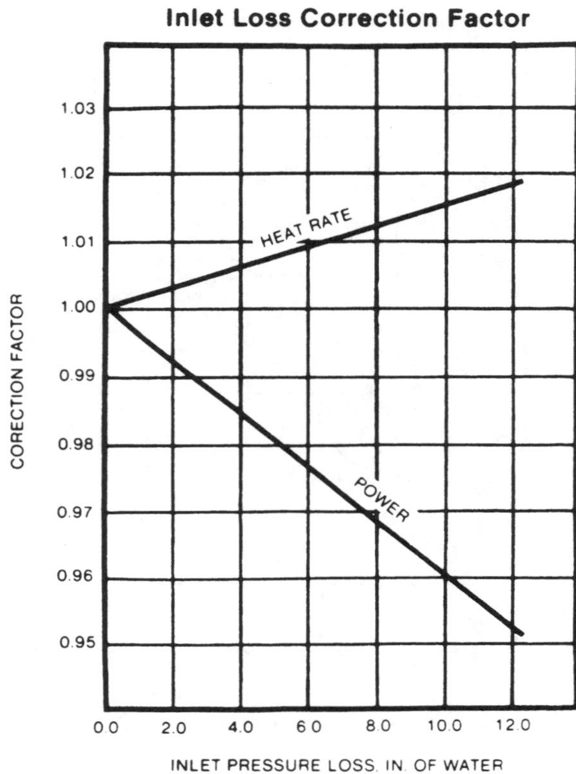

Figure 3. Inlet Loss Correction Factor.

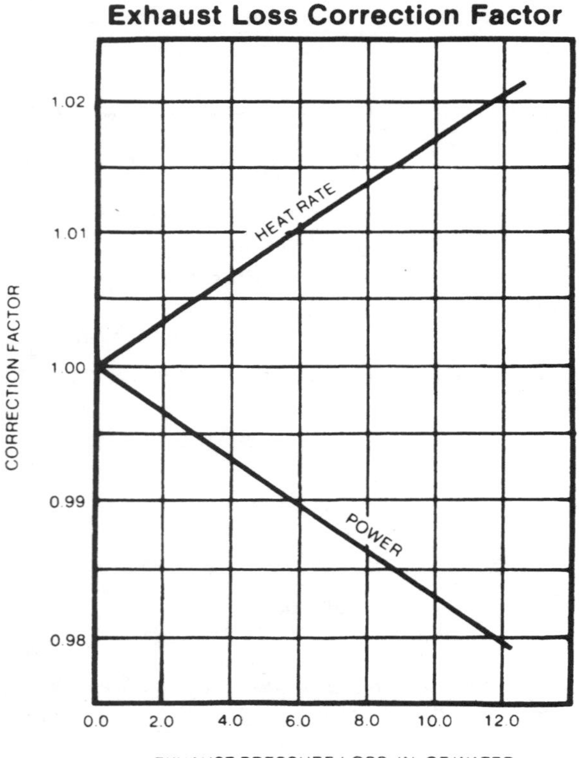

Figure 4. Exhaust Loss Correction Factor.

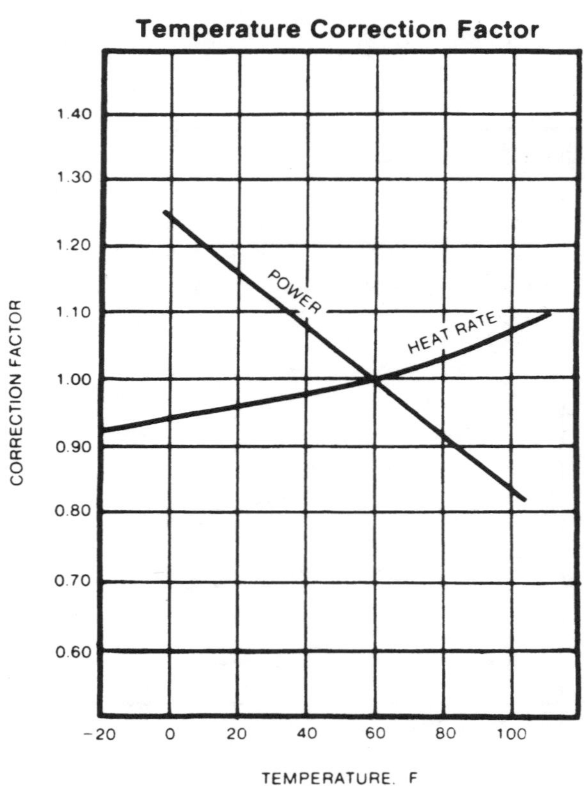

Figure 5. Temperature Correction Factor.

Gas Engines: Fuel Rates

Here are heat rates, for initial estimating, for gas engines.

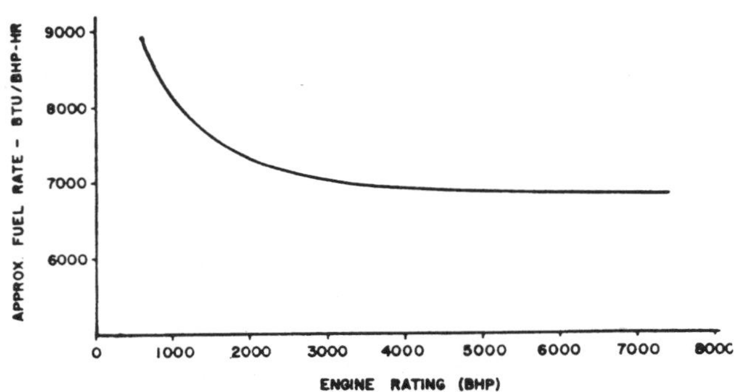

Figure 1. Approximate gas engine fuel rates.

Source

Evans, F. L. *Equipment Design Handbook for Refineries and Chemical Plants, Vol. 1, 2nd Ed.,* Gulf Publishing Co., 1979.

Gas Expanders: Available Energy

With high energy costs, expanders will be used more than ever. A quickie rough estimate of actual expander available energy is

$$\Delta H = C_p T_1 \left[1 - \left(\frac{P_2}{P_1} \right)^{(K-1)/K} \right] 0.5 \qquad (1)$$

where

ΔH = Actual available energy, Btu/lb
C_p = Heat capacity (constant pressure), Btu/lb °F
T_1 = Inlet temperature, °R
P_1, P_2 = Inlet, outlet pressures, psia
$K = C_p/C_v$

To get lb/hr-hp divide as follows:

$$\frac{2545}{\Delta H}$$

A rough outlet temperature can be estimated by

$$T_2 = T_1 \left(\frac{P_2}{P_1} \right)^{(K-1)/K} + \left(\frac{\Delta H}{C_p} \right) \qquad (2)$$

For large expanders, Equation 1 may be conservative. A full rating using vendor data is required for accurate results. Equation 1 can be used to see if a more accurate rating is worthwhile.

For comparison, the outlet temperature for gas at critical flow across an orifice is given by

$$T_2 = T_1 \left(\frac{P_2}{P_1} \right)^{(K-1)/K} = T_1 \left(\frac{2}{K+1} \right) \qquad (3)$$

The proposed expander may cool the working fluid below the dew point. Be sure to check for this.

The expander equation (Equation 1) is generated from the standard compressor head calculation (see Compressors, Horsepower Calculation) by:

1. Turning $[(P_2/P_1)^{(K-1)K} - 1]$ around (since work $= -\Delta h$)
2. Substituting $C_p = (1.9865/\text{Mol.wt.})[K/(K-1)] = (1.9865/1544)R[K/(K-1)]$
3. Cancelling 1.9865/1544 with 779 ftlb/Btu
4. Assuming $Z = 1$
5. Using a roundhouse 50% efficiency

Source

Branan, C. R., *The Process Engineer's Pocket Handbook, Vol. 1,* Gulf Publishing Co., 1976.

8
Separators/Accumulators

Liquid Residence Time

For vapor/liquid separators there is often a liquid residence (holdup) time required for process surge. Tables 1, 2, and 3 give various rules of thumb for approximate work. The vessel design method in this chapter under the Vapor/Liquid Calculation Method heading blends the required liquid surge with the required vapor space to obtain the total separator volume. Finally, a check is made to see if the provided liquid surge allows time for any entrained water to settle.

Table 1
Residence Time for Liquids

Service (times in minutes)	½ Full (Reference 1)	L_{HL} to L_{LL} (minimum) Reference 2*)	½ Full (Reference 3)	Miscellaneous
Tower reflux drum	See Table 2	5-based on reflux flow	5 to 10-based on total flow	—
Vapor-liquid separators	—	—	3–5	—
Product to storage	Depends on situation	Fractionator O.H. Prod.—2	—	—
Product to heat exchanger along with other streams	—	Fractionator O.H. Prod. —5	—	—
Product to heater	—	Fractionator O.H. Prod. —10	—	—
Furnace surge drums	—	—	—	10 min. 20 max.
Tower bottoms FRC control	—	—	—	5 min. 10 max.
LC control	—	—	—	3 min.

** This article deals only with reflux drums. Use only the larger vessel volume determined. Do not add two volumes such as reflux plus product. If a second liquid phase is to be settled, additional time is needed. For water in hydrocarbons, an additional 5 minutes is recommended.*

Table 2
Liquid Residence Rules of Thumb for Reflux Drums

½ Full Minutes (Reference 1)

Factor	Reflux Control	Instrument Factor w/ Alarm	Instrument Factor w/o Alarm	Labor Factor* G	Labor Factor* F	Labor Factor* P
Base reflux drum rules of thumb	FRC	½	1	2	3	4
	LRC	1	1½	2	3	4
	TRC	1½	2	2	3	4

The above are felt to be conservative. For tight design cut labor factors by 50%. Above based on gross overhead.

	Situation	Factor
Multipliers for overhead product portion of gross overhead depending on operation of external equipment receiving the overhead product	Under good control	2.0
	Under fair control	3.0
	Under poor control	4.0
	Feed to or from storage	1.25

	Situation	Factor
Multipliers for gross overhead if no board mounted level recorder	Level indicator on board	1.5
	Gauge glass at equipment only	2.0

**Labor factors are added to the instrument factors G = good, F = fair, P = poor.*

Table 3
Liquid Residence Rules of Thumb for Proper Automatic Control of Interface Level (References 3 and 4)

Flow (gpm)	Reservoir Capacity (gal/in. of depth)
100	2
200	3
600	4
800	8
1,000	10
1,500	15
2,000	24

Sources

1. Watkins, R. N., "Sizing Separators and Accumulators," *Hydrocarbon Processing,* November 1967.
2. Sigales, B., "How to Design Reflux Drums," *Chemical Engineering,* March 3, 1975.
3. Younger, A. H., "How to Size Future Process Vessels," *Chemical Engineering,* May 1955.
4. Anderson, G. D., "Guidelines for Selection of Liquid Level Control Equipment," Fisher Controls Company.

Vapor Residence Time

For vapor/liquid separators, this is usually expressed in terms of maximum velocity which is related to the difference in liquid and vapor densities. The standard equation is

$$U_{vapor\ max} = K[(\rho_L - \rho_v)/\rho_v]^{0.5}$$

where

U = Velocity, ft/sec
ρ = Density of liquid or vapor, lbs/ft^3
K = System constant

Figure 1 relates the K factor for a vertical vessel (K_v) to:

$$W_L/W_v\ (\rho_v/\rho_L)^{0.5}$$

where

W = Liquid or vapor flow rate, lb/sec

For a horizontal vessel $K_H = 1.25K_v$.

Figure 1 is based upon 5% of the liquid entrained in the vapor. This is adequate for normal design. A mist eliminator can get entrainment down to 1%.

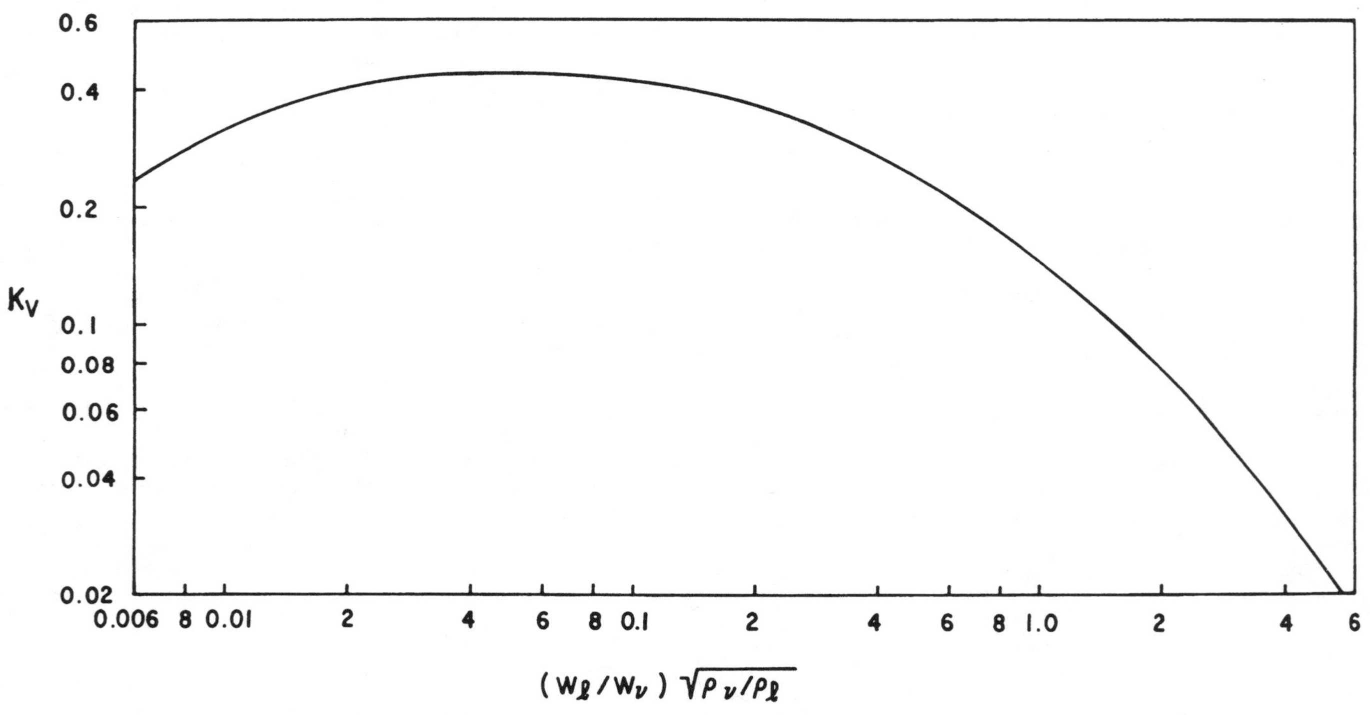

Figure 1. Design vapor velocity factor for vertical vapor-liquid separators at 85% of flooding.

An equation has been developed for Figure 1 as follows:

$X = \ln (W_L/W_V\ (\rho_v/\rho_L)^{0.5})$
$Y = K_v$ (Remember $K_H = 1.25K_v$)

Y = EXP (A + BX + CX^2 + DX^3 + EX^4 + FX^5)

A = −1.942936
B = −0.814894
C = −0.179390
D = −0.0123790
E = 0.000386235
F = 0.000259550

Sources

1. Watkins, R. N., "Sizing Separators and Accumulators," *Hydrocarbon Processing,* November 1967.
2. The equation was generated using FLEXCURV V. 2.0, Gulf Publishing Co.

Vapor/Liquid Calculation Method

A vessel handling large amounts of liquid or a large liquid surge volume will usually be horizontal. Also, where water must be separated from hydrocarbon liquid, the vessel will be horizontal. A vessel with small surge volume such as a compressor knockout drum will usually be vertical.

Vertical Drum

This method uses the separation factor given in the section titled Vapor Residence Time. The first three steps use equations and a graph (or alternate equation) in that section to get K_v and $U_{vapor\ max}$. Nomenclature is explained there.

1. Calculate separation factor = $W_L/W_v\ (\rho_v/\rho_L)^{0.5}$
2. Get K_v from graph or equation
3. Calculate $U_{vapor\ max}$
4. Calculate minimum vessel cross section
5. Set diameter based on 6-inch increments
6. Approximate the vapor-liquid inlet nozzle based on the following criteria:

$$(U_{max})\ \text{nozzle} = 100/\rho_{mix}^{0.5}$$
$$(U_{min})\ \text{nozzle} = 60/\rho_{mix}^{0.5}$$

where

U = max, min velocity, ft/sec
ρ_{mix} = mixture density, lbs/ft^3

7. Sketch the vessel. For height above center line of feed nozzle to top seam, use 36″ + ½ feed nozzle OD or 48″ minimum. For distance below center line of feed nozzle to maximum liquid level, use 12″ + ½ feed nozzle OD or 18″ minimum.
8. Select appropriate full surge volume in seconds. Calculate the required vessel surge volume.

$$V = Q_L\ (\text{design time to fill}),\ ft^3$$

where

Q_L = Liquid flow, ft^3/sec

9. Liquid height is

$$H_L = V\ (4/\Pi D^2),\ ft$$

10. Check geometry. Keep

$$(H_L + H_v)/D$$

between 3 and 5, where H_v is vapor height in feet.

For small volumes of liquid, it may be necessary to provide more liquid surge than is necessary to satisfy the L/D > 3. Otherwise this criteria should be observed. If the required liquid surge volume is greater than that possible in a vessel having L/D < 5, a horizontal drum must be provided.

Horizontal Drum

This method is a companion to the vertical drum method.

1. Calculate separation factor.
2. Look up K_H.
3. Calculate $U_{vapor\ max}$
4. Calculate required vapor flow area.

$$(A_v)_{min} = Q_v/U_{vapor\ max},\ ft^2$$

5. Select appropriate design surge time and calculate full liquid volume. The remainder of the sizing procedure is done by trial and error as in the following steps.
6. When vessel is at full liquid volume,

$$(A_{total})_{min} = (A_v)_{min}/0.2$$
$$D_{min} = \sqrt{4(A_{total})_{min}/\Pi},\ ft$$

7. Calculate vessel length.

$$L = \frac{\text{full liquid volume}}{(\Pi/4)D^2}$$

$D = D_{min}$ to the next largest 6 inches

8. If 5 < L/D < 3, resize.

If there is water to be settled and withdrawn from hydrocarbon, the water's settling time requirement needs to be checked. The water settling requirement, rather than other process considerations, might set the liquid surge capacity. Therefore, the liquid surge capacity we have previously estimated from tables might have to be increased.

Here is a quick check for water settling.

1. Estimate the water terminal settling velocity using:

$$U_T = 44.7 \times 10^{-8} (\rho_w - \rho_o) F_s/\mu_o$$

where

U_T = Terminal settling velocity, ft/sec
Fs = Correction factor for hindered settling
ρ_w, ρ_o = Density of water or oil, lb/ft^3
μ_o = Absolute viscosity of oil, lb/ft-sec

This assumes a droplet diameter of 0.0005 ft. F_s is determined from:

$$F_s = X^2/10^{1.82(1-x)}$$

where

X = Vol. fraction of oil

2. Calculate the modified Reynolds number, Re from:

$$Re = 5 \times 10^{-4} \rho_o U_T/\mu_o \qquad \text{(usually < 1.0)}$$

This assumes a droplet diameter of 0.0005 ft

3. Calculate U_S/U_T from:

$$U_S/U_T = A + B\ln Re + C\ln Re^2 + D\ln Re^3 + E\ln Re^4$$

where

U_S = Actual settling velocity, ft/sec
A = 0.919832
B = −0.091353
C = −0.017157
D = 0.0029258
E = −0.00011591

4. Calculate the length of the settling section as:

$$L = hQ/AU_S$$

where

L = Length of settling zone, ft
h = Height of oil, ft
Q = Flow rate, ft^3/sec
A = Cross-sectional area of the oil settling zone, ft^2

This allows the water to fall out and be drawn off at the bootleg before leaving the settling section.

Sources

1. Watkins, R. N., "Sizing Separators and Accumulators," *Hydrocarbon Processing,* November 1967.
2. Branan, C. R., *The Process Engineer's Pocket Handbook, Vol. 1,* Gulf Publishing Co., 1976.
3. The equations were generated using FLEXCURV V. 2.0, Gulf Publishing Co.

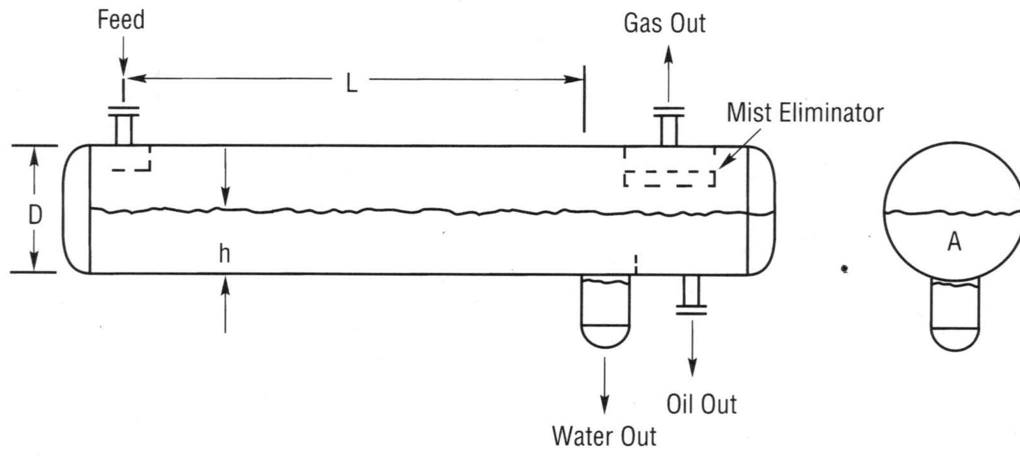

Liquid/Liquid Calculation Method

R. L. Barton ("Sizing Liquid-Liquid Phase Separators Empirically," *Chemical Engineering,* July 8, 1974, Copyright © (1974) McGraw-Hill, Inc., used with permission) provides the following quick method for sizing liquid-liquid phase separators empirically.

The separation of mixtures of immiscible liquids constitutes one of the important chemical engineering operations. This empirical design has proven satisfactory for many phase separations.

1. Calculate holdup time with the formula

   $T = 0.1\ [\mu/(\rho_b - \rho_t)]$

   where

   T = holdup time, hours
   μ = viscosity of the continuous phase, cp
   ρ_b = sp. gr. of the bottom phase
   ρ_t = sp. gr. of the top phase

2. Assign a length-to-diameter ratio of 5, and size a tank to accommodate the required holdup time.
3. Provide inlet and outlet nozzles at one end, and an internal flat cone (see Figure 1).

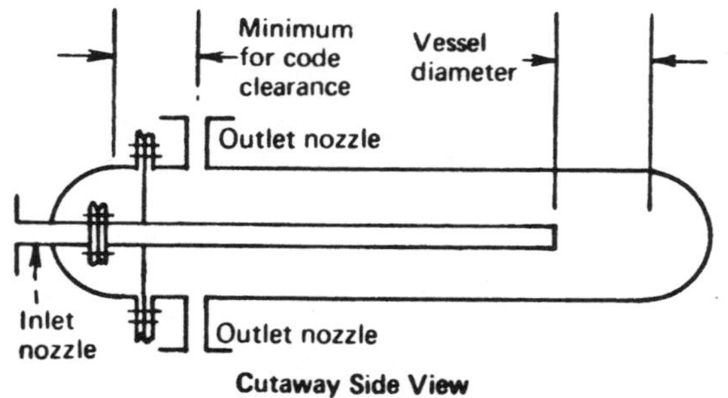

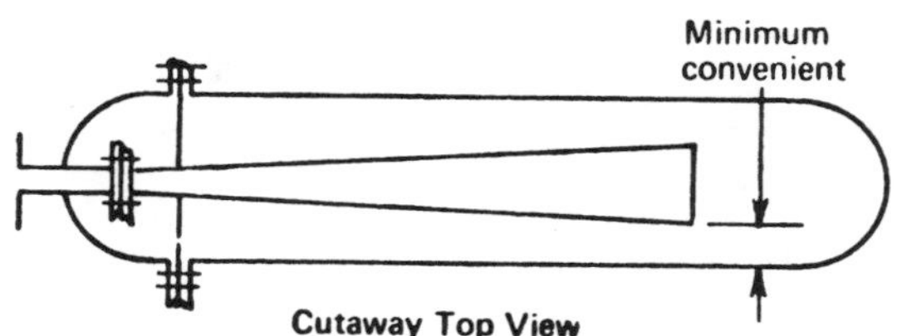

Notes:
1. For large vessels, a manway and internally dismantled cone may be used where more economical.
2. Inlet and outlet nozzles sized for pump discharge.
3. Gage glass and level instruments to be located at inlet-outlet end.
4. Mechanical design to suit for economy under operating conditions.

Figure 1. This is a recommended design for a liquid-liquid separator.

While this design procedure is empirical, there is some rationale behind it. The relation between viscosity and specific-gravity-difference of the phases corresponds to those of the equations for terminal settling velocity in the Stokes-law region and free-settling velocity of isometric particles. Also, the dimensions of the tank and cone recognize that the shape of turbulence created by nozzles discharging into liquids spreads at an angle whose slope is about 1 to 5.

This design is not good for emulsions.

Source

Barton, R. L., "Sizing Liquid-Liquid Phase Separators Empirically," *Chemical Engineering,* July 8, 1974.

Pressure Drop

Pressure Drop Across Mist Eliminator

Use 1″ H_2O pressure drop.

Pressure Drop Entering Plus Leaving Vessel

One velocity head for inlet and one half for outlet pipe velocity is close.

Source

Branan, C. R., *The Process Engineer's Pocket Handbook, Vol. 1,* Gulf Publishing Co., 1976.

Vessel Thickness

Equation 1 gives the required vessel thickness based on inside vessel radius. Equation 2 gives the thickness based on outside radius.

$$T = \frac{Pr_i}{SE - 0.6P} + C \qquad (1)$$

$$T = \frac{Pr_o}{SE + 0.4P} + C \qquad (2)$$

where

P = Pressure, psig
r = Radius, in.
S = Allowable stress, psi
E = Weld efficiency, fraction.
Use 0.85 for initial work.
C = Corrosion allowance, in.

Source

Branan, C. R., *The Process Engineer's Pocket Handbook, Vol. 1,* Gulf Publishing Co., 1976.

Gas Scrubbers

Fair Separation

$$G = 900\sqrt{\rho_V(\rho_L - \rho_V)}$$

Good Separation

$$G = 750\sqrt{\rho_V(\rho_L - \rho_V)}$$

where

G = Allowable mass velocity, lb/hr ft^2
ρ = Density, lb/ft^3

Source

Branan, C. R., *The Process Engineer's Pocket Handbook, Vol. 1,* Gulf Publishing Co., 1976.

Reflux Drums

Sigales has provided an optimizing method for reflux drums. Here are some useful comments from that article.

1. Reflux drums are usually horizontal because the liquid load is important.
2. When a small quantity of a second liquid phase is present, a drawoff pot (commonly called a bootleg) is provided to make separation of the heavy liquid (frequently water) easier. The pot diameter is ordinarily determined for heavy phase velocities of 0.5 ft/min. Minimum length is 3 ft for level controller connections. Minimum pot diameter for a 4 to 8 foot diameter reflux drum is 16 inches. For reflux drums with diameters greater than 8 feet, pot diameters of at least 24 inches are used. The pot must also be placed at a minimum distance from the tangent line that joins the head with the body of the vessel.
3. The minimum vapor clearance height above high liquid level is 20% of drum diameter. If possible this should be greater than 10 inches.

Source

Sigales, B., "How to Design Reflux Drums," *Chemical Engineering,* March 3, 1975.

General Vessel Design Tips

The process engineer gets involved in many mechanical aspects of vessel design such as thickness, corrosion allowance, and internals. Here are some pitfalls to watch for along the way:

1. Be sure to leave sufficient disengaging height above demisters,[1] otherwise a healthy derate must be applied.
2. For liquid/liquid separators, avoid severe piping geometry that can produce turbulence and homogenization. Provide an inlet diffuser cone and avoid shear-producing items, such as slots or holes.
3. Avoid vapor entry close to a liquid level. Reboiler vapor should enter the bottom of a fractionator a distance of at least tray spacing above high liquid level. Tray damage can result if the liquid is disturbed.
4. Avoid extended nozzles or internal piping that the operator cannot see, if at all possible.
5. Make sure items such as gauge glasses, level controls, or pressure taps do not receive an impact head from an incoming stream.
6. Use a close coupled ell for drawoff from gravity separators to eliminate backup of hydraulic head.
7. Check gravity decanters for liquid seal and vapor equalizing line (syphon breaker).
8. For gauge glasses, it is good to have a vent at the top as well as a drain at the bottom. These should be inline for straight-through cleaning.

Sources

1. "Performance of Wire Mesh Demisters®," Bulletin 635, Otto H. York Co., Inc.
2. Branan, C. R., *The Process Engineer's Pocket Handbook, Vol. 2,* Gulf Publishing Co., 1983.

9
Boilers

Power Plants

Battelle has provided a well written report that discusses power plant coal utilization in great detail. It gives a thermal efficiency of 80–83% for steam generation plants and 37–38% thermal efficiency for power generating plants at base load (about 70%). A base load plant designed for about 400 MW and up will run at steam pressures of 2400 or 3600 psi and 1000 °F with reheat to 1000 °F and regenerative heating of feedwater by steam extracted from the turbine. A thermal efficiency of 40% can be had from such a plant at full load and 38% at high annual load factor. The 3600 psi case is supercritical and is called a once-through boiler, because it has no steam drum. Plants designed for about 100–350 MW run around 1800 psi and 1000°F with reheat to 1000°F. Below 100 MW a typical condition would be about 1350 psi and 950°F with no reheat. Below 60% load factor, efficiency falls off rapidly. The average efficiency for all steam power plants on an annual basis is about 33%.

Source

Locklin, D. W., Hazard, H. R., Bloom, S. G., and Nack, H., "Power Plant Utilization of Coal, A Battelle Energy Program Report," Battelle Memorial Institute, Columbus, Ohio, September 1974.

Controls

Three basic parts of boiler controls will be discussed:

1. Level control
2. Firing control (also applies to heaters)
3. Master control

For steam drum level control, the modern 3-element system—steam flow, feedwater flow, and drum level—should be selected. Steam and feedwater flows are compared, with feedwater being requested accordingly and trimmed by the drum level signal. This system is better than having the drum level directly control feedwater, because foaming or changing steam drum conditions can cause a misleading level indication. Also, the 3-element controller responds faster to changes in demand.

The firing controls must be designed to ensure an air-rich mixture at all times, especially during load changes upward or downward. Steam header pressure signals the firing controls for a boiler. The signal to the firing controls comes from a master controller fed by the steam header pressure signal if multiple boilers are operating in parallel.

The firing controls that best ensure an air-rich mixture are often referred to as metering type controls, because gas flow and air flow are metered, thus the fuel-air ratio is controlled. The fuel-air ratio is the most important factor for safe, economical firing, so it is better to control it directly. Do not settle for low budget controllers that attempt to perform this basic job indirectly (such as controlling fuel-steam or other ratios).

The idea for safe control is to have the air lead the fuel on increases in demand and fuel lead the air on decreases in demand. On load increases, the air is increased ahead of the fuel. On load decreases, the fuel is decreased ahead of the air. This is accomplished with high- and low-signal selectors.

A high-signal selector inputs the air flow controller, which often adjusts forced draft fan inlet vanes. The high-signal selector compares the steam pressure and fuel flow signals, selects the highest, and passes it on to the air flow controller. For a load decrease, the steam line pressure tends to rise and its controller reduces the amount of firing air to the boiler, but not immediately. The high-signal selector picks the fuel flow signal instead of the steam pressure signal, which has decreased. The air flow will therefore wait until the fuel has decreased. On load increases, the steam pressure signal exceeds the fuel signal and the air flow is immediately increased.

The reader can easily determine how a low-signal selector works for the fuel flow controller. It would compare the signals from the steam pressure and the air flow. A flue gas oxygen analyzer should be installed to continuously monitor or even trim the air flow.

A master controller is necessary to control a single steam header pressure from multiple parallel boilers. The boilers are increased or decreased in load together. There is a bias station at each boiler if uneven response is desired. The "boiler master," a widely used device, can also be used to control other parallel units of equipment. Therefore, it should not be thought of as only a boiler controller if other applications arise, such as controlling parallel coal gasification units.

Source

Branan, C. R., *The Process Engineer's Pocket Handbook, Vol. 2,* Gulf Publishing Co., 1983.

Thermal Efficiency

Here is a graph that shows how thermal efficiency can be determined from excess air and stack gas temperature.

Source

GPSA Engineering Data Book, Gas Processors Suppliers Association, Vol. I, 10th Ed.

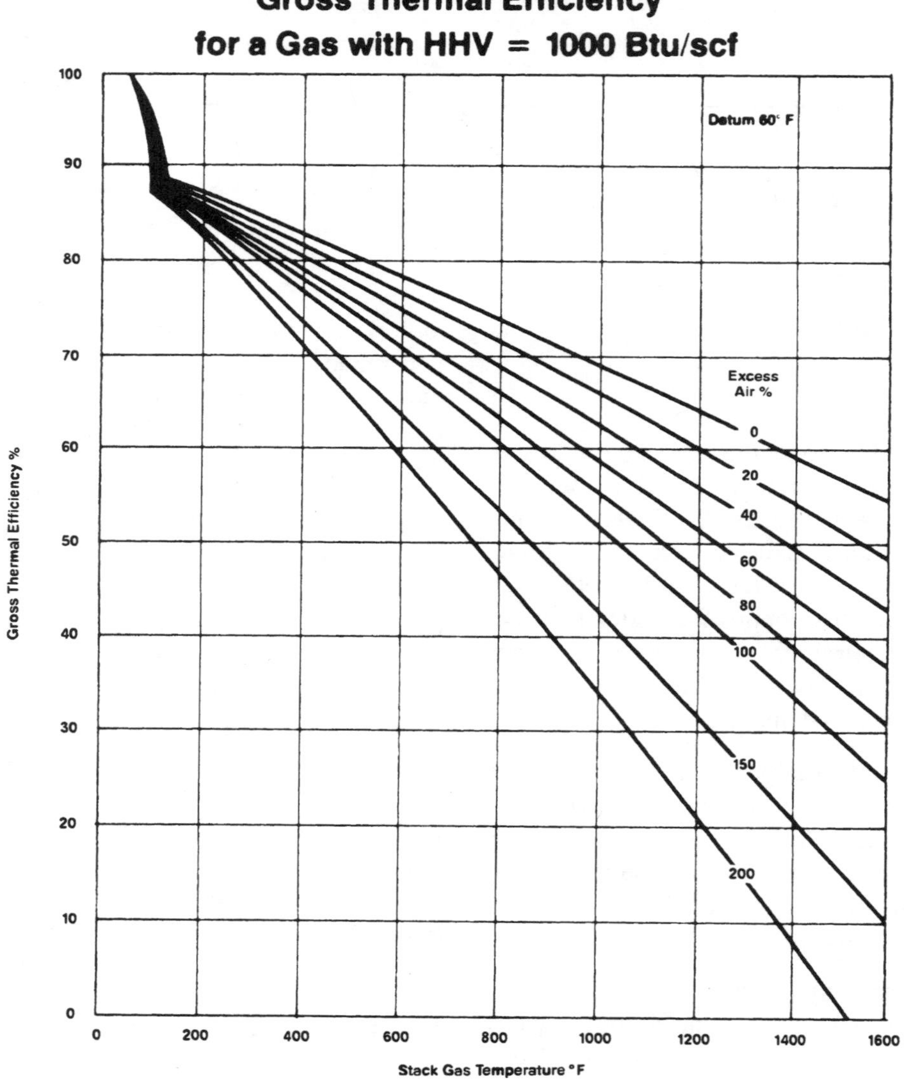

Figure 1. Thermal efficiency determination.

Stack Gas Enthalpy

Here is a handy graph for stack gas enthalpy with natural gas firing.

Source

GPSA Engineering Data Book, Gas Processors Suppliers Association, Vol. I, 10th Ed.

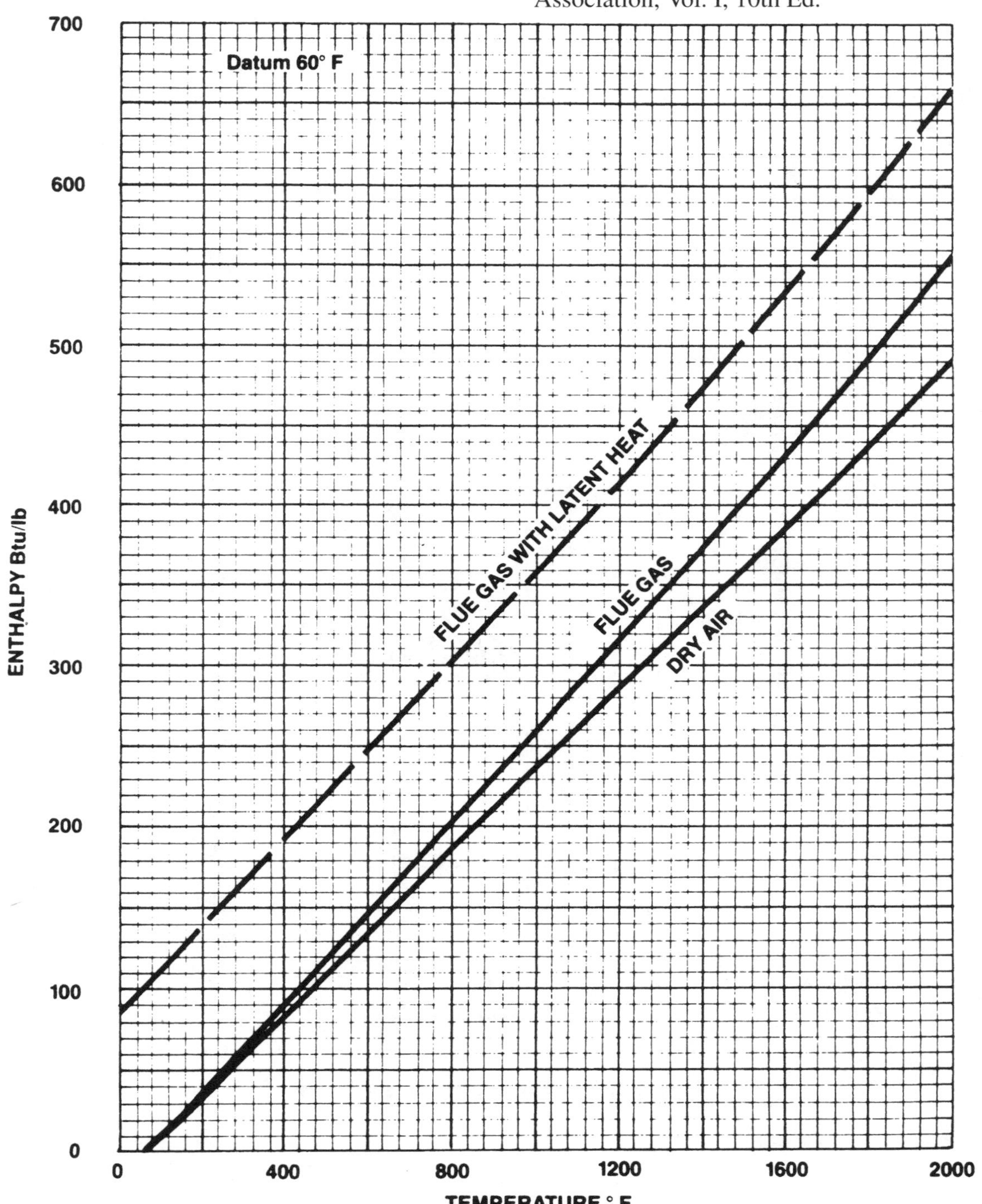

Figure 1. Typical Enthalpy of Combustion Gases for a Natural Gas Fuel and 20% Excess Air.

Stack Gas Quantity

Here is a handy graph for estimating stack gas quantity.

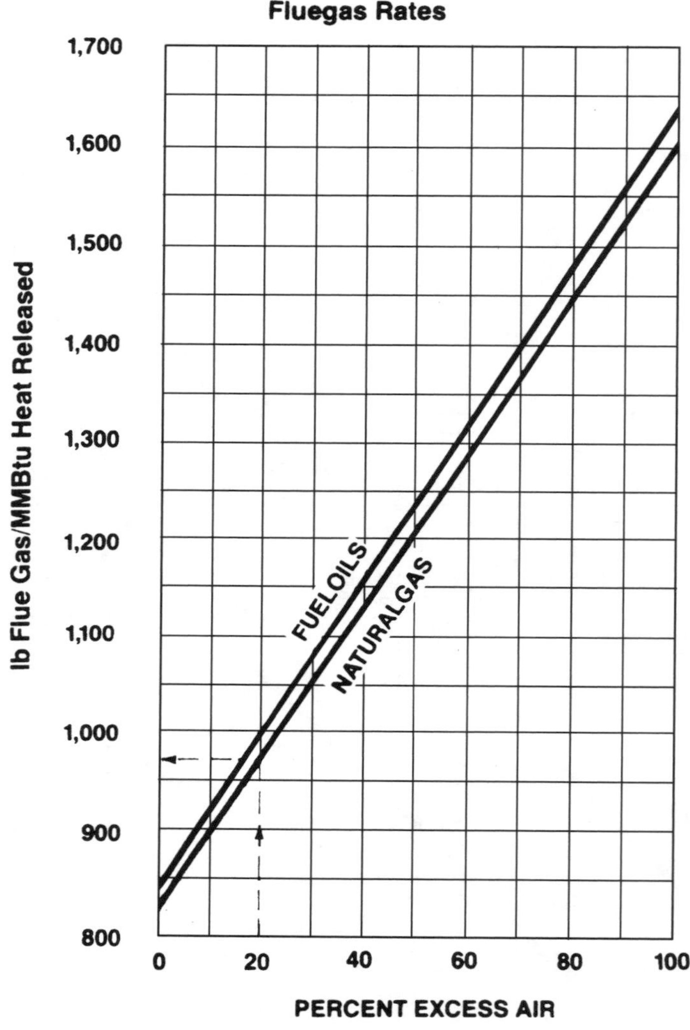

Figure 1. Stack gas quantity estimation.

Source

GPSA Engineering Data Book, Gas Processors Suppliers Association, Vol. I, 10th Ed.

Steam Drum Stability

Ellison has published an extremely important factor for steam drum design called the Drum-Level-Stability Factor. As manufacturers have learned how to increase boiler design ratings, the critria for steam drum design has lagged. The three historical steam drum design criteria have been:

1. The drum is sized for the required steaming rate.
2. The drum must be large enough to contain the baffling and separators required to maintain separation and steam purity.
3. The drum must extend some minimum distance past the furnace to mechanically install the tubes.

A fourth criterion needs to be:

4. The drum must have a water holding capacity with enough reserve in the drum itself, such that all the steam in the risers, at full load, can be replaced by water from the drum without exposing critical tube areas.

This fourth criterion can be met at a low steam drum cost. Only one percent of the cost of the boiler spent on the steam drum can provide it. The fourth criterion is met by requiring that the Drum-Level-Stability Factor (D.L.S.F.) be equal to 1.0 minimum. When this exists the steam drum level will be stable for wide and sudden operational changes.

The D.L.S.F. is defined as follows:

$$D.L.S.F. = V_a/V_m$$

where

V_a = The actual water holding capacity of the drum between the normal water level and the level at which tubes would be critically exposed, gal.

V_m = The minimum water holding capacity required to replace all of the steam bubbles in the risers, gal.

V_a is calculated based on Figure 1, which uses, as the "critical level," a height of one inch above the lower end of the drum baffle separating the risers from the downcomers.

Ellison has derived equations to simplify the calculation of V_m.

$$V_m = [(\%SBV)\ G\ (HR)]/[4\ (150{,}000)]$$

(The constants are based upon test data)

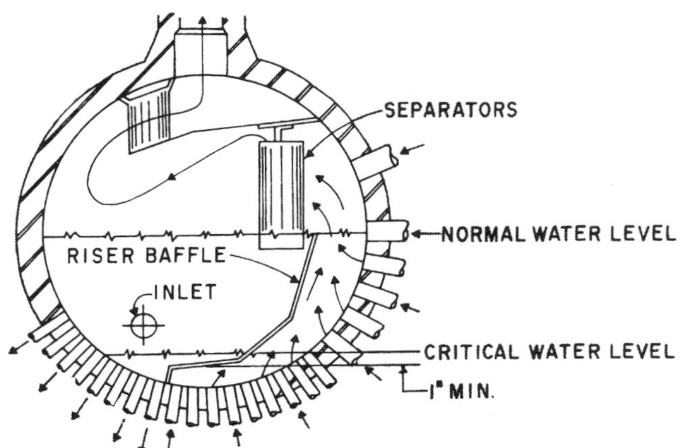

Figure 1. General steam drum configuration.

where

G = The volume of water required to fill the entire boiler to the normal water level, gal.

HR = The furnace heat release per square foot of effective projected radiant surface, BTU/ft^2

$$\%SBV = V_s/[(C - 1)\ V_W + V_S], \text{ dimensionless}$$

where

C = The average boiler circulation ratio at full load, lbs water-steam mixture circulated in a circuit per lb of steam leaving that circuit. (C is the average of all the circuits in the boiler.)

V_S, V_W = Specific volume of a pound of steam or water at saturation temperature and pressure of the steam drum at operating conditions.

Ellison gives the following example:

Given

Steam drum pressure	= 925 psig
C (from manufacturer)	= 18.5
G (from manufacturer)	= 6,000 gal
HR (from manufacturer)	= 160,000 BTU/EPRS
V_a	= 1,500 gal

Calculations

$V_W = 0.0214\ ft^3/lb$ (from steam tables)

$V_S = 0.4772\ ft^3/lb$ (from steam tables)

$$\%SBV = 0.4772/[(18.5 - 1)(0.0214) + 0.4772]$$
$$= 0.56 \text{ (or 56\% steam by volume)}$$
$$V_m = [(0.56)(6{,}000)(160{,}000)]/600{,}000$$
$$= 896 \text{ gal}$$
$$D.L.S.F. = V_a/V_m = 1500/896 = 1.67$$

This steam drum level would be very stable (D.L.S.F. well above 1.0).

Source

Ellison, G. L., "Steam Drum Level Stability Factor," *Hydrocarbon Processing,* May 1971.

Deaerator Venting

Knox has provided the following graphs for estimating the required vent steam from boiler feedwater deaerators. Vent steam rate depends upon the type of deaerator (spray or tray type) and the percentage of makeup water (in contrast to returning condensate). Low makeup water rates require relatively lower steam vent rates, but there is a minimum rate required to remove CO_2 from the returning condensate.

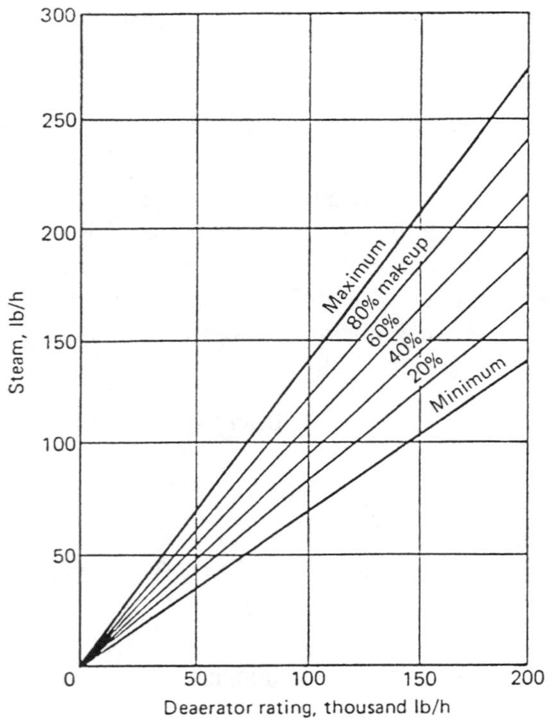

Figure 1. Steam vent rate vs. deaerator rating for a spray-type deaerator.

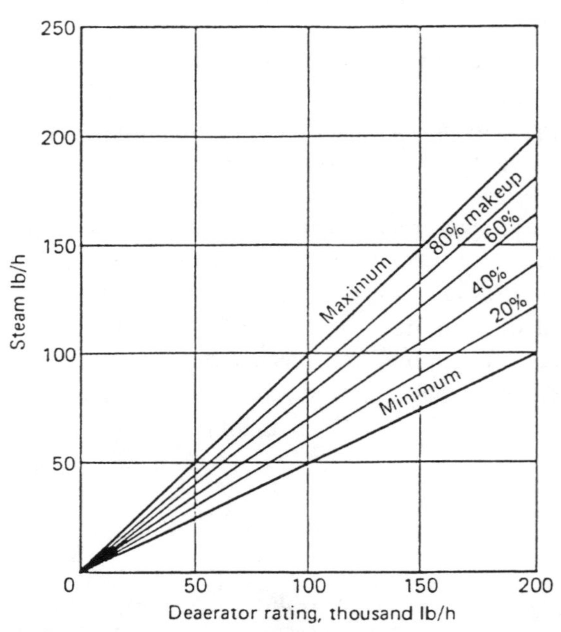

Figure 2. Steam vent rate vs. deaerator rating for a tray-type deaerator.

Source

Knox, A. C., "Venting Requirements for Deaerating Heaters," *Chemical Engineering,* January 23, 1984.

Water Alkalinity

Most water analysis results are rather easily interpreted. However, two simple and useful tests need explanation. These are the P and M alkalinity. The water is titrated with N/30 HCl to the phenolphthalein endpoint at pH 8.3. This is called the P alkalinity. Similar titration to the methyl orange end point at pH 4.3 is called the M alkalinity. They are reported as ppm $CaCO_3$.

This applies to waters having alkalinity caused by one or all of the following:

1. Bicarbonate (HCO_3^-)
2. Carbonate ($CO_3^=$)
3. Hydroxide (OH^-)

In natural waters, the alkalinity is usually caused by bicarbonate. Carbonate or hydroxide is rarely encountered in untreated water. The M alkalinity equals the sum of all three forms of alkalinity. The P alkalinity equals ½ the carbonate and all the hydroxide alkalinity. Table 1 shows what one can deduce from the P and M alkalinity.

Table 1
Comparison of P and M Alkalinities

	OH^-	$CO_3^=$	HCO_3^-
P = O	O	O	M
P = M	M	O	O
P = M/2	O	M	O
P < M/2	O	2P	M-2P
P > M/2	2P-M	2(M-P)	O

Source

Branan, C. R., *The Process Engineer's Pocket Handbook, Vol. 1,* Gulf Publishing Co., 1976.

Blowdown Control

The American Boiler Manufacturers' Association (ABMA) has established limits for boiler water composition. The limits are set to help assure good quality steam (for example, limiting silica in the steam to 0.02–0.03 ppm). Blowdown is normally based on the most stringent of these limits shown in Table 1.

Table 1
ABMA Limits For Boiler Water

Boiler Pressure (psig)	Total Solids (ppm)	Alkalinity (ppm)	Suspended Solids (ppm)	Silica (ppm)
0–300	3500	700	300	125
301–450	3000	600	250	90
451–600	2500	500	150	50
601–750	2000	400	100	35
751–900	1500	300	60	20
901–1000	1250	250	40	8
1001–1500	1000	200	20	2.5
1501–2000	750	150	10	1.0
over 2000	500	100	5	0.5

Source

Branan, C. R., *The Process Engineer's Pocket Handbook, Vol. 1,* Gulf Publishing Co., 1976.

Impurities in Water

Here is a table showing major water impurities, the difficulties they cause, and their treatment.

Source

GPSA Engineering Data Book, Gas Processors Suppliers Association, Vol. II, 10th Ed.

Table 1
Common Characteristics and Impurities in Water

Constituent	Chemical Formula	Difficulties Caused	Means of Treatment
Turbidity	None, usually expressed in Jackson Turbidity Units	Imparts unsightly appearance to water; deposits in water lines, process equipment, boilers, etc.; interferes with most process uses	Coagulation, settling, and filtration
Color	None	Decaying organic material and metallic ions causing color may cause foaming in boilers; hinders precipitation methods such as iron removal, hot phosphate softening; can stain product in process use	Coagulation, filtration, chlorination, adsorption by activated carbon
Hardness	Calcium, magnesium, barium and strontium salts expressed as $CaCO_3$	Chief source of scale in heat exchange equipment, boilers, pipe lines, etc.; forms curds with soap; interferes with dyeing, etc.	Softening, distillation, internal boiler water treatment, surface active agents, reverse osmosis, electrodialysis
Alkalinity	Bicarbonate (HCO_3^{-1}), carbonate (CO_3^{-2}), and hydroxyl (OH^{-1}), expressed as $CaCO_3$	Foaming and carryover of solids with steam; embrittlement of boiler steel; bicarbonate and carbonate produce CO_2 in steam, a source of corrosion	Lime and lime-soda softening, acid treatment, hydrogen zeolite softening, demineralization, dealkalization by anion exchange, distillation, degasifying
Free Mineral Acid	H_2SO_4, HCl, etc. expressed as $CaCO_3$, titrated to methyl orange end-point	Corrosion	Neutralization with alkalies
Carbon Dioxide	CO_2	Corrosion in water lines and particularly steam and condensate lines	Aeration, deaeration, neutralization with alkalies, filming and neutralizing amines
pH	Hydrogen Ion concentration defined as $\text{pH} = \log \frac{1}{(\text{H}^{+1}}$	pH varies according to acidic or alkaline solids in water; most natural waters have a pH of 6.0–8.0	pH can be increased by alkalies and decreased by acids
Sulfate	$(SO_4)^{-2}$	Adds to solids content of water, but, in itself, is not usually significant; combines with calcium to form calcium sulfate scale	Demineralization, distillation, reverse osmosis, electrodialysis
Chloride	Cl^{-1}	Adds to solids content and increases corrosive character of water	Demineralization, distillation, reverse osmosis, electrodialysis
Nitrate	$(NO_3)^{-1}$	Adds to solids content, but is not usually significant industrially; useful for control of boiler metal embrittlement	Demineralization, distillation, reverse osmosis, electrodialysis
Fluoride	F^{-1}	Not usually significant industrially	Adsorption with magnesium hydroxide, calcium phosphate, or bone black; Alum coagulation; reverse osmosis; electrodialysis
Silica	SiO_2	Scale in boilers and cooling water systems; insoluble turbine blade deposits due to silica vaporization	Hot process removal with magnesium salts; adsorption by highly basic anion exchange resins, in conjunction with demineralization; distillation
Iron	Fe^{+2} (ferrous) Fe^{+3} (ferric)	Discolors water on precipitation; source of deposits in water lines, boilers, etc.; interferes with dyeing, tanning, paper mfr., etc.	Aeration, coagulation and filtration, lime softening, cation exchange, contact filtration, surface active agents for iron retention
Manganese	Mn^{+2}	same as iron	same as iron
Oil	Expressed as oil or chloroform extractable matter, ppmw	Scale, sludge and foaming in boilers; impedes heat exchange; undesirable in most processes	Baffle separators, strainers, coagulation and filtration, diatomaceous earth filtration
Oxygen	O_2	Corrosion of water lines, heat exchange equipment, boilers, return lines, etc.	Deaeration, sodium sulfite, corrosion inhibitors, hydrazine or suitable substitutes

(table continued)

Table 1 Continued
Common Characteristics and Impurities in Water

Constituent	Chemical Formula	Difficulties Caused	Means of Treatment
Hydrogen Sulfide	H_2S	Cause of "rotten egg" odor; corrosion	Aeration, chlorination, highly basic anion exchange
Ammonia	NH_3	Corrosion of copper and zinc alloys by formation of complex soluble ion	Cation exchange with hydrogen zeolite, chlorination, deaeration, mixed-bed demineralization
Conductivity	Expressed as micromhos, specific conductance	Conductivity is the result of ionizable solids in solution; high conductivity can increase the corrosive characteristics of a water	Any process which decreases dissolved solids content will decrease conductivity; examples are demineralization, lime softening
Dissolved Solids	None	"Dissolved solids" is measure of total amount of dissolved matter, determined by evaporation; high concentrations of dissolved solids are objectionable because of process interference and as a cause of foaming in boilers	Various softening processes, such as lime softening and cation exchange by hydrogen zeolite, will reduce dissolved, solids; demineralization; distillation; reverse osmosis; electrodialysis
Suspended Solids	None	"Suspended Solids" is the measure of undissolved matter, determined gravimetrically; suspended solids plug lines, cause deposits in heat exchange equipment, boilers, etc.	Subsidence, filtration, usually preceded by coagulation and settling
Total Solids	None	"Total Solids" is the sum of dissolved and suspended solids, determined gravimetrically	See "Dissolved Solids" and "Suspended Solids"

Conductivity Versus Dissolved Solids

For a quick estimate of total dissolved solids (TDS) in water one can run a conductivity measurement. The unit for the measurement is mhos/cm. An mho is the reciprocal of an ohm. The mho has been renamed the Sieman (S) by the International Standard Organization. Both mhos/cm and S/cm are accepted as correct terms. In water supplies (surface, well, etc.) conductivity will run about 10^{-6} S/cm or 1 μS/cm.

Without any data available the factor for conductivity to TDS is:

TDS (ppm) = Conductivity (μS/cm)/2

However, the local water supplier will often supply TDS and conductivity so you can derive the correct factor for an area.

Table 1 gives conductivity factors for common ions found in water supplies.

Source

McPherson, Lori, "How Good Are Your Values for Total Dissolved Solids?" *Chemical Engineering Progress*, November, 1995.

Table 1
Water Quality: Conductivity Factors of Ions Commonly Found in Water

Ion	μS/cm per ppm
Bicarbonate	0.715
Calcium	2.60
Carbonate	2.82
Chloride	2.14
Magnesium	3.82
Nitrate	1.15
Potassium	1.84
Sodium	2.13
Sulfate	1.54

Silica in Steam

Figure 1 from GPSA shows how the silica content of boiler water affects the silica content of steam. For example at 1600 psia, 100 ppm silica in the boiler water causes 0.9 ppm silica in the steam. If this steam were expanded to 100 psia, following the steam line down (the saturated and superheated curves converge by the time 100 psia is reached), the solubility of silica decreases to 0.1 ppm. Therefore the difference (0.8 ppm) would tend to deposit on turbine blades.

The American Boiler Manufacturers' Association shoots for less than 0.02–0.03 ppm silica in steam by limiting silica in the boiler water. See the section entitled Blowdown Control.

Source

GPSA Engineering Data Book, Gas Processors Suppliers Association, Vol. II, 10th Ed.

Caustic Embrittlement

A boiler's water may have caustic embrittling characteristics. Only a test using a U.S. Bureau of Mines Embrittlement Detector will show whether this is the case. If the water is found to be embrittling, it is advisable to add sodium nitrate inhibitor lest a weak area of the boiler be attacked.

The amount of sodium nitrate required is specified as a $NaNO_3/NaOH$ ratio calculated as follows:

Ratio = (Nitrate) (2.14)/(M alkalinity – Phosphate)

where

Ratio = $NaNO_3/NaOH$ ratio
Nitrate = Nitrate as NO_3, ppmw
M alkalinity = M alkalinity as $CaCO_3$, ppmw
Phosphate = Phosphate as PO_4, ppmw

See the section on Water Alkalinity for an explanation of M alkalinity.

The U.S. Bureau of Mines recommends that the $NaNO_3/NaOH$ ratio be as follows:

Boiler Pressure psig	Ratio
0–250	0.20
250–400	0.25
400–700	0.40

Source

GPSA Engineering Data Book, Gas Processors Suppliers Association, Vol. II, 10th Ed.

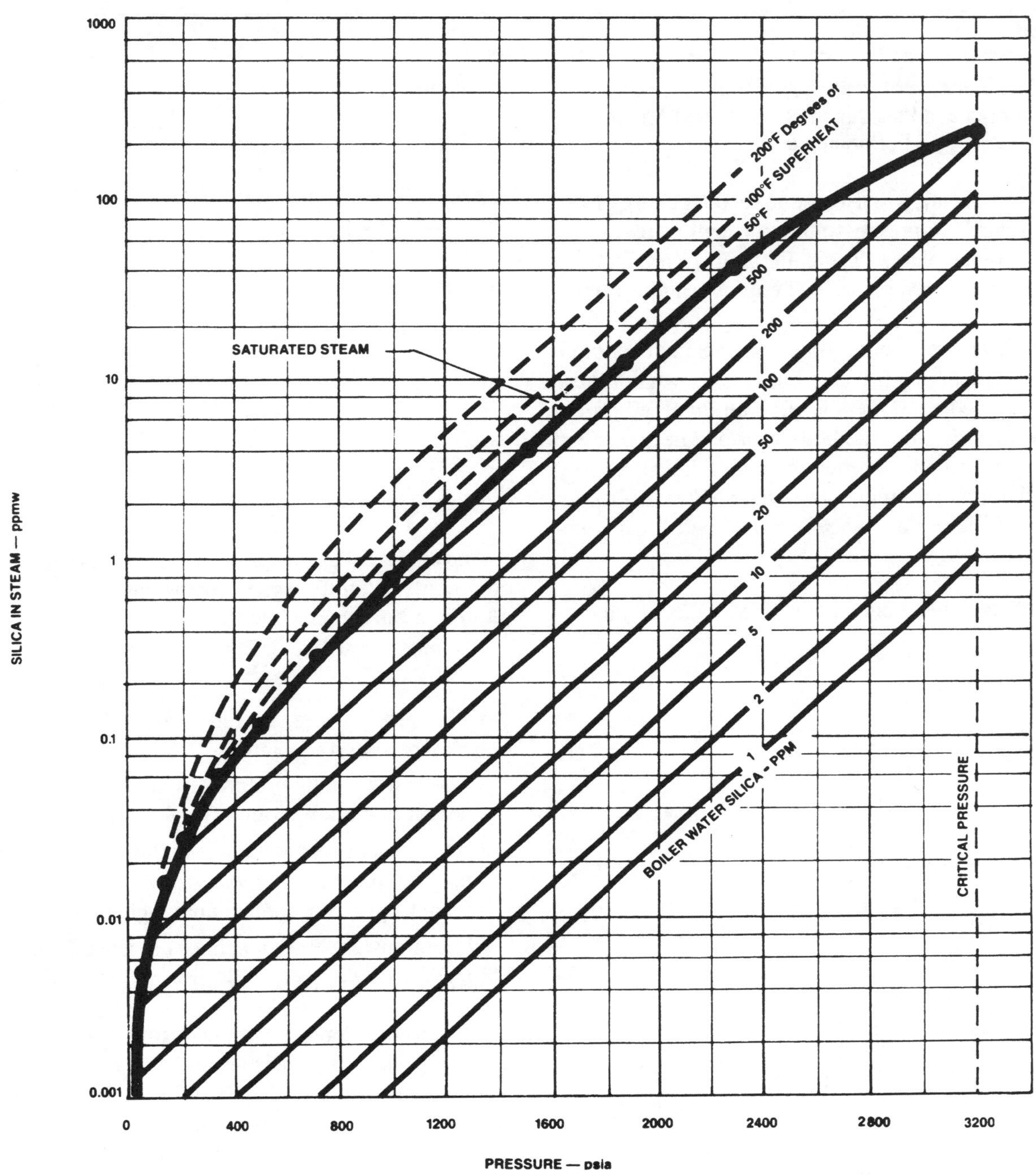

Figure 1. Relationships between boiler pressure, boiler water silica content, and silica solubility in steam.

Waste Heat

This Method Quickly Determines Required Heat Transfer Surface Area and How Changing Loads Will Affect Performance

Fire tube boilers are widely used to recover energy from waste gas streams commonly found in chemical plants, refineries and power plants. Typical examples are exhaust gases from gas turbines and diesel engines, and effluents from sulfuric acid, nitric acid and hydrogen plants. Generally, they are used for low-pressure steam generation. Typical arrangement of a fire tube boiler is shown in Figure 1. Sizing of waste heat boilers is quite an involved procedure. However, using the method described here one can estimate the performance of the boiler at various load conditions, in addition to designing the heat transfer surface for a given duty. Several advantages are claimed for this approach, as seen below.

Design Method. The energy transferred, Q, is given by[1].

$$Q = W_g C_p (T_1 - T_2)$$

$$= UA \frac{\{(T_1 - t_s) - (T_2 - t_s)\}}{\ln\left(\frac{T_1 - t_s}{T_2 - t_s}\right)} \quad (1)$$

where U, the overall heat transfer coefficient, is given by:

$$U = 0.95\ h_i. \quad (2)$$

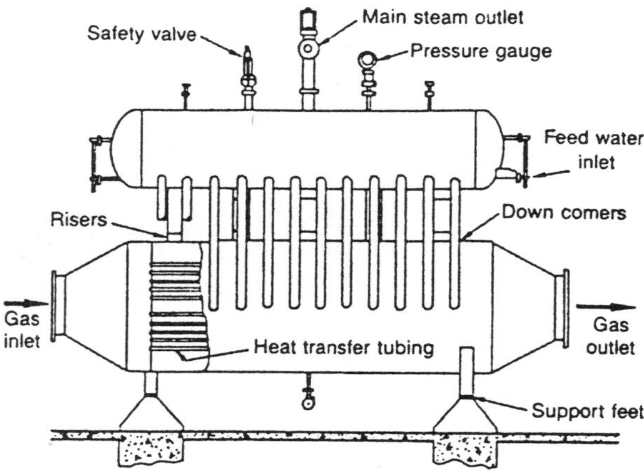

Figure 1. Typical boilers for waste heat recovery.

The inside heat transfer coefficient, h_i, is given by:

$$NU = (h_i d_i)/12k = 0.023\ (Re)^{0.8}\ (Pr)^{0.4} \quad (3)$$

From the previous equations by substituting $A = \pi d_i NL$, and $w = W_g/N$

$$\ln\left(\frac{T_1 - t_s}{T_2 - t_s}\right) = \frac{UA}{W_g C_p}$$

$$= \frac{0.62L\ k^{0.6}}{d_i^{\,0.8} w^{0.2} C_p^{\,0.6} \mu^{0.4}} = \frac{0.62L\ F(t)}{d_i^{\,0.8} w^{0.2}} \quad (4)$$

where $F(t) = (k^{0.6}/c_p^{\,0.6}\mu^{0.4})$

The above equation has been charted in Figure 2. F(t) for commonly found flue gas streams has been computed and plotted in the same figure.

Advantages of Figure 2 are:

- The surface area, A, required to transfer a given Q may be found (design).
- The exit gas temperature, T_2, and Q at any off design condition may be found (performance).
- If there are space limitations for the boiler (L or shell dia. may be limited in some cases) we can use the chart to obtain a different shell diameter given by $D_s = 1.3\ d\sqrt{N}$.
- Heat transfer coefficient does not need to be computed.
- Reference to flue gas properties is not necessary.

Thus, a considerable amount of time may be saved using this chart. The gas properties are to be evaluated at the average gas temperature. In case of design, the inlet and exit gas temperatures are known and hence, the average can be found. In case of performance evaluation, the inlet gas temperature alone will be known and hence, a good estimate of average gas temperature is $0.6\ (T_1 + t_s)$. Also, a good starting value of w for design is 80 to 150 lb/h.

Example 1. A waste heat boiler is to be designed to cool 75,000 lb/h of flue gases from 1,625°F to 535°F, generating steam at 70 psia. Tubes of size 2-in. OD and 1.8-in. ID are to be used. Estimate the surface area and geometry using 120 lb/h of flow per tube. Find Q and W_s if feed water enters at 180°F.

Solution: Calculate $(T_1 - t_s)/(T_2 - t_s) = (1{,}625 - 303)/(535 - 303) = 5.7$. Connect w = 120 with d_i = 1.8 and extend to cut line 1 at "A". Connect t = (1,625 + 535)/2 = 1,080 with $(T_1 - t_s)/(T_2 - t_s) = 5.7$ and extend to cut line 2 at "B". Join "A" and "B" to cut the L scale at 16. Hence, tube length is 16 ft. Number of tubes, N = 75,000/120 = 625. Surface area, A = 1.8 (16)(625)(π/12) = 4,710 ft². Note that this is only A design and several alternatives are possible by changing w or L. If L has to be limited, we can work the chart in reverse to figure w.

$$Q = 75{,}000\ (0.3)\ (1{,}625 - 535) = 24.52\ \text{MMBtu/h}$$

$$W_s = 24.52\ (10^6)/(1{,}181 - 148) = 23{,}765\ \text{lb/h}$$

(1,181 and 148 are the enthalpies of saturated steam and feed water.)

Example 2. What is the duty and exit gas temperature in the boiler if 60,000 lb/h of flue gases enter the boiler at 1,200°F? Steam pressure is the same as in example 1.

Solution: Connect w = (60,000/625) = 97 with d_i = 1.8 and extend to cut line 1 at "A". Connect with L = 16 and extend to cut line 2 at "B". The average gas temperature

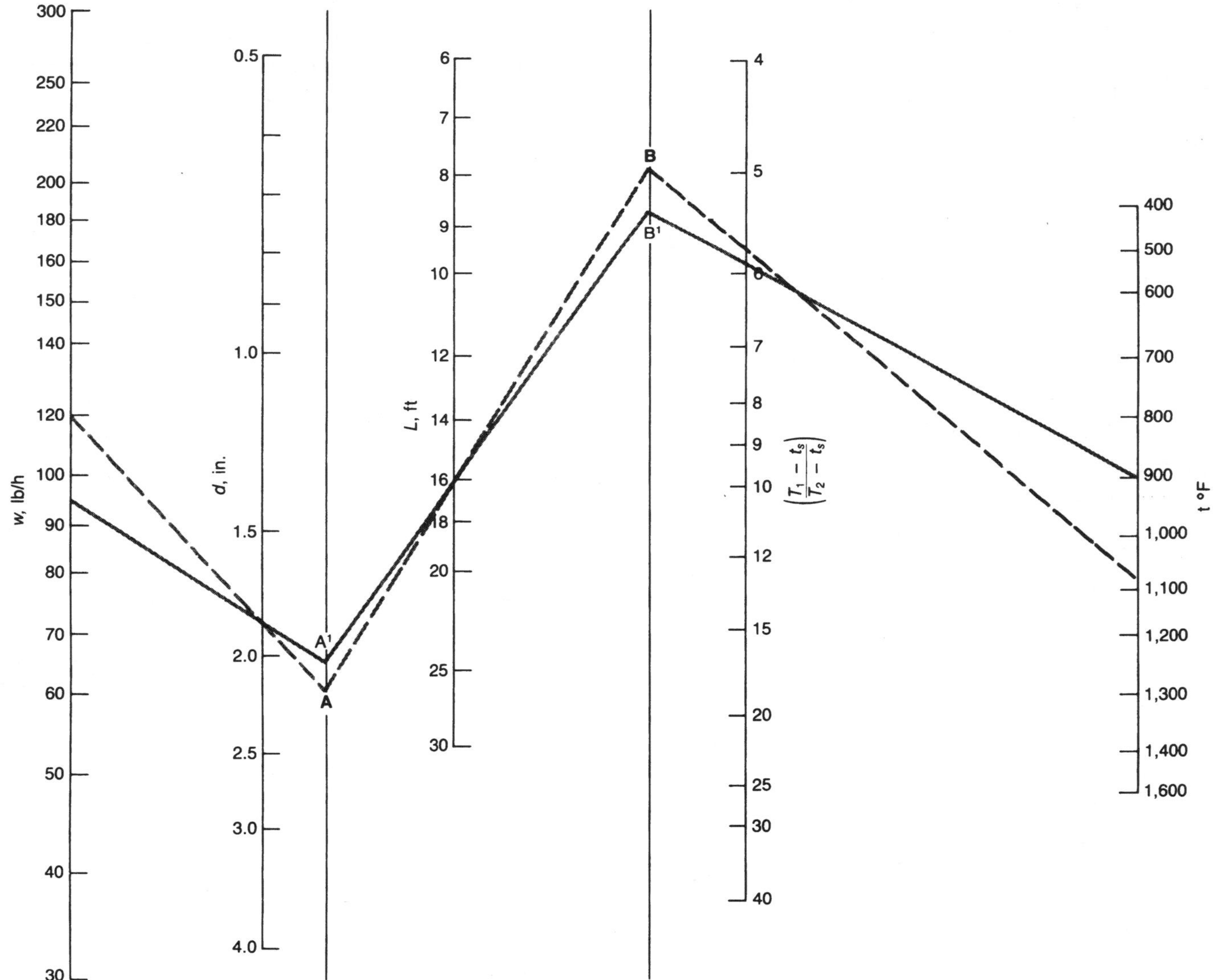

Figure 2. Performance evaluation chart.

is 0.6 (1,200 + 303) = 900°F. Connect "B" with t = 900 to cut $(T_1 - t_s)/(T_2 - t_s)$ scale at 5.9. Hence t = 455°F.

Assuming a C_p of 0.285, Q = 60,000 (0.285) (1,200 – 455) = 12.74 MMBtu/h. W_s = 12.74 (10^6)/(1,181 – 148) = 12,340 lb/h (C_p can be approximated by (0.22 + 7.2 (10^{-5})t) in the temperature range 600 to 1,500°F).

Nomenclature

A = Surface area ft^2, based on tube ID
C = Gas specific heat, Btu/lb°F
d = Tube inner diameter, in.
k = Gas thermal conductivity, Btu/ft-h°F
L = Tube length, ft
N = Total number of tubes in boiler
Pr = Gas Prandtl number
Q = Duty of the boiler, Btu/h
Re = Reynolds number
t_g, t_s = Average gas and saturation temperature of steam, °F
T_1, T_2 = Gas temperature entering and leaving the boiler, °F
U = Overall heat transfer coefficient, Btu/ft^{2}h °F (based on tube ID)
μ = Gas viscosity, lb/ft-h
W_g, W_s = Total gas and flow, lb/h
w = Flow per tube, lb/h

Sources

1. Ganapathy, V., "Size or Check Waste Heat Boilers Quickly," *Hydrocarbon Processing,* September 1984.
2. Ganapathy, V., Applied Heat Transfer, PennWell Books, 1982.

10
Cooling Towers

System Balances

To determine cooling water system flows, use a heat and material balance and a chloride balance (concentration ratio is usually calculated from chloride concentrations).

$$D = C\,(500)\ \text{lb/hr}\ (1)\ \text{Btu/lb °F}\ (\Delta T)\ \text{°F} = 500\,C\Delta T\ \text{Btu/hr} \qquad (1)$$

$$E \approx \frac{D\ \text{Btu/hr}}{1000\ \text{Btu/lb}\ (500)\ \text{lb/hr/GPM}} = \frac{C\Delta T}{1000} \qquad (2)$$

$$CR = \frac{Cl^-_B}{Cl^-_M};\ \text{also}\ MCl^-_M = BCl^-_B \qquad (3)$$

so

$$CR = M/B \qquad (4)$$

$$M = E + B\ \text{(overall material balance)} \qquad (5)$$

$$E = M - B = CR\,(B) - B = B\,(CR - 1) \qquad (6)$$

so

$$B = \frac{E}{CR - 1} \qquad (7)$$

$$M = E\left(\frac{CR}{CR - 1}\right) \qquad (8)$$

Use Equation 2 to get E, then Equation 7 to get B, and finally Equation 5 or 8 to get M.

where

D = Cooling system duty, Btu/hr
C = System circulation rate, GPM
ΔT = Cooling system temperature difference (hot return water minus cold supply water), °F
E = Cooling system evaporation rate, GPM
CR = Cooling system concentration ratio
Cl^- = Chloride concentration in the makeup or blowdown
M = Cooling system makeup rate, GPM
B = Cooling system total blowdown, GPM. This includes both planned blowdown plus cooling system windage (or drift) losses (of course any system leakage counts as part of "planned" blowdown)

To determine the required amount of planned blowdown, subtract windage losses from B. Use Table 1 for windage losses in lieu of manufacturer's or other test data.

When cooling systems are treated, chemicals are sometimes added in shots rather than continuously. Equation 9 gives a chemical's half life in a cooling system:

$$T_{1/2} = \frac{S}{B}(.693) \qquad (9)$$

where

$T_{1/2}$ = Half life, min.
S = System capacity, gal

Source

Branan, C. R., *The Process Engineer's Pocket Handbook, Vol. 1,* Gulf Publishing Co., 1976.

Temperature Data

Figure 1 taken from the *GPSA Engineering Data Book,* gives design wet and dry bulb temperatures for the continental United States.

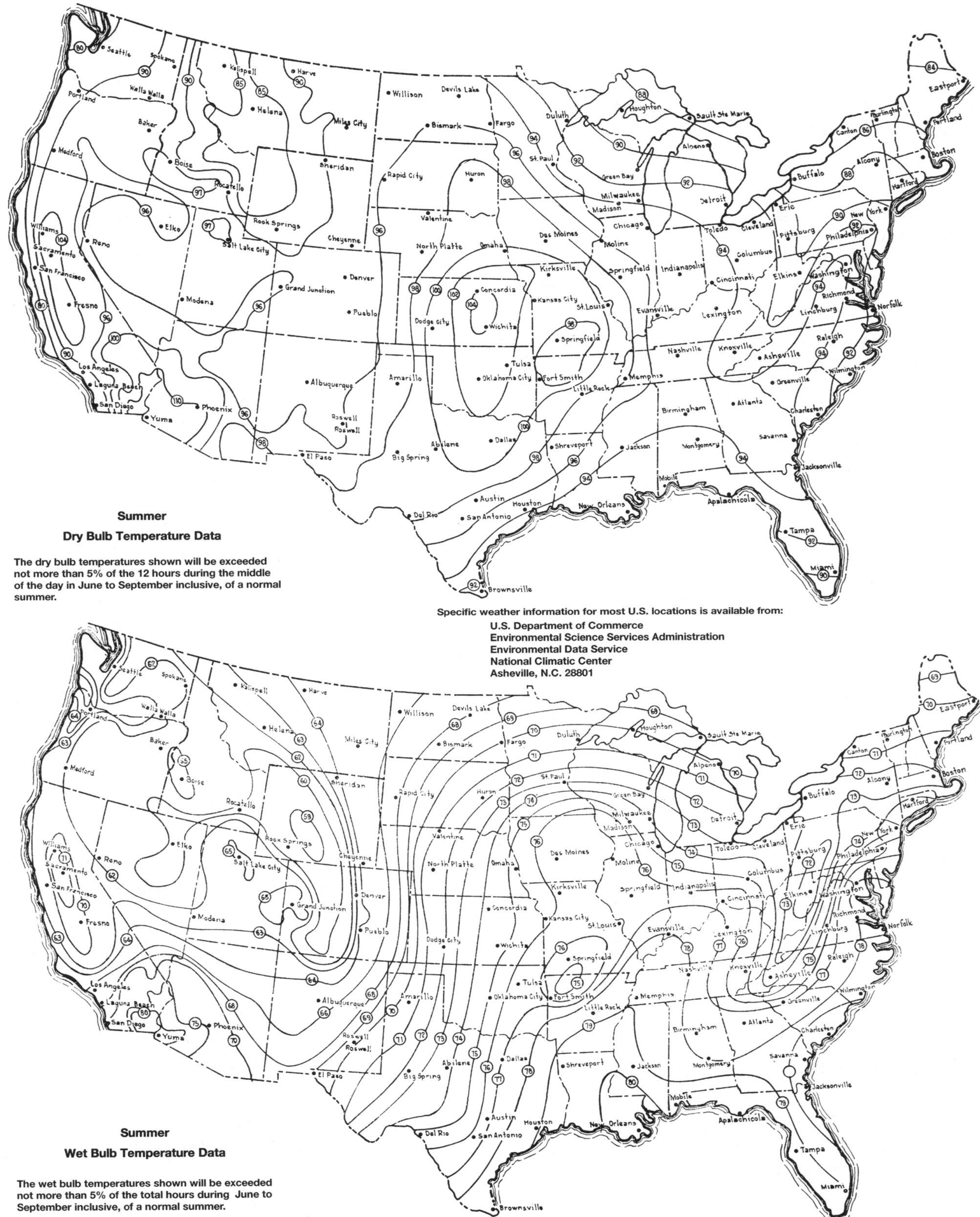

Summer

Dry Bulb Temperature Data

The dry bulb temperatures shown will be exceeded not more than 5% of the 12 hours during the middle of the day in June to September inclusive, of a normal summer.

Specific weather information for most U.S. locations is available from:

U.S. Department of Commerce
Environmental Science Services Administration
Environmental Data Service
National Climatic Center
Asheville, N.C. 28801

Summer

Wet Bulb Temperature Data

The wet bulb temperatures shown will be exceeded not more than 5% of the total hours during June to September inclusive, of a normal summer.

Figure 1. Dry bulb/wet bulb temperature data.

Source

GPSA Engineering Data Book, Vol. I, Gas Processors Suppliers Association, 10th Ed., 1987.

Performance

GPSA provides an extremely handy nomograph for cooling tower performance evaluation. The use of the nomograph is almost self-evident, especially with the examples that GPSA provides, condensed here.

Base data for examples:

Circulation = 1000 gpm
Hot Water = 105°F
Cold Water = 85°F
Wet Bulb = 75°F
Range = 105–85 = 20°F
Approach = 85–75 = 10°F

Example 1. Effect of Varying Wet Bulb Temperature

Change: Wet Bulb = 60°F
Procedure: Enter at 85°F Cold Water Temp.
Go Horizontally to 75°F Wet Bulb Temp.
Go Vertically down to 60°F Wet Bulb Temp.
Read 76°F new Cold Water Temp.

Example 2. Effect of Varying Range

Change: Range = 30°F
Procedure: Enter at 85°F Cold Water Temp.
Go Horizontally to 75°F Wet Bulb Temp.
Go Vertically up to 20°F Range
Go Horizontally to 30°F Range
Go Vertically down to 75°F Wet Bulb Temp.
Read 87.5°F new Cold Water Temp.

Example 3. Effect of Varying Circulation

Change: Circulation = 1500 gpm
Procedure: Enter at 85°F Cold Water Temp.
Go Horizontally to 75°F Wet Bulb Temp.
Go Vertically up to 20°F Range
Read Performance Factor = 3.1
Calculate new Performance Factor
3.1 (1500/1000) = 4.65
Go Horizontally left from 4.65 to 20°F Range
Go Vertically down to 75°F Wet Bulb Temp.
Read 90.5°F new Cold Water Temp.

Example 4. Effect of Varying Wet Bulb Temperature, Range, and Circulation

Changes: Wet Bulb Temp. = 60°F
Range = 25°F
Circulation = 1250 gpm
Procedure: Enter at 85°F Cold Water Temp.
Go Horizontally to 75°F Wet Bulb Temp.
Go Vertically up to 20°F Range
Read Performance Factor = 3.1
Calculate new Performance Factor
3.1 (1250/1000) = 3.87
Go Horizontally from 3.87 to 25°F Range
Go Vertically down to 60°F Wet Bulb Temp.
Read 82°F new Cold Water Temp.

Example 5. Effect of Varying Fan Horsepower

Change: Start from the conditions in Example 4 and raise fan horsepower from 20 to 25.
Procedure: Calculate new/old air flow ratio
$(25/20)^{1/3} = 1.077$
Calculate new Performance Factor
3.87/1.077 = 3.6
Go Horizontally from 3.6 to 25°F Range
Go Vertically down to 60°F Wet Bulb Temp.
Read 81°F new Cold Water Temp.

Example 6. Use the Increase in Fan Horsepower to Raise Circulation Rather Than Lower Cold Water Temp.

Procedure: Air flow ratio (Example 5) = 1.077
Calculate allowable increase in circulation
(holding 82°F Cold Water Temp.)
1250 (1.077) = 1346 gpm

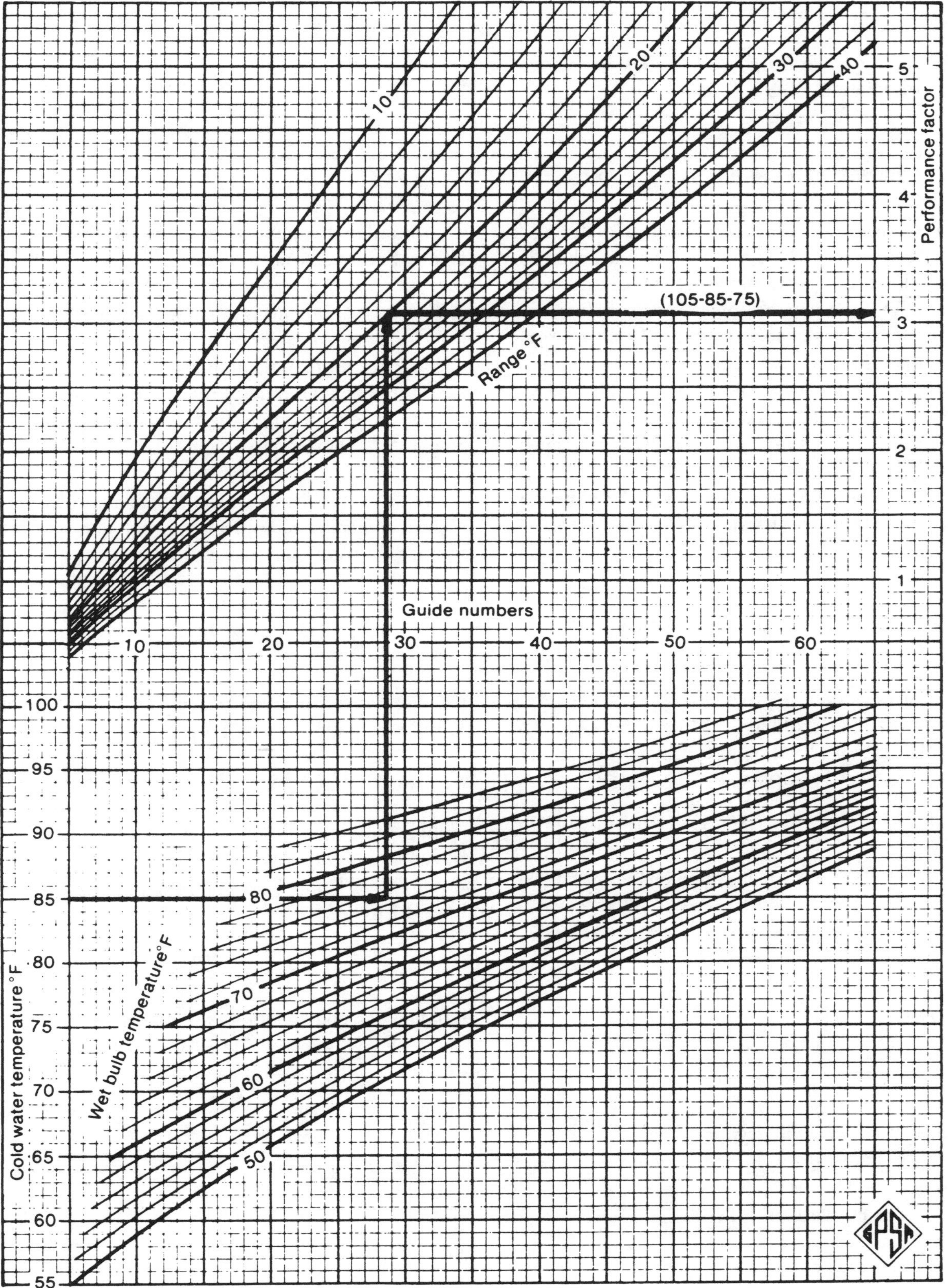

Figure 1. Performance characteristic nomograph.

Source

GPSA Engineering Data Book, Vol. I, Gas Processors Suppliers Association, 10th Ed., 1987.

Performance Estimate: A Case History

A temporary extra heat load was placed on an olefin plant cooling tower one winter. The operations people asked the question, "Will the cooling tower make it next summer with the extra load?" Of course the tower would deliver the heat removal, so the real question was, "What will the cold water temperature be next summer?"

The author rated the tower and then used the GPSA nomograph in the previous section titled Performance to determine the next summer's maximum cold water temperature. When the author proudly presented the results the operation people, of course, then asked, "Will the plant run at necessary production at your calculated cold water temperature?"

To simulate the next summer's condition the plant was run at the desired production rate and two cooling tower fans were turned off. It turned out that the cold water temperature rose to slightly above that predicted for the next summer. A thorough inspection of critical temperatures and the plant's operation indicated that the plant would barely make it the next summer. Process side temperatures were at about the maximum desired, with an occasional high oil temperature alarm on the large machines.

The decision was made to take a shutdown that spring to clean the water side of critical coolers. The following summer we were glad that we had done so.

Transfer Units

The section on Performance provides a handy nomograph for quick cooling tower evaluation. More detailed analysis requires the use of transfer units. The performance analysis then asks two questions:

1. How many transfer units correspond to the process requirement?
2. How many transfer units can the actual cooling tower or proposed new cooling tower actually perform?

The number of transfer units is given by:

$$N_{tog} = \int_{H_1}^{H_2} dH/(H_{sat} - H) \qquad (1)$$

where

H = Enthalpy of air, BTU/lb dry air
H_1 = Enthalpy of entering air, BTU/lb dry air
H_2 = Enthalpy of exit air, BTU/lb dry air
H_{sat} = Enthalpy of saturated air, BTU/lb

Equation 1 is normally integrated by graphical or numerical means utilizing the overall material balance and the saturated air enthalpy curve.

Khodaparast has provided an equation for H_{sat} that allows simpler evaluation of Equation 1:

$$H_{sat} = (-665.432 + 13.4608t - 0.784152t^2)/(t - 212) \qquad (2)$$

This equation is good if the air temperature is 50°F or above, the cooling tower's approach to the wet bulb temperature is 5°F or above, and N_{tog} is within a range of about 0.1 to 8.

Table 1
Equation 3 is quite accurate for calculating the number of transfer units for the example cooling tower

	N_{tog}	
G_s/L	Equation 5	Numerical Solution
0.75	4.500	4.577
0.85	3.341	3.369
0.95	2.671	2.682
1.05	2.231	2.234
1.20	1.793	1.791
1.40	1.423	1.419
2.0	0.883	0.878
5.0	0.306	0.304
10.0	0.147	0.146

Calculate N_{tog}

The calculation begins with the evaluation of several primary parameters:

$a = -(L/G_s) - 0.784152$
$b = -H_1 + (L/G_s)(t_1 + 212) + 13.4608$
$c = -665.432 + 212 [H_1 - (L/G_s) t_1]$
$d = 4ac - b^2$

If L/G_s is excessively small or large, the operating line will intersect the saturated-air enthalpy curve, and d will be negative. Therefore, the value of the integral depends on the sign of d, as follows:

$$N_{tog} = \left(\frac{L}{G_s}\right)\left[\frac{1}{2a}\ln\left(\frac{at_2^2 + bt_2 + c}{at_1^2 + bt_1 + c}\right) - \left(212 + \frac{b}{2a}\right)e\right] \quad (3)$$

where:

$$e = \frac{2}{\sqrt{d}}\left[\operatorname{atan}\left(\frac{2at_2 + b}{\sqrt{d}}\right) - \operatorname{atan}\left(\frac{2at_1 + b}{\sqrt{d}}\right)\right]$$

for $d > 0$;

$$e = \frac{1}{\sqrt{-d}} \times \ln\left[\frac{\left(2at_2 + b - \sqrt{-d}\right)\left(2at_1 + b + \sqrt{-d}\right)}{\left(2at_2 + b + \sqrt{-d}\right)\left(2at_1 + b - \sqrt{-d}\right)}\right]$$

for $d < 0$; and

$$e = \frac{1}{a}\left[\frac{1}{\left(t_1 + \frac{b}{2a}\right)} - \frac{1}{\left(t_2 + \frac{b}{2a}\right)}\right]$$

if ever $d = 0$.

Example

To compare the results of the correlation presented in this article and an exact numerical solution, let us consider the case where air with a wet-bulb temperature of 70°F is used to cool water from 120°F to 80°F. Table 1 summarizes the results for different air-to-water flow rate ratios

Nomenclature

C_p = heat capacity of water, Btu/lb • °F
G_s = air flow rate on a dry basis, lb/h
H = enthalpy of air on a dry basis, Btu/lb dry air
H_1 = enthalpy of entering air, Btu/lb dry air
H_2 = enthalpy of exiting air, Btu/lb dry air
H_{sat} = enthalpy of saturated air at 1 atm, Btu/lb
L = water flow rate, lb/h
N_{tog} = number of transfer units
t = water temperature, °F
t_1 = temperature of exit water, °F
t_2 = temperature of inlet water, °F

Literature Cited

1. Treybal, R. E., "Mass Transfer Operations," 3rd ed., McGraw-Hill, p. 247 (1981).
2. Perry, R. H., "Perry's Chemical Engineers' Handbook," 6th ed., McGraw-Hill, p. 12–14 (1984).

Source

Calculate N_{tog} and Table 1, from "Predict the number of Transfer Units for Cooling Towers," by Kamal Adham Khodaparast, *Chemical Engineering Progress,* Vol. 88, No. 4, pp. 67–68 (1992).

S E C T I O N T W O

Process Design

11 Refrigeration

Types of Systems

The following table shows the three most used refrigeration systems and approximate temperature ranges.

Table 1
Types of Refrigeration Systems

	Approx. Temp. Range, °F	Refrigerant
1. Steam—Jet	35° to 70°	Water
2. Absorption		
Water—Lithium Bromide	40° to 70°	Lithium Bromide solution
Ammonia	–40° to +30°	Ammonia
3. Mechanical Compression (Reciprocating or centrifugal)	–200° to +40°	Ammonia, halogenated hydrocarbons, propane, ethylene, and others

The most common light hydrocarbon refrigerant cooling temperature ranges are:

Methane	–200 to –300°F
Ethane and ethylene	–75 to –175°F
Propane and propylene	+40 to –50°F

Source

Ludwig, E. E., *Applied Process Design for Chemical and Petrochemical Plants, Vol. 3, 2nd Ed.,* Gulf Publishing Co., p. 201.

Estimating Horsepower per Ton

This quick but accurate graph shows the design engineer how much horsepower is required for mechanical refrigeration systems, using the most practical refrigerant for the desired temperature range.

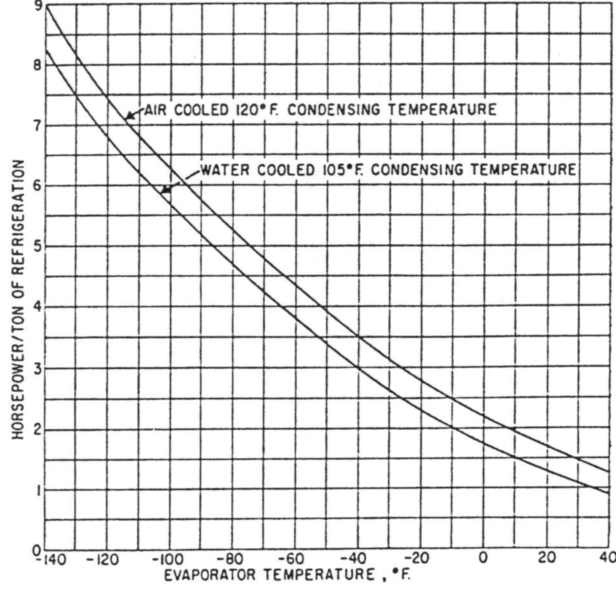

Figure 1. Horsepower requirements per ton of refrigeration.

Example. A water-cooled unit with an evaporator temperature of -40°F will require 3 horsepower/ton of refrigeration. A ton of refrigeration is equal to 12,000 BTU/hr.

Here are equations for these curves in the form:

$$y = A + Bx + Cx^2 + Dx^3 + Ex^4$$

where

y = horsepower/ton refrigeration
x = evaporator temperature, °F

Condenser temperature °F	A	B	C	D	E
105	1.751	–2.686e-2	1.152e-4	3.460e-8	1.320e-9
120	2.218	–2.882e-2	1.036e-4	3.029e-7	3.961e-9

Equations were generated using FLEXCURV V.2, Gulf Publishing Co.

Source

Ballou, Lyons, and Tacquard, "Mechanical, Refrigeration Systems," *Hydrocarbon Processing,* June 1967, p. 127.

Horsepower and Condenser Duty for Specific Refrigerants

Technical literature has graphs for a number of often used refrigerants. Here is a set of graphs for horsepower and condenser duty per 10^6 BTU/hr of refrigeration duty. Parameters are evaporator and condensing temperature. Single-stage, two-stage, and three-stage systems are included for propane, propylene, ethane, and ethylene.

Example. A 50 ton two-stage propane refrigeration system has a 40°F evaporator temperature and a 120°F refrigerant condensing temperature. Find the gas horsepower requirement and condenser duty.

Using the two-stage propane graphs at 40°F evaporator temperature and 120°F condensing temperature, the gas horsepower requirement is 100 hp/10^6BTU/hr. A 50 ton unit is equivalent to $50 \times 12{,}000 = 600{,}000$ BTU/hr, so the gas horsepower is $100 \times 0.6 = 60$. The condenser duty read from the companion two-stage propane graph is $1.25 \times 0.6 = 0.75$ MMBTU/hr.

If equations are desired, the curves can be fitted accurately using polynomials. Here are the equations for the single-stage propane horsepower graphs, for example.

$$y = A + Bx + Cx^2 + Dx^3 + Ex^4$$

where

x = evaporator temperature, °F

y = horsepower/10^6 BTU/hr refrigeration duty

Condenser temperature °F	A	B	C	D	E
60	87.8	–1.786	9.881e-3	–7.639e-5	—
70	103.4	–1.916	1.101e-2	–6.944e-5	—
80	121.3	–2.121	1.253e-2	–6.250e-5	—
90	138.3	–2.294	1.530e-2	–7.986e-5	—
100	160.8	–2.545	1.570e-2	–6.345e-5	—
110	183.0	–2.816	1.741e-2	–6.376e-5	—
120	208.4	–3.047	2.009e-2	–1.054e-4	2.422e-7
130	240.9	–3.447	2.285e-2	–1.220e-4	3.205e-7
140	279.9	–3.939	2.657e-2	–1.390e-4	3.451e-7

Equations were generated using FLEXCURV V.2, Gulf Publishing Co.

(text continued on page 177)

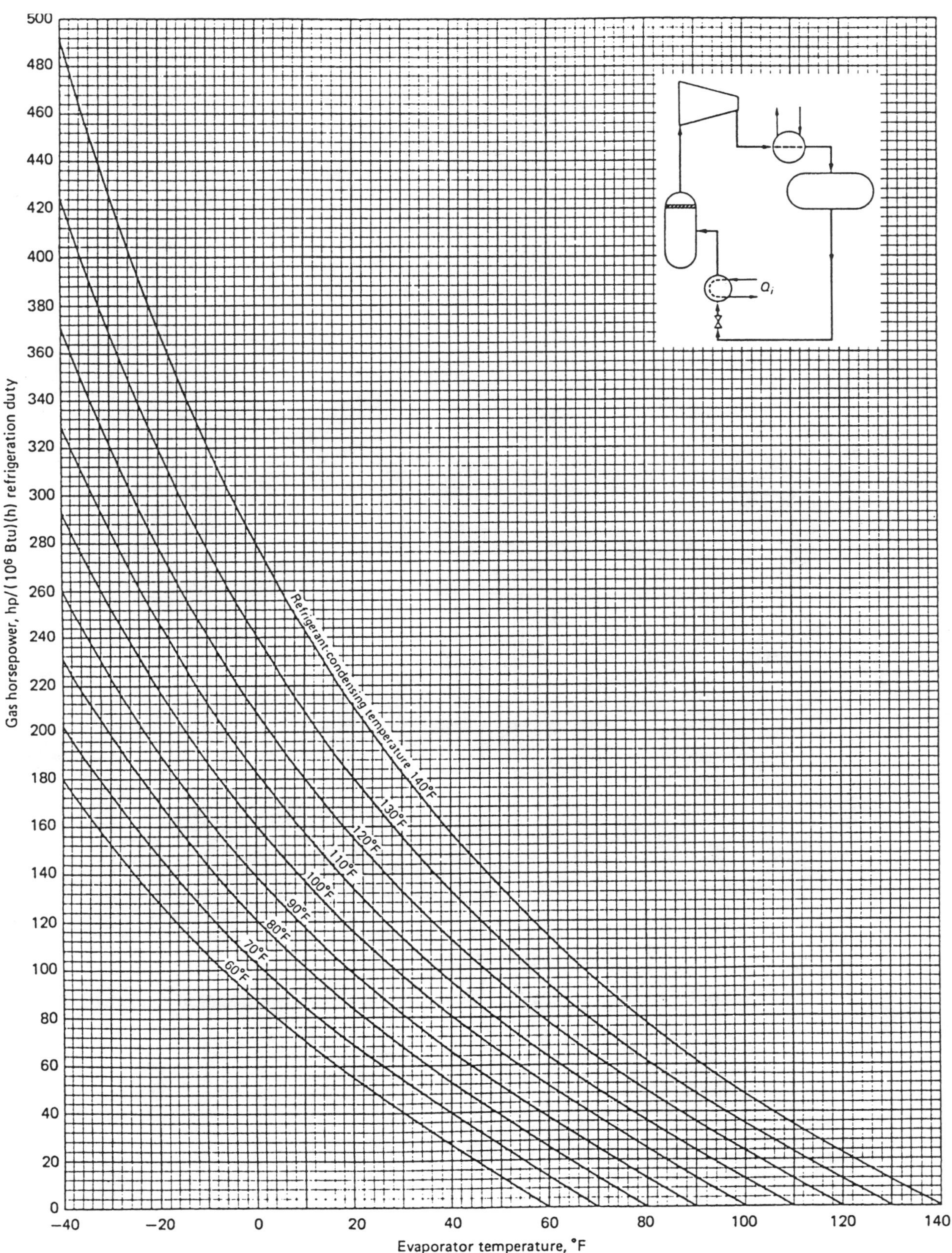

Figure 1. Gas horsepower for single-stage propane refrigeration system.

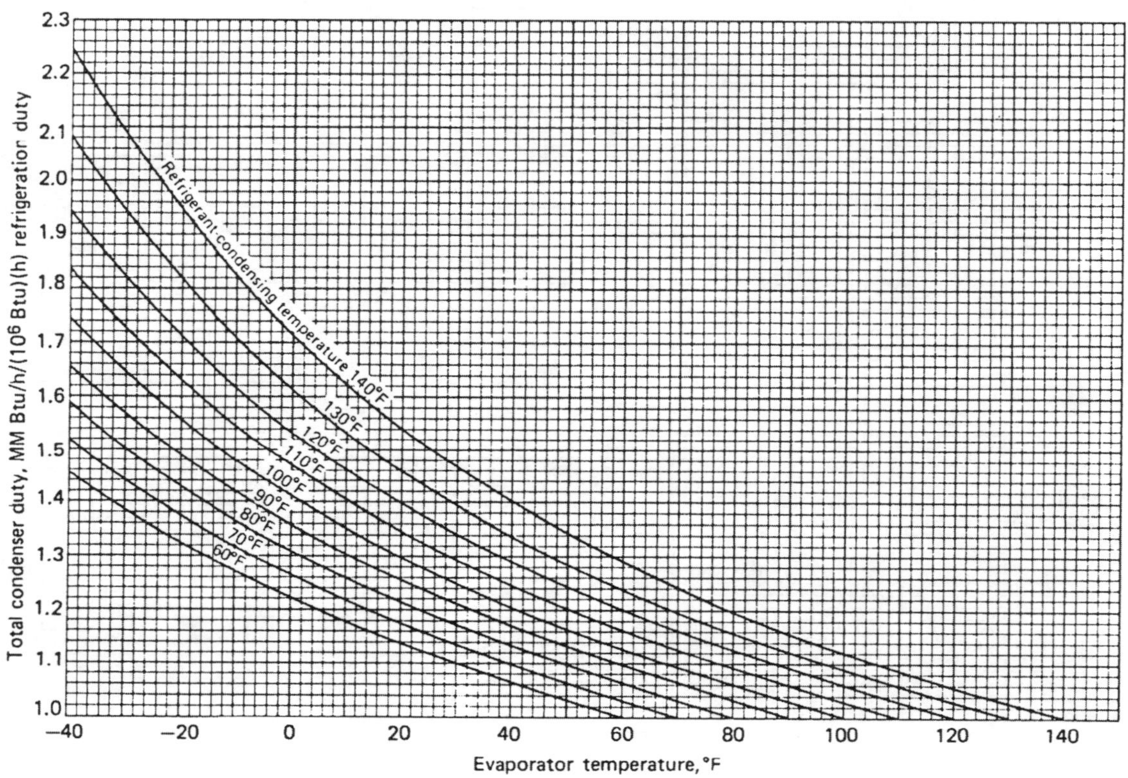

Figure 2. Condenser duty for single-stage propane refrigeration system.

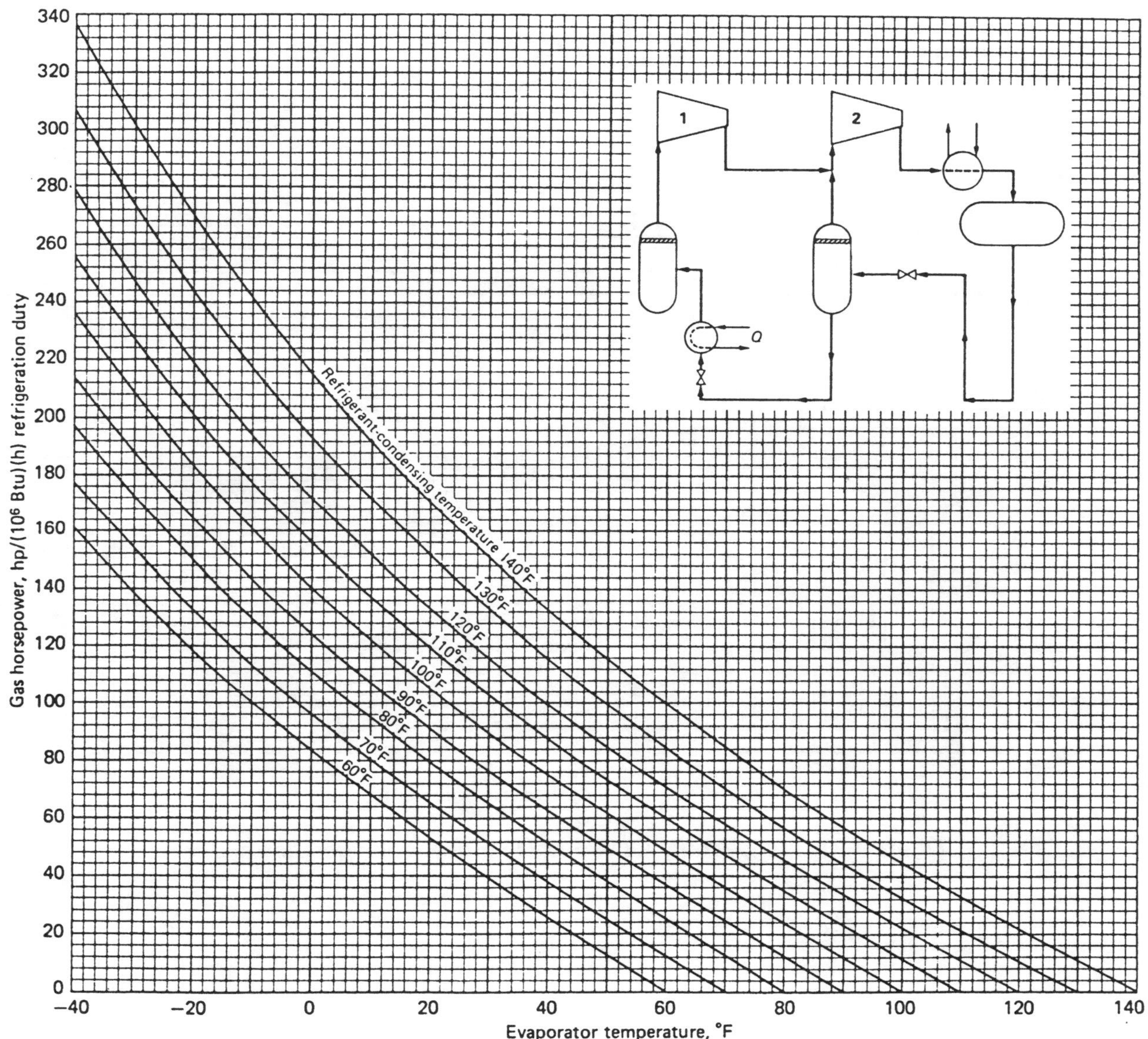

Figure 3. Gas horsepower for two-stage propane refrigeration system.

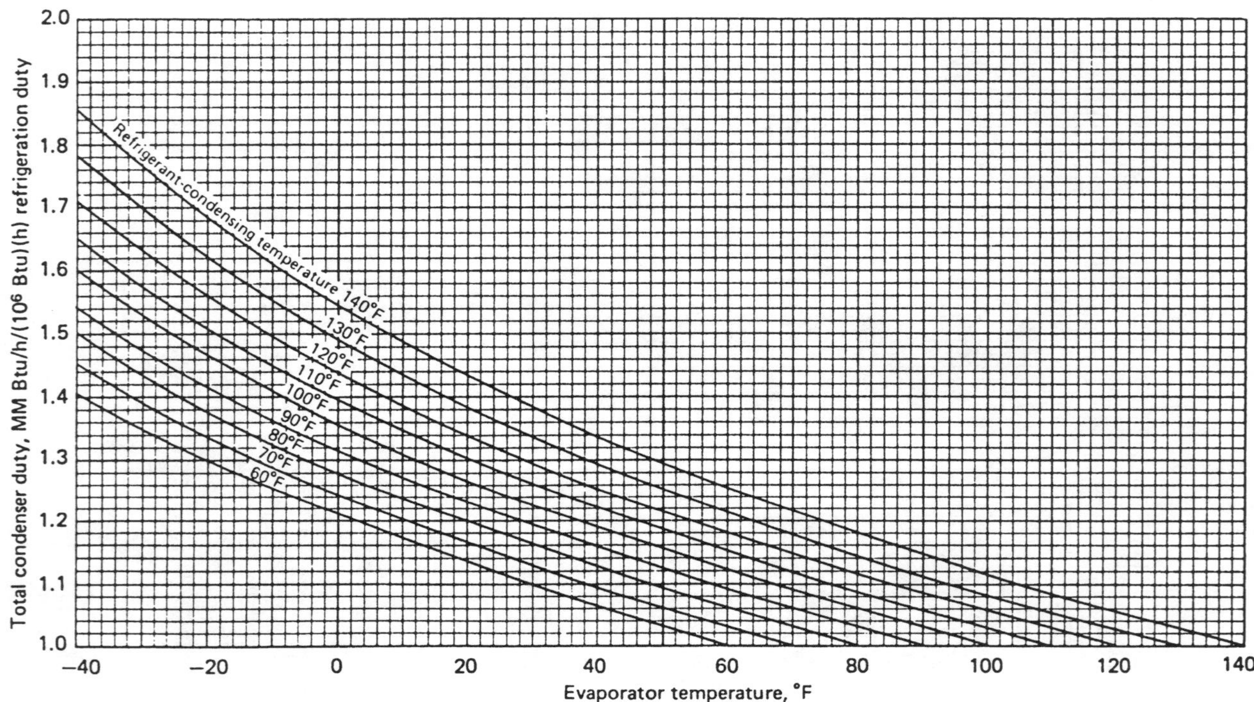

Figure 4. Condenser duty for two-stage propane refrigeration system.

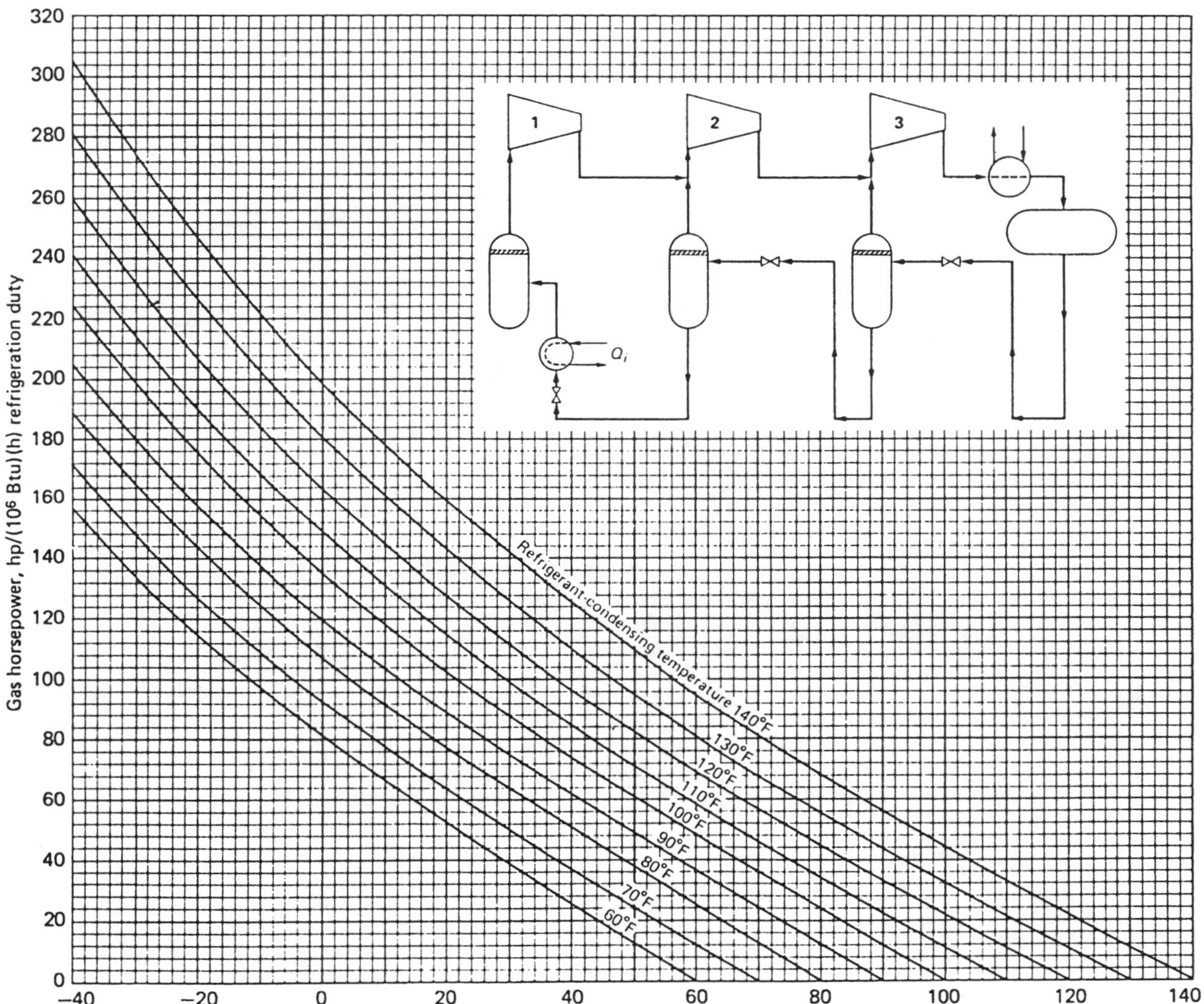

Figure 5. Gas horsepower for three-stage propane refrigeration system.

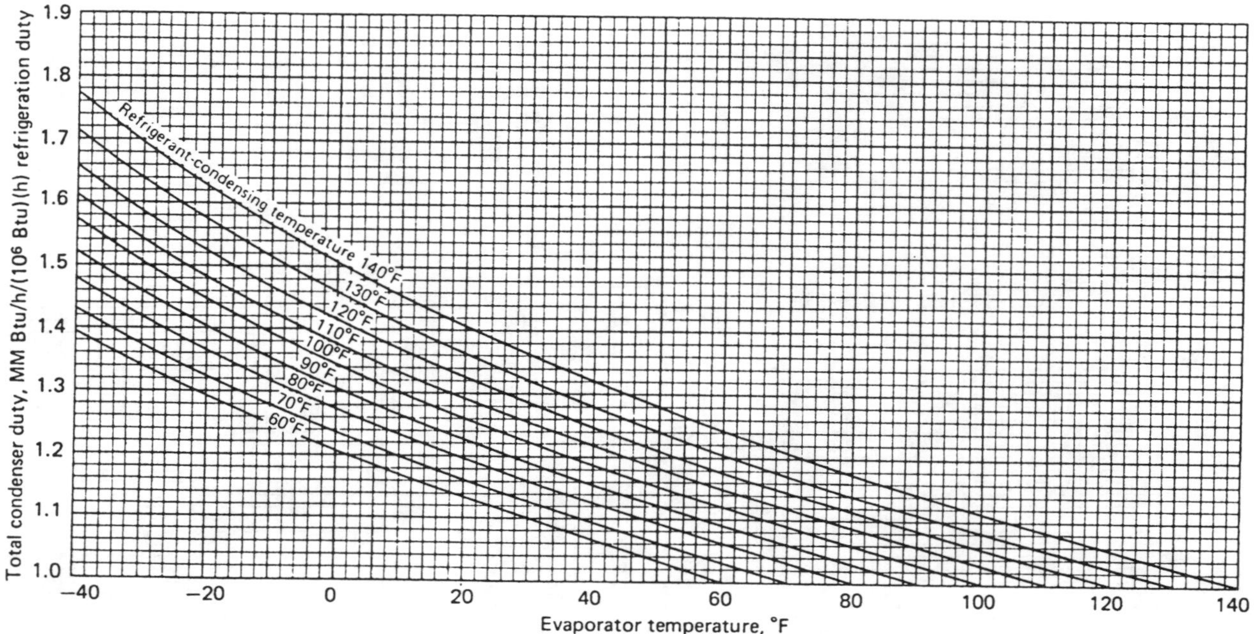

Figure 6. Condenser duty for three-stage propane refrigeration system.

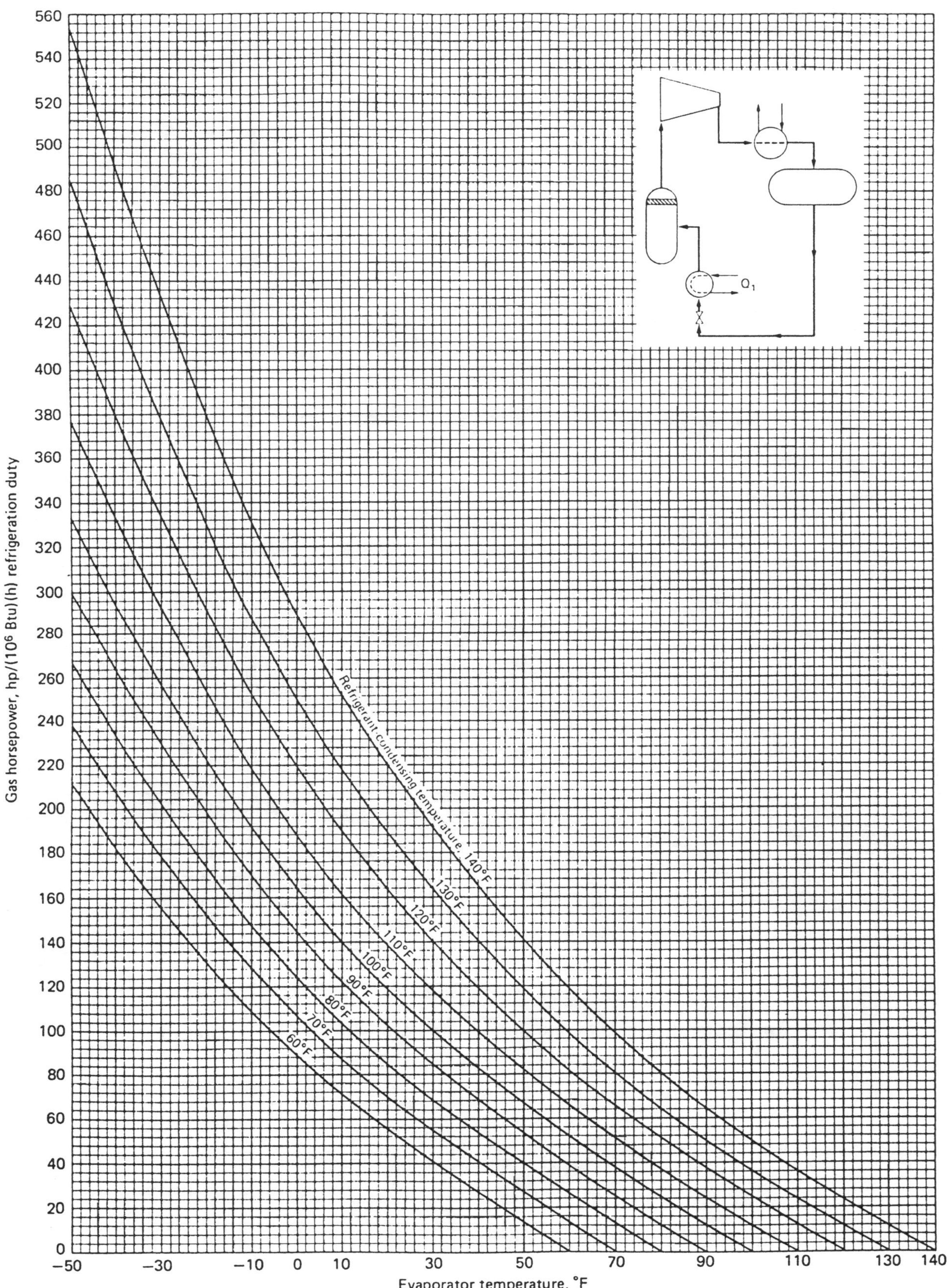

Figure 7. Gas horsepower for single-stage propylene refrigeration system.

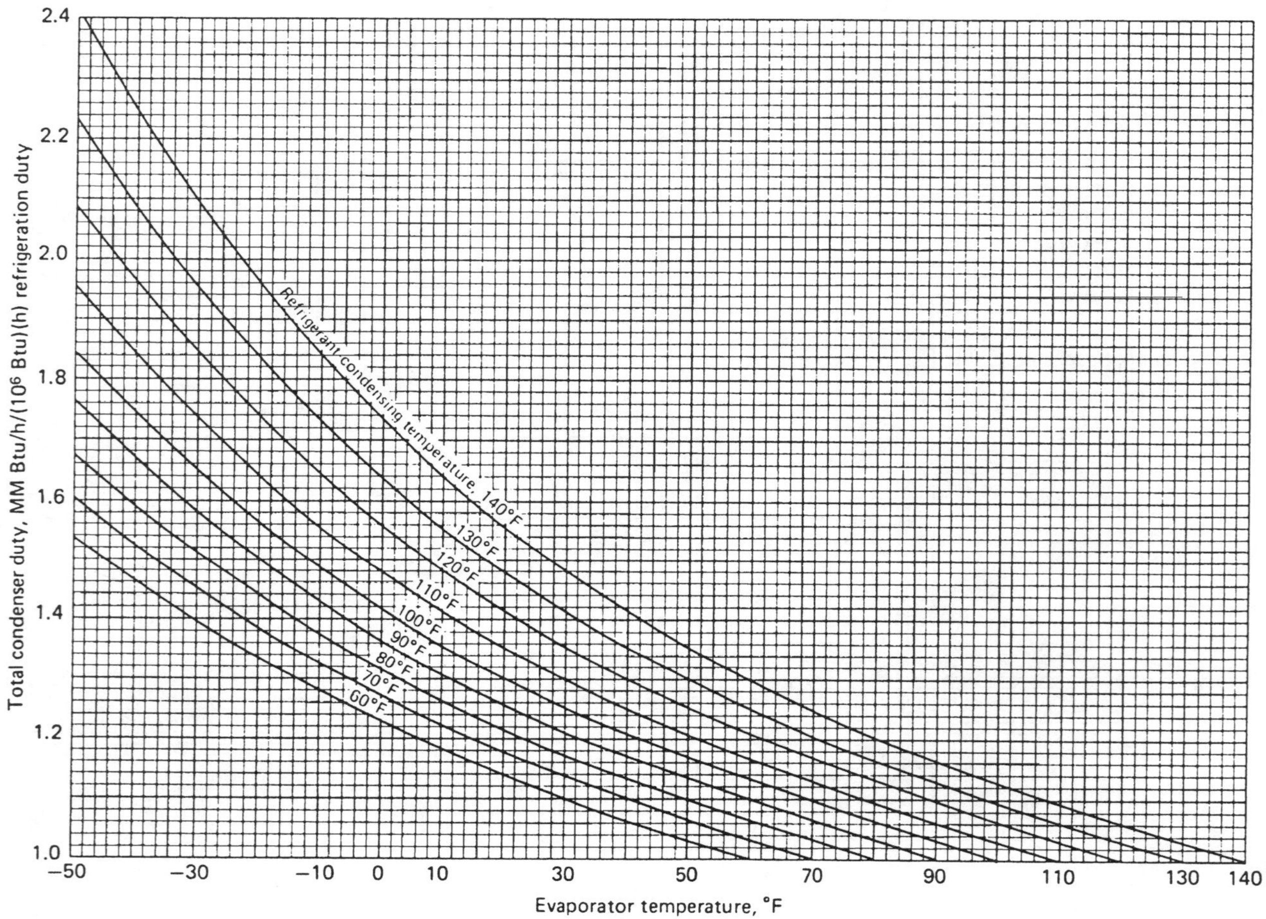

Figure 8. Condenser duty for single-stage propylene refrigeration system.

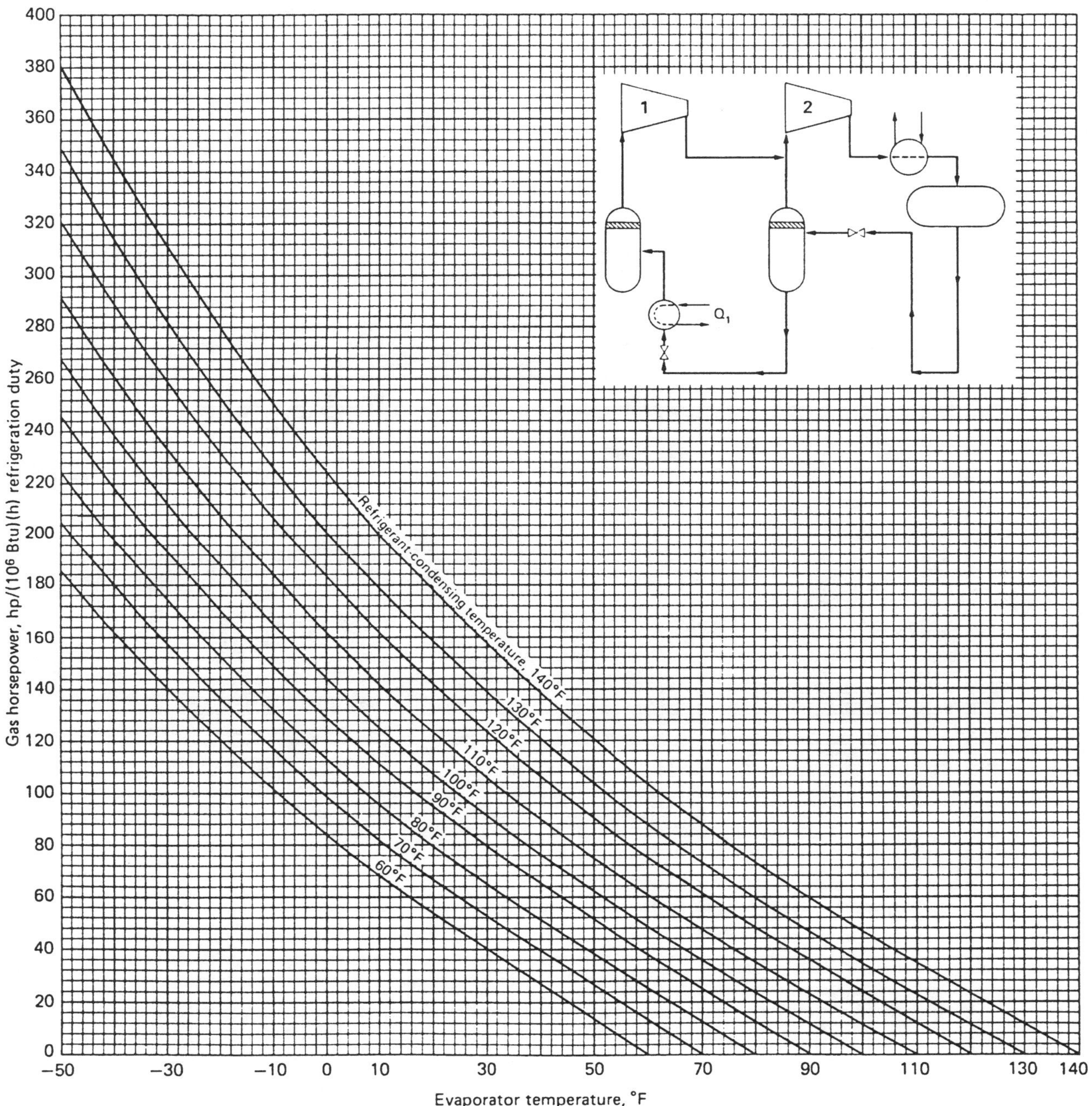

Figure 9. Gas horsepower for two-stage propylene refrigeration system.

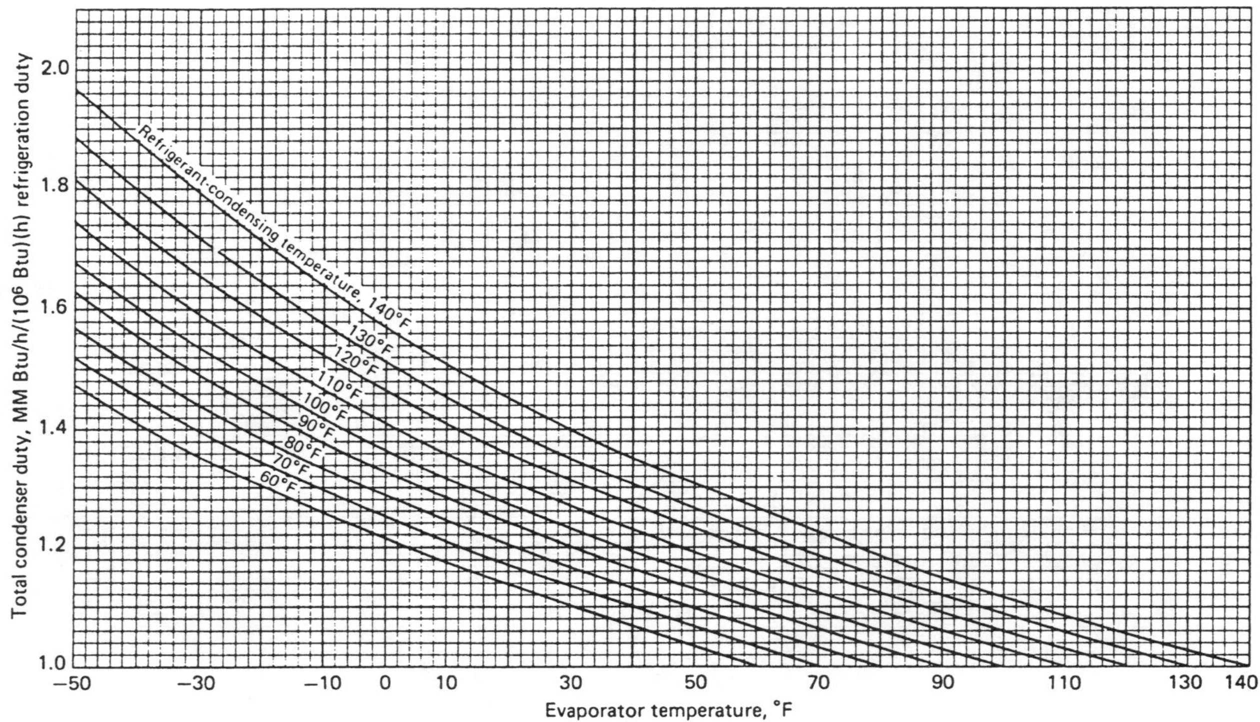

Figure 10. Condenser duty for two-stage propylene refrigeration system.

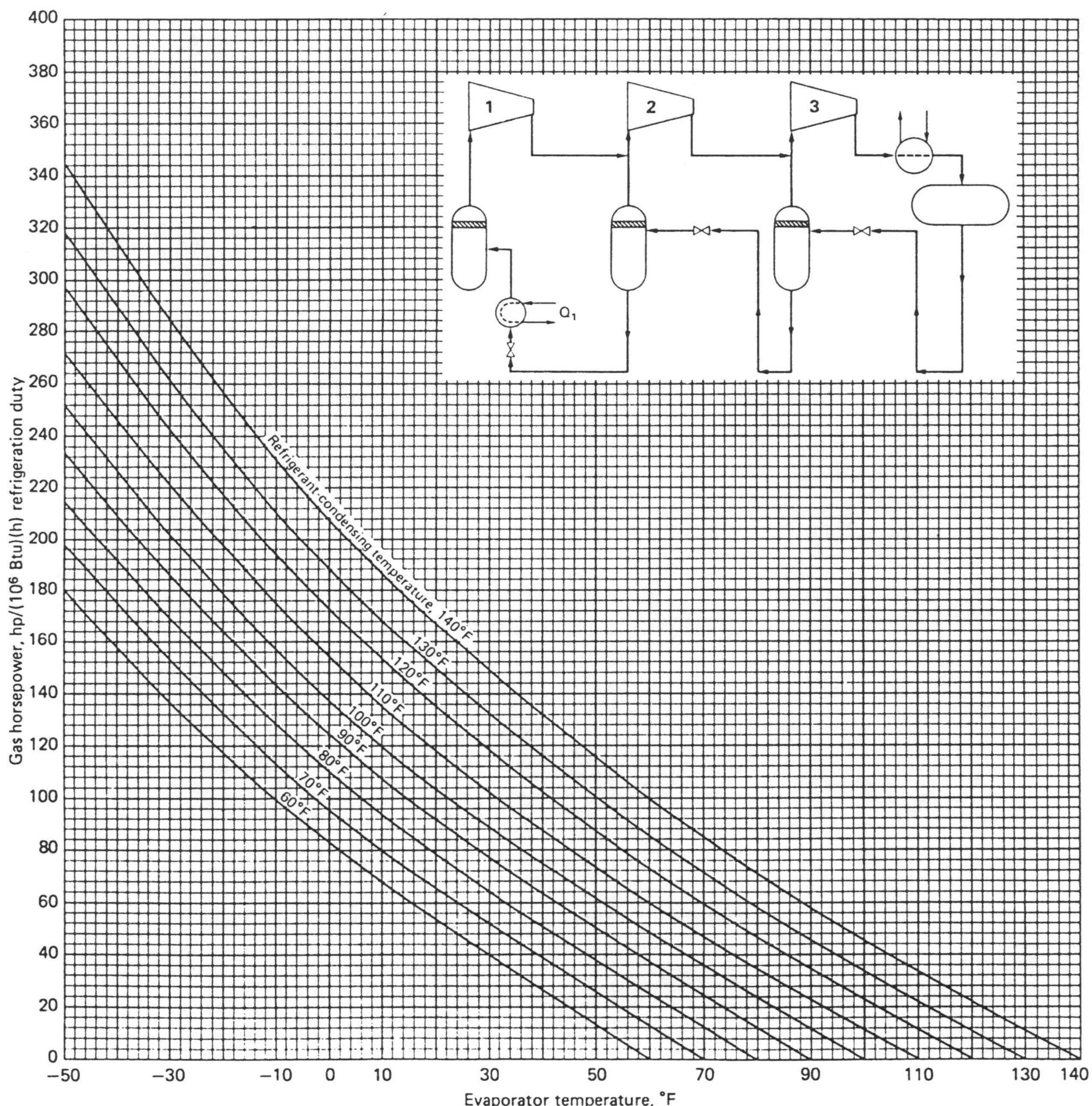

Figure 11. Gas horsepower for three-stage propylene refrigeration system.

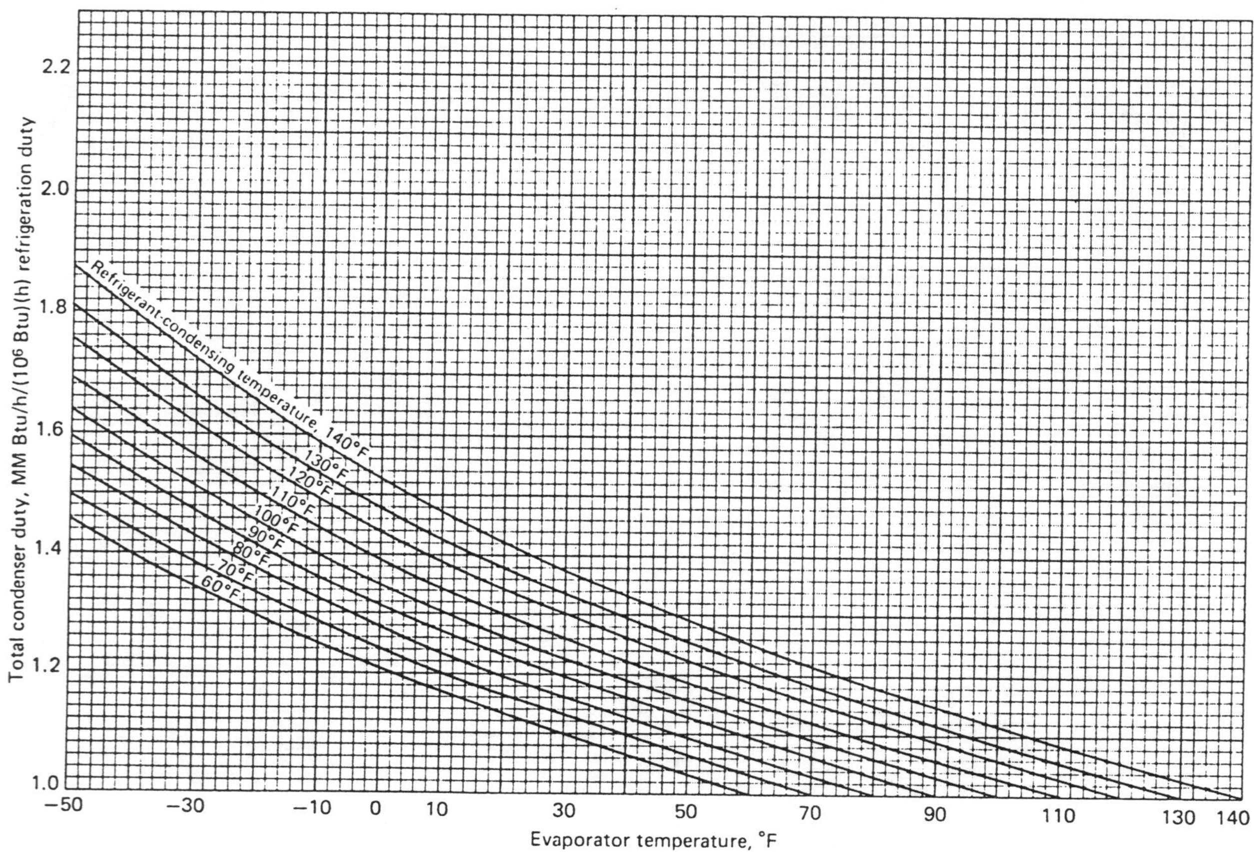

Figure 12. Condenser duty for three-stage propylene refrigeration system.

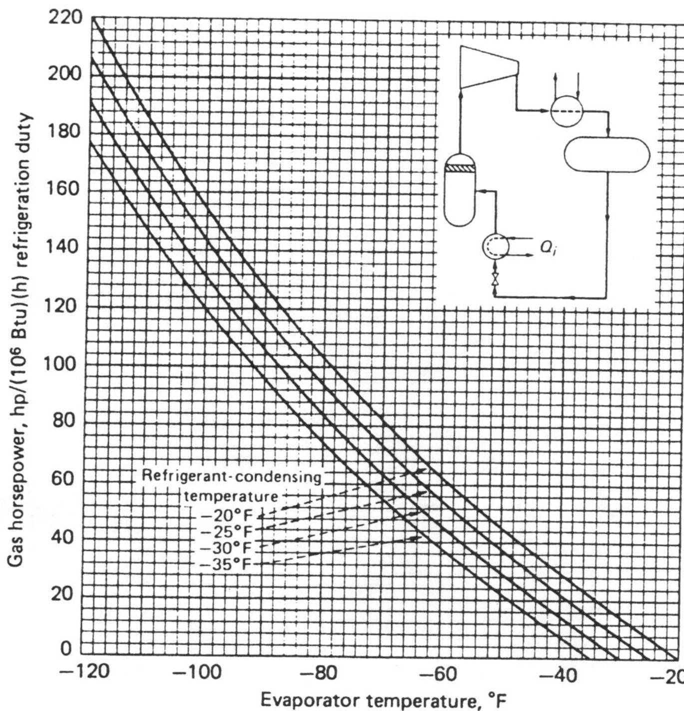

Figure 13. Gas horsepower for single-stage ethane refrigeration system.

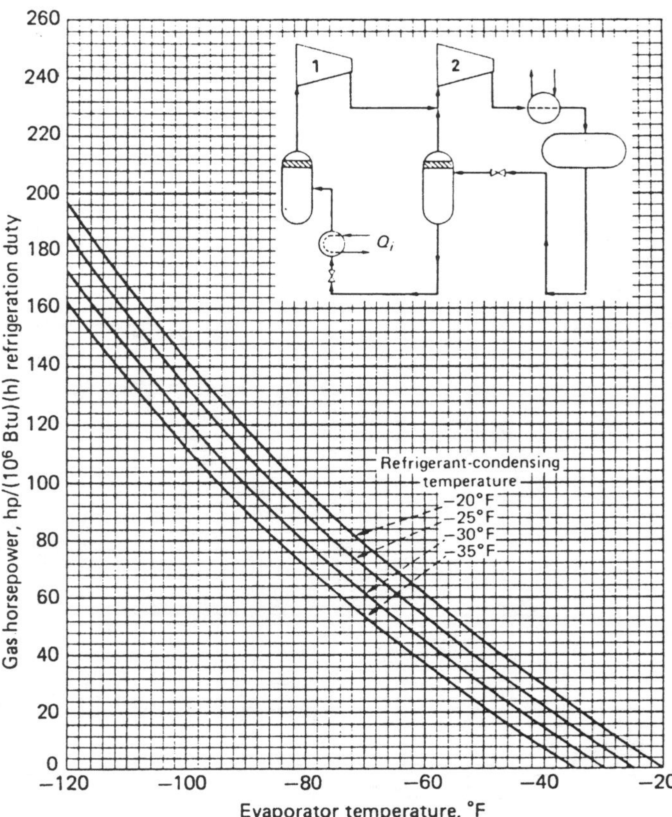

Figure 15. Gas horsepower for two-stage ethane refrigeration system.

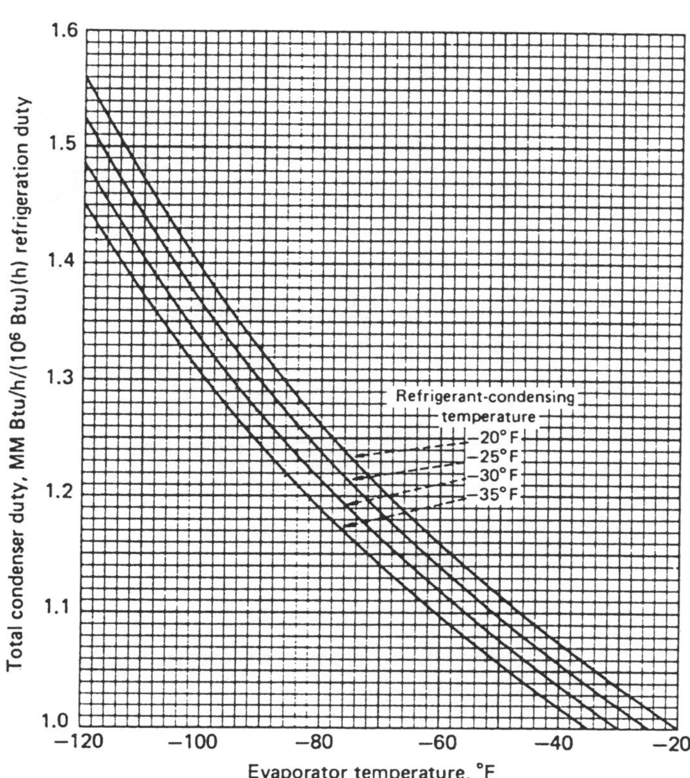

Figure 14. Condenser duty for single-stage ethane refrigeration system.

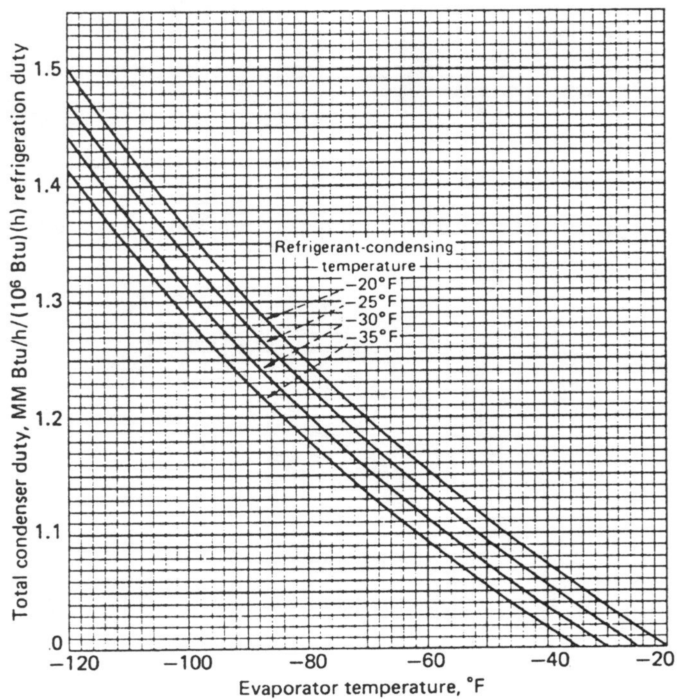

Figure 16. Condenser duty for two-stage ethane refrigeration system.

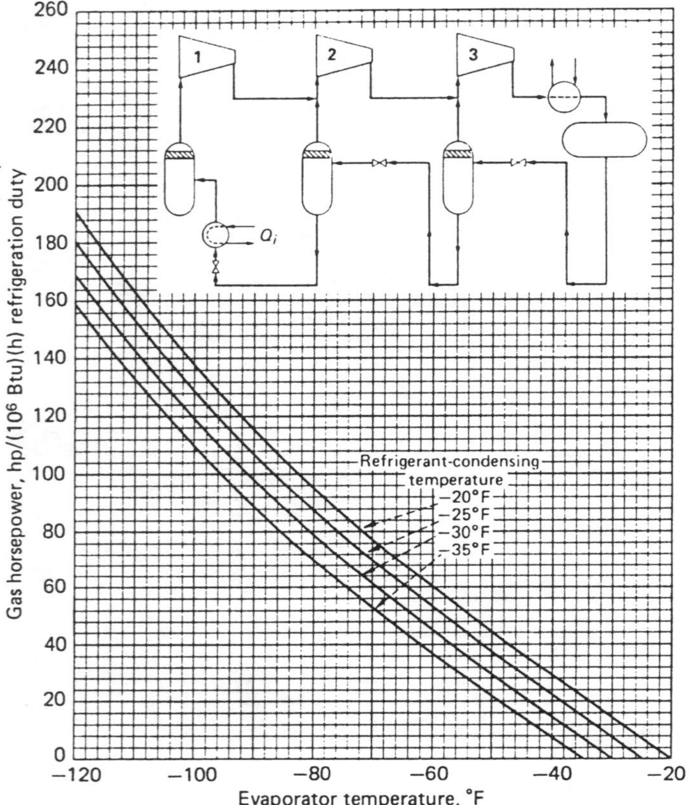

Figure 17. Gas horsepower for three-stage ethane refrigeration system.

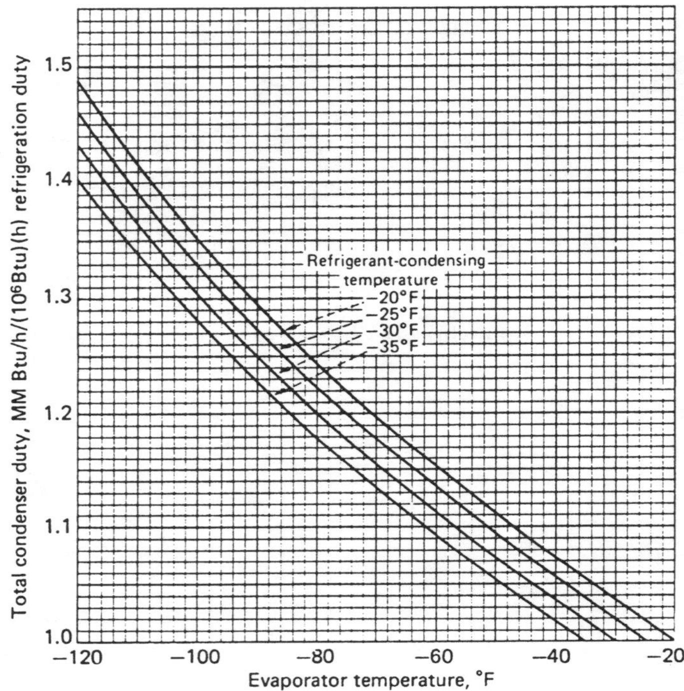

Figure 18. Condenser duty for three-stage ethane refrigeration system.

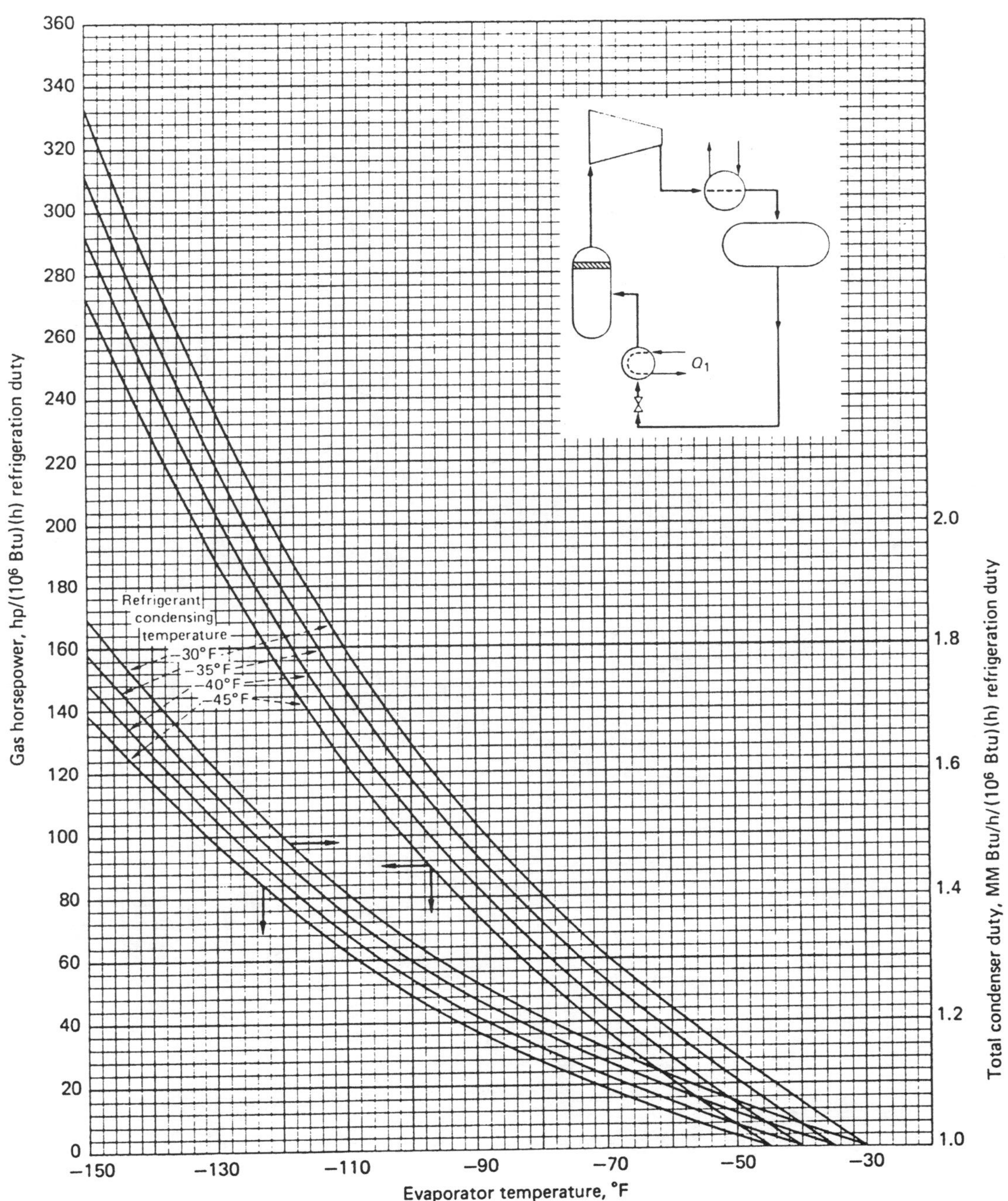

Figure 19. Single-stage ethylene refrigeration system.

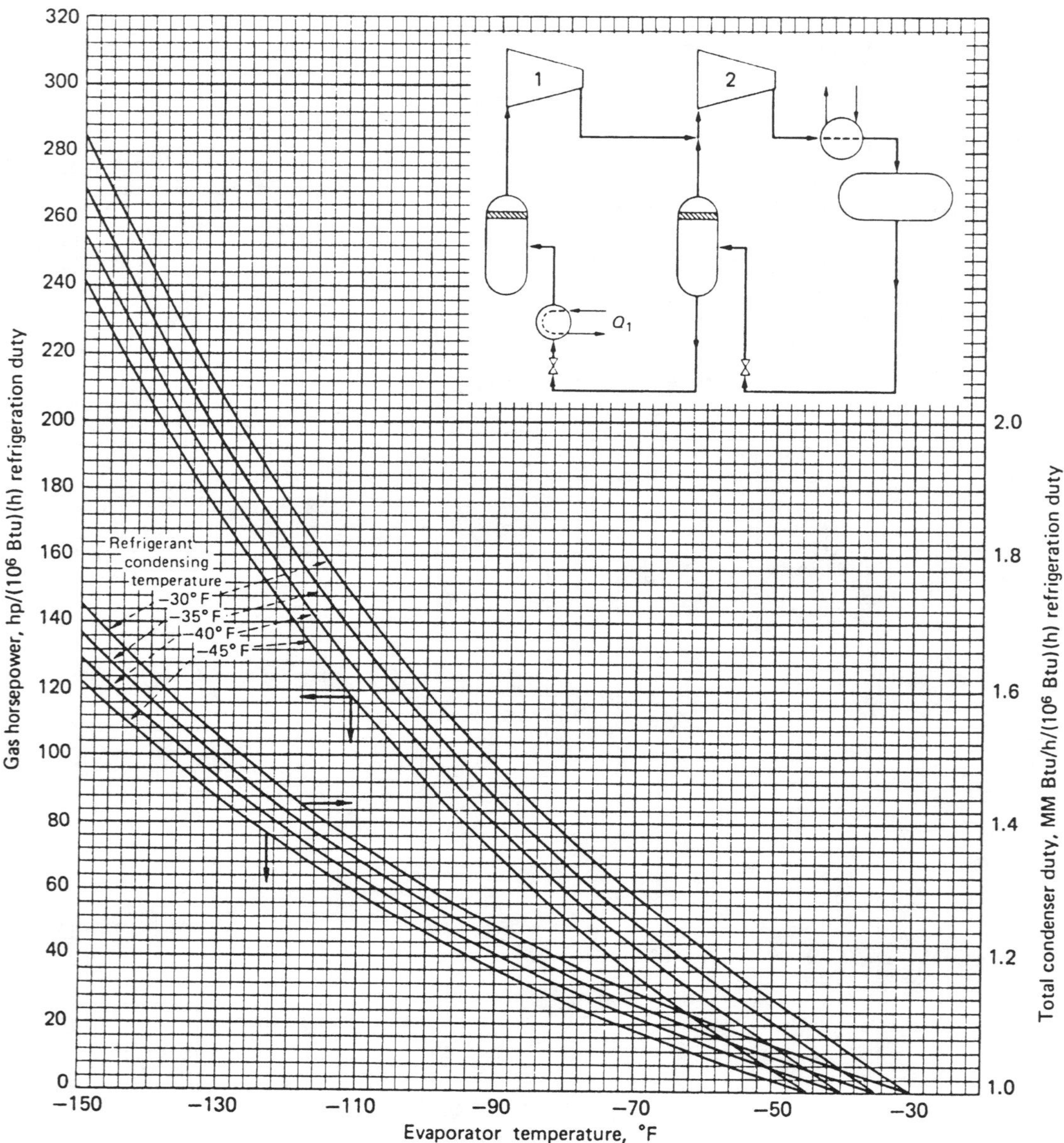

Figure 20. Two-stage ethylene refrigeration system.

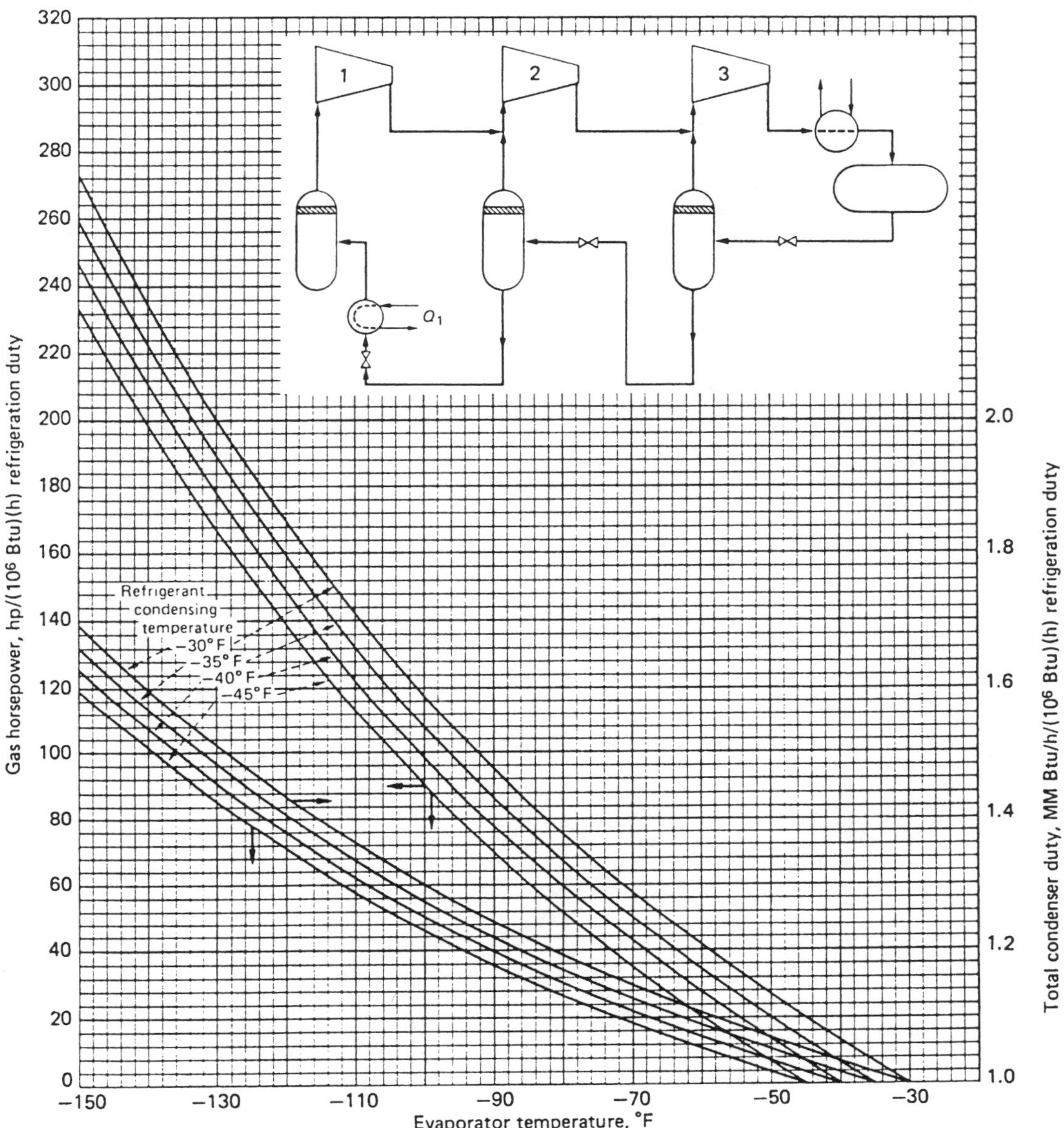

Figure 21. Three-stage ethylene refrigeration system.

(text continued from page 160)

Source

Mehra, Y. R., "Hydrocarbon Refrigerants series," parts 1, 2, 3, and 4, *Chemical Engineering,* December 18, 1978, January 15, 1979, February 12, 1979, and March 26, 1979.

Refrigerant Replacements

As you know, most countries are phasing out certain refrigerants to lessen damage to the ozone layer. The chemicals being phased out are chloroflurocarbons (CFCs) and hydrochloroflurocarbons (HCFCs). Replacements are hydroflurocarbons (HFCs) and certain blends. The DuPont web site (www.dupont.com) gives the handy Table 1 of recommended replacement refrigerants for various applications.

DuPont also provides a computer program to simulate performance of different refrigerants in your system. It is called the "DuPont Refrigeration Expert" (DUPREX).

See the "Properties" section for a description of physical data available.

Source

E. I. DuPont de Nemours and Company web site (www. dupont.com), "Selection of DuPont Refrigerants," reprinted by permission.

Table 1
DuPont Refrigerants

Application	CFC/HCFC Refrigerants	Retrofit	New System
Air conditioning in buildings and industrial temperature control	R-11	SUVA® 123	SUVA® 123
	R-12	SUVA® 134a SUVA® MP39	SUVA® 134a
	R-22	SUVA® 9000	SUVA® 9000 SUVA® 9100
Split and window A/C systems	R-22	SUVA® 9000	SUVA® 9000 SUVA® 9100
Air and marine A/C systems	R-114, R-12B1	SUVA®124	SUVA®124
Automobile air conditioning	R-12	SUVA®134a SUVA® MP52	SUVA®134a
Fresh food storage, above 0°C	R-12	SUVA® MP39	SUVA® 134a
Domestic refrigerators, drink cooler, commercial and restaurant non-frozen chilled foods storage	R-12	SUVA® MP39	SUVA® 134a
Light commercial refrigeration	R-22	SUVA® HP62	SUVA® HP62 SUVA® HP62

Table 1 (Continued)

Application	CFC/HCFC Refrigerants	Retrofit	New System
Frozen Food storage below -18°C	R-502	SUVA® HP80 SUVA® HP81	SUVA® HP62
Refrigerated Transport	R-12	SUVA® MP66	SUVA® 134a
Low Temperature Transport	R-502	SUVA® HP80	SUVA® HP62
Medium Temperature commercial refrigeration	R-12, R-500	SUVA® MP39 SUVA® MP66	SUVA® 134a
Low to Medium Temperature commercial refrigeration	R-22	SUVA® HP62	SUVA® HP62
Very low temperature	R-13B1	SUVA® 9100	SUVA®9100
	R-13	HFC-23	HFC-23
	R-503	SUVA®95	SUVA®95

**The choice of a SUVA® refrigerant will depend on the application as well as on the type of CFC to replace.*
This table is intended as a guide, to cover the situations most likely to be encountered. The equipment owner may request the use of the permanent or "New System" refrigerant in a retrofit, usually to ensure the lowest ODP.
Retrofitting directly to a SUVA® HFC Refrigerant can of course be done, but takes longer and costs more than the simple change to a SUVA® Blend.

Ethylene/Propylene Cascaded System

The following information was used in olefin plant case studies to determine if the ethylene/propylene cascaded refrigeration systems had enough horsepower for various plant operations. The propylene was condensed against cooling water at 110°F and the ethylene was condensed against propylene at –20°F. For comparison, the horsepower requirements for each refrigerant alone are also shown.

The data is presented in the following equation form:

$y = A + Bx + Cx^2$

where

y = horsepower/ton refrigeration

x = evaporator temperature, °F

Example. With the example cascaded system at an evaporator temperature of –100°F, the horsepower requirement is 6.2 hp/ton refrigeration. A ton of refrigeration is equal to 12,000 BTU/hr.

Propylene alone at 110°F condensing temperature

A	B	C	Range, °F
1.98	–1.991e-2	1.691e-5	–40 to 110

Ethylene alone at -20°F condensing temperature

A	B	C	Range, °F
4.67	–1.256e-2	7.778e-5	–150 to –60

Cascade with ethylene condensing at -20°F and propylene at 110°F

A	B	C	Range, °F
3.47	–2.289e-2	4.444e-5	–150 to –60

Equations were generated using FLEXCURV V.2, Gulf Publishing Co.

Source

The horsepower per ton raw data is from a private source.

Steam Jet Type Utilities Requirements

Steam and cooling water requirements for barometric steam jet refrigeration units are shown in the following graphs for given available cooling water temperature and delivered chilled water temperature. The graphs are for 100 PSIG motivating steam. For 30–50 PSIG steam, the quantity required will increase by a factor of about 2 for 40°F chilled water and a factor of 1.5 for 55°F chilled water.

Example. To produce 20 tons of refrigeration while delivering 50°F chilled water, the steam consumption depends upon the quantity and temperature of the cooling water. If one has 140 gpm of 85°F cooling water the y-axis is 7 gpm/ton of refrigeration. The steam consumption on the x-axis is about 17 lb/hr steam per ton of refrigeration. A ton of refrigeration is equal to 12,000 BTU/hr.

Source

Ludwig, E. E., *Applied Process Design for Chemical and Petrochemical Plants, Vol. 3, 2nd Ed.*, Gulf Publishing Co., pp. 208, 209.

(text continued on page 182)

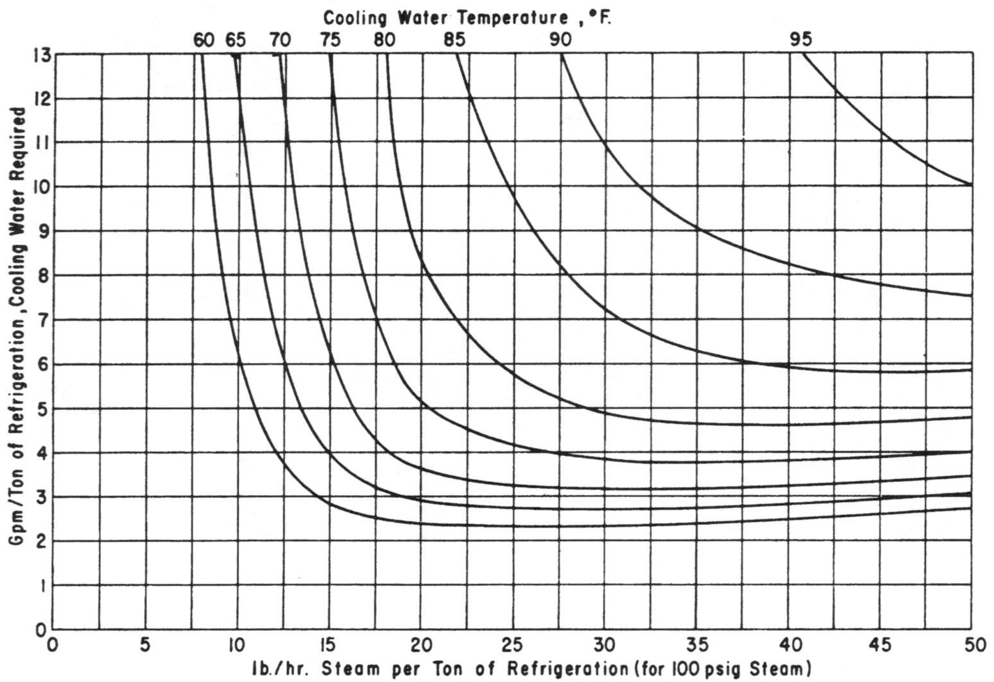

Figure 1. lb/hr steam per ton of refrigeration (for 100 psig steam) for 40°F chilled water.

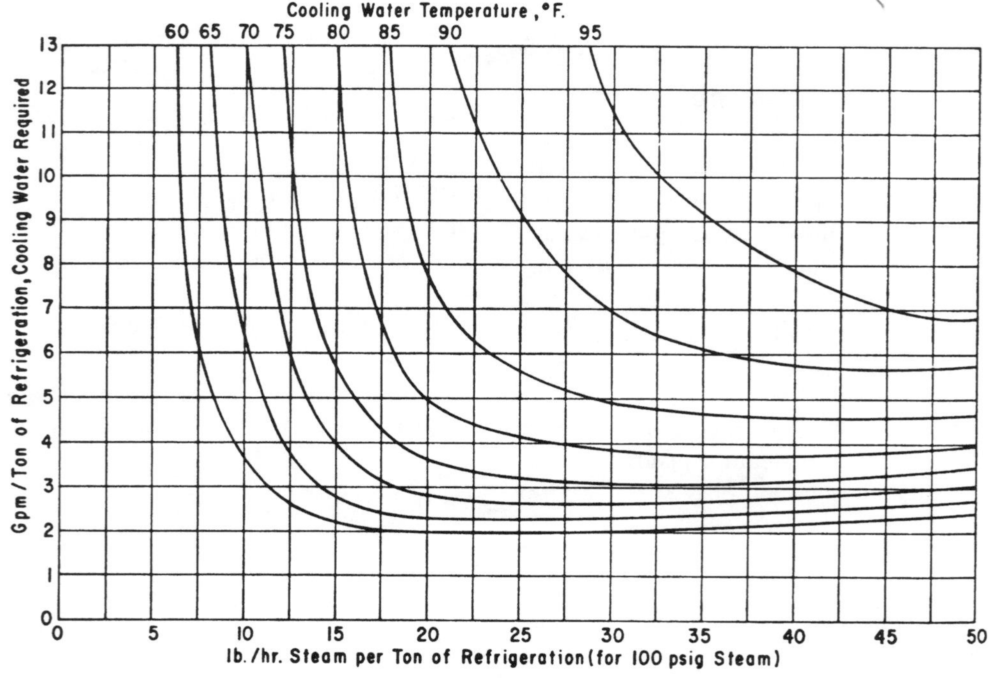

Figure 2. lb/hr steam per ton of refrigeration (for 100 psig steam) for 45°F chilled water.

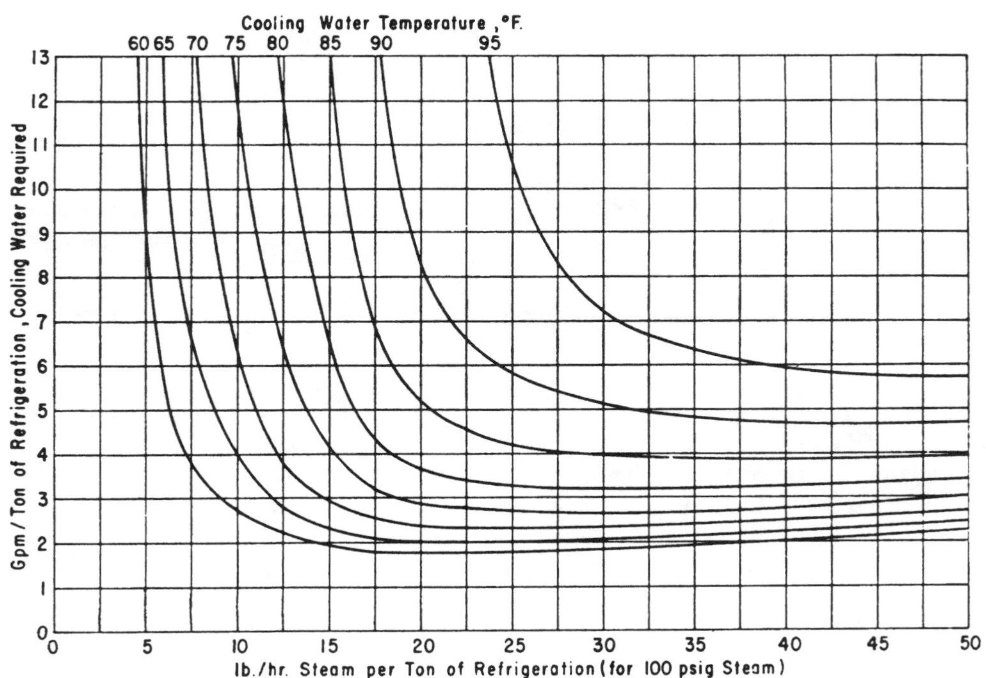

Figure 3. lb/hr steam per ton of refrigeration (for 100 psig steam) for 50°F chilled water.

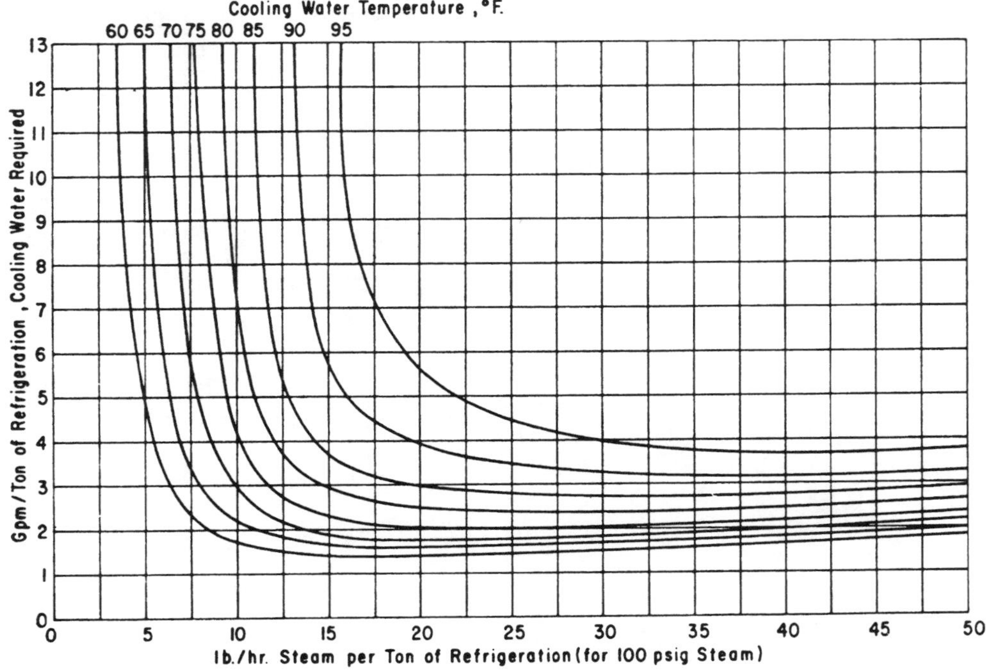

Figure 4. lb/hr steam per ton of refrigeration (for 100 psig steam) for 60°F chilled water.

(*text continued from page 179*)

Ammonia Absorption Type Utilities Requirements

Steam and cooling water requirements for ammonia absorption refrigeration systems are shown in Table 1 for single-stage and two-stage units. The tables are based upon cooling water to the condenser of 85°F with 100°F condensing temperature. Water from the condenser is used in the absorbers.

Example. For an evaporator temperature of −10°F, a steam rate (300°F saturated temperature in the generators) of 33.6 lb/hr/ton refrigeration is required. Also, 5.4 gpm cooling water/ton refrigeration, assuming a 7.5°F rise through the condenser, are required in this system.

Source

Ludwig, E. E., *Applied Process Design for Chemical and Petrochemical Plants, Vol. 3, 2nd Ed.,* Gulf Publishing Co., p. 214, Table 11-2.

Table 1
Steam and Cooling Water Required for Ammonia Absorption Refrigeration Systems

Single-stage

Evap. Temp., °F	Steam Sat. Temp., °F. Req. In Generators	Btu Per Min. Req. In Generator Per Ton Refrig. (200 BTU/ Min.)	Steam Rate Lb/Hr/Ton Refrig.	Water Rate Thru Cond. (7.5°F. Temp. Rise), GPM/Ton
50	210	325	20.1	3.9
40	225	353	22.0	4.0
30	240	377	23.7	4.1
20	255	405	25.7	4.3
10	270	435	28.0	4.6
0	285	467	30.6	4.9
–10	300	507	33.6	5.4
–20	315	555	37.3	5.9
–30	330	621	42.5	6.6
–40	350	701	48.5	7.7
–50	370	820	57.8	9.5

Two-stage

Steam Sat. Temp., °F., Req. In Generators	Btu Per Min. Req. In Generator Per Ton Refrig.	Steam Rate, Lb/Hr/Ton Refrig.	Water Rate Thru Cond. (7.5°F. Temp. Rise), GPM/Ton
175	595	35.9	4.3
180	625	37.8	4.5
190	655	40.0	4.6
195	690	42.3	4.9
205	725	44.7	5.3
210	770	47.5	5.7
220	815	50.6	6.3
230	865	54.0	6.9
240	920	58.0	7.8
250	980	62.3	9.0
265	1050	67.5	11.0

12
Gas Treating

Introduction

A study of gas treating can be very confusing because so many processes exist. The discussions in this chapter attempt to explain the differences among some of the most popular processes. Understanding of the different processes will aid in process selection.

Gas treating is defined here as removal of H_2S and CO_2. Other sulfur compounds are discussed where applicable. Dehydration and sulfur production are not included, except for discussing sulfur production in the Stretford Process and for selective H_2S removal. H_2S must be removed from natural gas and process streams for health reasons and prevention of corrosion. Natural gas pipeline specifications require no more than ¼ grain/100 SCF. This is equivalent to 4 ppmv or 7 ppmw (for a 0.65 specific gravity gas). By comparison, the human nose can detect 0.13 ppmv and the threshold limit value for prolonged exposure is 10 ppmv.

CO_2 must be removed for prevention of corrosion and because it lowers the heating value of natural gas. In some treatment situations it is desired to selectively remove H_2S and "slip" most of the CO_2. If the CO_2 can be tolerated downstream, a more economical treating plant can often be provided that removes only the H_2S. Sometimes, selective H_2S removal is practiced to enrich the H_2S content of the acid gas stream feeding a sulfur recovery plant. The concentration of CO_2 should be kept below 80% in the acid gas feeding a sulfur recovery plant. In a coal gasification plant design, power and steam were produced using low BTU gas-fired turbines. CO_2 was slipped to the turbines because the CO_2 mass flow contained significant high pressure energy.

In summary, some of the reasons for selective H_2S removal are:

- Economical design (capital and operating costs)
- H_2S enrichment for sulfur recovery
- Special cases such as energy recovery

Sources

1. *GPSA Engineering Data Book,* Gas Processors Suppliers Association, Vol. II, 10th Ed.
2. Maddox, R. N., *Gas and Liquid Sweetening,* Campbell Petroleum Series, 1977.

Gas Treating Processes

There are so many gas treating processes that one hardly knows where to begin. That makes selection of a process for a given situation difficult. Economics will ultimately decide, but some initial rough screening is required to eliminate inordinate study. I have attempted to condense the mass of information in the literature into some brief, user-friendly guidelines for the rough cut.

The selected popular processes are grouped as follows:

Reaction type: MEA, DEA, MDEA, DGA, Stretford (also produces sulfur)
Physical solvent type: Fluor solvents (propylene carbonate example), Selexol
Physical/chemical type: Sulfinol
Carbonate type: Potassium carbonate
Solution batch type: Lo-Cat, Chemsweet
Bed batch type: Iron Sponge, Mol Sieve

Figure 1 should be used in the process selection as one of the important parameters. The processes listed within the graph are in preference order.

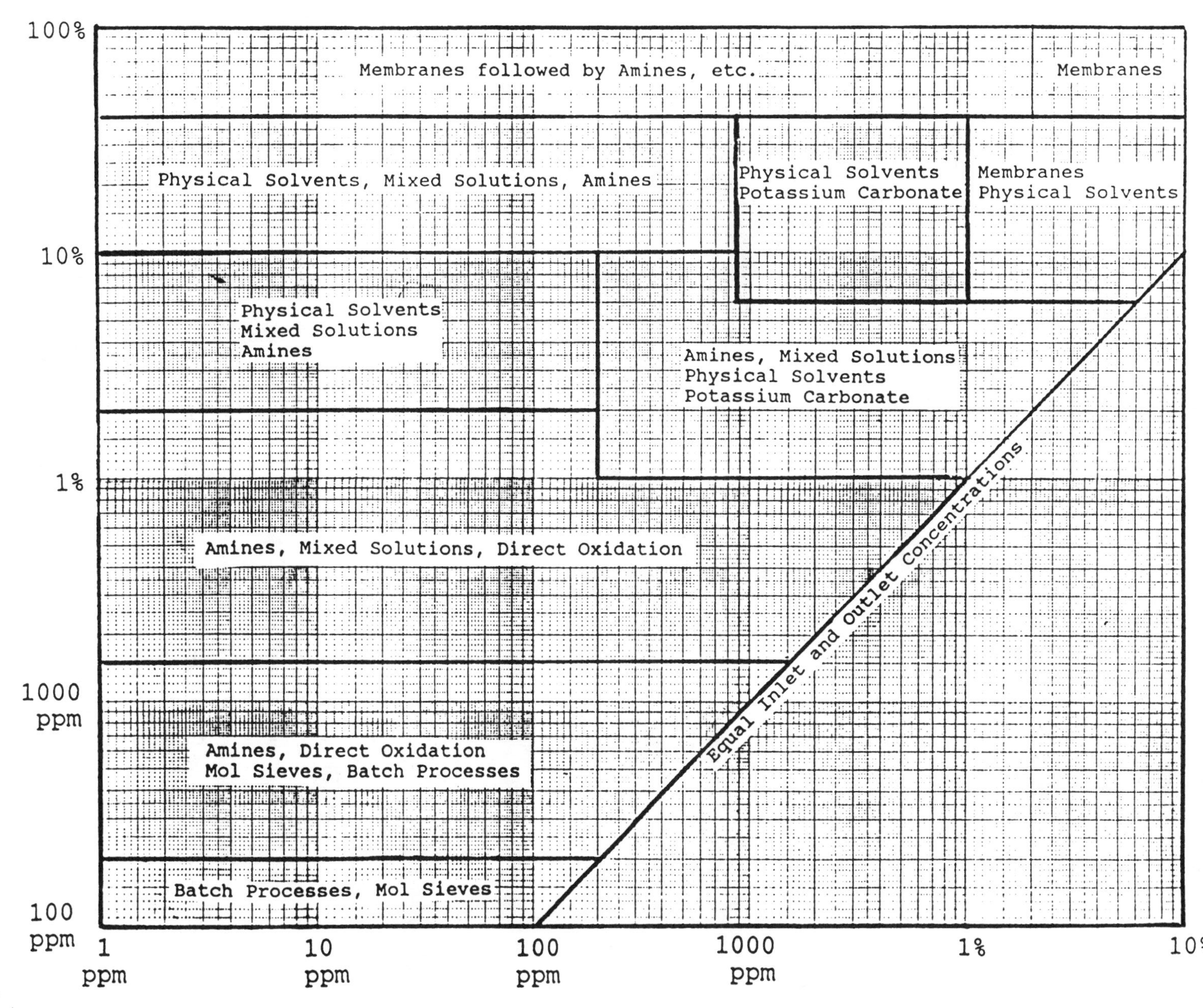

Figure 1. Guide for selecting gas sweetening processes.

Source

Ball, T., Ballard, D., and Manning, W., "Design Techniques For Amine Plans," Presented at PETRO ENERGY, October 1988.

Reaction Type Gas Treating

The amines (MEA, DEA, MDEA, DGA) are the most popular treating solutions. At one time MEA, and later MEA and DEA, dominated the market. Amines in general should be considered for:

- Low acid gas partial pressures (product of system pressure and concentration of acid gases—H_2S and CO_2 in the feed) of roughly 50 psi and below. Another way of looking at selection based upon acid gas concentration is Figure 1[3] in the previous section entitled Gas Treating Processes.
- Low acid gas concentrations in the gas product of roughly 4–8 ppmv (0.25–0.50 grains/100 scf)
- Heavier hydrocarbons present (provide better filtration for DGA)
- For CO_2 removal with no H_2S, a CO_2 partial pressure of 10–15 psi[4]

If you plan to operate an amine plant, be sure to get a copy of Don Ballard's classic article.[5]

Among the amines, MEA is preferred for low contactor pressures and stringent acid gas specifications such as H_2S well below 0.25 grains/100 scf, and CO_2 as low as 100 ppmv[1]. MEA is degraded by COS and CS_2 with the reactions only partially reversible with a reclaimer.

DEA is preferred when system pressure is above 500 psi.[1] The 0.25 grains/100 scf is more difficult to produce with DEA. COS and CS_2 have few detrimental effects on DEA. This, and high solution loadings, provide advantages over MEA. DEA will typically be used for refinery and manufactured gas streams that have COS and CS_2.[2]

MDEA is preferred for selective H_2S removal and lack of degradation from COS and CS_2.

DGA is preferred for cold climates and high (50–70 wt%) solution strength for economy. By comparison, solution strength for MEA is 15–25 wt%, and for DEA, 25–35 wt%.[1] Provide good filtration for DGA because it has a greater affinity for heavy hydrocarbons than other amines. The feed gas must have at least 1% acid gas for DGA to provide savings over MEA.[2] The author understands that DEA was picked for treating coal seam gas with 7–8% CO_2 (this level is not considered high CO_2).

The Stretford Process sweetens and also produces sulfur. It is good for low feed gas concentrations of H_2S.[4] Economically, the Stretford Process is comparable to an amine plant plus a Claus sulfur recovery plant. Usually, the amine/Claus combination is favored over Stretford for large plants.[4] Stretford can selectively remove H_2S in the presence of high CO_2 concentrations. This is the process used in the coal gasification example in the Introduction.

Nomenclature

MEA = Monoethanol amine
DEA = Diethanol amine
MDEA = Methydiethanol amine
DGA = Diglycol amine ® Jefferson Chemical Co., Fluor patented process (Econamine)

Sources

1. *GPSA Engineering Data Book,* Gas Processors Suppliers Association, Vol. II, 10th Ed.
2. Maddox, R. N., *Gas and Liquid Sweetening,* Campbell Petroleum Series, 1977.
3. Ball, T., Ballard, D., and Manning, W., "Design Techniques for Amine Plants," presented at PETRO ENERGY, October 1988.
4. Tennyson, R. N., and Schaap, R. P., "Guidelines Can Help Choose Proper Process for Gas-Treating Plants," *Oil and Gas Journal,* January 10, 1977.
5. Ballard, Don, "How To Operate An Amine Plant," *Hydrocarbon Processing,*" April 1966.

Physical Solvent Gas Treating

The physical solvent types of gas treatment are generally preferred when acid gases in the feed are above 50–60 psi.[1,3] This indicates a combination of high pressure and high acid gas concentration. Heavy hydrocarbons in the feed discourage physical solvents, but not COS and CS_2 which do not degrade the solvents. Usually, physical solvents can remove COS, CS_2, and mercaptans.[1] Physical solvents are economical because regeneration occurs by flashing or stripping which require little energy.[1]

Selexol®, licensed by the Norton Company, uses the dimethyl ether of polyethylene glycol. A Selexol® plant can be designed to provide some selectivity for H_2S.[2] For example, the plant can be designed to provide pipeline quality gas (0.25 grains H_2S/100 scf) while slipping 85% of the CO_2.[1]

The Fluor solvent, propylene carbonate, is used primarily for removal of CO_2 from high pressure gas streams.[2] The author is familiar with a plant using propylene carbonate with 15–20% CO_2 in the feed at about 800 psi. The CO_2 off gas stream was used for enhanced oil recovery. Propylene carbonate loses economic incentive below about 12% acid gas in the feed.[4]

Sources

1. *GPSA Engineering Data Book,* Gas Processors Suppliers Association, Vol. II, 10th Ed.
2. Maddox, R. N., *Gas and Liquid Sweetening,* Campbell Petroleum Series, 1977.
3. Tennyson, R. N., and Schaap, R. P., "Guidelines Can Help Choose Proper Process for Gas-Treating Plants," *Oil and Gas Journal,* January 10, 1977.
4. Private communication with an operating company.

Physical/Chemical Type

The Sulfinol® process from Shell Development Company is a good example of the physical/chemical type of process. It blends a physical solvent and an amine to obtain the advantages of both. The physical solvent is Sulfolane® (tetrahydrothiophene dioxide) and the amine is usually DIPA (diisopropanol amine). The flow scheme is the same as for an amine plant.[1]

Advantages of the Sulfinol® Process:

- The physical solvent Sulfolane®, like other physical solvents, has higher capacity for acid gas at higher acid gas partial pressures. At these higher partial pressures the Sulfinol® process has lower circulation rates (higher solution loading) and better economy than MEA.[2] Sulfinol® is demonstratively advantageous against MEA when the H_2S/CO_2 ratio is greater than 1:1[1] while at high acid gas partial pressures.
- In the solution, the amine DIPA is meanwhile able to achieve pipeline quality gas (0.25 grains H_2/100 scf).
- COS, CS_2, and mercaptans are removed. CO_2 slightly degrades DIPA, but reclaiming is easy.[1]
- Low corrosion/carbon steel[2]
- Low foaming[2]
- Low vapor losses[2]

Disadvantages Include:

- The same disadvantages with heavy hydrocarbons in the feed as other physical solvents
- High priced chemicals and process royalty, but losses are low and the licensee receives many engineering services.[2]

Sources

1. *GPSA Engineering Data Book,* Gas Processors Suppliers Association, Vol. II, 10th Ed.
2. Maddox, R. N., *Gas and Liquid Sweetening,* Campbell Petroleum Series, 1977.

Carbonate Type

Hot (230–240°F) potassium carbonate treating was patented in Germany in 1904[2] and perfected into modern commercial requirements by the U. S. Bureau of Mines. The U. S. Bureau of Mines was working on Fischer-Tropsch synthesis gas at the time.[2] Potassium carbonate treating requires high partial pressures of CO_2. It therefore cannot successfully treat gas containing only H_2S.[1]

Natural gas streams have been economically treated with potassium carbonate. Medium to high acid gas concentrations combined with the high pressures of natural gas transmission lines yield the high partial pressures of acid gases required.

For stringent specifications of outlet H_2S or CO_2, special designs or a two-stage process may be required.[1] Maddox[2] states that a two-stage process would typically be used for acid gas feed concentrations of 20–40%. Maddox[2] gives the following rules of thumb for process flow scheme selection:

- Single-stage process will remove CO_2 down to 1.5% in the treated gas
- A split stream cooling modification will produce 0.8% CO_2
- Two-stage process or two-stage process with cooler will go below 0.8% CO_2

Maddox[2] shows how the major process concerns of corrosion, erosion, and column instability must be met in the design and operation of a hot carbonate process. These items will impact the capital and operating/maintenance costs.

Various processes attempt to improve on the basic potassium carbonate process by using activators to increase the rate of CO_2 absorption such as the Catacarb, Benfield, and Giammarco—Vetrocoke processes.

Sources

1. *GPSA Engineering Data Book,* Gas Processors Suppliers Association, Vol. II, 10th Ed.
2. Maddox, R. N., *Gas and Liquid Sweetening,* Campbell Petroleum Series, 1977.

Solution Batch Type

The Lo-Cat® process can be used to sweeten or convert H_2S to sulfur. It removes H_2S only and will not remove CO_2, COS, CS_2, or mercaptans. Iron is held in dilute solution (high circulation rates) by the common chelating agent EDTA (ethylene diamine tetra acidic acid). The iron oxidizes the H_2S to sulfur. The solution is circulated batchwise to an oxidizer for regeneration.

Chemsweet from C. E. Natco is another H_2S-only process. It uses a water dispersion of zinc oxide and zinc acetate to oxidize H_2S and form zinc sulfide. The process can handle H_2S up to 100 grains/100 scf at pressures from 75 to 1400 psig. Mercaptan concentrations above 10% of the H_2S concentration can cause problems by forming zinc mercaptides. The resulting sludge can cause foaming.

Source

GPSA Engineering Data Book, Gas Processors Suppliers Association, Vol. II, 10th Ed.

Bed Batch Type

The iron sponge process is very old (introduced in England in the mid-19th century) and very simple. It removes only H_2S and mercaptans. It is good only for streams containing low H_2S concentrations at pressures of 25 to 1200 psig.[1] Hydrated iron oxide containing water and of proper pH is supported on wood chips or other material. Water is injected with the gas.

Regeneration with air can be done with continuous or periodic addition of small amounts of air. Both must be done carefully because of exothermic reaction. Regeneration is never complete, so the beds must be eventually changed out. This must be done carefully because of the pyrophoric (spontaneously combustible) nature of the iron sulfide. The entire bed is wetted first.

Mol sieve processes can be developed to do almost anything desired: remove H_2S, sweeten and dehydrate at the same time, remove CO_2, remove mercaptans, etc. Regeneration is done by switching beds and sending hot gas to the regenerating bed. Two, three, or four towers can be used to allow time for processing, regeneration, cooling, etc. Design is complicated and so it is necessary to contact the manufacturers (the Linde Division of Union Carbide or the Davison Chemical Company Division of W. R. Grace and Co.) when considering an application.

There is a Gas Research Institute (GRI) evaluation of a solid-based scavenger (Sulfa Treat®). Full-scale evaluations have been conducted at a production plant in central Texas. The reference is GRI-95/0161.

Sources

1. *GPSA Engineering Data Book,* Gas Processors Suppliers Association, Vol. II, 10th Ed.
2. Maddox, R. N., *Gas and Liquid Sweetening,* Campbell Petroleum Series, 1977.

13 Vacuum Systems

Vacuum Jets

Jet design is normally handled by the vendor. However, the process engineer must specify the system into which the jets are incorporated. He must also supply the vendor with operating conditions which include

1. Flows of all components to be purged from the system (often air plus water vapor).
2. Temperature and pressure entering the jets and pressure leaving if not atmospheric.
3. Temperature and pressure of steam available to drive the jets.
4. Temperature and quantity of cooling water available for the intercondensers. Also cooling water allowable pressure drop for the intercondensers.

In addition, the process engineer must be aware of good design practices for vacuum jets.

The vendor will convert the component flow data into an "air equivalent." Since jets are rated on air handling ability, he can then build up a system from his standard hardware. The vendor should provide air equivalent capability data with the equipment he supplies. Determination of air equivalent can be done with Equation 1.

$$ER = F\sqrt{0.0345\,(MW)} \qquad (1)$$

where

ER = Entrainment ratio (or air equivalent). It is the ratio of the weight of gas handled to the weight of air which would be handled by the same ejector operating under the same conditions.
MW = Gas mol. wt.
F = 1.00, for MW 1 – 30
F = 1.076 – 0.0026 (MW), for MW 31 – 140

Equation 1 will give results within 2% of the Reference 1 entrainment ratio curve.

The effect of temperature is shown by Equations 2 and 3.

$$ERTA = 1.017 - 0.00024T \qquad (2)$$

$$ERTS = 1.023 - 0.00033T \qquad (3)$$

where

ERTA = The ratio of the weight of air at 70°F to the weight of air at a higher temperature that would be handled by the same ejector operating under the same conditions.
ERTS = Same as above for steam
T = Gas temperature, °F

This information is based on Reference 1.

The vendor should also supply steam consumption data. However, for initial planning the process engineer needs to have an estimate. Use the following equations to calculate the horsepower required to compress noncondensing components from the jet inlet pressure and temperature to the outlet pressure.

$$HP = \frac{WH_{poly}}{E_p 33{,}000}$$

$$HP = \frac{WH_{AD}}{E_A\ 33{,}000}$$

where

HP = Gas horsepower
W = Flow, lb/min
H_{poly} = Polytropic head
H_A = Adiabatic head
E_P = Polytropic efficiency
E_A = Adiabatic efficiency

$$H_{poly} = \frac{ZRT_1}{(N-1)/N}\left[\left(\frac{P_2}{P_1}\right)^{(N-1)/N} - 1\right]$$

$$H_{AD} = \frac{ZRT_1}{(K-1)/K}\left[\left(\frac{P_2}{P_1}\right)^{(K-1)/K} - 1\right]$$

where

Z = Average compressibility factor; using 1.0 will yield conservative results
R = 1,544/mol. wt
T_1 = Suction temperature, °R
P_1, P_2 = Suction, discharge pressures, psia
K = Adiabatic exponent, C_p/C_v
N = Polytropic exponent, $\frac{N-1}{N} = \frac{K-1}{KE_p}$

For process water vapor handled by the jets with intercondensing, calculate horsepower for the first stage only. After the first stage the condenser will bring the system to the same equilibrium as would have occurred without the process water vapor. Use an adiabatic efficiency of 7% for cases with jet intercondensers and 4% for noncondensing cases. Estimate the steam consumption to be the theoretical amount which can deliver the previously calculated total horsepower using the jet system steam inlet and outlet conditions. These ballpark results can be used until vendor data arrive. This procedure will give conservative results for cases with high water vapor compared to the Ludwig[2] curves for steam consumption.

Following are some general rules of thumb for jets:

1. To determine number of stages required, assume 7:1 compression ratio maximum per stage.
2. The supply steam conditions should not be allowed to vary greatly. Pressure below design can lower capacity. Pressure above design usually doesn't increase capacity and can even lower capacity.
3. Use Stellite or other hard surface material in the jet nozzle. For example 316 s/s is insufficient.
4. Always provide a suitable knockout pot ahead of the jets. Water droplets can quickly damage a jet. The steam should enter the pot tangentially. Any condensate leaves through a steam trap at the bottom. It is a good idea to provide a donut baffle near the top to knock back any water creeping up the vessel walls.
5. The jet barometric legs should go in a straight line to the seal tank. A 60°–90° slope from horizontal is best.

Sources

1. *Standards for Steam Jet Ejectors, 3rd Ed.,* Heat Exchange Institute, New York, N.Y.
2. Ludwig, E. E., *Applied Process Design for Chemical and Petrochemical Plants, Vol. I,* Gulf Publishing Co.
3. Jackson, D. H., "Selection and Use of Ejectors," *Chemical Engineering Progress,* 44, 347 (1948).
4. Branan, C. R., *The Process Engineer's Pocket Handbook, Vol. 1,* Gulf Publishing Co., 1976.

Typical Jet Systems

Figure 1 provides a convenient comparison of capacity and suction pressure ranges for typical commercial steam jet systems handling any noncondensable gas such as air. Figure 1 is based on each of the designs using the same

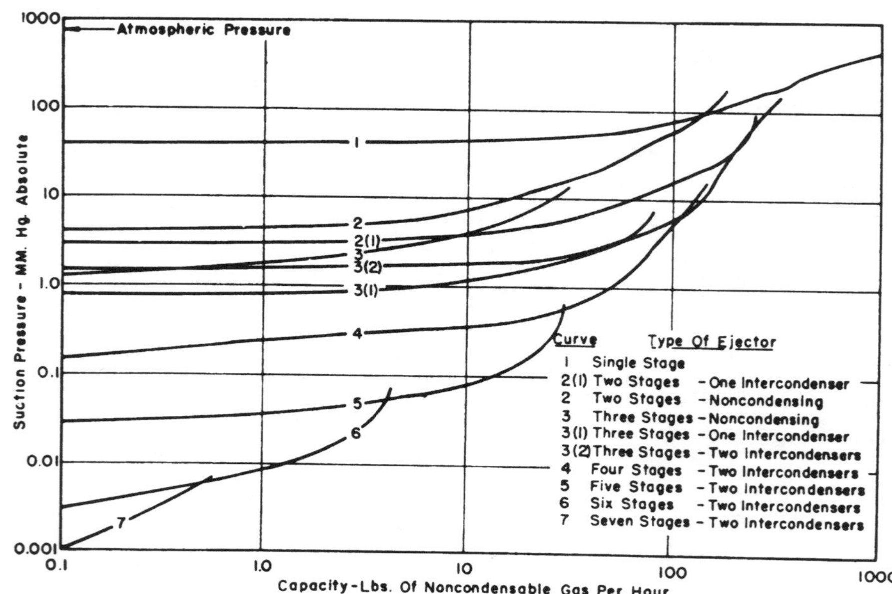

Figure 1. A wide range of pressures can be achieved by using various combinations of ejectors and condensers. The same steam consumption is used for each design here. Note: Curves are based on 85°F condensing water. If warmer water is used, curves shift to the left—cooler water, shift right.

amount of motive steam at 100 psig. Each point on each curve represents a point of maximum efficiency. Therefore, the curves represent a continuum of designs rather than a single design. A single design normally wouldn't operate over the wide ranges shown. For example, a reasonable range for good efficiency would be 50–115% of the design capacity.

Sources

1. Evans, F. L., *Equipment Design Handbook for Refineries and Chemical Plants, Vol. I, 2nd Ed.,* Gulf Publishing Co., 1979.
2. Berkley, F. D., "Ejectors Have Wide Range of Uses," *Petroleum Refiner,* December 1958, p. 95.

Steam Supply

Steam jets require a constant-pressure steam supply for best performance. Pressure below design will usually lower performance and pressure above design usually doesn't increase capacity and can even lower capacity. Lieberman has recommended an intelligent way to handle the problem of variable steam pressure, shown in Figure 1. The steam to the jet is controlled at a pressure slightly below the lowest supply pressure so the jet has a constant steam pressure. Lieberman points out that there is hardly any steam penalty in providing this better operating system. For example, a jet designed for 130 psig steam will use only about 15% more steam than one designed for 150 psig steam.

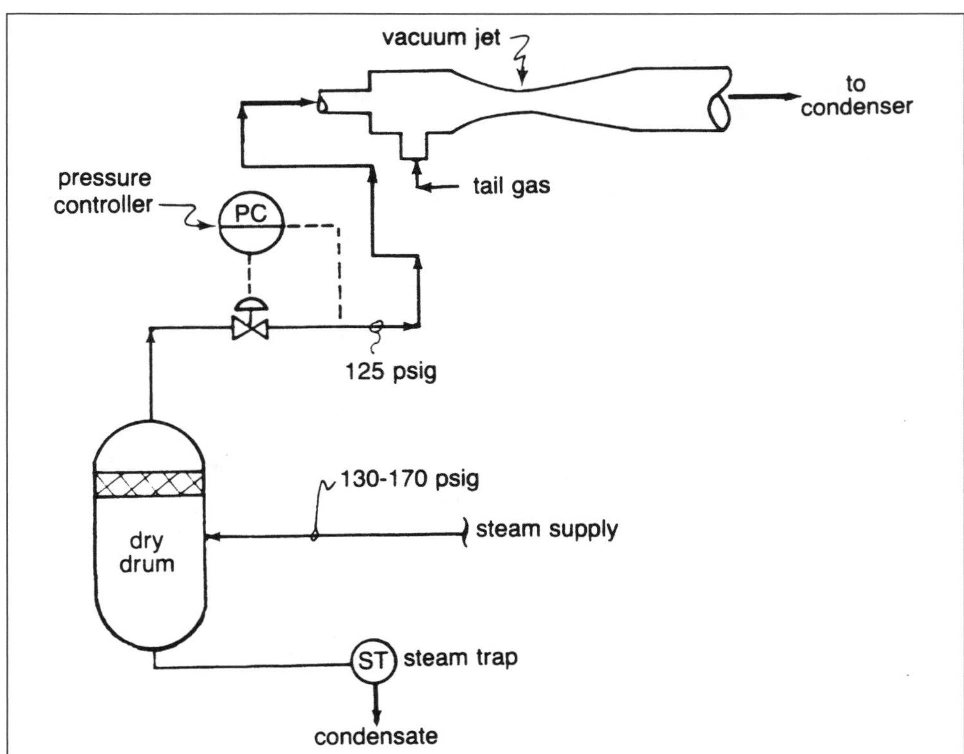

Figure 1. Wet steam of variable pressure will ruin a vacuum steam jet's performance.

Source

Lieberman, N. P., *Process Design For Reliable Operations, 2nd Ed.,* Gulf Publishing Co., 1988.

Measuring Air Leakage

Air is usually the basic load component to an ejector, and the quantities of water vapor and/or condensable vapor are usually directly proportional to the air load. Unfortunately, no reliable method exists for determining precisely the optimum basic air capacity of ejectors. It is desirable to select a capacity which minimizes the total costs of removing the noncondensable gases which accumulate in a process vacuum system. An oversized ejector costs more and uses unnecessarily large quantities of steam and cooling water. If an ejector is undersized, constant monitoring of air leaks is required to avoid costly upsets.

Experience with a similar system is the most reliable guide when sizing an ejector. Air leakage in an existing system having a multistage condensing ejector may be measured by bleeding steam to the first-stage suction through a manual or controlled valve to control the system pressure. Then, observe the pressure at the last stage and refer to the performance curve to estimate the air load, or measure the volume of vent gases from a surface after-condenser.

To determine the amount of air leak in an existing system, estimate the total volume of the system. Operate the ejector to secure a pressure somewhat less than 15 inches Hg abs. Then isolate the ejector from the system. Measure the time required for a rise in pressure in the vessel (say 2 inch Hg). It is essential that the absolute pressure does not rise above 15 inches Hg abs during this time. The following formula will then give the leakage:

$$W_L = 0.15\ V\ (\Delta p)/t$$

where

W_L = Leakage, lb/hr
V = System volume, cu. ft.
Δp = Pressure rise, in. Hg
t = Time, min.

If the volume of the system is not known, the leakage can still be determined, but two tests will be required. First, the test described above must be run. Then a known air leak must be introduced to the system. This can be done by means of a calibrated air orifice. A second test is made, obtaining a new pressure rise and time. The unknown leak is then given by

$$W_l = W'/[\Delta_p{}'t)/(\Delta_p t')]-1$$

where

W' = Known leak, lbs/hr
$\Delta p'$ = Second pressure rise, in. Hg
t' = Second time, min.

If the air leakage is greater than the load for which the jet was designed, the alternatives are to correct the leaks or to use a larger ejector. The ejector must be large enough to handle not only the leaks but the normal load from the process.

For moderately tight, small chemical processing systems (say 500 cu. ft.), an ejector air capacity of 10 lbs/hr is adequate. For large systems, use 20 lbs/hour. For very tight, small systems, an air capacity of 2–5 lbs/hr is reasonable.

The designer should consider the ejector price and its operating costs when he specifies the air capacity. Larger capacities may be economical with single-stage ejectors, for example.

A reverse philosophy in sizing ejectors is occasionally applicable to a system in which even a small quantity of air leakage will upset the operation or contaminate the product. In such a system it may be desirable to install an ejector having a deliberately limited air handling capacity, so that the system cannot be operated until an injurious rate of air leakage is corrected.

Source

Evans, F. L., *Equipment Design Handbook for Refineries and Chemical Plants, Vol. 1, 2nd Ed.*, Gulf Publishing Co., 1979.

Time to Evacuate

To estimate the time required for an ejector to evacuate a system from atmospheric pressure down to the design pressure, assume that the average air handling capacity during the evacuation period is twice the design air handling capacity. Assume also that the actual air leakage into the system during the evacuation is negligible. The approximate evacuation time is

$T_e = 2.3\ V/C_a$

where

T_e = Time to evacuate a system from atmospheric pressure to the design pressure of an injector, min.
V = System volume, vapor space, cu. ft.
C_a = Ejector design air capacity, lbs/hr

If this approximate evacuation period is too long, it may be shortened by adding a larger last stage to the ejector or by adding a noncondensing ejector in parallel with the primary ejector. One noncondensing ejector may be used as an evacuation and spare ejector serving several adjoining systems. The evacuation performance is specified by indicating the system volume, the desired evacuation time, and the absolute pressure to which the system must be evacuated.

Source

Evans, F. L., *Equipment Design Handbook for Refineries and Chemical Plants, Vol. 1, 2nd Ed.,* Gulf Publishing Co., 1979.

Design Recommendations

First, one must estimate air or other gas leakage into the vacuum system. Of course every effort is made to keep it as tight as possible. The author is aware of possible leak points being sealed with polystyrene, which produces an excellent seal. When tests cannot be made, one must use rules of thumb. Many such rough estimating techniques exist.

A close friend with years of experience in designing and operating vacuum systems claims that a system can be made very tight when designed properly. He advocates purging instrument leads with nitrogen. The rate is 1 SCFM maximum per lead. He has designed and successfully operated a number of vacuum systems with provisions for the required instrument purge times a safety factor of about 4. This factor is a contingency for such things as sudden load changes which could bring in fractionator feed that has been more poorly stripped.

Other recommendations are

1. For good control, design the pressure drop for the control valve between the fractionating system and the jet system for sonic velocity (approximately 2:1 pressure ratio). This means that the jets' suction must be designed for half the absolute pressure of the evacuated system.
2. In even a large fractionator system, under the following conditions:
   a. Properly designed condenser
   b. Vacuum tight
   c. Sonic velocity across the control valve
   d. Well stripped feed

   the control valve trim rarely exceeds ¾″ and usually runs ½″.
3. Set the pressure controller for low proportional band.
4. For applications having column temperature control above the feed point, put the measuring elements for the temperature and pressure controllers on the same tray. This will make for good composition control at varying column loads (varying column differential pressure). In vacuum systems a slight pressure change will produce large equilibrium temperature changes.
5. For big vapor lines and condensers (frequent in vacuum systems) always insulate the line, condenser, and top of column. Rain or sudden cold fronts will change column control otherwise. It is possible to have more surface in the overhead line than in the condenser.
6. Never use screwed fittings in any vacuum system, regardless of size.
7. Installation of a pressure controller measurement tap is shown in Figure 1.
8. Avoid liquid traps in vacuum system piping by never going up after having gone horizontal.
9. Put the vacuum system control valves at the highest point of a horizontal run and the control valve bypass in the same horizontal plane. This is in compliance with item 8.

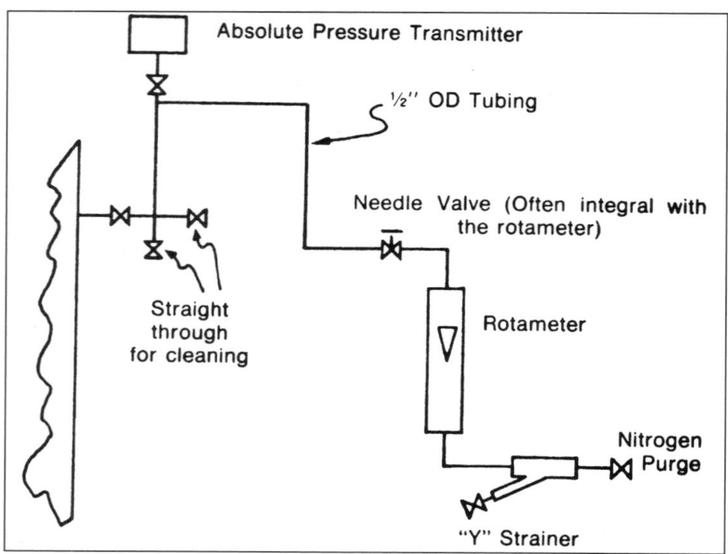

Figure 1. Shown here is a vacuum measurement installation.

Source

Branan, C. R., *The Process Engineer's Pocket Handbook, Vol. 1,* Gulf Publishing Co., 1976.

Ejector Specification Sheet

Here is a specification sheet for steam jet and liquid jet ejectors. It contains data needed by the manufacturer to design and quote an ejector for a specific application.

Steam Jet Ejector

Application ______

Absolute pressure to be maintained by the ejector ______

Fluids to be handled by the ejector ______

Noncondensable gases, #/hr. ______ M.W. ______

Water vapor, #/hr. ______

Condensable vapors other than water vapor:

#/hr. ______ M.W. ______

Temperature of load fluid, F. ______

Steam pressure available at the ejector, psig ______

Steam temperature at the ejector, F. ______

Cost of steam (if known), $/1,000 lbs. ______

Is cooling water available for intercondensing? ______

Cooling water source? Cooling tower ☐

City supply ☐ Well ☐

Cooling water temperature at the ejector, F. ______

Cooling pressure at the ejector, psig ______

Cost of cooling water (if known), $/1,000 gal. ______

Is cooling water relatively clean? ______

Turbidity (if known) ______

If evacuation time to establish the required absolute pressure in the system is important, then state net volume to be evacuated—cu. ft. ______

Time allowed for evacuation, minutes ______

Will system to be evacuated be wet or dry? ______

If wet, state liquid present and quantity to be removed by evaporation.

Altitude at installation if abnormally high ______

Discharge pressure final stage ______

Type of intercondenser preferred?

Surface ☐ Direct contact ☐

Is an aftercondenser required?

Type: Surface ☐ Direct contact ☐

Is economy of operation to be stressed? ______

Is low initial cost of ejector to be stressed? ______

Any limitation on steam supply available that might affect ejector design? ______

Any limitation on water supply available that might affect ejector design? ______

Is installation to be inside or outside? ______

Is A.S.M.E. Code construction required on inter and aftercondensers? ______

Are there any other construction codes required? ______

Is there any special painting required? ______

Is there any special orientation of components to be considered? ______

Is there any space limitation to be considered? ________
Is there any weight limitation to be considered? ________
Size of suction connection from system to ejector. ______
Is there any special supporting structure to be furnished? ________
Accessory equipment to be furnished with ejector: ____
Steam valves, strainers, and interconnecting steam piping ________
Condensate pumps ________
Discharge head required ________
Electric power available: cycles ________
volts ________ phase ________
Electric motors to be: Drip proof □ Splash proof □ TEFC □ Explosion proof □
Steam purifier ________
Liquid level controls for intercondenser hotwell ______
Traps ________
Spare parts ________
Other equipment ________
Material: Diffuser and suction chamber:
Standard ________ Special ________
Steam nozzle: Standard ________ Special ________
Inter and after condenser-shell:
Standard ________ Special ________
(If shell & tube) Tubes: Adm. ______ Special ______
(If shell & tube) Tube sheet: Steel ____ Special ______
Standard materials are chosen to resist ordinary corrosion of steam and water. If corrosive vapors or liquids are involved, state under special material customer has found best for his operation. Is export packing required?

Water or Liquid Jet Ejector

Suction pressure to be maintained ________
Discharge pressure ________
Total load fluid #/hr. ________ Temp. F. ________
Load fluid: Liquid #/hr. ________ M.W. ________
Non-condensables #/hr. ________ M.W. ________
Condensable vapor #/hr. ________ M.W. ________
Propelling liquid ________
Propelling liquid pressure, psig ________
Propelling liquid temperature (Maximum), F ________
Properties of propelling liquid (if other than water) ______
____
Specific gravity ________
Specific heat, Btu/#—F. ________
Vapor pressure at inlet, temperature, psia ________
Materials of Construction
Suction chamber ________ Diffuser ________
Steam nozzle ________ Steam chest ________
Accessories
Propelling liquid valve ________
Propelling liquid strainer ________
Other equipment (if any) ________

Source

Evans, F. L., *Equipment Design Handbook for Refineries and Chemical Plants, Vol. 1, 2nd Ed.*, Gulf Publishing Co., 1979.

14
Pneumatic Conveying

Types of Systems

This chapter considers conveying solids with air. Materials conveyed with air vary greatly in properties, often requiring special considerations. This chapter will, therefore, give only means for preliminary rough sizing and selection of equipment. Often the entire pneumatic conveying system for a plant is bought as a package from a vendor specializing in such equipment. Even then it is helpful for the process engineer to be familiar with rough equipment sizing methods. Such knowledge can help him plan such things as plot area and utility draws in the initial design phase. The knowledge may also be helpful in later startup or troubleshooting. The methods described are primarily those of References 1, 2, and 3.

The following are the general types of systems and their uses:

1. Negative (vacuum) system; normally used when conveying from several pickup points to one discharge point.
2. Positive-pressure system; normally used when conveying from one pickup point to several discharge points.
3. Pressure-negative (push-pull) combination system; normally used when conveying from several pickup points to several discharge points.
4. Venturi, product systems, or blow tanks will not be discussed here.

The final choice is determined by economics or some special material demand.

Negative System

The negative system usually sucks on a cyclone and/or filter/receiver mounted above the receiving storage hopper or bin. Solids are usually sent down to the hopper with a rotary air lock feeder. Air is sucked into the transfer pipe or duct at the pickup end of the system. A variety of feeders can be used to introduce solids into the flowing air stream such as rotary air lock feeders, pan-type manifolds under rail car hoppers, paddle-type rail car unloaders, screw conveyors, etc.

Positive Pressure System

In the positive pressure system air is blown into the pickup duct often up to a cyclone with atmospheric vent. Usually solids are introduced to the conveying air stream with a rotary air lock feeder. In sending the solids down to the storage bin from the cyclone a simple spout connection can be used instead of the rotary air lock feeder required in the negative pressure system. The positive system doesn't need a vacuum vessel at each receiving location. Also, in the positive pressure system conveying to a number of hoppers, a simple bag-type cloth can serve as the filter. So, for conveying from one pickup location to several receiving locations, the positive pressure system is often cheaper than the negative system.

Pressure-Negative System

The pressure-negative system is ideal for unloading rail cars, and also for conveying light dusty materials, because the suction at the inlet aids the product in entering the conveying line. Positive systems are poorer than negative for handling such materials since feeding into the pressure line can be difficult, and can present a dust collection problem at the pickup location because of blowback or leakage of air through the rotary valve. Also, the author has noted that the negative system works better in instances when there are lumps at the pickup end. The positive pressure system tends to pack at the lump while the negative system will often keep solids moving around the lump and gradually wear it away.

For the pressure-negative system, a single blower can be used for both the negative and positive sides of the system. However, the lack of flexibility with a single blower usually dictates the need for separate blowers for the negative and positive sides.

Sources

1. Fisher, John, "Practical Pneumatic Conveyer Design," *Chemical Engineering,* June 2, 1958.
2. Gerchow, Frank J., "How to Select A Pneumatic Conveying System," *Chemical Engineering,* February 17, 1975.
3. Perkins, Don E., Wood, Jim E., "Design and Select Pneumatic Conveying Systems," *Hydrocarbon Processing,* March 1974.

Differential Pressures

If the system differential pressure requirements are low, sometimes a fan can be used. A fan is limited to a maximum of about 65 inches of water (just over 2 psi) in vacuum service or 77 inches of water in pressure service (just under 3 psi). To estimate fan horsepower use the equations found in Chapter 6 in the section entitled Horsepower Calculation. In lieu of manufacturer's data use an adiabatic efficiency of 50% for initial work.

For higher differential pressures (or lower differential pressures where it is preferred not to use a fan), rotary positive-displacement blowers are used. These are excellent for conveying systems since they provide

1. Reasonably constant volume at variable pressure discharge.
2. Vacuums of about 8 psi (some can go to 11 psi with water injection) and pressure differentials of 15 psi.
3. A low slippage with improved efficiency.

The higher pressure differentials allow longer and sometimes smaller lines than for fans. To estimate blower horsepower use the equations found in Chapter 6 in the section entitled Horsepower Calculation. In lieu of manufacturer's data use an adiabatic efficiency of 80% for initial work.

Even though 15 psi is possible from a blower, most positive-pressure systems are limited to 10–12 psi differential pressure because of limits of the rotary air lock feeder valves (deflections in shaft and bearings and increased blowback air).

Sources

1. Fisher, John, "Practical Pneumatic Conveyer Design," *Chemical Engineering,* June 2, 1958.
2. Gerchow, Frank J., "How to Select A Pneumatic Conveying System," *Chemical Engineering,* February 17, 1975.
3. Perkins, Don E. and Wood, Jim E., "Design and Select Pneumatic Conveying Systems," *Hydrocarbon Processing,* March 1974.

Equipment Sizing

To rough out line sizes and pressure drop for fan or blower sizing, use the following quickie method:

1. Arbitrarily assume air velocity of 5,000 ft/min (good for 90% of conveying situations).
2. Use Table 1.
3. Calculate pressure drop. This is in two parts:

   a. Material losses

      E_1 = Acceleration losses
      E_2 = Lifting energy
      E_3 = Horizontal losses
      E_4 = Bends and elbows

   b. Air losses

4. First do material losses in ft-lb/min

   $E_1 = MU^2/2g = 108M$ @ 5,000 ft/min
   $E_2 = M(H)$
   $E_3 = M(L)(F)$
   $E_4 = MU^2/gR\ (L)\ (F)\ (N) = 342\ (M)\ (F)\ (N)$ for 48″ radius 90° ell. Assume this to be the case.

where

M = Solids conveyed, lb/min
U = Velocity, ft/min
$2g = 2.32 \times 10^5$ ft/min^2
H = Vertical lift, ft
L = Duct horizontal length, ft
R = 90° ell radius, ft
F = Coefficient of friction and tangent of solids "angle of slide" or "angle of repose." Use 0.8 in lieu of solids data for initial estimating
N = Number of 90° ells. For 45°, 30° etc., express as equivalent 90° ells by direct ratio (Example: A 30° ell is 0.33 of a 90° ell)

5. From Table 1 and solids rate, estimate duct size and flow in SCFM (standard cubic feet per minute).
6. Express material losses in inches of water:

$$\frac{\text{ft} - \text{lb} / \text{min}}{\text{ft}^3 / \text{min} \times 5.2} = \text{in. } H_2O$$

7. Calculate air losses.
   a. Calculate equivalent length of straight pipe by adding to actual length of straight pipe an allowance for conveying type 90° ells of 1 ft of pipe/in. of diameter. (Example: 4″ 90° ell = 4 ft of pipe)
   b. Assume the following losses for other items in inches of water:

Duct entry loss	1.9
Y branch	0.3
Cyclone	3.0
Collector vessel	3.0
Filter	6.0

8. Add material and air losses.
9. Calculate fan or blower horsepower as explained earlier.
10. Be sure to use the atmospheric suction pressure at the site. Normal blower ratings are given at sea level.

Table 1
Capacity Range

Duct Dia. in.	Flow SCFM at 5,000 ft/min	Friction Loss, in. H_2O/100 ft	Usual Capacity Thousands of lbs/hr Negative	Positive
4	440	11.0	2–6	12–40
5	680	8.0	3–10	15–60
6	980	6.3	4–15	20–80
8	1,800	4.5	15–30	30–160

Sources

1. Fisher, John, "Practical Pneumatic Conveyer Design," *Chemical Engineering,* June 2, 1958.
2. Gerchow, Frank J., "How to Select A Pneumatic Conveying System," *Chemical Engineering,* February 17, 1975.
3. Perkins, Don E., Wood, Jim E., "Design and Select Pneumatic Conveying Systems," *Hydrocarbon Processing,* March 1974.

15
Blending

Single-Stage Mixers

Portable or fixed mixers up to 5 hp normally use propellers and run at either direct drive speeds of 1,150 or 1,750 rpm or at single reduction gear drive speeds between 300 and 420 rpm. They may either be clamped on the rim of open tanks or mounted with a fixed assembly for open or closed tank operation. These mixers are the most economical, and are usually used in tanks without baffles. They are rugged and long-lasting.

Side entering mixers are used for blending purposes. The side entering propeller type mixer is economical and establishes an effective flow pattern in almost any size tank. Because the shaft seal is below the liquid level, its use in fluids without corrosive and erosive properties is usually ideal.

Top entering mixers are heavy duty equipment. They are usually fixed to a rigid structure or tank mounting. Either radial flow or axial flow turbines may be used. Speeds vary from 50 to 100 rpm and usually require a double set of helical gearing or a single set of worm gears to achieve these low speeds. Therefore, they are more expensive than single reduction mixers.

Slow speed close-clearance impellers are used when mixing high viscosity materials. Helical or anchor type close-clearance impellers are used in this application at speeds from 5 to 20 rpm. Table 1 compares the power required and cost for conventional axial flow turbines and the helical type.

Table 1
Axial Flow Turbine Compared to Helical Impeller

Impeller Type	Horsepower (Op. cost & mix. heat added)	N rpm	Torque	Initial cost
Axial flow turbine	1	1	1	1
Helical impeller	¼	⅛	2	3

D/T = 0.5
½-inch radial clearance
Blend time—equal
Heat transfer coefficient—equal

A lineblender is a mixer placed directly in process piping when mixing times of several seconds are required. Agitation in a lineblender is sufficient to disrupt the flow pattern through the pipe so that one or two stages of mixing are accomplished. Designers must pay particular attention to the pressure drop through this type of mixer when selecting pumps for the piping system.

Source

Evans, F. L., *Equipment Design Handbook For Refineries and Chemical Plants, Vol. 1, 2nd Ed.,* Gulf Publishing Co., 1979.

Multistage Mixers

These mixers are specified when one liquid must be dissolved in another, a solid and a liquid must be mixed, a high viscosity liquid must be reacted, a light liquid must be extracted from a mixture of heavy and light liquids, or when gas must be absorbed in a liquid. To select the proper mixer, certain fluid properties must be known:

1. Required—specific gravity of components and the mixture.
2. Fluid viscosity—For Newtonian fluids (a constant viscosity at all impeller speeds) approximate viscosities up to 5,000 centipoises are satisfactory. Above 5,000 centipoises, estimating errors of 20 to 50% can mean undersizing or oversizing the agitator.

   For non-Newtonian fluids, viscosity data are very important. Every impeller has an average fluid shear rate related to speed. For example, for a flat blade turbine impeller, the average impeller zone fluid shear rate is 11 times the operating speed. The most exact method to obtain the viscosity is by using a standard mixing tank and impeller as a viscosimeter. By measuring the power response on a small scale mixer, the viscosity at shear rates similar to that in the full scale unit is obtained.
3. The phase to be dispersed must be known.
4. For solid-liquid systems, settling velocities of the 10, 50, and 90 percent by weight fractions of particle size distribution must be available from calculations or measurements.

5. With gases, flow rates must be available at standard temperature and pressure as well as actual temperature and pressure. The range of gas flow must be given, as well as whether the mixer is to be operated at full horsepower for all gas ranges or operated with the gas on.

Source

Evans, F. L., *Equipment Design Handbook For Refineries and Chemical Plants, Vol.1 , 2nd Ed.,* Gulf Publishing Co., 1979.

Gas/Liquid Contacting

Dispersing gas in liquid involves absorption or desorption of the gas into or out of the liquid, plus a chemical reaction. A key element is the appearance of the gas/liquid dispersion because it indicates the type of splashing or foaming that will occur on the surface and the volume of gas that will be contained in the mixture. As the gas expands going through the vessel, it gives up energy. This energy can be calculated, and is related to the superficial gas velocity.

If the gas is to be intimately dispersed in the liquid, the input power to the mixer will be several times the gas power level for the gas. In other words, the mixer power level must be compared to the gas power level to be certain that the required degree of mixing is achieved.

For gas/liquid mass transfer, a data point is taken at a particular power level and gas rate. This point is obtained from a similar application or from laboratory or pilot plant data.

The gas/liquid process may be batch liquid, continuous liquid, or continuous multistage liquid. If the process is multistage liquid flow, the gas flow must be specified as cocurrent or countercurrent.

Source

Evans, F. L., *Equipment Design Handbook for Refineries and Chemical Plants, Vol. 1, 2nd Ed.,* Gulf Publishing Co., 1979.

Liquid/Liquid Mixing

An example of liquid/liquid mixing is emulsion polymerization, where droplet size can be the most important parameter influencing product quality. Particle size is determined by impeller tip speed. If coalescence is prevented and the system stability is satisfactory, this will determine the ultimate particle size. However, if the dispersion being produced in the mixer is used as an intermediate step to carry out a liquid/liquid extraction and the emulsion must be settled out again, a dynamic dispersion is produced. Maximum shear stress by the impeller then determines the average shear rate and the overall average particle size in the mixer.

For liquid/liquid extraction, data on mass transfer rate of the system at typical operating conditions are required. Also required are an applicable liquid/liquid equilibrium curve and data on chemical reactions occurring after mass transfer in the mixer.

Source

Evans, F. L., *Equipment Design Handbook For Refineries and Chemical Plants, Vol. 1, 2nd Ed.,* Gulf Publishing Co., 1979.

Liquid/Solid Mixing

Suspended solids are often the process objective that requires a specific degree of uniformity. Five guides to better liquid/solid mixing are:

1. Complete uniformity implies 100% uniformity of the suspended solid in the liquid. The upper layer of liquid in the mixer tank is very difficult to bring to 100% suspension. The problem is getting particles with settling velocities above 6 ft per minute suspended uniformly in the upper part of the tank because the primarily horizontal flow pattern cannot keep high settling velocity solids in suspension.
2. Complete off-bottom suspension means moving all particles off the tank bottom.

3. Complete motion on the tank bottom refers to all particles suspended off the bottom or rolling on the bottom.
4. A fillet is a stagnant deposit of solids commonly found at the outside periphery of the bottom where it joins the tank. However, it could occur anywhere in the tank where the flow pattern is stagnant. Filleting is permitted, but no progressive fillet buildup is allowed. The reason is that it costs less to allow some solids to settle into fillets than to provide additional mixer horsepower to eliminate fillets.
5. Height of suspension is the height to which solids are suspended in the tank. It is commonly expressed as the percent of solids of each particle size fractions at various liquid heights off bottom, or as the particle size distribution in samples taken from various points.

In continuous flow operations, a slurry is continuously added and withdrawn from a single-stage mixer. Progressive buildup must not occur in fillets, and the discharge composition to the mixer must equal the inlet composition. Discharge from the tank occurs at a particular draw-off point, the only point in the tank that requires the mixture to have the same composition as the inlet.

If there is sufficient agitation for uniform particle size in each stage, residence time of the solids will be the same as the average retention time of the total mixture passing through the tank. If there is not a complete uniformity of particles in the mixture, the average concentration of solids within the tank will be lower than the concentration in the feed and discharge. Therefore, the solids residence time will be less than the calculated average retention time.

The larger the ratio of impeller diameter to tank diameter, the less mixer power required. Large, slow speed impellers require a lower horsepower for a given pumping capacity, and solid suspension is governed by the circulation rate in the tank.

It might appear that the most economical mixer would have a large impeller and low mixer horsepower. However, the torque required to turn large impellers is often greater than for smaller impellers, even though the power required for larger impellers is less. In other words, the mixer cost is governed largely by the torque required to drive the mixer.

Source

Evans, F. L., *Equipment Design For Refineries and Chemical Plants, Vol. 1, 2nd Ed.,* Gulf Publishing, 1979.

Mixer Applications

Applications for mixers in blending operations vary from small laboratory mixers to large petroleum tanks of several million gallons. Three common mixing operations are:

1. Blend in similar viscosity materials to a specified blend point.
2. Blend in low and high viscosity materials to a specified blend point.
3. Promote temperature uniformity by controlling the length of time a reaction mass is away from a cooling or heating surface.

When two-phase mass transfer is required to supply reactants by mixing for a chemical reaction, the most important factor to consider is whether the mass transfer controls the operation or whether the chemical reaction controls it. This can be done by increasing the mixer speed to a point where mass transfer effects become very high and the operation is limited by the chemical reaction.

Only data on thermal properties of the fluid are necessary to calculate mixer side coefficients in most systems. For non-Newtonian fluids, the heat transfer surface must be estimated and a suitable viscosity obtained. As the Reynolds number decreases, the transition area is approached. In going to close-clearance impellers, the mechanism of heat transfer changes from forced convection to conduction through a stagnant film. Measurements indicate that the heat transfer coefficient is given by conduction through a film having a thickness of one-half the radial clearance. For one-half inch radial clearance in an organic fluid, a coefficient of 4 to 5 Btu/hr °F ft^2 is obtained regardless of impeller speed and fluid viscosity.

If a scraper is added, it will double or triple a coefficient while also doubling or tripling the impeller's power consumption.

Source

Evans, F. L., *Equipment Design For Refineries and Chemical Plants, Vol. 1, 2nd Ed.,* Gulf Publishing Co., 1979.

Shrouded Blending Nozzle

This section discusses a simple low cost tank blending method. It is referred to as a shrouded blending nozzle system, and it works quite well. The shroud causes the jet nozzle to educt a large quantity of surrounding fluid, improving blending. A simplified diagram of the nozzle portion is shown in Figure 1.

Make the nozzle point upward at an angle such that a straight line projected from the nozzle would hit the liquid surface ½ to ⅔ of the way across the tank diameter. The idea is to promote top to bottom turnover. However, tilt the nozzle slightly to the left to promote a slight swirl effect. Aim the nozzle at a point about ⅓ of a radius off-center.

The following rules of thumb apply primarily to the situation where a final tank requires blending after, and perhaps during, transfer from rundown tanks.

1. Base tank size 10,000 barrels.
2. Use about 25 hp circulating pump.
3. Provide roughly 50–75 feet of head or a little higher.
4. For size other than 10,000 barrels, ratio directly for horsepower.
5. Circulation rate in GPM can be calculated from horsepower and head.

Often in refrigerated storage, light hydrocarbons are involved that have a large gravity change with temperature. This condition makes such storage relatively easy to stratify. Be sure not to put in too large a blending pump for refrigerated storage. Not only does it waste blending pump horsepower, but extra heat is added to the fluid which must be removed by the refrigeration system (a double penalty). Operators often tend to leave blending systems running full time.

Provide a means of filling the tank to operating level prior to operating the blending nozzle system. Damage could result from discharging the high velocity jet into an empty tank.

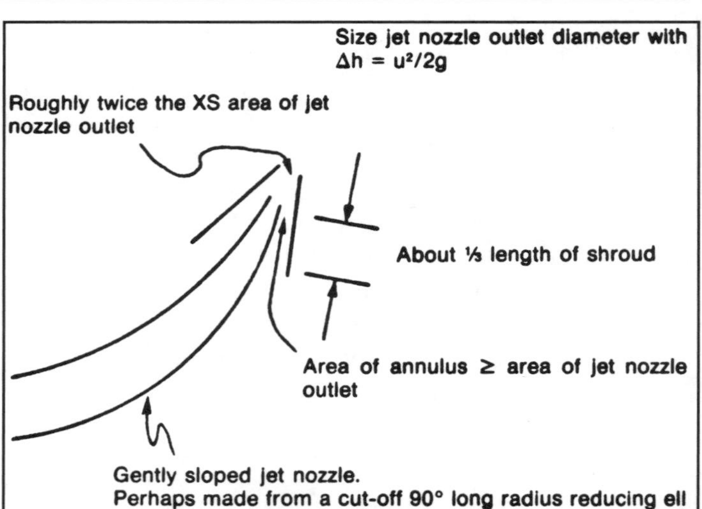

Figure 1. Blending can be improved by a shrouded blending nozzle system because the jet nozzle educts a large quantity of surrounding fluid.

When circulating the blending system and running down into the tank at the same time, it may be possible to direct the rundown stream into the circulating pump suction for additional blending in the pump.

Nomenclature

Δh = Head loss in feet of flowing fluid
u = Velocity in ft/sec
g = 32.2 ft/sec^2

Source

Branan, C. R., *The Process Engineer's Pocket Handbook, Vol. 1,* Gulf Publishing Co., 1976.

SECTION THREE

Plant Design

16 Process Evaluation

Introduction

One of the most important activities within a company is project screening. One losing venture can negate the gain of several winners. Therefore, the project evaluators must be careful to obtain the best and most complete information possible so that the economics will accurately model the future "real world" situation. Possible model inaccuracies must be covered by risk analysis. This work is difficult and demanding, but the people who engage in it can take credit for helping to keep the company's "bottom line" intact and the nation's economy healthy. In this business, the bottom line *is* the real world.

Study Definition

In problem solving and process studies, the front-end work is important. Proper definition and direction developed at the beginning will assure meeting corporate goals in a timely fashion. This section establishes the type of study example to be used for illustrating principles and philosophy, and presents the basic items necessary to begin such a study.

Presentation and Evaluation Aspects

Discussions in this section center around studies involving a proposed new plant or major addition to an existing plant. Practical applications address situations where a commercially proven or extensively tested process is offered as a relatively detailed design package by a licensing organization to an operating company for evaluation. Many of the principles developed apply to other situations, such as evaluation of fundamental processes in the research or semiworks stages.

Following development of the study direction, the evaluation describes the efforts of obtaining and validating process information, and then discusses equipment specifications and a cost estimate of the feasibility or budget type; i.e., with plant costs factored from major material. Finally, project economics and financing complete the evaluation.

Some fundamental questions should be answered prior to launching into the study details.

Who Should Do the Evaluation?

Usually, the evaluation is done by in-house company personnel, a contractor, or a combination. For capital cost estimation it is difficult to beat a contractor. He will, of necessity, have an excellent cost-estimation group that is familiar with a wide range of processes and is on a first-name basis with many equipment vendors, who provide quotes.

Equally important, the contractor's cost-estimation team will have established a sharply honed, systematic approach including all the correct checklists, forms, and report formats designed to preclude oversights. The team will have developed the discipline, as a result of competitive bidding, to produce the kind of final report that fits in correctly with any type of economic evaluation.

Contractors are also effective in the process definition stage. Such things as checking the licensor's package for completeness, firming up utility balances, and validation of input data are their bread and butter. An in-house engineering team would work well alongside a contractor for this evaluation stage. Similar comments could be made for developing equipment final specifications.

Operating costs are best done in-house, because company personnel are familiar with corporate philosophies of staffing, maintenance, control laboratory operations, administrative requirements, and many other support aspects of running the business. If adequately staffed, the in-house study group should handle operating costs rather than try to teach a contractor company requirements.

Likewise, *project economics* and *financing studies* should be done in-house, and the reasons are even stronger in this case than for operating costs. Each company, whether using private enterprise or regulated utility-type economics, has small distinctions within its detailed methods of economic analysis. It is important that the corporation maintain consistency among all its evaluation studies. Also, because confidentiality must be maintained in economics/financing, this is definitely not the place to bring in an outsider for help.

In summary, the in-house personnel should be assigned certain parts of the study, and help can be sought in the other parts if in-house staffing or expertise needs support.

What Should the Plant Size Be?

This parameter must be decided before proceeding into the evaluation. Popular commercial processes have established sizes, as shown in Reference 1, however, a company might opt for a size change, such as desiring two trains instead of the standard single train olefin plant. Bucking a standard trend must be evaluated carefully. Often, compressors set the maximum or standard plant size; vendors have developed standard packages for full-sized olefin plants; or half-sized trains may require development and greater expense.

The total plant, whether single or multiple trains, should normally be of the established commercial maximum size. Otherwise, without maximum economy of scale, it will be difficult to compete in the marketplace.

Sometimes, processes lend themselves to several limited-sized trains of equipment. Two prime examples are coal gasification and LNG plants. A Lurgi coal gasification plant train is limited by gasifier size. The maximum size has been increased significantly with the Mark V, 15-ft-diameter gasifiers first installed in 1980 (Reference 5). Modern mixed refrigerant LNG plant trains are typically limited by main exchanger size. In 1968, at Skikda, Algeria, Technip of France employed axial compressors for the first time in hydrocarbon service and at a power rating of 110,000 bhp (Reference 2). For this design, the compressor set the train size limit.

For a fundamentally new process, plant size is a variable requiring extensive study outside the scope of this handbook. Reference 4 addresses determination of such a project's "core of viability."

What Side Studies Are Required?

Side studies are often required for major decisions, and a good example is a study of gas versus steam turbines for LNG plants (Reference 3). Indications are that higher capital investment is associated with gas turbine drivers than with steam turbines for liquefaction train service, but that the added capital investment for gas turbines shows good payout in improved feedstock utilization if natural gas cost is high. Conversely, gas turbines generate lower capital investment in the utility area.

Another example is a study of coal fired boilers versus low-BTU gas turbines for power generation in a new coal gasification plant. Such a study is complicated and the decision reached might be irreversible since environmental considerations are involved. Smaller scale side studies might involve such things as packed versus trayed fractionation towers, sparing philosophy, air versus water cooling, or package versus field-erected boilers. In any event, required major side studies should be decided early.

Who Should Do the Side Studies?

Side studies should be done by in-house company personnel if possible. If communications are not perfect, it is easy for a contractor to provide a study on the wrong basis. Company design and operating and economic philosophy often are not fully communicated to the contractor with resulting waste of time. In any event, as discussed previously, in-house personnel must do economics. Therefore, use of a contractor is limited to the same portions of the side study as for the overall evaluation.

References

1. Corrigan, T. E., and Dean, M. J., "Determining Optimum Plant Size," *Chemical Engineering,* August 14, 1967, p. 51.
2. Dolle, J., and Gilbourne, D., "Startup of the Skikda LNG Plant," *Chemical Engineering Progress,* January 1976, p. 39.
3. DiNapoli, R. N., "LNG Costs Reflect Changes in Economy and Technology," *Oil and Gas Journal,* April 4, 1983, p. 138.
4. Hackney, J. W., "How to Explore the Profitability of Process Projects," *Chemical Engineering,* April 18, 1983, p. 99.
5. Schad, M. K., and Hafke, C. F., "Recent Developments in Coal Gasification," *Chemical Engineering Progress,* May 1983, p. 45.
6. Branan, C. and Mills, J., *The Process Engineer's Pocket Handbook, Vol. 3.* Houston, Texas: Gulf Publishing Co., 1984.

Process Definition

In this study phase, we determine exactly what the licensor is proposing, validate his claims, and do major side studies for decisions on process changes or additions. The necessity to complete this phase prior to detailed equipment specification is self-evident.

Establishing Proposal Details

The gap must be bridged between the licensor's standard proposal package and the operating company's specific needs. The operating company must also make sure, however, that the basics are included in the standard package. The licensor may have already included site specific items with his standard package, but loose ends will often remain.

Closed Heat and Material Balance. One would suppose that this most basic item would automatically be included in the licensor's package. Indeed, a heat and material balance of sorts is usually included. The operating company must make sure that the balances are meticulously done for overall stream flows and by component. Inconsistent or haphazard roundoff can give ridiculous results for streams whose flows are obtained by difference. Such streams can be small and appear insignificant, but are of great importance in determining such things as waste disposal or air pollution control. An example would be a side stream of gas liquor from a coal gasification plant.

The operating company must underwrite the emissions associated with the plant through environmental impact reporting. Such accounting has become an important part of the design. Rather than pass over the need for a closed heat and material balance at the study stage of a project, it is better to get this job done as early as possible. Persistence is sometimes required.

Utility Balances. The operating company should also require a balance for each plant utility. The most involved of the utility balances is usually the supply/demand steam tabulation showing all levels of steam and condensate and their interactions. The *steam balance* is almost always required at this stage for any required side studies. The steam balance influences many design parameters, such as boiler size and contingency, treated water makeup rates, blowdown disposal rates, chemicals usage, and surface condenser size.

An *electrical balance* is also important as it affects many plant areas. Often, onsite generation versus purchased power is studied. Even with total onsite generation, there is often a community power backup. This must be factored into the cost estimate, especially if the plant has to share costs for major new community generation or feeder equipment. Geographic location of electrical loads will influence location of subs and control centers and, thus, costs.

Balances or summaries for the other utilities, such as cooling water and fuel gas, are also needed at this stage.

Preliminary Process Flowsheet. This will show major equipment and lines, preliminary equipment details (vessel diameter, number of trays, pump flow and driver horsepower, etc.), major instrumentation, and, it is hoped, have a material balance at the bottom of each drawing with flows keyed to a numbering system on the diagram. The process flowsheets should cover both the "process" and "utility" sides of the plant.

Process Specification Sheets. Typical process specification sheets for major material are contained in the Appendix. An initial completed set of these from the licensor will be invaluable for any side studies. For heat exchangers in particular, the specification sheet becomes the base case against which a vast theory of ratio relationships can be applied for minor design excursions. For example, with shell and tube heat exchangers, the shell and tube side film coefficients vary with the 0.6 and 0.8 powers of flow rates, respectively. Reference 1 discusses the extensive use of such heat exchange ratios for shell and tube heat exchanger design. Reference 1 also has handy charts for cooling tower rerate. It must be clear when specification sheets are discussed at this study stage that the process type is being referred to rather than types relating strictly to mechanical and metallurgical integrity.

In addition to being useful for side studies, the process specification sheets can be used as a file to capture the latest thinking for each individual unit of major material for cost estimation and final design. Vendor quotes are expedited if data is provided to them on standard-type process specification sheets.

Site Conditions. If the licensor's package is at the custom design stage, it is important to review the site conditions to be used in the calculations. Weather data is available for most proposed sites or at least for areas not far away. Decisions are necessary as to how to apply the weather data; for example, whether to use 95% or 99% temperature extremes (95% is typical), or 50 or 100-year snowfall extremes. For remote areas, supplementary weather data collection may be specified, including certain temporarily installed equipment. Other environmental data collection may also be necessary. These activities may have costs large enough for inclusion in feasibility cost estimates.

Seismic zone basis must be specified for structural design. Soil data is important, especially for cases where extensive use of foundation piling is required with major cost impact. Availability of aggregate or natural pond stabilization materials near the site will not be considered for early cost estimates, but can be kept in mind for future planning if the project is given the green light.

Critical Analytical Specifications. The material balance will show percentage quantities in process streams. Many critical specifications will be in parts per million or smaller. For an LNG plant, feed gas treating removes H_2S and CO_2 and drying with mol sieves removes water down to low levels; otherwise, plugging would occur in the colder parts of the process. Without the initial feed conditioning, the process would quickly shut down and expensive equipment damage would occur, such as ruptured tubes in locations difficult to repair. Even slight upsets in the initial feed conditioning section demand prompt action by operators to save the train.

In Lurgi coal gasification, an example of extremely important treating is in the sulfur removal step ahead of methanation where the catalyst is poisoned by even small traces of any sulfur compound. The sulfur removal step is a relatively high capital and operating cost item.

It is of paramount importance to make sure that the operating company has a full understanding of all analytical product and raw materials specifications within the plant at the earliest study stage. Such specifications have major impact on capital and operating costs.

Plant Service Areas. A major portion of the plant is lumped under this heading for this discussion. The items previously discussed are so basic that they should be included in the licensor's proposal at every stage of his design. The items in the plant service areas category might not be included in the licensor's proposal in his early design stages, but they still must be defined prior to cost estimation. This implies considerable work for the operating company, but work that must be done. This section is not intended to be a complete checklist (such as the lists in Reference 2 for cost estimation and economics), but rather a discussion of some of the more important plant service subgroups.

Air Pollution and Waste Disposal. Some provisions should be included in the licensor's package for air pollution and waste disposal; even foreign licensors understand that U.S. rules are strict. However, the operating company will have considerable checking to do in this area. State or local rules are often more strict than national rules, and local conditions can portend considerable expense with problems such as high existing emissions, geographical smog influences, or nearby community aquifers. An example of a major waste disposal expense is phenol removal from coal gasification gas liquor. This process requires a sizable and expensive plant area.

In-house or outside environmental experts should be brought into the study at an early stage. By having these people in "on the ground floor," feasibility analysis, cost estimates, and environmental impact studies will be more accurate with fewer surprises later in the project.

Secondary Systems. It is easy for secondary systems to become orphans even well into the final design. Such things as plant air, instrument air, nitrogen, closed cooling water, pump gland cooling, service water, steam tracing, and hydraulic oil are secondary systems, and the list can go on almost forever. Such systems are important to the operation of the plant and extend into many plant areas. Since people often tend to take such systems for granted, it is difficult to generate the kind of interest that these areas deserve until well into the project.

Since the studies under discussion have cost estimates factored from major material, secondary systems at this stage have relatively smaller impact than later when more definitive cost estimates are done. At this stage, it is well to gain a feel for the completeness of the licensor's design with respect to secondary systems.

Contractors often tend to purchase certain secondary systems as complete packages, such as an instrument air system complete with compressors, receivers, driers and instrumentation, or a nitrogen plant. While this may be cost effective, problems can be created for the operating company, such as having the package secondary system shown simply as a square on the drawings without details and having no process specification sheets for the major material within the purchased package. Both of these problems hamper cost comparison among vendors and troubleshooting once the plant is built.

Another problem often noticed once the plant is built is incompatibility of materials (heat exchangers, tubing fittings, pipe valves and fittings, etc.) within the purchased secondary systems package with materials in the main plant. Quite often, equipment such as condensers and evaporators in a package refrigeration unit is lighter duty design than in the main system. Startup of packaged systems is discussed in Reference 15.

Safety Systems. Major expenditures here include the flare system (the flare structures and large lines extending throughout the plant) and the firewater system (high-capacity pumps and extensive piping). Safety systems, fortunately, are usually given particular attention. At this study phase, the main thrust should be to check the completeness of licensor equipment lists for cost estimation purposes.

Offsites Other Than Utilities. Utilities have already been discussed. The operating company will play a major role in deciding other offsites. Storage and loading requirements, rail siding facilities, and buildings such as maintenance shops, control and satellite laboratory facilities, and administrative accommodations are the type of facilities that depend heavily on the operating company's way of doing business. The operating company will have its own unique operating and maintenance philosophies, as well as philosophies on control laboratory operation, amount of product storage required, and personnel accommodations. Therefore, the operating company, perhaps with the help of a contractor, should complete these lists and specifications for cost estimating purposes.

Land. At this point, it should be possible to determine how much land will be required. Do not forget to include things like holding ponds, environmental buffer area, camp facilities, and rights of way or easements. It may be well to allow some contingency in initial cost estimates for land requirement increases, as more definitive design information becomes available.

Validating Input

Having extracted a full disclosure from the licensor, the operating company owes it to itself to spend some time sharpshooting major claims and specifications. Even if everything is perfect, a better understanding of the process is developed.

Basic Yield Data. This is a good place to start asking questions. If the process uses a catalytic reaction, do the yields represent new catalyst or catalyst regenerated a number of times? For a thermal reaction like an olefin plant steam cracker, questions might be asked about onstream time between decokings. Therefore, how much contingency is there in the specified number of crackers required?

For a foreign, commercially-proven, coal gasification design, one might ask, "How will it do on my coal (yields, ash handling, etc.)?" A set of yields might have already been stated as being representative of a particular company's coal. Ask how these yields were obtained, vis-á-vis commercial units charging similar coal, or pilot plant or bench scale runs. Having discussed their data thoroughly, it can be accepted at face value or arrangements can be made for a commercial run on your coal at an overseas location where these plants exist. This has been done by at least one company that felt the expense was worthwhile.

If the yields are accepted without full-scale testing, questions can and should arise as to how much contingency exists in the yields (since after all, they were obtained by correlations of similar coals, or perhaps by small-scale tests for your coal, for example). For at least one study, initially presented yields of this sort were found to represent a conservative case and upon request, yields were revealed that were closer to licensor expectations with no contingency. Some design contingency must be provided, but to do this intelligently, any yield contingency must be identified.

For a new process, the basis used for reactor scaleup should be discussed with the licensor. Scaleup of other equipment may also require discussions.

Thermodynamics. Along with stated yields goes heat requirements for the reactor. The thermodynamics for this operation should be checked, as the author once did for a proposed ethyl-benzene dehydrogenation process. Ethylbenzene and steam were fed to the reactor, and unreacted ethylbenzene and steam exited the reactor together with the sought product, styrene, and eight side products.

To make the necessary thermodynamic calculations, plausible reaction equations are written and balanced for production of the stated molar flows of all reactor products. Given the heat of reaction for each applicable reaction, the overall heat of reaction can be determined and compared to that claimed. However, often the individual heats of reaction are not all readily available. Those that are not available can be determined from heats of combustion by combining combustion equations in such a way as to obtain the desired reaction equations by difference. It is a worthwhile exercise to verify this basic part of the process.

Fractionation and Absorption. It is a good idea to do a quick check of fractionation and absorption column separations to see if they appear reasonable. Fractionating and absorption columns are major cost items and major users of utilities. Fractionator and absorber sizing can be checked in Chapter 3.

Commercial computer services are available to do rigorous distillation calculations. Perhaps the licensor will provide copies of rigorous computer runs to validate his balances. Alternately, the operating company can make such runs. For highly non-ideal systems, literature data for binary pairs may have to be sought. In some cases, laboratory equilibrium data may have to be obtained in-house or contracted out to one of several organizations or universities that are in this business.

If money and time are short and the system is fairly ideal, commercial computer services available to the company might have shortcut options available in lieu of the more expensive rigorous routines. Finally, methods will be recommended for checking the separations by hand.

To determine the reasonableness of the top and bottom compositions of a fractionation column, a Hengstebeck plot is fast and easy (Reference 4). First, select a heavy key component and determine the relative volatility (α) of all column components to the heavy key. The α can be α_{feed} or perhaps more accurately $\alpha = (\alpha_{top}\ \alpha_{bottom})^{0.5}$. Plot ln D/B versus ln α and the component points should fall close to a straight line. If a fairly straight line does not result, the compositions are suspect. A nomenclature table is provided at the end of this chapter.

A more quantitative and lengthy method, but still very useful for checking of the type required here is the Smith-Brinkley method (Reference 5). It uses two sets of separation factors for the top and bottom parts of the column for a fractionator or reboiled absorber and one overall separation factor for a simple absorber. The method is tailor-made for analysis of a column design or a field installed column. The Smith-Brinkley method starts with the column parameters and calculates the resulting product compositions unlike other methods that require knowing the compositions to determine the required reflux.

For a distillation column component i:

$$f = \frac{(1 - S_n^{N-M}) + R(1 - S_n)}{(1 - S_n^{N-M}) + R(1 - S_n) + hS_n^{N-M}(1 - S_m^{M+1})}$$

where:

$f_i = (BX_B/FX_F)_i$
$S_{ni} = K_iV/L$
$S_{mi} = K'_iV'/L'$
h = Correlating factor defined by Equations 1 or 2; if the feed is mostly liquid, use Equation 1, and if mostly vapor, Equation 2.

$$h_i = (K'_i/K_i)\,(L/L')\,[(1 - S_n)/(1 - S_m)]_i \qquad (1)$$

$$h_i = (L/L')\,[(1 - S_n)/(1 - S_m)]_i \qquad (2)$$

The effective top and bottom section temperatures are used to determine K_i and K'_i. These are used along with the effective top and bottom section molar liquid and vapor rates to determine S_n and S_m. Section temperatures are the average of the top and feed design temperatures for the top section and bottom and feed design temperatures for the bottom section, or averaged column profile data, if available. The molar flows can be obtained assuming equal molar overflow or with averaged column profile data, if available.

If it is desired to estimate the overhead and bottoms molar flows of a single component, this can be done without performing the full calculation. This is another advantage of the Smith-Brinkley method.

For a reboiled absorber component i:

$$f = \frac{(1 - S_n^{N-M}) + q_s(S_n^{N-M} - S_n)}{(1 - S_n^{N-M}) + hS_n^{N-M}(1 - S_m^{M+1})}$$

The calculations are handled similarly to a distillation column. The only new element introduced is q_s, the fraction of the component in the lean oil.

For a simple absorber component i:

$$f = \frac{(1 - S^N) + q_s(S^N - S)}{1 - S^{N+1}}$$

Pretreating. Pretreating for acid gas (H_2S, CO_2) or water removal (glycol, molecular sieves) is always important, but more so for processes such as LNG because of the extremely low temperatures. Some commercial computer services provide rigorous treating calculations for acid gas removal using various standard treating methods, such as MEA or DEA. If the licensor cannot provide computer solutions, it will be worthwhile to run some design cases.

Most companies have extensive treating and drying equipment. The design case can be compared to existing field equipment as a rough check on capability and possible oversights. For overall quick design checking of amine and glycol plants, References 8 and 9 are most helpful. Reference 7 provides a good overview of typical design parameters for molecular sieve driers. Branan's[6] Chapter 9 gives a quick overview of absorption design in general.

Heat Exchangers. Several aspects of heat exchanger design should be reviewed at this stage of the study since they impact on workability and cost.

A check should be made to be sure that the licensor has not included any temperature cross situations in shell and tube heat exchanger design. In Figure 1, the colder fluid being heated emerges hotter than the outlet temperature of the other fluid. This is an impossible real world situation, or close enough to impossible to be undesirable for the plant design.

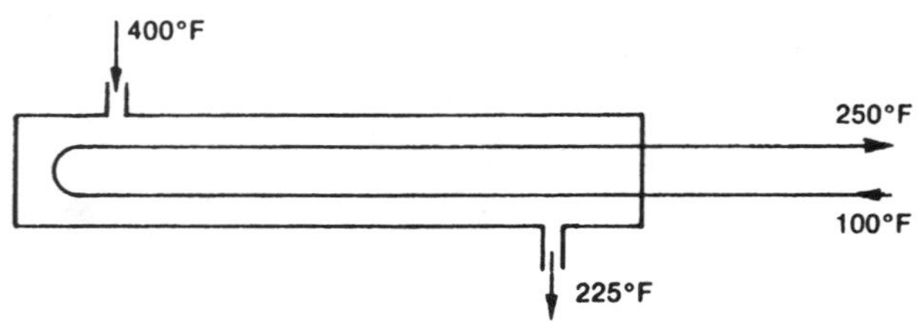

Figure 1. Shown here is an example of a temperature cross.

Another desirable check is to note exchanger types picked for each service. For fixed tubesheet shell and tube heat exchanger design, be sure an expansion joint is included for cases where large differences exist between shellside and tubeside temperatures. Branan[6] gives insight into analyzing this situation. The added cost of expansion joints, if required, can be added to later estimates. Also, for fixed tubesheet units be sure that the shellside fluid is clean enough for this design. If not, the more expensive removable bundle design may be required. Also, for frequent shellside cleaning requirements, the more expensive square tube pitch, instead of triangular, may be required, to provide straight-through cleaning lanes.

Be sure to look closely at fouling resistances used for heat exchanger rating. They play a large part in calculating the heat exchanger coefficient and, thus, in setting required heat transfer surface. Reference 10 is a good source for checking fouling resistances.

Metallurgy is another cost versus operability factor for heat exchangers. For example, for seawater service some companies specify 90/10 Cu-Ni tubes as minimum and do not allow the cheaper, but more prone to corrosion, aluminum-brass. One can even consider going on up to 70/30 Cu-Ni, or to one of the modern, high-performance stainless steels (Reference 14), or even to titanium, if superior resistance to corrosion is required and money is available. Whatever material is specified, however, good operating and maintenance practices are necessary, such as not exceeding maximum tubeside velocities specified for the particular material:

	ft/sec
70/30 Cu-Ni	14
90/10 Cu-Ni	8–10
Aluminum-Brass	7–8

Stagnant seawater must not be left in the unit during shutdown periods. This must be worked out at this stage because cost is impacted.

Other examples of metallurgy decisions are red brass versus admiralty tubes with fresh water on the tubeside and suspected stress corrosion cracking conditions on the shellside, and stainless steel versus carbon steel with chlorides present. A good metallurgist should be brought in when these kinds of decisions are needed.

For condensers, especially refrigerated units with high temperature difference, a check should be made for fogging tendency. Branan's[6] Chapter 7 shows how to do this.

Compressors and Expanders. Process engineers can become involved in reviewing compressor and expander design. The outlet temperatures are one such design feature. For compressors, the outlet temperature for most multistage operations is designed to stay below 250–300°F. High temperature requires a special and more costly machine. Mechanical engineers will be watching the design outlet temperature for this reason. The process engineer should make sure that the outlet temperature stays within any process limits. Sometimes, for example, temperature is limited to protect against polymerization as in olefin or butadiene plants.

For expanders, the process engineer should be alert for outlet temperatures in the dew point range. For changes in design parameters, such as for proposed alternate modes of operation, the process engineer is best equipped to track this tendency.

It is not a bad idea for the process engineer to familiarize himself with compressor surge controls. The interaction of the compressor surge controls with downstream process control valves can become a problem area later, and this study phase is not too early to put such items on a checklist. An LNG plant example comes to mind where such an operating problem existed.

Cooling Towers and Air-Cooled Heat Exchangers. Recirculation is the enemy of cooling towers and air-cooled heat exchangers. Humid air recirculation to the inlet of cooling towers and hot air recirculation to the inlet of air-cooled heat exchangers must be defined in the design stage.

For cooling towers, one specifies the required cold water temperature and heat duty. Usually, the 95% summer hours maximum wet bulb temperature for the area is the starting point. To this, an allowance is added for recirculation by raising the wet bulb temperature (say, 1–3°F). After the design air wet bulb inlet temperature is set, the cold water approach temperature difference to this wet bulb temperature is specified (often, 10°F).

For example, if 72°F is the wet bulb temperature not exceeded for 95% of the summer hours for the area, a conservative design air wet bulb inlet temperature would be 75°F. Then, for a 10°F approach, the design cold water temperature would be 85°F.

This then is the design condition having no contingency (unless some contingency is felt to exist in the 3°F added to the wet bulb temperature). Cooling tower deterioration and fouling can occur over a few years so some contingency is needed just to keep even. In addition, some contingency may be desired for increases in process duty. The author has seen enough examples of plants with higher than design cold water temperature over the years to feel that a contingency of 20–30% in tower heat duty is not out of line.

For air-cooled heat exchangers, recirculation is more difficult to define or relate to standard practices. Banks of air-coolers are installed in a variety of configurations so, for a particular proposed installation, a study by specialists may be required. This study of recirculation would probably be done later in the project even though the results impact costs. There is a practical limit to the number of front-end studies and this is one that can be deferred until geometry is better defined.

Use of Vendors. During the validation phase of the study, vendors can be effectively employed. Often, vendors will provide quite detailed studies free of charge for goodwill or in hopes of a later sale. Contractors use vendor help routinely for process designs or studies. Credit is usually given in the contractor's presentation of results for any vendor participation.

Performing Side Studies

Management might make certain decrees for the major study: keep costs to a minimum; maximize air-cooling; make sure we meet or exceed air quality standards; do not provide spares or bypasses—if we have a problem, we will shut down a train; all large compressors will be steam driven; provide at least two spare boilers; all pneumatic instrumentation. The list can go on and on. It is a good practice to document all such decisions lest they be forgotten, especially when new managers are named.

A contractor helping with the study will definitely document such decisions. In fact, a contractor's most difficult task is extracting major decisions like this from the operating company's management. He will be diligent in documenting those that are handed to him on a silver platter.

Management decisions might not be very definitive. For example, how does one maximize air-cooling? Normally, a study will reveal the optimum crossover temperature between air- and water-cooling (often, air-cooling will be shown to be economical above 140°F). What management really wants in this case is air-cooling wherever it can be used economically.

Some side studies are very common and are done on most new projects, while others depend upon the local circumstances. Some of the more common are:

- Air- versus water-cooling optimum temperature
- Package versus field erected boilers
- Gas versus steam turbines
- Trays versus packing for fractionation or absorption
- What equipment should be spared?
- Electric versus steam driven pumps
- Electronic versus pneumatic instrumentation
- Grounded Y versus ungrounded delta electrical system
- Self-generated versus purchased power

In many designs, a combination of the choice items are used such as air- and water-cooling, steam and electric driven pumps and electronic and pneumatic instrumentation. As much as possible, these type decisions should be made by results of side studies rather than simple edict.

"Classic" Side Studies and Their Results. Let's discuss briefly the "classic" side studies and even predict what the final results might be! For *air versus water cooling,* the optimum temperature will probably lie in the range 120–150°F, so time can be saved by building initial cases therein.

For *package versus field erected boilers,* field erected are usually selected when conservative design and high reliability are required. Field erected are more expensive, but feature lower heat density, larger steam drums, more elaborate steam drum separation internals, and generally more heavy duty or robust design. If package boilers are selected, full water wall design is strongly recommended. Refractory applied at the factory often does not stand shipment well and is difficult to cure correctly onsite. Also, controls should be of the metering rather than positioning type as explained in Reference 11. Reference 13 discusses specification of package boilers. A boiler design alternate would be modular construction which is starting to be used.

Gas versus steam turbines can involve a major side study. The results can be different for the process and utility sides of the plant as shown in Reference 12. For the gas turbine case, simple cycle versus waste heat boilers can be studied. Usually, waste heat boilers will win out unless the plant is in a cheap gas country. If gas turbines are selected for power generation, black start capability is usually a good investment.

Trays versus packing for fractionation can be a standoff if large numbers of tray passes are allowed for large fractionation towers by the operating company. Often, company preferences will decide. The licensor may also have preferences based on field experience.

Usually, *major pumps are spared,* even if the philosophy of shutting down an entire train is adopted. Other spares or bypasses are often left out for such a philosophy, but standard bypasses (control valves, major block valves, etc.) are normally included otherwise. The utility area will have normal sparing (for example, three ⅔-capacity electric generators), even if train shutdown philosophy dictates the process area.

For pumps, it is common to have steam driven primary units with electrical spares having inherent fast startup capability. Mixed utilities often allow continuation of operations, after a fashion, when one utility fails.

Instrumentation is rapidly becoming more electronic. However, many users prefer pneumatic, and computer compatibility is available with either; although electronic interface with computers is generally preferred. One coal gasification company prefers pneumatic because they feel the inherent corrosive atmosphere around such plants is not kind to electronic equipment.

Many companies are realizing the advantages of ungrounded delta circuits over grounded Y. Operating people like the delta because it allows the plant to keep operating if one phase goes to ground. It is far better for the operating management to be able to choose the time for an orderly shutdown than to have the plant suddenly "come down around your ears". Equipment damage and other losses are lower for the orderly shutdown.

Self-generated versus purchased power often comes out in favor of self-generated, especially when the sale of power from a cogeneration cycle is possible, with a diesel driven emergency generator and with purchased power backup. Battery banks with inverters supply critical lighting and instrumentation during power failures. Automatic-on gas turbine packages supply emergency power for some hotels and hospitals, but the author is not familiar with any process applications.

Identifying Bottlenecks

The total plant or train main process bottleneck will probably be identified by the licensor, such as the gasifier for a coal gasification train, the main exchanger for a mixed refrigerant LNG plant train, or the cracked gas compressors for an olefin plant. First and foremost, be sure that the licensor has not made the utility area a bottleneck. This can never be allowed since overloaded utilities could repeatedly shut the entire complex down on a crash basis, adversely impacting economics.

Once the main bottleneck is defined as a process area item, it is not a bad idea to identify secondary bottlenecks. If the second bottleneck were to be a fractionating column shell, it might be advantageous to increase the diameter slightly if the main bottleneck statement is felt to be conservative. If a heat exchanger were the second bottleneck, space might be left for another unit later. Identification of secondary bottlenecks is a worthwhile exercise, even if the search does not produce design changes. At least, better understanding of the process is developed with possible dividends in later operations.

Nomenclature

B = Bottoms molar rate
D = Distillate molar rate
F = Feed molar rate
f_i = Smith-Brinkley ratio of the molar rate of component i in the bottoms to that in the feed
h_i = Smith-Brinkley correlating factor defined by Equations 1 and 2
K_i = Equilibrium constant of component i in top section equals Y/X

K'_i = Equilibrium constant of component i in bottom section equals Y/X
L = Effective liquid molar rate in column top section
L′ = Effective liquid molar rate in column bottom section
M = Total equilibrium stages below the feed stage
N = Total equilibrium stages including reboiler and partial condenser
q_s = Fraction of the component in the lean oil of a simple or reboiled absorber
R = Actual reflux ratio
S = Smith-Brinkley overall stripping factor for component i
S_{mi} = Smith-Brinkley stripping factor for component i in bottom section
S_{ni} = Smith-Brinkley stripping factor for component i in top section
V = Effective vapor molar rate in column top section
V′ = Effective vapor molar rate in column bottom section
X = Mol fraction in the liquid
Y = Mol fraction in the vapor
α_i = Relative volatility of component i versus the heavy key component

References

1. GPSA Engineering Data Book, Gas Processors Suppliers Association, Latest Revision.
2. Perry, R. H., and Chilton, C. H., *Chemical Engineers' Handbook.* New York: McGraw-Hill, Inc, 1973.
3. Branan, C., *The Fractionator Analysis Pocket Handbook.* Houston, Texas: Gulf Publishing Co., 1978.
4. Hengstebeck, R. J., *Trans. A.I.Ch.E.* 42, 309, 1946.
5. Smith, B. D., *Design of Equilibrium Stage Processes,* New York: McGraw-Hill, Inc., 1963.
6. Branan, C., *The Process Engineer's Pocket Handbook, Volume 1.* Houston, Texas: Gulf Publishing Co., 1976.
7. Barrow, J. A., "Proper Design Saves Energy," *Hydrocarbon Processing,* January 1983, p. 117.
8. Ballard, D., "How to Operate an Amine Plant," *Hydrocarbon Processing,* April 1966.
9. Ballard, D., "How to Operate a Glycol Plant," *Hydrocarbon Processing,* June 1966.
10. Standards of Tubular Exchanger Manufacturers Association (latest edition) Tubular Exchanger Manufacturers Association, Inc. (TEMA), 707 Westchester Ave., White Plains, New York 10604.
11. Branan, C., *The Process Engineer's Pocket Handbook, Volume 2—Process Systems Development.* Houston, Texas: Gulf Publishing Co., 1983.
12. DiNapoli, R. N., "LNG Costs Reflect Changes in Economy and Technology," *Oil and Gas Journal,* April 4, 1983, p. 138.
13. Hawk, C. W. Jr., "Packaged Boilers for the CPI," *Chemical Engineering,* May 16, 1983, p. 89.
14. Redmond, J. D., and Miska, K. H., "High Performance Stainless Steels for High-Chloride Service"—Part I, *Chemical Engineering,* July 25, 1983, p. 93. Part II, *Chemical Engineering,* August 22, 1983, p. 91.
15. Gans, M., Kiorpes, S. A., and Fitzgerald, F. A., "Plant Startup—Step By Step By Step," *Chemical Engineering,* October 3, 1983, p. 99 (Inset: Some afterthoughts on startups).
16. Edmister, W. C., and Lee., B. I., *Applied Hydrocarbon Thermodynamics, Volume 1,* second edition. Houston, Texas: Gulf Publishing Co., 1984.
17. Branan, C. and Mills, J., *The Process Engineer's Pocket Handbook, Vol. 3.* Houston, Texas: Gulf Publishing Co., 1984.

Battery Limits Specifications

The licensor's design package has been digested, validated, changed where necessary, and completed where necessary. Major decisions are made. The effort now involves finalizing equipment specifications, particularly sizing of high cost items, and generating a complete battery limits major material list with process specification sheets as complete as possible for this study stage.

Reactors

The licensor's basis for sizing has already been discussed and agreed to or changed. For an olefin plant, the number of steam crackers of the licensor's standard size is firm. For a new process, reactor scaleup methods have been agreed to. For a coal gasification plant, gasifier size, number and yields are set. Therefore, there is little left to do in this study phase other than finalizing the process specification sheets.

Fractionation and Absorption

Methods for quick sizing trayed fractionation and absorption column diameter have been reduced here to equations to facilitate programming for calculators or computers. Three methods are discussed and it is not a bad idea to compare results with all three.

Souders-Brown. The Souders-Brown method (References 1, 2) is based on bubble caps, but is handy for modern trays since the effect of surface tension can be evaluated and factors are included to compare various fractionator and absorber services. These same factors may be found to apply for comparing the services when using valve or sieve trays. A copy of the Souders-Brown C factor chart is shown in Reference 2.

The developed equation for the Souders-Brown C factor is

$$C = (36.71 + 5.456T - 0.08486T^2) \ln S - 312.9 + 37.62T - 0.5269T^2$$

Correlation ranges are:

C = 0 to 700
S = 0.1 to 100
T = 18 to 36

A nomenclature table is included at the end of this chapter.

The maximum allowable mass velocity for the total column cross section is calculated as follows:

$$W = C\,[\rho_v\,(\rho_L - \rho_v)]^{1/2}$$

The value of W is intended for general application and to be multiplied by factors for specific applications as follows:

Absorbers: 0.55
Fractionating section of absorber oil stripper: 0.80
Petroleum column: 0.95
Stabilizer or stripper: 1.15

The top, bottom, and feed sections of the column can be checked to see which gives the maximum diameter.

The Souders-Brown correlation considers entrainment as the controlling factor. For high liquid loading situations and final design, complete tray hydraulic calculations are required.

Reference 2 states that the Souders-Brown method appears to be conservative for a pressure range of 5 to 250 psig allowing W to be multiplied by 1.05 to 1.15, if judgment and caution are exercised.

F Factor. The use of the F factor for fractionating column diameter quick estimation is shown in Reference 3. The developed equation for the F factor is:

$$F = (547 - 173.2T + 2.3194T^2)\,10^{-6}P + 0.32 + 0.0847T - 0.000787T^2$$

Correlation ranges are:

F = 0.8 to 2.4
P = 0 to 220
T = 18 to 36

The F factor is used in the expression $U = F/(\rho_V)^{0.5}$ to obtain the allowable superficial vapor velocity based on free column cross-sectional area (total column area minus the downcomer area). For foaming systems, the F factor should be multiplied by 0.75.

For estimating downcomer area, Reference 3 gives a correlation of design liquid traffic versus tray spacing. The developed equation is:

$$DL = 6.667T + 16.665$$

Clear liquid velocity (ft/s) through the downcomer is then found by multiplying DL by 0.00223. The correlation is not valid if $\rho_L - \rho_V$ is less than 30 lb/ft^3 (very high pressure systems). For foaming systems, DL should be multiplied by 0.7. Reference 3 recommends segmental downcomers of at least 5% of total column cross-sectional area regardless of the area obtained by this correlation. For final design, complete tray hydraulic calculations are required.

Smith-Dresser-Ohlswager

Smith[4] uses settling height as a correlating factor, which is intended for use with various tray types. The correlation is shown in Figure 1. Curves are drawn for a range of settling heights from 2 to 30 inches. Here, U is the vapor velocity above the tray not occupied by downcomers.

The developed equations for the curves are (Y subscript is settling height in inches):

$Y = A + BX + CX^2 + DX^3$

	A	B	C	D
Y_{30}	−1.68197	−.67671	−.129274	−.0046903
Y_{24}	−1.77525	−.56550	−.083071	+.0005644
Y_{22}	−1.89712	−.59868	−.080237	+.0025895
Y_{20}	−1.96316	−.55711	−.071129	+.0024613
Y_{18}	−2.02348	−.54666	−.067666	+.0032962
Y_{16}	−2.19189	−.51473	−.045937	+.0070182
Y_{14}	−2.32803	−.44885	−.014551	+.0113270
Y_{12}	−2.47561	−.48791	−.041355	+.0067033
Y_{10}	−2.66470	−.48409	−.040218	+.0064914
Y_{8}	−2.78979	−.43728	−.030204	+.0071053
Y_{6}	−2.96224	−.42211	−.030618	+.0056176
Y_{4}	−3.08589	−.38911	+.003062	+.0122267
Y_{2}	−3.22975	−.37070	−.000118	+.0110772

Smith recommends obtaining the settling height (tray spacing minus clear liquid depth) by applying the familiar Francis Weir formula. For our purposes of quickly checking column diameter, a more rapid approach is needed.

For applications having 24-in. tray spacing, the author has observed that use of a settling height of 18 in. is good enough for rough checking. The calculation yields a superficial vapor velocity that applies to the tower cross section not occupied by downcomers. A downcomer area of 10% of column area is minimum, except for special cases of low liquid loading.

For high liquid loading situations and final design, complete tray hydraulic calculations are required.

Packed Columns

For packed columns see the rules of thumb in Chapter 3, Fractionators, in the Packed Columns section.

Heat Exchangers

In the Process Definition section, heat exchanger design was validated by eliminating the possibility of temperature cross conditions, checking application of types, sharpshooting fouling factors, establishing metallurgy and investigating fogging tendencies. For this study phase, a final check will be made of the licensor's supplied process specification sheets for agreement of overall heat transfer coefficients and resulting heat transfer surface specifications preparatory to cost estimation. For reboilers and other boiling applications, the heat transfer surface will be set by allowable heat flux.

Typical film coefficients can be used to build rough overall heat transfer coefficients. This should suffice in most cases to establish that the design is within ballpark accuracy. Later, for final design, certain critical services will be checked in detail. Typical film resistances for shell and tube heat exchangers and overall heat transfer coefficients for air cooled heat exchangers are shown in Chapter 2, Heat Exchangers.

If exchanger shell diameter is in doubt, see Chapter 2, Heat Exchangers, the Shell Diameter section. In addition, this book provides a rough rating method for air-cooled heat exchangers.

Vessels

For rough sizing check of vapor/liquid separators and accumulators, see the Fluor method in Chapter 8, Separators/Accumulators—Vapor/Liquid calculation method.

Vessel thickness is quickly checked using standard equations:

$$T = \frac{Pr_i}{SE - 0.6P} + C$$

$$T = \frac{Pr_o}{SE + 0.4P} + C$$

Pumps and Drivers

Horsepower is conveniently checked by using:

$$HP = \frac{GPM(\Delta P)}{1,715(Eff)}$$

Pump efficiency can be approximated using the equation (Reference 6):

$$\begin{aligned} Eff = 80 - 0.2855F + 3.78 \times 10^{-4}FG - 2.38 \times 10^{-7}FG^2 \\ + 5.39 \times 10^{-4}F^2 - 6.39 \times 10^{-7}F^2G + 4 \times 10^{-10}F^2G^2 \end{aligned}$$

Ranges of applicability: $F = 50 - 300$ ft
$G = 100 - 1000$ GPM

For flows in the range 25–99 GPM, a rough efficiency can be obtained by using this equation for 100 GPM and then subtracting 0.35%/GPM times the difference between 100 GPM and the low flow GPM.

Compressors and Drivers

Compressor horsepower is best determined using the horsepower calculation in Chapter 6. For refrigeration compressors, the horsepower can be approximated another way that may prove to be simpler. The compressor horsepower per ton of refrigeration load depends upon the evaporator and condenser temperatures. See the section titled Estimating Horsepower per Ton in Chapter 11.

Molecular Sieve Driers

Reference 8 is valuable for doing a quick check of a molecular sieve drier system giving an example of a typical installation.

Nomenclature

C = Souders-Brown allowable mass velocity factor or corrosion allowance, in.
DL = Design downcomer liquid traffic, gpm/ft^2
E = Weld efficiency, fraction (use 0.85 for initial work)
Eff = Pump efficiency
F = Factor for fractionation allowable velocity or packed column packing factor or pump developed head, ft.
G = Fractionator vapor rate, lb/hr or packed column gas rate, lbs/ft^2 sec or pump flow, gpm
GPM = Pump flow, gpm
K = Vessel allowable velocity factor
K_H = Horizontal vessel allowable velocity factor
K_V = Vertical vessel allowable velocity factor
L = Fractionator liquid rate, lb/hr or packed column liquid rate, lbs/ft^2 sec
P = Pressure, psia for fractionator F factor correlation or psig for vessel thickness calculation
r_i = Inside radius, in.
r_o = Outside radius, in.
S = Surface tension, dynes/cm for Souders-Brown correlation or allowable stress, psi for vessel thickness calculation
T = Tray spacing, in., or vessel thickness, in.
U = Velocity, ft/sec
W = Maximum allowable mass velocity, lb/(ft^2 total cross section) (hr)
W_L = Liquid rate, lb/sec
W_V = Vapor rate, lb/sec
X = Abscissa of Smith-Dresser-Ohlswager correlation, or abscissa of Norton Generalized Pressure Drop Correlation, or Fluor vessel sizing separation factor or evaporator temperature, °F
Y = Ordinate of Smith-Dresser-Ohlswager correlation, or ordinate of Norton Generalized Pressure Drop Correlation, or K_V, or horsepower per ton of refrigeration.
ρ_G, ρ_V = Gas or vapor density, lb/ft^3
ρ_L = Liquid density, lb/ft^3
ν = Liquid viscosity, centistokes

References

1. Souders, M. and Brown, G., *Ind and Eng Chem.,* Vol. 26, p. 98 (1934), copyright The American Chemical Society.
2. Ludwig, E. E., *Applied Process Design for Chemical and Petrochemical Plants, Vol. 2,* second edition. Houston, Texas: Gulf Publishing Co., 1979.
3. Frank, O., "Shortcuts for Distillation Design," *Chemical Engineering,* March 14, 1977, p. 111.
4. Smith, R., Dresser, T., and Ohlswager, S., "Tower Capacity Rating Ignores Trays," *Hydrocarbon Processing and Petroleum Refiner,* May 1963, p. 183.
5. Branan, C., *The Process Engineer's Pocket Handbook, Volume 2—Process Systems Development.* Houston, Texas: Gulf Publishing Co., 1983.
6. Branan, C., *The Process Engineer's Pocket Handbook,* Volume 1. Houston, Texas: Gulf Publishing Co., 1976.
7. Ballou, D. F., Lyons, T. A., and Tacquard, J. R., "Mechanical Refrigeration Systems," *Hydrocarbon Processing,* June 1967, p. 119.
8. Barrow, J. A., "Proper Design Saves Energy," *Hydrocarbon Processing,* January 1983, p. 117.
9. *GPSA Engineering Data Book,* Gas Processors Suppliers Association, latest edition, 41.
10. Smith, E. C., Hudson Products Corp., *Cooling with Air-Technical Data Relevant to Direct Use of Air for Process Cooling.*
11. Branan, C. and Mills, J., *The Process Engineer's Pocket Handbook Vol. 3.* Houston, Texas: Gulf Publishing Co., 1984.

Offsite Specifications

This discussion of offsites is subdivided into Utilities and Other Offsites. The utility portion interacts with the process area, while the other offsites have minor interaction with the process area, if any. In addition, the process area may have utility generation, such as waste heat boilers. It is convenient to discuss all utility generation as one package pointing out special considerations for the process area units along the way. The goal for this study phase is the same as for battery limits specification: complete major material list and process specification sheets.

Utilities

A utility area superintendent once made the remark to me, "Others have only one process unit to worry about, but in utilities, everybody's problem is my problem." Utilities have a way of being taken for granted until a problem develops. In the study phase of a project, it is well to attach great importance to utilities, making sure that this portion of the capital estimate is large enough to provide reliability and sufficient spare capacity.

Steam Pressure Levels. Steam pressure levels have risen over the years to achieve higher efficiencies. While large central power stations run 1,800 to 3,600 psig (with the 3,600 psig being supercritical), steam generation in process plants tends to run in the 900 to 1,500 psig range. The decision setting the plant's highest steam level should result from a study of capital versus operating cost. Higher pressure boilers have higher capital cost. Energy savings at the higher pressures tend to lower operating costs, but water treating costs are usually higher at higher steam pressures (both capital and operating costs). The American Boiler Manufacturers' Association (ABMA) has established limits for boiler water composition to help assure good quality steam. The limits shown in Table 1 are more stringent for higher pressure. Boiler blowdown is normally based on the most stringent of the limits for a given pressure operation. Feed water treating must produce purer water as pressure rises, otherwise ridiculous or impossible blowdown projections could result. For example, at 1,500 psig operation and above, silica removal is normally required since the 2.5-ppm-boiler-water maximum may already exist in the makeup water.

For package units, the manufacturer may propose limits tighter than ABMA; also these specifications might include boiler feedwater limits. The author has seen 1 ppm total solids specified for boiler feedwater at 900 psig operation for package boilers. Such a limit will probably require complete feedwater demineralization as well as return condensate polishing (demineralization). Certainly, this information must be factored into any decision on package versus field erected boilers and included in cost estimates.

Table 1
ABMA limits for boiler water

Boiler Pressure (psig)	Total Solids (ppm)	Alkalinity (ppm)	Suspended Solids (ppm)	Silica (ppm)
0–300	3500	700	300	125
301–450	3000	600	250	90
451–600	2500	500	150	50
601–750	2000	400	100	35
751–900	1500	300	60	20
901–1000	1250	250	40	8
1001–1500	1000	200	20	2.5
1501–2000	750	150	10	1.0
over 2000	500	100	5	0.5

Once the highest steam level is set, then intermediate levels must be established. This involves having certain turbines exhaust at intermediate pressures required of lower pressure steam users. These decisions and balances should be done by in-house or contractor personnel having extensive utility experience. People experienced in this work can perform the balances more expeditiously than people with primarily process experience. Utility specialists are experienced in working with boiler manufacturers on the one hand and turbine manufacturers on the other. They have the contacts as well as knowledge of standard procedures and equipment size plateaus to provide commercially workable and optimum systems. At least one company uses a linear program as an aid in steam system optimization.

Sometimes, conventional techniques do not produce a satisfactory steam balance for all operating modes. Options are available for steam drives for flexibility, such as extraction and induction turbines. Extraction turbines are widely used. In these, an intermediate pressure steam is removed or "extracted" from an intermediate turbine stage with the extraction flow varying as required over preset limits. Induction turbines are not as widely used as extraction turbines, but are a very satisfactory application if instrumented properly. Intermediate pressure steam is added or "inducted" into an intermediate stage from an outside source. Protection must be provided to prevent the induction steam from entering during shutdown or certain offload conditions.

The lowest pressure steam is usually set to allow delivery of 10–20 psig steam to boiler feedwater deaerators. Excess low-pressure steam could be a design problem. If so, low-pressure turbines can be designed, but their application requires large piping and inlet nozzles. A good application would be electrical generation or a noncritical drive application located close to the low-pressure steam source to avoid long runs of large piping and pressure drop. Some designs vent small amounts of excess low-pressure steam to a special condenser for condensate recovery.

If steam is generated at intermediate pressure levels, such as in waste heat boilers cooling process streams at too low a temperature level for highest pressure steam generation, a good technique is to blowdown stagewise. The blowdown from high-pressure boilers with their stringent boiler water limits is often satisfactory makeup for the lower pressure boilers. The lowest pressure blowdown might have enough heat content to make it economical to flash for deaerator heating steam.

To compare the values of steam at various pressures for design studies or accounting once the plant is built, the ΔB method is useful. The maximum available energy in a working fluid can be determined from

$$\Delta B = \Delta H - T_o \Delta S$$

where: ΔB = Maximum available energy, Btu/lb
ΔH = Enthalpy difference between source and receiver, Btu/lb. For a typical condensing steam turbine, it would be the difference between the inlet steam and the liquid condensate.
T_o = Receiver temperature, °R
ΔS = Entropy difference between the source and receiver, Btu/lb °F.

Comparing ΔB's for steam pressure levels using their intended drivers' receiver conditions gives an indication of the relative steam values.

It is good to limit the number of steam levels to the minimum absolutely required.

Boilers. Boiler efficiency will determine how much fuel is used, an important operating cost parameter. Pin the supplier down on just what efficiency is quoted and whether the basis is higher or lower fuel heating value. Once the efficiency claim is clear, establish how the supplier plans to achieve the efficiency. What stack gas temperature is required for the efficiency to hold and does the design appear to have the ability to consistently achieve this temperature? What percentage excess air is required for efficiency maintenance?

The controls are important, especially if low-percentage excess air is projected. Provide enough capital in the estimate to include metering type combustion controls, as mentioned earlier, and modern 3-element feedwater makeup controls. Stack oxygen monitoring should also be included.

It is important to make sure that enough spare boiler capacity is included in the design. Frequency of maintenance and required inspections by authorized agencies are factors. Also, required process operating factor is important since boiler maintenance and/or inspection can be carried out concurrently with process turnarounds. At this study stage, the estimate should be on the conservative side, with a generous sparing philosophy. Spare deaerator and boiler feedwater pump capacity should also be considered at this stage.

Process Steam Generation. Steam generated in the process sections of the plant may be at the highest plant pressure level or an intermediate level. Also, the process area may have fired boilers, waste heat boilers or both. There may be crossties between utility and process generated steam levels. Enough controls must be provided to balance far-ranging steam systems and protect the most critical units in the event of boiler feedwater shortage situations.

It is important for crosstied systems that a sufficient condensate network is provided for balancing the mix of condensate return and makeup treated water as required. The author has seen a system designed with process area and utility area fired boilers of the same pressure. Periodically, the utility area was required to supply makeup steam to balance a shortage in the process area, but no provisions were made to return equivalent condensate from the process to the utility area. The earlier such a mistake is caught, the better.

The cost estimate should include provisions for any required satellite boiler water analysis laboratories. The central control lab cannot normally handle analyses of widely spread boilers satisfactorily. The designers, while remembering satellite water laboratory facilities for the utilities area, might overlook similar facilities for the steam generation in the process area.

Electrical Power. The major considerations for the electrical generation area are reliability, backup supply, and spare capacity. A typical setup for a large plant would be three or four ½- to ⅔-capacity electric generators with a combination of steam and gas turbine drivers. The gas turbine driver or drivers would have black start capability. A diesel-driven emergency generator would supply critical equipment on an emergency bus plus 25–50% excess capacity. There would perhaps be a tie to a community power supply for partial additional backup. Critical services such as key lighting and instrumentation are backed up by a UPS (uninterruptible power supply) system of batteries charged by an inverter. A load shedding system would drop preselected electrical services in steps if turbogenerator overloading was indicated by falling current cycles.

An electrical engineer would best chair discussions leading to decisions and provide the complete cost estimation package. The package may include such things as air-conditioned motor control centers, undervoltage protection for motors, any costs to the company for community backup electrical supply, and forced cooling for large motors. The process engineer can provide input for decisions on assignment of equipment to the emergency bus and UPS lists and electrical versus steam drives for normally operated versus spare pumps and other equipment.

Instrument and Plant Air Systems. A typical setup for a large plant could include three to four 50% instrument air compressors and two 100% plant air compressors, with steam drives for normally operated units and electrical drives for spares. Common practice would provide an interconnection to allow makeup from plant air into instrument air, but not vice versa, and two sets (two 100% driers per set—one onstream and one regenerating) of 100% instrument air driers. Two main receivers on instrument air near the compressors with several minutes holdup time and satellite receivers at process trains would be likely and proper for feasibility cost estimating.

Cooling Water System. A list of cooling duties will be available at this point so the cost estimate for this system can be factored or estimated based on a similar operating system. For a more definitive estimate based on initial or detailed layout, it is probably best to use a contractor or consultant skilled in these designs. If a cooling tower is involved, the groundwork will already have been set. This basis can be passed along on specification sheets provided in the Appendix to a vendor for quotes.

Calculations for determining system makeup rates and chemical treating effective half life are presented in the section on Cooling Tower design.

Fuel System. An adequate knockout vessel should be provided for natural gas entering the plant as fuel or feed gas. Hydrocarbon liquids can and will enter the fuel system otherwise. Double-pressure letdown plus heating to preclude hydrates is also typically specified.

If liquid streams are to be used as fuel for firing boilers or heaters, special burners, nozzles, or "guns" can be used, but a fuel vaporization system featuring a knockout/blending drum followed by a centrifugal separator has proven to be a good system in an Algerian LNG plant where byproduct gasoline is fired.

Other Offsites

Flare Systems. There is a good chance that the operating company will not have anyone experienced in flare system design. For feasibility cost estimates, rough estimates can be made by comparison with existing plants or a vendor can be contacted for budget cost estimates for the flare stacks and associated knockout drum, burner tip, igniter, and molecular seal.

If in-house personnel are required to provide a flare system piping layout, many good literature articles are available. Reference 2 has simplified the procedure by allowing the calculations to begin with the outlet (atmospheric pressure) and work back towards the source; thus overcoming tedious trial and error required by methods that require beginning at the source.

It may be well to include capital for control valves for at least the larger manifolds for relief valve support. A control valve set at a pressure slightly lower than the corresponding relief valve will handle many of the relieving situations, especially on startup, saving the relief valve from having to open. Control valves tend to reseat better than relief valves, so the control valve may pay its way in leakage savings and relief valve maintenance savings.

Firewater systems. These systems are best laid out by contractors or other specialists. National Fire Protection Association (NFPA) rules will spell out required coverage, typical pump size, and other standard items. A small jockey pump will maintain system pressure at all times.

Rail Facilities. Here, specialists are needed as much as if not more than anywhere else. Certainly, railroad representatives must be in on any design considerations. There are strict limits on such things as turning radius, percentage grade (for example, one degree maximum versus 16 degrees for coal slurry pipelines, Reference 3), and spotting requirements. For hydrocarbon loading operations, an emergency shutdown system (ESD) should be provided in case of fire.

Keep in mind rail tariff breaks for numbers of cars in a string so that the ability to handle the required numbers can be provided. If the operation is large enough, as for large coal mining operations, unit trains (100 cars) are used.

Buildings. Company philosophies on operating and maintenance as well as control and satellite laboratory operations and administrative requirements will set building requirements. The licensor will make suggestions, but the operating company will have to take the lead in setting up these requirements.

Not only the buildings themselves, but an allowance for equipment and stocks within the buildings must be included in the cost estimates, such as maintenance shop power equipment, control laboratory analytical equipment and reagents and, largest of all, the thousands of warehouse equipment spares, fittings, and supplies. For feasibility cost estimates, such items are factored from experience.

Storage of catalyst and chemicals must be decided—some will be indoors and some perhaps outdoors. Tankage will be provided for treating chemicals, such as MEA, for holding process inventory for turnarounds.

Waste Disposal

Wastewater Treating. Earlier we discussed the importance of bringing environmental experts into the project early. Wastewater treating can be part of the battery limits process area in the case of phenol extraction from coal gasification gas liquor, but this is an exception.

Sanitary and process wastewater treating facilities are both required. For process wastewater treating, API separators might be used to remove oil as from refinery processing units or tanker ballast discharge. In such cases, emulsion-breaking facilities recover slop oil for rerunning from API hydrocarbon product. Thickeners or filters may also be required for particulates removal. Considerable land area can be required for these facilities and associated sludge disposal.

One unique application reused community tertiary-treated wastewater as plant cooling water in a major chemical complex in Odessa, Texas.

Wastewater Deep Well Injection. This is an alternate wastewater disposal procedure and requires some treating prior to injection, such as filtration or pH adjustment. Permits for this procedure require long lead times. Reference 4 gives prediction methods for refinery wastewater, generation.

Incineration. If land or ocean disposal is not available for waste sludges, incineration may be used. Reference 5 discusses this technique, including dewatering prior to incineration, since wastewater treating sludge can have more than 99% water.

Other Items

"Other Items" is considered for this discussion to be a catchall category including "everything else." Cranes and other mobile equipment, product storage, and holding ponds are some examples. Again, the operating company needs to take the lead in making sure the list is complete enough for the cost estimate accuracy required. The cost estimators cannot be expected to design the plant.

References

1. Branan, C., *The Process Engineer's Pocket Handbook, Volume 1.* Houston, Texas: Gulf Publishing Co., 1976.
2. Mak, H. Y., "New Method Speeds Pressure-Relief Manifold Design," *The Oil and Gas Journal,* November 20, 1978, p. 166.
3. Santhanam, C. J., "Liquid CO_2-Based Slurry Transport Is On the Move," *Chemical Engineering,* July 11, 1983, p. 50.
4. Finelt, S. and Crump, J. P., "Predict Wastewater Generation," *Hydrocarbon Processing,* August 1977, p. 159.
5. Novak, R. G., Cudahy, J. J., Denove, M. B., Sandifer, R. L., and Wass, W. E., "How Sludge Characteristics Affect Incinerator Design," *Chemical Engineering,* May 9, 1977.
6. Musser, E. G., *Designing Offsite Facilities by Use of Routing Diagrams.* Houston, Texas: Gulf Publishing Co., 1983.
7. Branan, C. and Mills, J., *The Process Engineer's Pocket Handbook, Vol. 3.* Houston, Texas: Gulf Publishing Co., 1984.

Capital Investments

Capital estimates are the heart and soul of all project justification and feasibility studies. Inaccuracies can cause serious harm to a company as well as to the engineer making the mistake. With the flow of a project from its inception to construction, many cost estimates are made. Each new estimate is based on more data, and should be more accurate.

Management usually wants a single accurate cost number, not a range of costs. Thus, an engineer making the first, order of magnitude estimate, usually has very little data, resources, or time; yet, he must produce an estimate that tends to become cast in concrete as it is reviewed and passed up in the organization. Unlike marketing forecasts or operating cost estimates where errors may never be discovered or publicized, the next cost estimate, made after additional engineering, shows the original estimate's inaccuracies. This difference is obvious to everyone from top management to the bookkeepers. Again, as the project progresses, and new, more accurate estimates are made, the spotlight shifts to the second most recent estimate. Reports explaining the differences are usually required.

This means the engineer should strive to do the best job he can on each estimate, and apply adequate factors to keep the cost conservative. All project costs seem to grow as the project matures. Few projects reduce in scope or cost as more information is developed. A Rand Corporation study documents these cost increases as plans mature (Reference 1).

There is a natural tendency for an engineer, or group of engineers, to start believing an estimate after they have worked with it for some time. As each new estimate is made, or an existing one reviewed, the engineer should keep an open mind on every number and assume it is wrong until it is confirmed or changed.

All cost estimates are based on historical costs accumulated from previous projects. This history can be in-house, from vendors, or from the literature. The accuracy of an estimate depends on how completely the project is defined, and on how well the costs from previous projects have been analyzed and correlated. If your company does not have good (or any) past project records, the literature abounds with correlations of cost data, as discussed later. However, this data must be used very carefully.

Process Contingencies

One way to achieve better early estimates is to add a process contingency to each estimate to provide for inevitable future changes. This is in addition to the normal contingency needed to cover errors in estimating techniques. Table 1 lists recommended process contingencies for each type of estimate. To use these process factors, make the best possible estimate with the information available, then apply the process contingency. Many skilled estimators make allowances for unknowns, consciously or unconsciously, as the estimate progresses. This has the effect of a process contingency, and the two should not be additive.

Table 1
Recommended Process Contingencies

Type of Estimate	Process Contingency	Data Available
Order of Magnitude	50%	Lab. Data, Plant Size Raw Materials
Feasibility (factored)	30%	Process Flowsheets, Equipment Size, Regional Location
Project Control (factored and take off)	5%	Mechanical Flowsheets, Preliminary Plot Plans, Location, Off Site Definition
Detailed (material take off)	0%	Most of Engineering

Making a Factored Estimate

Early in the life of a project, information has not been developed to allow definitive cost estimates based on material takeoff and vendor quotes for equipment. Therefore, it is necessary to estimate the cost of a facility using shortcut methods. The first step is to develop or check flowsheets, major equipment sizes, and specification sheets as described in earlier chapters. From the equipment specification sheets, the cost of each piece of equipment is estimated, using techniques discussed later. Once the major equipment cost has been estimated, the total battery limit plant cost can be quickly estimated using factors developed on a similar project.

Correlating Plant Costs

Each engineer should analyze and correlate the costs from at least one project. This gives a feel for various accounts and gives the values found in the literature more meaning. To do this, cost data during project construction should be kept according to a logical set of accounts. As discussed, most estimators do early cost estimates by determining the cost of the major equipment, then multiplying the equipment cost by factors to get the total plant cost. Therefore, costs should be accumulated by major equipment groupings, and then subdivided into accounts such as pipe, valves, electrical, and instruments.

Factors

As an example of correlating plant costs, the cost data from a nylon intermediate plant constructed from stainless steel was accumulated and converted to factors as shown in Table 2.

The question can be asked, "Why estimate the concrete, steel, piping, electrical, etc., when the total plant can be estimated from major equipment in one step?" It is better to show all of the accounts because later estimates will be made this way and the preliminary estimate can be used to check the new estimate. Further, an experienced estimator gets a feel for each account, which allows recognition of errors in early estimates.

Additional information on factors and their use can be found in References 3, 4, 5, and 6.

Table 2
Typical Factors as % of Total Cost of Battery Limits

Account Number	Account Name	Stainless Plant	Recommended Vol. II***
Direct Costs			
1	Major equipment	42.0	35.4
2	Earth work	0.7	1.1
3	Concrete	2.2	3.0
4	Control bldgs.	0.4	0.7
5	Pipe, valves & fittings	13.2	16.4
6	Steel work	5.0	3.4
7	Electrical	4.4	3.4
8	Instrumentation	4.4	3.0
9	Service tankage	*	1.0
10	Small equipment, chem. & cat.	*	0.4
11	Painting	1.1	1.4
12	Insulation & scaffolding	2.9	3.3
13	Payroll taxes & insurance	1.5	1.8
14	Misc. field exp.	3.0	4.4
Indirect costs			
15	Expendable supp.	*	0.2
16	Sales tax	1.3	per state
17	Equipment use	3.1	1.2
18	Field supervision	*	1.2
19	Field overhead	*	0.5
20	Engineering	8.6	7.8
21	Overheads	2.6	2.8
22	Contingency	3.4	7.8**
Total Battery Limits		100.0	100.0

**Included in other accounts.*
***Contingency can vary from 5 to 25%, depending on project status.*
****Reference 2.*

Offsite Estimating

Once the battery limits have been summed, the offsite (the difficult part) must be estimated. Since there is so much variation from site to site, and between "grassroots" plants and construction at existing sites, the use of factors is not recommended.

Site preparation and soil analysis are very important for grassroots plant estimating. If the stage of the project is such that no site has been selected, a generous allowance for site preparation should be included. Once the site has been selected, this phase of the estimate should be firmed at once. If soil conditions are less than ideal, an estimate of the added cost for piling, compacting, or whatever the soil conditions require must be included.

The following items are usually included in the offsite estimate:

Land

1. Site preparation
2. Grading
3. Roads
4. Sanitary sewers
5. Fencing

Utilities

1. Boilers
2. Cooling towers
3. Water supply and treating
4. Offsite electrical
5. Instrument air
6. Plant air
7. Sanitary water system

8. Fuel systems—gas, liquid, solid
9. Flare system

Storage and handling for:

1. Feed stocks
2. Products
3. Chemicals and catalysts
4. Purchased utilities
5. Fuels

Transportation facilities:

1. Railroad
2. Access roads
3. Docks

Waste disposal

1. Liquid waste disposal—evaporation, deep well injection (don't forget filtration and treating), mixing with river or ocean water (don't forget treating)
2. Solids disposal—land fill, hauling, etc.
3. Gas wastes—sulfur recovery and incineration

Personnel facilities

1. Construction camp
2. Buses
3. Offices
4. Showers

Interconnecting piping

Additional checklists of items to include in capital cost estimates can be found in References 3 and 7.

Total Offsite Costs. The offsite costs can range from 20% to 50% of the total cost of the project. If a preliminary built-up estimate of the offsites is less than 30% of the total costs, it should be suspect. Unless the offsites are very well defined, it would be better to use a factor of 50% to 75% of battery limits estimate as the offsite figure.

Utilities

Utilities are very expensive and highly variable from plant to plant. Great care must be exercised to get the proper steam and electrical loads, not only in the process areas, but also in the offsite areas to make sure the cost estimate for the utilities is complete.

Water supply can be quite inexpensive, or very expensive, depending on the plant location. In one coal gasification study, river water was to be used. However, the river was thirty miles away, full of silt, and so unreliable that several days storage of water at the plant site was necessary. The cost of raw water became a significant cost item!

Estimating Major Equipment Costs

Once the sizes of the major equipment items are known, there are several ways to get the cost of each. By far the most accurate is to get quotations from vendors. This is difficult to do in the early stages of a project because many equipment details are not known. Also, the time between making the preliminary study and purchasing the equipment is so long that the quotation ages, and vendors cannot afford to prepare quotations on every preliminary study. However, many vendors are very cooperative in providing verbal prices or "estimating quotations." Often, the vendor is the only source for accurate cost information.

The usual estimating technique is to collect equipment pricing information from other projects and correlate this data by size, weight, pressure rating, and/or materials of construction. Each piece must be adjusted for inflation to bring all costs to one base time. Adjusting costs for inflation is discussed later under the heading, "Construction Cost Indexes."

As an example of this technique, the estimated equipment costs for a large coal gasification project have been correlated and programmed for a computer. Thus, it is very easy to get the cost of any one piece, or of many pieces of equipment, for a coal gasification or hydrocarbon processing project once the specification sheets are completed.

Estimating Vessels

Historical data on similar vessels and fractionation towers can best be used by correlating the costs of this equipment vs. weight. Many methods can be found in the literature for estimating the weight and costs of vessels and fractionators (References 8, 9, 10, and 11). Make sure the estimated weight is complete; including skirt, ladders and platforms, special internals, nozzles, and manholes.

Once the weight has been determined, the cost is obtained by multiplying by a $/lb figure. Up-to-date numbers for the type of vessel or fractionator being estimated can be obtained from a vendor, in-house historical data, literature, or estimating books such as Reference 12. Make sure the cost reflects the materials of construction to be used.

Other methods for estimating the cost of vessels and fractionators can be used, but weight is usually the best. The cost of fractionators can be correlated as a function of the volume of the vessel times the shell thickness, with an addition for the cost of trays based on their diameter (Reference 13). Fractionator costs can also be correlated based on the volume of the vessel with the operating pressure as a parameter. This requires a great deal of data and does not give as good a correlation as weight. Hall et al. (Reference 14) presents curves of column diameter vs. cost.

Estimating Heat Exchangers

The price of air-cooled exchangers should be obtained from vendors if possible. If not, then by correlating in-house historical data on a basis of $/ft^2 of bare surface vs. total bare surface. Correction factors for materials of construction, pressure, numbers of tube rows, and tube length must be used. Literature data on air coolers is available (Reference 15), but it should be the last resort. In any event, at least one air-cooled heat exchanger in each project should be priced by a vendor to calibrate the historical data to reflect the supply and demand situation at the expected time of procurement.

The price of a shell and tube exchanger depends on the type of exchanger, i.e., fixed tube, U-tube, double tube sheets, and removable bundles. The tube side pressure, shell side pressure, and materials of construction also affect the price. If prices cannot be obtained from vendors, correlating in-house data by plotting $/ft^2 vs. number of ft^2 with correction factors for the variables that affect price will allow estimating with fair accuracy. If not enough in-house data is available to establish good correlations, it will be necessary to use the literature, such as References 16, 17, and 18.

Pumps and Drivers

Pumps and their drivers are sometimes obtained from two vendors—the pump manufacturer and the driver manufacturer. There is, however, a big advantage in buying from a single source, thereby fixing responsibility with one vendor for matching and coupling the driver and pump. However, the estimating numbers for pump and driver can often be obtained from different vendors more easily than from the pump manufacturer.

In-house correlation of pumps should be made using gpm vs. cost with head as a parameter. These should result in step functions, since one size pump with different impellers can serve several flow rates and heads. Different correlations should be made for each type of pump. The price of a vertical multistage pump may be quite different from a horizontal splitcase multistage pump. A sump pump with a 15-ft drive shaft would cost more than a single-stage horizontal pump with the same gpm and head.

For in-house correlations, the cost of electric motors should be correlated vs. horsepower with voltage, speed, and type of construction as correction factors or parameters. Correction factors for explosion proof or open drip-proof housings could be developed if most of the data is for TEFC (totally enclosed fan cooled) motors. Similarly, correction factors could be developed for 1,200 rpm and 3,600 rpm with 1,800 rpm as the base.

For steam turbines the cost should be correlated vs. horsepower with steam inlet and outlet pressure as secondary variables.

Several sources (References 19, 20, 21, and 22) are available for estimating pumps and drivers to check in-house correlations or to fill in where data is not available. Care must be exercised in using construction cost indexes to update the literature data. It would be wise to calibrate the indexes and literature data by getting vendor prices on a few of the larger, more expensive pumps, and 5% or 10% of the common types of pumps in the project being estimated.

Storage Tanks

Cone roof and floating roof tanks are usually correlated using $ vs. volume, with materials of construction as another variable. The cost of internal heat exchangers, insulation, unusual corrosion allowance, and special internals should be separated from the basic cost of the tank in the correlations.

Cone roof storage tanks could be correlated using $/lb of steel vs. weight, but the roof support for larger tanks is difficult to estimate, as is the overall thickness.

Pressure storage tanks should be correlated using $/lb vs. weight, much the same as other pressure vessels. Materials of construction, of course, would be another variable. Special internals, insulation, and internal heat exchangers should again be separated from the base cost of the tank. The weight of supports, ladders, and platforms should be estimated and added to the weight of the tank.

Literature estimating techniques for tankage can be found in Reference 17.

High-Pressure Equipment

This equipment presents problems in estimating preliminary costs, since there is seldom enough information in-house to make good correlations. Vendors are by far the best source of costs. Guthrie (Reference 23) discusses the complexities of estimating high-pressure equipment and presents some cost data.

Process Furnaces

Due to the great variation in pressures, flux rates, materials of construction, heat recovery, burner configuration, etc., correlation of process heaters is difficult even with large amounts of data. For similar furnaces, heat absorption vs. cost gives the best correlation. It is again recommended that vendor help be obtained for estimating process furnaces, unless data on similar furnaces is available. Data can be found in References 24 and 25.

Compressors

Compressors are usually high-cost items, but easily correlated by brake horsepower vs. $/horsepower. Variations in engine-driven reciprocating compressor prices can be caused by the type of driver, the speed (the slower the speed the more costly, but the more reliable), the total discharge pressure, and the size.

The driver on centrifugal compressors should be correlated separately. The difference in the initial cost of an electric motor and gas turbine would overshadow the cost correlation of compressors. Literature data can be found in Reference 25.

Construction Cost Indexes

Once the cost of each piece of major equipment is known, it must be adjusted by construction cost indexes. Due to inflation and changing competitive situations, the price of equipment changes from year to year (Reference 26). Fortunately, there are several indexes that help in estimating today's costs based on historical data. Some of these indexes are: Nelson Refinery Construction Cost Index (Reference 27); *Engineering News Record* (ENR) Construction Index (Reference 28); Marshal and Stevens Equipment Index (Reference 29); and *Chemical Engineering* Plant Index (Reference 29).

For feasibility studies on chemical plants, the *Chemical Engineering* Plant Index is recommended. However, if extensive civil work is being estimated, the ENR Index will be applicable.

Since the adjustments for inflation are so large, it is important to fix the date for historical data as closely as possible. For instance, a historical cost estimate from a vendor or contractor for equipment to be delivered in two years would have escalation built in, so the index should be for two years later, when the equipment was expected to be manufactured. However, data based on purchased equipment delivered on a certain date should use the index for the date the equipment was manufactured.

United States Location Factors

The productivity of labor can make enough difference in the cost of plants to justify an adjustment for location, even in early estimates.

Field labor usually runs 25% to 50% of the cost of a project. If one location has 20% lower labor productivity, the cost of the project may increase by 5% to 10%. Contractors have a good feel for the expected productivity in a given area. Such productivity adjustments are usually made with the U.S. Gulf Coast as unity.

Costs in a Foreign Country

Preliminary estimates of costs in foreign countries are difficult due to rapidly changing conditions overseas. For the many companies that do not have construction experience in foreign countries, or in the one being considered, help from a contractor who has built plants in that country is necessary in adjusting U.S. costs. If literature data is needed, References 30 and 31 give some data on foreign construction.

"0.6" Factor Estimates

An alternative to a factored estimate, in some cases, can be a scaled estimate if the battery limits cost of a similar plant is known, but the size is different. The cost of the new plant, C_n, is equal to the known plant cost, C_k, times the ratio of the two plants' capacity raised to a fractional power. That is:

$$C_n = C_k \left(\frac{V_n}{V_k} \right)^F$$

where: V_n = Capacity of the new plant
V_k = Capacity of the known plant
F = Factor, usually between 0.4 and 0.9, depending on the type of plant

A factor of 0.6 is often used in lieu of literature or historical data, so this estimating technique is commonly referred to as the "0.6" factor method.

To build in conservatism, different factors can be used when going up in size, and when going down in size. For instance, if an ammonia plant has been built for $20 million that produces 250 tons per day, and two studies for plants producing 200 TPD and 300 TPD are being prepared, the estimates could be done as follows:

Cost of 200 TPD plant $= 20 \times (200/250)^{0.58}$
$= \$17.6$ million
Cost of 300 TPD plant $= 20 \times (300/250)^{0.62}$
$= \$22.4$ million

Exponent factors can also be used to estimate individual pieces of equipment using prices of similar equipment of a different size. Here, also, the factor varies with the type of equipment and the units chosen for capacity.

Literature values for "0.6" factor exponents can be found in References 3 and 32.

Start-Up Cost Estimates

The start-up costs are usually included in the capital cost of the project. For tax purposes they are often amortized over 5 to 10 years.

Estimating start-up costs can be as simple as choosing 10% of the fixed capital cost, or as complicated as estimating each step of the start-up. If time permits, a detailed estimate is recommended. Not only does it make the start-up costs more accurate, but also starts the planning for operating the plant.

Start-Up Planning

This section discusses some of the items considered in the start-up costs for a coal gasification project, along with the methods of estimating each item.

The management staff (superintendent, plant engineer, operations supervisor, maintenance manager) was assumed to be hired and functioning twelve months before start-up. About one-half of the first six months of their time was charged to engineering, the rest to start-up planning. The last six months before start-up were assumed to be 85% start-up planning and 15% engineering.

The five senior shift supervisors and two maintenance foremen were assumed to be hired nine months before start-up. Their time was divided into 30% checking engineering and construction of the project, 35% operations planning, and 35% training. Five additional shift supervisors were assumed to be hired six months before start-up. Their time was assumed to be 100% training and planning for start-up.

Operator training was planned to take place four months prior to the first heat-up. Half of the operators were to be on the payroll five months, and the other half for three months. As part of the training, each operator was to have spent four weeks in the plant inspecting construction, becoming familiar with the equipment, and helping with hydrostatic testing and internal cleaning.

Since the coal gasification project was very large, the initial heat-up and utilities commissioning was planned to take four weeks. During this time, coal, purchased power, and consumable supplies would be used. The amounts increased from zero on day −30 to 15% on day 0. Day 0 was the day coal was first to be fed to the gasifiers. From day 0 to day +30, the consumption of raw materials was assumed to increase rapidly to 50% of design. The production of pipeline gas was expected to start by day +10, but the start-up expenses were planned with zero credit for gas sales until day +30. From day +30 to day +120, all operating expenses were assumed to increase from 50% to 80%, and the revenue from gas and byproduct sales was assumed to increase from 0 to 70% of design. At this point, the start-up was to be declared finished and the normal operation of the plant commenced. Due to possible continuing start-up problems, the profitability of the project would probably have started lower than projections, but increased to better than projections as operating experience increased. Making allowance for this is discussed in this chapter under Economics.

The expenses associated with each phase of the start-up were estimated to develop the start-up expenses for this first of a kind project. This exercise helped greatly in developing the operating costs.

Total Plant Cost

Regardless of the estimating method, a process contingency should be added to the total plant cost for feasibility studies. As discussed earlier, this contingency depends on the status of the project. For most factored estimates, on a first of a kind process that is fairly well defined, the process contingency should be 30%.

References

1. The Rand Corp., "New Study Provides Key to Better Cost Estimates," *Chemical Engineering,* Feb. 9, 1981, p. 41.
2. Branan, C., *The Process Engineer's Pocket Handbook, Vol. 2,* Gulf Publishing Co., Houston, Texas, 1976, p. 28.
3. Perry, R. H., and Chilton, C. H., *Chemical Engineers' Handbook,* 5th Edition, McGraw-Hill, Inc., New York, 1973, section 25, p. 16.
4. Gallagher, J. T., "Rapid Estimation of Plant Costs," *Chemical Engineering,* Dec. 18, 1967, pp. 89–96.
5. Desai, M. B., "Preliminary Cost Estimating of Process Plants," *Chemical Engineering,* July 27, 1981, pp. 65–70.
6. Viola, J. K., Jr., "Estimating Capital Costs Via a New, Shortcut Method," *Chemical Engineering,* April 6, 1981, pp. 80–86.
7. Hackney, J. W., "How to Appraise Capital Investments," *Chemical Engineering,* May 15, 1961, pp. 145–ff.
8. Mulet, A., Corripio, A. B., and Evans, L. B., "Estimating Costs of Distillation and Absorption Towers," *Chemical Engineering,* Dec. 28, 1981, pp. 77–82.
9. Jordans, A., "Simple Weight Calculations for Steel Towers and Vessels," *Hydrocarbon Processing,* August 1981, p. 146.
10. Prater, N. H., and Antonacci, D. W., "How to Estimate Fractionating Column Costs," *Petroleum Refiner,* July 1960, pp. 119–126.
11. Phadke, P. S., and Kulkarni, P. D., "Estimating the Costs and Weights of Process Vessels," *Chemical Engineering,* April 11, 1977, pp. 157–158.
12. Richardson Rapid System, "Process Plant Construction Estimating Standards," Richardson Engineering Service, Inc. (latest edition) San Marcos, California.
13. Miller, J. S., and Kapella, W. A., "Installed Cost of Distillation Columns," *Chemical Engineering,* April 11, 1977, pp. 129–133.
14. Hall, R. S., Matley, J. and McNaughton, K. J., "Current Costs of Process Equipment," *Chemical Engineering,* April 5, 1982, pp. 80–116.
15. Clerk, J., "Costs of Air vs. Water Cooling," *Chemical Engineering,* Jan. 4, 1965, pp. 100–102.
16. Fanaritis, J. P. and Bevevino, J. W., "How to Select the Optimum Shell and Tube Heat Exchanger," *Chemical Engineering,* July 5, 1976, pp. 62–71.
17. Corripio, A. B., Chrien, K. S., and Evans, L. B., "Estimate Costs of Heat Exchangers and Storage Tanks Via Correlations," *Chemical Engineering,* Jan. 25, 1982, pp. 125–127.
18. Purohit, G. P., "Estimating Costs of Shell and Tube Heat Exchangers," *Chemical Engineering,* Aug. 22, 1983, p. 56.
19. Corripio, A. B., Chrien, K. S., and Evans, L. B., "Estimate Costs of Centrifugal Pumps and Electric Motors," *Chemical Engineering,* Feb. 22, 1982, pp. 115–118.
20. Guthrie, K. M., "Pump and Valve Costs," *Chemical Engineering,* Oct. 11, 1971, pp. 151–159.
21. Olson, C. R. and McKelvy, E. S.,"How to Specify and Cost Estimate Electric Motors," *Hydrocarbon Processing,* Oct. 1967, pp. 118–125.
22. Steen-Johnsen, H., "How To Estimate Cost of Steam Turbines," *Hydrocarbon Processing,* Oct., 1967, pp. 126–130.
23. Guthrie, K. M., "Estimating the Cost of High-Pressure Equipment," *Chemical Engineering,* Dec. 2, 1968, pp. 144–148.
24. Von Wiesenthal, P. and Cooper, H. W., "Guide to Economics of Fired Heater Design," *Chemical Engineering,* April 6, 1970, pp. 104–112.
25. Pikulik, A. and Diaz, H. E., "Cost Estimating for Major Process Equipment," *Chemical Engineering,* Oct. 10, 1977, pp. 106–122.
26. Farrar, G. L., "How Indexes Have Risen," *Oil and Gas Journal,* July 6, 1981, pp. 128–129.
27. "Nelson Construction and Operating Cost Indexes," *Oil and Gas Journal,* First issue each month.

28. "Construction and Building Indexes," *Engineering News-Record,* continuing.
29. "Economic Indicators," *Chemical Engineering,* continuing.
30. Bridgwater, A. V., "International Construction Cost Location Factors," *Chemical Engineering,* Nov. 5, 1979, pp. 119–121.
31. Miller, C. A., "Converting Construction Cost From One Country to Another," *Chemical Engineering,* July 2, 1979, pp. 89–93.
32. Guthrie, K. M., "Capital and Operating Costs," *Chemical Engineering,* June 15, 1970, pp. 140–156.
33. Page, J., *Conceptual Estimating Manual for Process Plant Design and Construction,* Gulf Publishing Co., Houston, Texas, 1984.
34. Branan, C. and Mills, J., *The Process Engineer's Pocket Handbook, Vol. 3,* Gulf Publishing Co., 1984.

Operating Costs

Unlike capital cost estimates, operating cost estimating errors are difficult to see, even in hindsight. If more operators, utilities, or raw materials are needed during actual operations than projected in the early feasibility studies, no one seems to notice. This lack of feedback on estimating operating costs means the engineer doing the early studies must do an especially careful and complete job. The estimator should also undertake an operating audit as soon as possible to validate the estimating techniques.

Estimating Form

A systematic method of developing operating costs should be devised. A table or form that everyone becomes familiar with helps to make all studies uniform and complete. Reference 1 presents a good form and checklist.

Building Up the Operating Cost

The trick to getting a good estimate of the operating costs is being thorough. Accounting plant cost statements help, but of necessity, many costs are lumped into one category so it is difficult to determine all of the items to be included in a first of a kind, or grassroots, plant. Some of the expense items are discussed in the following sections, but each project will have its own unique costs that must be estimated.

Raw Materials

The amount of raw materials needed is supplied by the licensor or process developer. However, unless similar plants are in operation, allowances for unforeseen impacts on raw material usages must be included. Such items as off-spec material, spills, plant upsets, and operator inexperience can increase the quantity of raw materials used.

Once the volume of raw material is set, the price must be estimated. In some studies, a captive source is available with a set transfer price. In other studies, contracts for raw materials will be far enough along to establish the price. However, in some studies, contacts with vendors and the literature is the only source of raw material prices.

The *Chemical Marketing Reporter* tabulates the current list price for many chemicals. These prices, like "off the cuff" estimates from vendors, tend to be conservative. The *European Chemical News* also publishes chemical prices for many places in the world. These prices are sometimes better to use for raw materials.

After the volume and price of the raw materials have been set, the freight estimate must be made. A call to the local rail, barge, or truck office by the traffic department or the engineer will get a price. Here again, the price tends to be conservative for preliminary studies since several months or years elapse before serious freight rates are negotiated.

Operating Labor

The engineer estimating the operating labor must visualize the plant operation, degree of automation, and the labor climate for the project being estimated. For most hydrocarbon processing plants, each control room should have at least one operator with no outside duties. For very large control rooms, more than one such operator may be needed.

The number of outside operators will depend on the layout of the equipment, and the number of operating levels that have pumps, valves, or other equipment needing attention. In remote locations, operators should never work alone. A not-so-good alternative to a second "buddy" operator might be a closed circuit TV in the control room, or a good voice communication system that allows a second person to make sure the remote operator is okay. Certain units, such as boilers and water treating units, tend to be manned with their own operator, even if the amount of equipment does not justify an operator. Laboratory personnel, loading rack operators, guards, and daylight operators must be considered.

When the number of operators per shift has been estimated, multiply the number by five to get the total operating personnel to cover all shifts. This provides for days off, some training, and sick leave. The price per person must be estimated for the project location. Fringes, taxes, and other overheads must be added.

When the engineer has decided on the number of operators and estimated their cost, it is wise to consult an old operating hand who has worked shift on a similar plant to review the number of operators and other assumptions.

Maintenance Labor and Materials

Maintenance labor is usually estimated as a percentage of the capital cost of the project, or as a percentage of the maintenance materials, which in turn is estimated as a percentage of the project cost. For many petrochemical and refinery projects, the maintenance materials can be estimated as 2%–3% of the current cost of the project. The ratio of labor to materials is around 50-50. Exact ratios for the type of project being estimated should be obtained from historical data if available. Get relatively current percentages for materials and labor because the numbers change with economic conditions and the learning curve on the plant. An older plant with all the bugs worked out usually requires less maintenance than a new plant. Reference 2 presents data on the maintenance cost as a percentage of current replacement cost, percentage of original cost, and percentage of depreciated value for many chemical, refining, and other companies.

Supplies and Expenses

This account covers everything from chemicals and catalysts to paper clips. Safety supplies, protective clothing, and tools can usually be estimated knowing the number of operators. When all known items have been estimated, an allowance for unforeseen things should be added.

Supervision

All salaried employees working directly with the plant operation should be included in the estimate of supervision costs. The plant manager's secretary and other supervision support personnel are sometimes included as supervision even though they are not salaried nor in a direct supervisory position. The main thing is to make sure all employee costs are included in some category, but not duplicated in some other category.

Administrative Expenses

The allocation of home office expenses shows up in this category. It has been argued that home office expenses do not increase with the addition of a new plant or unit. However, a full allocation of home office overhead should be made for each study, since old units are phasing out or losing profitability. The new units must carry their share. Further, as discussed next in the section on Economics, each new unit must stand on its own and contribute to the overall company's profitability. The cost of yield clerks, plant accountants, plant personnel representatives, plant transportation department personnel, etc., must be estimated and included here.

Utilities

The cost of utilities is one of the most significant, yet difficult chores encountered in estimating operating costs. As discussed earlier, the amount of utilities required for both the process and the offsite areas must be estimated as accurately as possible. If utilities are generated in the project, the utilities required to operate the utility area must be included. Any increase in the project requires re-estimating the utilities consumed in the utility area. This can result in a trial and error calculation to get the total cost of utilities.

The cost of fuel for generating steam and electricity is usually a major one. The efficiency from fuel to finished utility must be determined. The choice of fuel can become a major study in many circumstances. Future availability and price must be considered.

Water costs can also be substantial in many locations. The raw water cost, as well as the everyday cost of gathering, transporting, settling, treating, and demineralizing must be estimated. Do not forget potable, fire, process, utility, and cooling tower water treating costs. Each treating cost must be estimated and included.

If utilities are supplied to the new project from some other source, the cost and amount must be determined. If purchased from a second party, the cost will be determined by contract and can be estimated by discussions with the vendor. If utilities are transferred from an affiliated source, the cost must include a profit to the supplying entity. Some estimators use a lower return on utility plants than on a new hydrocarbon processing unit, since the utility can be used for some alternate plant if the new one shuts down for any reason. However, the preferred analysis allows a high enough utility transfer price to provide the same return on the utilities as the new unit being studied. This can require a trial and error approach, especially if the utilities are a significant part of the selling price of the product.

Waste Disposal Costs

Historical data on disposal costs are often inadequate for estimating operating costs for a new unit, since environmental laws have changed extensively. In the early stages of planning for a plant, many of the trace impurities and their disposal problems may not be defined, especially for a first of a kind unit. Every effort should be made to define the waste products and their compositions. The operating costs for the ultimate disposal of every waste stream must be estimated. If not enough is known, then an allowance for the costs of the most expensive disposal method must be included.

Wastewater disposal costs depend on the area, and of course, the quality of the wastewater. In one plant, the wastewater was sold to an oil company for secondary oil recovery. However, the cost of treating and filtering the wastewater far exceeded the revenues from the oil company.

Gas waste stream disposal operating costs can be significant. Incineration with or without a catalyst usually requires additional fuel. A heat balance provides the amount. If sulfur must be removed before releasing the waste gas, all operating costs associated with its removal and recovery must be included.

Solid waste disposal has been in the limelight and will continue to be a political, emotional, and technical problem. Due to the attention, solid waste disposal problems must be solved with a sledge, even though a tack hammer would do the job. Make sure there is enough operating money included so that the ultimate disposal of solids is satisfactory for all time.

Local Taxes and Insurance

Current costs for taxes and insurance should be obtained for the proposed project site.

Estimating Operating Costs from Similar Plants

If overall operating costs on similar plants are available, a reasonable estimate can be made for the new project. The closer the plants resemble each other, the more accurate this method.

Several rules of thumb have been developed for estimating operating costs. For instance, Grumer (Reference 3) suggests that the raw material costs can be estimated as a percentage of the sales price of the finished product. Correlations have been made that indicate the overall operating costs can be related to the lbs of products, i.e., the total operating costs are a constant cents/lb of finished product. This obviously will work only when history can be used to get the recent cost on a similar plant.

Hidden Operating Costs

Every project has its share of hidden operating costs, those items that cannot be foreseen and estimated. As an example, the catalyst in a first of a kind ethylbenzene process did not operate in the plant as the research data indicated it should. This caused all facets of the operating costs to be much higher than the estimates. However, after the bugs were worked out, the cost of the catalyst dropped to near zero and the rest of the operating costs were below original projections. In another project, unexpected trace impurities in the product caused extra expenses for rerun, and ultimately increased utilities for additional fractionation. Further, operators can increase or decrease the operating costs, depending on their ability and attitude.

To compensate for some of these unexpected costs, a 10%–20% contingency should be added to the operating costs.

Adjusting Operating Costs for Location

Every location has advantages and disadvantages that impact on the operating costs. Data must be collected for the proposed site and used to adjust historical data to get a proper estimate. Such costs as transportation, labor, and purchased utilities can be determined. However, such items as operating labor productivity, local customs, weather, and local management attitudes can only be estimated.

The estimator's main objective should be to include all the known costs and keep the contingency for unknown and unexpected items to a reasonable level.

References

1. Perry, R. H. and Chilton, C. H., *Chemical Engineers' Handbook,* Fifth Edition, McGraw-Hill, Inc., New York, section 25, pp. 26–28.
2. Annon., "Maintenance Spending Levels Off," *Chemical Week,* July 6, 1983, p. 25.
3. Grumer, E. L., "Selling Price vs. Raw Material Cost," *Chemical Engineering,* April 24, 1967, pp. 12–14.
4. Branan, C., and Mills, J., *The Process Engineer's Pocket Handbook, Vol. 3,* Gulf Publishing Co., 1984.

Economics

The goal of all economic evaluations should be to eliminate all unprofitable projects. Some borderline, and perhaps a few good projects, will be killed along with the bad ones. However, it is much better to pass up several good projects than to build one loser.

Every company should develop a standard form for their economic calculations, with "a place for everything and everything in its place." As management becomes familiar with the form, it is easier to evaluate projects and choose among several based on their merits, rather than the presentation.

Components of an Economic Study

Most economic studies for processing plants will have revenue, expense, tax, and profit items to be developed. These are discussed in the following sections.

Revenues. The selling price for the major product and the quantity to be sold each year must be estimated. If the product is new, this can be a major study. In any case, input is needed from sales, transportation, research, and any other department or individual who can increase the accuracy of the revenue data. The operating costs may be off by 20% without fatal results, but a 20% error in the sales volume or price will have a much larger impact. Further, the operating costs are mostly within the control of the operator, while the sales price and volume can be impacted by circumstances totally beyond the control of the company.

Depending on the product and sales arrangement, the revenues calculation can take from a few man-hours to many man-months of effort. For instance, determining the price for a synthetic fuel from coal can be done in a very short time, based on the cost of service. However, if the price is tied to the price of natural gas or oil, the task becomes very difficult, if not impossible. On the other hand, determining the sales growth and selling price for a new product requires a great deal of analysis, speculation, market research, and luck, but projections can be made.

Most revenue projections should be on a plant net back basis with all transportation, sales expense, etc., deducted. This makes the later evaluation of the plant performance easier.

Operating Costs. The operating costs, as developed in the previous section, can be presented as a single line item in the economics form or broken down into several items.

Amortization. Items of expense incurred before start-up can be accumulated and amortized during the first five or ten years of plant operations. The tax laws and corporate policy on what profit to show can influence the number of years for amortization.

Such items as organization expense, unusual prestart-up expenses, and interest during construction, not capitalized, can be amortized. During plant operations this will be treated as an expense, but it will be a noncash expense similar to depreciation, which is discussed later.

Land Lease Costs. If there are annual payments for land use, this item of expense is often kept separate from other operating costs, since it does not inflate at the same rate as other costs. If the land is purchased, it cannot be depreciated for tax purposes.

Interest. The interest for each year must be calculated for the type of loan expected. The interest rate can be obtained from the company's financial section, recent loan history, or the literature. The *Wall Street Journal* publishes the weekly interest rates for several borrowings. The company's interest rate will probably be the banks' prime rate, or slightly higher.

The calculation of interest is usually based on the outstanding loan amount. For example, semi-annual payments are expected, the amount of the first year's interest (I) would be:

$$I = \left(\frac{i}{2}\right)P + \left(\frac{i}{2}\right)(P - PP) \tag{1}$$

where i = Annual interest rate
P = Original principal amount
PP = First principal payment

The equation for any annual payment would be:

$$I_y = \sum_{n=1}^{n=N} \left(\frac{i}{N}\right)(P_{y,n-1} - PP_{y,n-1}) \tag{2}$$

where I_y = Interest for any year
$P_{y,n-1}$ = Principal or loan balance during the previous period
n = Period for loan payment
N = Total number of payments expected each year
y = Year during the loan life
$PP_{y,n-1}$ = Principal payment made at the end of the previous period

Debt Payments. The method of making principal and interest payments is determined by the terms of the loan agreement.

Loans can be scheduled to be repaid at a fixed amount each period, including principal and interest. This is similar to most home mortgage payments. The size of level payments can be calculated using the formula:

$$I + PP = P\,\frac{\left(\frac{i}{n}\right)\left(1+\frac{i}{n}\right)^{n\times y}}{\left[\left(1+\frac{i}{n}\right)^{n\times y}\right]-1} \tag{3}$$

In this type of loan, the principal payment is different each period, so the calculation of interest using Equation 2 must take into account the variable principal payment. Even though both interest and principal are paid together, they must be kept separate. Interest is an expense item for federal income taxes, while principal payments are not.

Alternatively, the debt repayment can be scheduled with uniform principal payments and varying total payments each period. The principal payment can be calculated by the formula:

$$PP_{y,n-1} = PP = \frac{P}{n\times y} \tag{4}$$

In some loans a grace period is allowed for the first six to twelve months of operation, where no principal payments are due. The economics must reflect this.

Depreciation. This is a noncash cost that accumulates money to rebuild the facilities after the project life is over. There are usually two depreciation rates. One is the corporation's book depreciation, which reflects the expected operating life of the project. The other is the tax depreciation, which is usually the maximum rate allowed by tax laws. To be correct, the tax depreciation should now be called "Accelerated Capital Cost Recovery System" (ACRS).

Economics can be shown using either depreciation rate. If book depreciation is used, a line for deferred income tax must be used. This is a cash available item that becomes negative when the tax depreciation becomes less than the book depreciation. At the end of the book life, the cumulative deferred income tax should be zero.

The 1984 allowable ACRS tax depreciation is shown in Table 1. Most chemical plants have a five-year tax life.

Federal Income Tax. The calculation of the federal income tax depends on the current tax laws. State income taxes are often included in this line item, since many are directly proportional to the federal taxes.

Table 1
ACRS Schedule in % Per Year of Depreciable Plant

Year	Years for Recovery			
	3	5	10	15
1	25	15	8	5
2	38	22	14	10
3	37	21	12	9
4		21	10	8
5		21	10	7
6			10	7
7			9	6
8			9	6
9			9	6
10			9	6
11–15				6

The federal tax laws are complicated and ever-changing. It is recommended that expert tax advice be obtained for most feasibility studies. If no current tax advice is available, many tax manuals are published each year that can be purchased or borrowed from a library.

Current tax laws allow deductions from revenues for sales expenses, operating costs, amortizations, lease costs, interest payments, depreciation, and depletion.

The net income before federal income tax (FIT) is calculated by subtracting all of the deductions from the gross revenues. If there is a net loss for any year, it can be carried forward and deducted the next year from the net income before FIT. If not all of the loss carried forward can be used in the following year, it can be carried forward for a total of seven years.

At times, tax laws allow a 10% investment tax credit for certain qualifying facilities. Thus, 10% of the qualifying investment can be subtracted from the first 85% of the FIT due. If the taxes paid in the first year are not enough to cover the investment tax credit, it can sometimes be carried forward for several years.

Exhaustible natural deposits and timber might qualify for a cost depletion allowance. Each year, depletion is equal to the property's basis, times the number of units sold that year, divided by the number of recoverable units in the deposit at the first of the year. The basis is the cost of the property, exclusive of the investment that can be depreciated, less depletion taken in previous years.

At times, mining and some oil and gas operations are entitled to an alternate depletion allowance calculated from a percentage of gross sales. A current tax manual should be consulted to get the percentage depletion allowance for the mineral being studied.

Percentage depletion for most minerals is limited to a percentage of the taxable income from the property receiving the depletion allowance. For certain oil and gas producers the percentage depletion can be limited to a percentage of the taxpayer's net income from all sources.

The procedure for a feasibility study is to calculate both the cost and percentage depletion allowances, then choose the larger. Remember, either type reduces the basis. However, percentage depletion deducted after the basis reaches zero.

The cash taxes due are calculated by multiplying the book net income before tax, times the tax rate, then adjusting for deferred tax and tax credits, if any.

Economics Form

As stated at the start of this chapter, it is desirable to establish a form for reporting economics. Table 2 shows such a form. By following this form, the procedure can be seen for calculating the entire economics.

As can be seen in Table 2 all revenues less expenses associated with selling are summed in Row 17. All expenses including noncash expenses such as depreciation, amortization, and depletion are summed in Row 30. The net profit before tax, Row 32, is obtained by subtracting Row 30 from Row 17 and making any inventory adjustment required. Row 34 is the cash taxes that are to be paid unless offset by investment or energy tax credits in Row 36. The deferred income tax is shown in Row 35. The deferred tax decreases the net profit after tax in the early years and increases the net profit after tax in later years. The impact on cash flow is just the other way around as discussed later. Row 37, profit after tax, is obtained as follows:

Row 32 (profit before tax)
−Row 34 (income tax)
−Row 35 (deferred tax)
+Row 36 (tax credits)
=Row 37 (profit after tax)

Row 33, loss carried forward, is subtracted from Row 32 before calculating the income tax.

The cash flow, Row 39, is calculated as follows:

Row 37 (profit after tax)
+Row 35 (deferred tax)
+Row 29 (depletion)
+Row 28 (amortization)

+Row 27 (depreciation)
−Row 38 (principal payments)
=Row 39 (cash flow)

If one of the rows is a negative number, such as deferred tax, the proper sign is used to obtain the correct answer. Any required inventory adjustment is also made.

Table 2
Economic calculation example

	Periods/Year = 1	
Fiscal years	1992	1993
12 Sales revenue	239,093	248,657
13 Byproduct revenue	17,031	17,713
14 Advert'g & market'g	(1,638)	(1,703)
15 Shipping & freight	(1,310)	(1,363)
16 Other sales expense	(983)	(1,022)
17 Income from sales	252,194	262,282
18 Raw material cost	76,075	78,738
19 Operating labor	26,508	27,701
20 Maintenance labor	23,984	25,063
21 Maint'nce material	37,869	39,573
22 Utilities	25,246	26,382
23 Administr. & general	9,467	9,893
24 Unchanging expense	5,000	5,000
25 Temporary expense	0	0
26 Interest	7,614	7,188
27 Depreciation	5,500	5,500
28 Amortization	3,639	2,139
29 Depletion	12,621	19,541
30 Total Cost	233,524	246,719
31 Inventory adjust't	0	0
32 Profit before tax	18,671	15,563
33 Loss carried forw'd	12,621	4,031
34 Income tax	0	7,835
35 Deferred tax	2,783	(2,530)
36 Tax credits	0	7,835
37 Profit after tax	15,888	18,093
38 Principal payments	4,105	4,531
39 Cash flow	36,325	38,212

Economic Analysis

Once the cash flow has been calculated, various yardsticks can be applied to determine the desirability of the project. The years for payout (payback) is often used. This is obtained by adding the cash flow each year, from the first year of construction, until the sum of the negative and positive cash flows reaches zero. The payout is usually reported in years to the nearest tenth, i.e., 5.6 years.

Each company has its own cutoff period for payout, depending on the economy, the company position, alternate projects, etc. One company used five years as the maximum payout acceptable for several years. As conditions changed, this was reduced to three years, then one year. It was hard to find projects with one year payout after taxes.

Some companies feel they should earn interest on their equity as well as get a payout. To calculate this, the cash flows are discounted back to year 0 using some arbitrary interest (discount) rate, then the positive discounted cash flows are summed each year of operation until they equal the sum of the discounted negative cash flows. This yardstick also works but consistency must be maintained from one project to another. The discount (interest) rate and payout both become variables that must be set and perhaps changed from time to time depending on economic conditions inside and outside the company.

A simple return on equity can be used as a yardstick. Utilities are required by regulators to set their rates to obtain a given rate of return. This return is usually constant over the years and from project to project, so management must choose among projects on some basis other than profit.

In free enterprise projects, the return on equity will probably vary from year to year so an average would have to be used. However, use of an average would consider return in later years to be as valuable as income in the first year of operation, which is not true.

The best way to evaluate free enterprise projects is to use discounted cash flow (DCF) rate of return, sometimes called internal rate of return.

This calculation can be expressed as follows:

$$SDCF = \sum_{y=1}^{y=N} \frac{CF_y}{(1+i)^y} = 0 \tag{5}$$

where SDCF = Sum of the equity discounted cash flows
CF_y = Cash flow, both negative and positive for each year y
i = Discount rate in %/100 that results in the sum being zero
N = Total number of years

Since this calculation uses trial and error to find a discount rate at which all discounted negative and positive cash flows are equal, it can be tedious without a computer. Table 3 demonstrates this calculation for a simple set of cash flows. The DCF rate of return results in a single number that reflects the time value of money. It is certainly one of the yardsticks that should be used to measure projects.

Sensitivity Analysis

Once the economic analysis has been completed, the project should be analyzed for unexpected as well as expected impacts on the economics. This is usually done through a set of "what if" calculations that test the project's sensitivity to missed estimates and changing economic environment. As a minimum, the DCF rate of return should be calculated for ±10% variations in capital, operating expenses, and sales volume and price.

Table 3
Discounted cash flow calculations

Discount Rate (%)	Discounted Cash Flows			
Year	0	20	30	25.9
1	−20	−17	−15	−16
2	−20	−14	−12	−12
3	−10	−6	−5	−5
4	10	5	3	4
5	15	6	4	4
6	20	7	4	5
7	25	7	4	5
8	30	7	4	5
9	40	8	4	5
10	50	8	4	5
Totals	+140	+11	−5	00

The impact of loss of tax credits, loss carry forward and other tax benefits should be studied. Such things as start-up problems that increase the first year's operating costs and reduce the production should also be studied. With a good economic computer program, such as described earlier, all of these sensitivity analyses can be made in an afternoon.

Additional reading on economics can be found in the References.

References

1. Perry, R. H. and Chilton, C. H., *Chemical Engineers Handbook,* Fifth Edition, McGraw-Hill, Inc., New York, section 25, p. 3.
2. Constantinescu, S., "Payback Formula Accounts for Inflation," *Chemical Engineering,* Dec. 13, 1982, p. 109.
3. Horwitz, B. A., "How Does Construction Time Affect Return?" *Chemical Engineering,* Sept. 21, 1981, p. 158.
4. Kasner, E., "Breakeven Analysis Evaluates Investment Alternatives," *Chemical Engineering,* Feb. 26, 1979, pp. 117–118.
5. Cason, R. L., "Go Broke While Showing A Profit," *Hydrocarbon Processing,* July, 1975, pp. 179–190.
6. Reul, R. I., "Which Investment Appraisal Technique Should You Use," *Chemical Engineering,* April 22, 1968, pp. 212–218.
7. Branan, C. and Mills, J., *The Process Engineer's Pocket Handbook, Vol. 3,* Gulf Publishing Co., 1984.

Financing

Most feasibility studies involve debt and equity. In the section on Economics it was assumed that there was a debt and interest to be paid. The amount of debt, or debt to equity ratio, varies widely from company to company. Some companies assume 100% equity and require the project to meet their set criteria on the entire investment. In this case all borrowings are considered to be at the corporate level, which then provides 100% of the funds to each project. In other companies, particularly utilities, the debt at the project level can be as high as 75% or more.

Some energy and other projects that obtain government loan guarantees could truly treat the equity as their total investment in the project.

Debt

Banks are not in business to take risks. They rent money and do everything they can to insure the return of their principal as well as the interest. Elaborate rating systems have been developed to measure each company's ability to repay its loans. One criterion is the debt to equity ratio. The higher the debt the more risk in a loan, and the higher the interest rate.

Some projects obtain project financing. That is, the parent company does not become liable for the loan. Some have been based on back to back contracts for construction, raw materials, and sales. These usually have the backing of a large sponsoring company with a reputation of paying its debts. Some projects have obtained project financing based on federal loan guarantees. Large natural gas projects have been financed based on FPC or FERC certificates, gas purchase contracts, and acceptable construction contracts. This package essentially guarantees the repayment of all debt.

In trying to finance large synthetic fuel projects based on first of a kind processes in the U.S. during a highly inflationary time, the question of guaranteeing project completion was raised by the banks. FERC certificates do not guarantee loan repayment until after the project is in operation. The banks perceived a risk of the cost going so high that the sponsoring companies either could not or would not provide the necessary funds to finish the project. Certainly, the engineering and construction contractor could not afford to guarantee completion through a lump sum price. Government guaranteed loans solved this and other risk problems in a few of these synthetic fuel projects.

Additional reading on project financing can be found in Reference 1.

Cost of Money

The money used in a project must be returned by the project with interest whether it comes from debt or equity. The recent swings in the cost of money have brought to everyone's attention the complexity of setting return and interest rates for feasibility studies. If a project is ready to be built and the loans have been negotiated, the interest rate can be determined accurately. Establishing the interest rate for a feasibility study on a project that will not be built for one or more years in the future is difficult and becomes a policy decision that should be set by management. The same rate should be used on all studies to keep the yardstick the same length.

Lenders: Their Outlook and Role

The lender has an obligation to his depositors to see that all loans are collected. To fulfill this obligation, the lenders sometimes become intimately involved with a company's operation if they feel the risk of a bad debt is too large. They can however, be very helpful to a company in evaluating feasibility studies by injecting an outside view. Their expertise should be used.

Equity Financing

Equity financing comes 100% from the owners. They are in business to take risks but must realize a higher return to make up for this higher risk. Several ways are open to get equity funds. An asset may be sold, common or preferred stock may be sold, or retained earnings may be accumulated. Whatever the source, management must make sure the project is feasible, will pay the debt, and will make a profit.

Leasing

Tax and economic conditions can make leasing a feasible method of financing hydrocarbon processing facilities. The operating company leases the plant from an investor.

The analysis on whether to buy or lease should be done very carefully. The assumption of risk, impact on the balance sheet, tax consequences, etc., must be studied.

Davis (Reference 2) lists some of the reasons for leasing, as well as some of the pitfalls.

References

1. Castle, G. R., "A Banker's Approach to Project Finance," *Hydrocarbon Processing,* March, 1978, pp. 90–98.
2. Davis, J. C., "Rent-A-Plant Plans Spread to the CPI," *Chemical Engineering,* Sept. 15, 1975, pp. 78–82.
3. Branan, C. and Mills, J., *The Process Engineer's Pocket Handbook, Vol. 3,* Gulf Publishing Co., 1984.

17
Reliability

RELIABILITY

The process engineer doesn't find a large volume of material written on plant reliability in general. Reliability is the concern of several groups of people:

- The original designers
- The operating people
- The maintenance department
- The technical service people

We are concerned, in this discussion, with original design for reliability. Norman Lieberman's excellent book[1] laments the dying art of process design born of field imparted discernment. We must be careful that computerized standard design methods do not produce "cookie cutter" unreliable plants.

The best and most effective time to provide for reliability is in the initial design. The importance of initial design is illustrated by a study[2] undertaken by Sohio at their Toledo refinery. Their first listed major finding from the study was as follows:

1. "A disproportionately high percentage (39%) of all failure producing problems encountered during the 11 year life of the processes were revealed during the first year of operation. This high failure rate during the early life of a new unit is felt to be characteristic of most complex processes. From the standpoint of reliability control, these early failures could potentially be prevented by improvements in design, equipment selection, construction, and early operator training."

Since the process engineer plays a large role in initial design, what are some things he/she can do to help assure reliability?

1. Understand existing operations of similar process units before embarking on a new design.[1] Lieberman gives the following ways to become familiar with an existing process:
   - Inspect the equipment in the field
   - Talk to the operators
   - Examine operating data
   - Crawl through towers during turnarounds
2. Include tech service (design) engineers in turnaround inspections.[3]

   Look familiar? This suggestion is the title of an article but is also the last listed suggestion by Lieberman.[1] Headings in the article are:
   - Why and when should process engineers inspect equipment?
   - What items should a process engineer take to inspect equipment?
   - What should the process engineer look for?
   - What should process engineers do with inspection data?
3. Reflect the unique nature of every process application.[1] Lieberman gives typical things that a successful process design must consider:
   - Local climate
   - Associated processes
   - Variabilities of feedstocks
   - Character of the unit operators
   - Environmental constraints
   - Heat and material balances
4. Define requirements.[4]

   Kerridge has provided an excellent article on the interface between the operating company and the contractor to define all requirements in complete and standardized detail. This includes "who is responsible?" for every deliverable. The operating company and contractor must work as a team. An example of one area that needs to be reviewed often with the contractor is the provision of secondary systems as packages, perhaps from a third party. Such systems can easily become orphans. This problem is discussed in the Process Definition section of Chapter 16.
5. Provide requirements for specific processes. Every commercial process has unique requirements. One of the best examples that I am familiar with is the requirements outlined for an amine gas treating plant in Don Ballard's timeless article.[5] Such advice as:
   - The temperature of the lean amine solution entering the absorber should be about 10°F higher than the inlet gas temperature to prevent hydrocarbon condensation and subsequent foaming.
   - The reboiler tube bundle should be placed on a slide about six inches above the bottom of the shell to provide good circulation.

- About two percent of the total circulation flow should pass through the carbon towers.

is invaluable to the design. Seek out advice from experts on a particular process. These can be in-house or consultants.

References

1. Lieberman, N. P., *Process Design For Reliable Operations, 2nd Ed.,* Gulf Publishing Co., 1988.
2. Cornett, C. L., and Jones, J. L., "Reliability Revisited," *Chemical Engineering Progress,* December 1970.
3. Miller, J. E., "Include Tech Service Engineers In Turnaround Inspections," *Hydrocarbon Processing,* May 1987.
4. Kerridge, A. E., "For Quality, Define Requirements," *Hydrocarbon Processing,* April 1990.
5. Ballard, Don, "How To Operate An Amine Plant," *Hydrocarbon Processing,* April 1966.

18
Metallurgy

Embrittlement

Paul D. Thomas, Consulting Metallurgist, Houston

Brittleness is ". . . the quality of a material that leads to crack propagation without appreciable plastic deformation."[1]

Embrittlement is ". . . the reduction in the normal ductility of a metal due to a physical or chemical change."[1]

With these definitions in mind, a systematic classification has been made. The various types of embrittlement found in refineries and petrochemical plant equipment, susceptible steels, basic causes, and common remedies are listed in the accompanying table.

Embrittlement originates in the following four ways:

Intrinsic steel quality refers to the metallurgical and chemical properties of steel products (plate, pipe, tubes, structurals, castings, forgings) supplied to the fabricator for conversion into process equipment. Factors related to deoxidation, controlled finishing temperatures in rolling, and cleaning up of surface defects are included.

Fabrication and erection. Embrittlement problems associated with forming, welding and heat treatment are included in this section, although in some instances the heat treatment is done by the steel manufacturer.

Metallurgical problems in service. Long exposure to high temperatures in service, to aging temperatures, or to alternating stresses can lead to different types of embrittlement. The changes in metallurgical microstructure range from gross (sigma) to subtle (aging), while the early stages of grain distortion in fatigue give no easily detectable clue, even microscopically.

Corrosion problems in service. The wide variety of corrosive factors that are present in refining and petrochemical processing (e.g. acids, caustics, chlorides, sulfides, sulfates) give rise to a wide variety of corrosive attacks, some of which act:

1. Directly on the steel (intergranular corrosion)
2. In concert with the stress environment (stress corrosion cracking, corrosion fatigue)
3. By indirect embrittlement (reaction by-product, atomic hydrogen diffusing into the lattice of the steel)

Impact testing. The common measure of brittleness, or the tendency towards embrittlement, is an impact test. The most frequently used tests are the pendulum-type, simple-beam Charpy test and some form of drop weight test. The term "impact" does not relate strictly to either shock loading of the test piece or to shock loading of vessels, piping, and structurals during fabrication, handling, transportation, erection, start up, operation, and/or shutdown of a unit. Of more importance is the respectable degree of correlation that impact test results have been found to have with failure behavior of steel in the presence of:

Stress concentrations, occurring at

1. Sharp corners in design
2. Points of restraint imposed by design
3. Notch-type defects in:
   a. The steel comprising the unit
   b. Welds joining the steel components
4. Any points of triaxial loading

Temperature changes, entailing

1. Low temperature effects, including ductile-to-brittle transition temperatures
2. Embrittling effects incurred
   a. At operating temperatures
   b. During cooling from operating to ambient temperatures

Cracks related to

1. Initial stage
2. Propagation stage to failure
   a. Type of failure
   b. Rate of failure
3. Aging (between ambient and 450°F)
4. Graphitization
5. Intergranular precipitation at elevated temperatures
6. Transgranular cracking at elevated temperatures

Table 1 should be regarded only as a general guide. Its conciseness does not permit either absolute completeness or detailed consideration of specific problems; some of these problems are being discussed in more detail in the other articles. However, the table does outline:

- The various forms in which embrittlement can show up in petroleum refining and petrochemical processing
- The basic causes of the embrittlement
- The common remedies

Care must be exercised however in applying the common remedies across the board for economic and technical reasons:

Economic. The degree or extent of the embrittlement problem may require little or nothing in the way of special materials and/or extra precautions (e.g., aging, forming in fabrication).

Technical. The effects of any corrective or precautionary steps must be evaluated with consideration given to ef-

(text continued on page 253)

Table 1
Embrittlement in Petroleum Refining and Petroleum Operations

Types of Embrittlement	Temperature Range	Manifestations	Typical Service, Equipment Structures Involved	Susceptible Steels	Basic Cause(s)	Common Remedies
Intrinsic Steel Quality						
Aging and Strain-aging	500/900°F 210/480°C	Increased hardness; lowered ductility and impact resistance; higher D-B transition temperatures	Piping, valves, lines, particularly in cold weather (below 40°F/5°C approx.)	Rimmed, capped, and semikilled low carbon steels	Nitrogen in steel, insufficient aluminum to take up nitrogen (Al N2); strain-aging is aging intensified by cold working.	Use fully killed or fine grained steels in services below 750°F in critical locations.
High Ductile-to-Brittle Transition Temperature (DBT)	–20/+180°F –29/+80°C (2)	Low notch toughness; high notch sensitivity; rapid, running failures below DBT.	Carbon steel piping, valves, structurals and vessels operating at cryogenic and ambient temperatures.	Carbon steels generally, unless particular precautions are taken in deoxidation, control of rolling temperatures, and heat treatment	Coarse grain, high finishing temperature, cold working, presence of notches and sharp corners (stress raisers generally).	Use fully killed or fine grain steel, controlled rolling temperatures; high Mn/C ratios; eliminate sharp corners in design, remove defects from steel; heat treat steel. For cryogenic operations use high nickel alloy steels or austenitic stainless steels, depending on temperature.
Notch sensitivity Associated with Defects in Steel	Any temperature but aggravated by low temperatures.	Visually observable breaks in surface; laminations; indications on non-destructive testing charts.	Any equipment subject to applied and residual stresses in fabrication, erection and service.	Any steel (defects are imperfections of sufficient extent to warrant rejection of the piece of steel). No steel is entirely free of imperfections.	Metallurgical; deoxidation, rolling, damage, inadequate croppage of ends; defects form notches or planes of internal, directional weakness.	Apply operational controls in mill, including careful removal of surface defects at intermediate or final stages; crop sufficiently to remove piped steel.
Fabrication and Erection Cold Working in Forming	Ambient	Warping, residual stresses in pressing, cutting; possible surface damage; lowered ductility; higher DBT.	Any plate or pipe being pressed, cut to fit, or bent.	Rimmed, capped and semikilled most affected but all types and grades have some degree of susceptibility.	Cold working and grain distortion in working, uneven working, build up and irregular distribution of residual stresses.	Use stress relief or full heat treatment.

(table footnotes found on page 253)

Table 1 (continued)
Embrittlement in Petroleum Refining and Petroleum Operations

Types of Embrittlement	Temperature Range	Manifestations	Typical Service, Equipment Structures Involved	Susceptible Steels	Basic Cause(s)	Common Remedies
Fabrication and Erection (Cont.) Poor Weldability		a. Underbead cracking, high hardness in heat-affected zone. b. Sensitization of nonstabilized austenitic stainless steels.	a. Any welded structure. b. Same	a. Steel with high carbon equivalents (3), sufficiently high alloy contents. b. Nonstabilized austenitic steels are subject to sensitization.	a. High carbon equivalents (3), alloy contents, segregations of carbon and alloys. b. Precipitation of chromium carbides in grain boundaries and depletion of Cr in adjacent areas.	a. Use steels with acceptable carbon equivalents (3); preheat and postheat when necessary; stress relieve the unit. b. Use stabilized austenitic or ELC stainless steels.
High Alloy Content in Weld Beads		Cracking in service, particularly in wet sulfide service when Rockwell hardness exceeds C 22 (4).	Any welds made with bonded high alloy—or low alloy-fluxes unless welding current/temperature is closely controlled.	Rods/electrodes used with bonded fluxes.	Irregular reduction of alloys (particularly Si and Mn) from the bonded flux, causing hard spots in weld.	Use neutral flux or chemically compatible rod; alternately minimize alloy reduction from bonded flux. Keep Rockwell hardness C22 or under in weld bead if moist H_2S involved in service (4).
Welding Defects		Visual, radiographic and/or ultrasonic indications.	Any welded joints.	Basic welding problem; laminated steel can cause trouble.	Electrode manipulation.	Control of welding speeds, procedures, careful inspection and nondestructive testing to locate defects for cutting out or repair.
Temper Embrittlement in Heat Treatment	650/1100°F 340/600°C	Very low impact test values and higher D-B transition temperatures.	Heat treated alloy piping and bolting.	Heat treated alloy steels.	Slow cooling through, or prolonged time in range of 650/1100°F; high chromium, manganese and phosphorus accentuate the temper brittleness.	Add about .50% molybdenum; quench or otherwise rapidly cool from tempering temperature; do not prolong tempering treatment.
Metallurgical Factors in Service Graphitization	850/1100°F 450/600°C	Breakdown of iron carbide into iron and carbon, with graphite particles precipitating and tending to form into a "chain," particularly alongside welds, thus embrittling affected section.	Reactors, regenerators power piping components operating at or above 850°F (maximum breakdown at 1,025°F, 550°C).	Fine grain and fully killed carbon and carbon-molybdenum steels.	Inhibiting nitrogen taken up by aluminum as Al N_2 in fully killed and fine grain steels (nitrogen in steel stabilizes iron carbides.)	Add chromium or avoid use of fully killed and fine grain steels unless chromium is added.

Table 1 (continued)
Embrittlement in Petroleum Refining and Petroleum Operations

Types of Embrittlement	Temperature Range	Manifestations	Typical Service, Equipment Structures Involved	Susceptible Steels	Basic Cause(s)	Common Remedies
Metallurgical Factors in Service (Cont.)						
885°F Embrittlement (475°C)	700/1050°F 370/656°C	Room temperature brittleness after exposure to temperatures between about 700 to 1,050°F.	Ferritic chromium stainless steels.	400 Series ferritic chromium stainless steels, over 13% Cr and any 400 Series martensitic chromium stainless steels low in carbon content (high Gr/C ratio).	Precipitation of a complex chromium compound, possibly a chromium-phosphorus compound.	Do not use ferritic chromium steels at temperatures above about 700°F (370°C); keep carbon up in martensitic chromium steels and limit Cr to 13% max.
Sigma Phase Embrittlement	800/1600°F 425/870°C	Brittleness at ambient and high temperatures (1,600°F max.)	Tubes, pipe, piping components, C.C. regenerators operating within the effective temperature range, 1,100/1,600°F for short times.	Austenitic steels.	Precipitation of iron-carbon phase along grain boundaries, and within grains.	Employs alloy combinations that will minimize chances of Sigma forming.
Fatigue		Development of cracks at the base damage marks, material defects, weld defects, or sharp corners and other notches present in design.	Any unit, structure or component that is subject to vibration or other repeated stresses in the presence of a notch.	Any grade of steel can be subject to fatigue when operating under repeated stresses in the presence of a notch, microscopic or macroscopic in size.	Fatigue progresses at the base of a notch in three stages: (1) Stress concentration; (2) Cracks initiate under repeated stresses; (3) Cracks propagate to failure.	Eliminate notches in design or minimize notch effect as much as possible; carefully inspect welds and base material; eliminate or minimize vibrations, stress levels.
Carbide Precipitation, Sensitization	800/1600°F 425/870°C	Low impact values at cryogenic temperatures (under –320°F). Room temperature embrittlement is nominal.	Pipe and piping components, trays and vessels.	Austenitic stainless steels, except the stabilized (Types 321, 347) and extra low carbon (304L, 316L) grades.	Precipitation of chromium carbides into grain boundaries, with depletion of chromium in contiguous grain boundary areas.	Use stabilized or extra low carbon austenitic stainless steels.
Corrosive Factors in Service Intergranular Corrosion (attends carbide precipitation if electrolyte present)	800/1600°F 425/870°C	Sensitization of some stainless steels to selective attack on adjacent grain boundaries by corrosive solutions, followed by "sugaring." Most damaging effect of sensitizing.	Pipe and piping components, trays and vessels.	Austenitic stainless steels, except the stabilized (types 321, 347) and extra low carbon (304L, 316L) grades.	Precipitation of chromium carbides into grain boundaries, with depletion of chromium in contiguous grain boundary areas resulting in selective, galvanic attack on Cr- depleted areas, in the presence of aqueous phases.	Use stabilized or extra low carbon stainless steels, or if steel has become sensitized in service apply full solution heat treatment to regular nonstabilized grades.

(table footnotes found on page 252)

Table 1 (continued)
Embrittlement in Petroleum Refining and Petroleum Operations

Types of Embrittlement	Temperature Range	Manifestations	Typical Service, Equipment Structures	Susceptible Steels	Basic Cause(s)	Common Remedies
Corrosive Factors in Service (Cont.)						
Stress Corrosion Cracking (SCC) including Caustic Embrittlement	Up to maximum of H_2O liquid phase.	Branched cracks, transgranular except for caustic embrittlement and for improperly heat treated steel, both of which give intergranular cracks. (Intercrystalline penetration by molten metals is also considered SCC).	Stainless steels in chlorides and caustic solutions. Other steels in caustic nitrates and some chloride solutions. Brass in aqueous ammonia and sulfur dioxide.	Almost any grade of steel under some chemical and physical environments.	Stresses—dynamic or static—concentrate at bases of small corrosion pits, and cracks form with vicious circle of additional corrosion and further crack propagation until failure occurs. Stresses may be dynamic, static, or residual.	Use material which is not susceptible to SCC; thermally stress relieve susceptible materials. Consider the new superaustenitic stainless steels.
Corrosion Fatigue		Development of cracks at the roots of corrosion pits, under the influence of repeated stresses.	Any unit, structure, or component that is subject to vibration, or any other fluctuating stress in presence of a corrosive environment.	Any grade of steel can be subject to corrosion fatigue when operating under conducive fluctuating stress and corrosive conditions.	Corrosion fatigue progress through three stages: (a) formation of corrosion pits with stress concentration; (b) Cracks initiate at base of pits under repeated stresses; (c) Cracks propagate to failure.	Eliminate or minimize corrosive conditions (inhibitors, coatings). Eliminate vibrations or minimize frequency and range of stress variations.
High temp. Hydrogen attack a. Steels without chromium additives.	a. Temperature to Pressure Relationship.	a. Formation of intergranular cracks with blisters forming in any laminated areas; reduction of ductility and hardness.	a. Hydrogenation, dehydrogenation, and synthesis plants.	a. Carbon, C-Mn and C-Mo steels.	a. Decarburization of steel by hydrogen; formation of intergranular cracks and blisters; these reduce strength and promote brittleness. Blisters develop at laminations.	a. Refer to Nelson curves (5); Use Cr-Mo steels in high temperature zones and low carbon or C-Mo steels in lower temperature zones.
b. Chromium steels.	b. Same.	b. Low impact values.	b. Same as a.	b. Cr-Mo steels in range of 1 to 6% Cr operating outside limits of Nelson curves (5).	b. Entrapment of hydrogen in lattice as steel cools down from operating temperature.	b. Observe Nelson curves (5); cool vessel down at rate sufficiently slow to reduce hydrogen in lattice to safe limits.

Table 1 (continued)
Embrittlement in Petroleum Refining and Petroleum Operations

Types of Embrittlement	Temperature Range	Manifestations	Typical Service, Equipment Structures	Susceptible Steels	Basic Cause(s)	Common Remedies
Hydrogen attack at ambient temperatures a. Hydrogen Blistering	a. Ambient to boiling	a. Numerous scattered blisters of varying sizes developing on surface of steel; fissure cracks developing in castings.	a. Gas plants, acidic water lines, acid storage tanks, acid cleaning and pickling operations; amine treating plants; any services involving reducing acids; ammonium hydrosulfide in stream.	a. Rimmed, capped, and semikilled carbon and low alloy steels when Rockwell hardness is under C 22.	a. Diffusion of atomic hydrogen (H) from reacting surface of steel into the crystal structure, concentrating in voids, inclusions and minor laminations. The atomic H changes to nondiffusible molecular H_2, building up high localized pressures.	a. Eliminate water from the systems; use fully killed steel; use inhibitors and/or resistant coatings and linings, if feasible; neutralize acid waters.
b. Hydrogen Sulfide Attack on softer steels (4).	b. Ambient to boiling	b. Blistering and corrosion pitting. Pitting promotes corrosion fatigue.	b. Any equipment handling hydrogen sulfide with any water present.	b. Any carbon or low alloy steel with Rockwell under C 22. (4) Rimmed, capped and semikilled steels most susceptible to blistering.	b. Same as immediately above plus the stimulation and acceleration of H diffusion by the presence of the S ion H_2S is source of both H and S.	b. Keep Rockwell hardness under C 22 (4) and pH high; Use inhibitors and/or coatings where feasible. Time or heating up will remove atomic hydrogen, but will not reduce blisters.
c. Hydrogen Sulfide Embrittlement (Sulfide Stress Cracking) on steels over Rockwell C 22. (4)	c. Ambient to about 150°F (66°C)	c. Brittle fracture under dynamic or static stresses.	Pipe and piping components and any other equipment handling sour gas, oil and/or water wherein H_2S and H_2O (liquid phase) are present up to about 150°F, where sulfide stress cracking slows down perceptibly.	c. Any carbon, low alloy and many stainless steels with Rockwell hardness over C 22. (4)	c. Embrittlement by atomic H diffusion into crystal structure, exact mechanism uncertain. Sulfur expedites absorption of atomic H into grain structure.	c . Keep Rockwell hardness under C 22 (4) if feasible; use inhibitors and/or resistant coatings where feasible; time or heating up will permit H to diffuse out but will not relieve any areas when H_2 has concentrated.

Notes:

(1) "Temperature Range" refers to the range of operating temperatures, heat treatmet temperatures, or temperatures developed in welding which can produce the type of embrittlement shown.

(2) Codes set up their controls on impact values of steel at –20°F (–29°C) and below but brittleness can well be a problem at ambient and higher temperatures.

(3) "Carbon Equivalent" (CE) is an approximate measure of weldability expressed in terms of the sum of carbon content and the alloy contents divided by applicable factors to r equivalence in carbon in effectiveness in hardening—and thereby cracking. Commonly used formulas with commonly accepted but rather arbitrarily set maximums are:

(a) $C + (Mn/4) = 0.60$

(b) $C + (Mn/6) = 0.45$

Denominator factors for nickel and chromium are given as 20 and 10 respectively in Linnert, Welding Metallurgy, Volume 2, Third Edition, American Welding Society (page 394).

(4) Rockwell C 22 is the commonly selected limit above which sulfide embrittlement and resultant sulfide stress cracking become problems. The change, however, is not that abrupt but the critical "gray band" is about C 20 to 25, with the point of change affected by mechanical, physical and chemical environmental factors.

(5) "Nelson Curves" have been published in API Publication 941, First Edition, July 1970.

(text continued from page 248)

fects on subsequent associated units (e.g., inhibitors). Some compromise material or precautions may be necessary. Some forms of embrittlement are temporary (e.g., hydrogen embrittlement, with or without the influence of sulfides), or are present at ambient temperature but not at operating temperature (e.g., 885°F embrittlement and embrittlement of high alloy 25/20 and 35/20 after service at high temperatures).

For these reasons and because brittleness does not reduce to a formula, a competent metallurgist and a corrosion engineer who are familiar with the overall operation as well as the specific unit should be called in by the design and process engineers to evaluate the problems in their full scope.

Literature Cited

1. "Metals Handbook, Vol. 1 Properties and Selection of Metals," 8th Edition, American Society for Metals, Metals Park (Novelty), Ohio.

Stress-Corrosion Cracking

When an alloy fails by a distinct crack, you might suspect stress-corrosion cracking as the cause. Cracking will occur when there is a combination of corrosion and stress (either externally applied or internally applied by residual stress). It may be either intergranular or transgranular, depending on the alloy and the type of corrosion.

Austenitic stainless steels (the 300 series) are particularly susceptible to stress-corrosion cracking. Frequently, chlorides in the process stream are the cause of this type of attack. Remove the chlorides and you will probably eliminate stress-corrosion cracking where it has been a problem.

Many austenitic stainless steels have failed during downtime because the piping or tubes were not protected from chlorides. A good precaution is to blanket austenitic stainless steel piping and tubing during downtime with an inert gas (nitrogen).

If furnace tubes become sensitized and fail by stress corrosion cracking, the remaining tubes can be stabilized by a heat treatment of 24 hours at 1,600°F. In other words, if you can't remove the corrosive condition, remove the stress.

Chemically stabilized steels, such as Type 304L have been successfully used in a sulfidic corrodent environment but actual installation tests have not been consistent.

Straight chromium ferritic stainless steels are less sensitive to stress corrosion cracking than austenitic steels (18 Cr-8 Ni) but are noted for poor resistance to acidic condensates.

In water solutions containing hydrogen sulfide, austenitic steels fail by stress corrosion cracking when they are quenched and tempered to high strength and hardness (above about Rockwell C24).

Hydrogen Attack

Cecil M. Cooper, Los Angeles

Atomic Hydrogen will diffuse into and pass through the crystal lattices of metals causing damage, called hydrogen attack. Metals subject to hydrogen attack are: the carbon, low-alloy, ferritic and martensitic stainless steels, even at subambient temperatures and atmospheric pressure. Sources of hydrogen in the processing of hydrocarbons include the following:

- Nascent atomic hydrogen released at metal surfaces by chemical reactions between the process environment and the metal (corrosion or cathodic protection reactions)
- Nascent atomic hydrogen released by a process reaction such as catalytic desulfurization
- Dissociation of molecular hydrogen gas under pressure at container metal surfaces

Types of damage. Hydrogen attack is characterized by three types of damage, as follows:

Internal stresses with accompanying embrittlement may be temporary only, during operation. Embrittlement is caused by the presence of atomic hydrogen within the metal crystal lattices, with ductility returning once the source of diffusing hydrogen is removed. But such embrittlement becomes permanent when hydrogen atoms combine in submicroscopic and larger voids to form trapped molecules which, under extreme conditions, may build up sufficient stresses to cause subsurface fissuring.

Blistering and other forms of local yielding. Diffusing atomic hydrogen combines to form molecular hydrogen gas in all voids until local yielding or cracking results.[1] Such voids include laminations, slag pockets, shrinkage cavities in castings, unvented annular spaces in duplex tubes, unvented metal, plastic or ceramic linings, unvented voids in partial-penetration welds and subsurface cracks in welds.

Decarburization and fissuring. Diffusing hydrogen combines chemically with carbon of the iron carbides in steels to form methane at hydrogen partial pressures above 100 psia at temperatures above 430 to 675°F, depending upon hydrogen pressures. This reaction begins at the metal surface and progresses inward, causing both surface and subsurface decarburization and the buildup of the gas in voids and grain boundaries, and resulting in eventual subsurface fissuring.

Conditions for hydrogen damage. Only the first two types of damage described above occur under aqueous-phase (wet) conditions.[2] When the source of diffusing hydrogen is corrosion of the containing equipment, high-strength, highly-stressed parts, such as carbon steel or low-alloy gas compressor rotor blades, exchanger floating-head bolts, valve internals and safety valve springs are particularly susceptible to embrittlement.[3] Experience has confirmed, however, that if yield strengths are limited to 90,000 psig, maximum, and working stresses are limited to 80 percent of yield, cracking failures are minimized. Welds and their heat affected zones should be stress relieved, hardnesses should be limited to Rockwell C22, or less, and they should be free from hard spots, slag pockets, subsurface as well as surface cracks, and other voids.

Tests of such steels as ASTM-A302, A212, and T-1, at elevated pressures have revealed no appreciable effects of hydrogen on the ultimate tensile strength of unnotched specimens, but have shown losses as high as 59 percent in the ultimate tensile strength of notched specimens.[4] Special quality control measures are necessary in the fabrication of such containers to minimize notch effects, and design pressures should be lowered 30 to 50 percent below the pressures considered safe for the storage of inert gases. The safest procedure would seem to be to use vented, fully austenitic stainless steel liners in such containers, and austenitic stainless steel auxiliary parts for which linings are not practical.

The use of fine-grained, fully-killed carbon steels (such as ASTM-A516 plate) are not justifiable as a first choice for aqueous-phase hydrogen services. This is because these premium grades are equally subject to embrittlement and subsurface fissuring as the lower grades. In the absence of blistering, such emrittlement is not easily discernible. Although the lower grades may blister, this blistering serves as its own warning that steps should be taken to protect the equipment from hydrogen damage.

When corrosion is the source of diffusing hydrogen, the elimination of this reactive hydrogen is necessary to stop its diffusion into the metal. This can be done either by eliminating the corrosion, or by eliminating chemicals in the process stream that "poison" the corroding metal surface: that is, by eliminating chemical agents which retard or inhibit the combining of nascent hydrogen atoms at the corroding surface into molecular hydrogen which cannot enter the metal.

Corrosion can be eliminated by the use of corrosion resistant barriers and/or alloys. This becomes necessary in situations in which measures to eliminate "poisoning agents" are not permissible or effective. Ganister linings (one part lumnite cement and three parts fine fire clay) over expanded metal mesh, and to a limited extent, plastic coatings have been used successfully where the normal operating pH is maintained above 7.0. Austenitic strip or sheet linings and metallurgically bonded claddings are used extensively, and are usually adequate for the most severe operating conditions. Surface aluminizing of small parts such as relief valve springs has been reported to be effective in protecting against hydrogen embrittlement.

Compounds such as hydrogen sulfide and cyanides are the most common metal surface poisoners occurring in process units subject to aqueous-phase hydrogen attack. In many process units, these compounds can be effectively eliminated and hydrogen diffusion stopped by adding ammonium polysulfides and oxygen to the process streams which converts the compounds to polysulfides and thiocyanates, provided the pH is kept on the alkaline side.

Nonaqueous (dry), elevated-temperature hydrogen attack can produce all three types of damage previously described. Since iron carbides in steels give up their carbon to form methane at lower temperatures and pressures than do the carbides of molybdenum, chromium, and certain other alloying elements, progressively higher operating temperatures and hydrogen partial pressures are permissible by tying up, or stabilizing, the carbon content of the steel with correspondingly higher percentages of these alloying elements.[5] The fully austenitic stainless steels are satisfactory at all operating temperatures.

To avoid decarburization and fissuring of the carbon and low-alloy steels, which is cumulative with time and, for all practical purposes irreversible, the limitations of the Nelson Curves should be followed religiously, as a minimum.[5] Suitable low-alloy plate materials include ASTM-A204-A, B, and C and A387-A, B, C, D, and E, and similarly alloyed materials for pipe, tubes, and castings, depending upon stream temperatures and hydrogen partial pressures, as indicated by the Nelson Curves.

Stainless steel cladding over low-alloy, or carbon steel base metal is usually specified for corrosion resistance, where necessary in elevated-temperature hydrogen services. This poses the additional hazards of voids created by unbonded spots in roll-bonded cladding, sensitization of the cladding to stress corrosion cracking by the final stress relief heat treatment of the vessel, and cracking of the cladding welds by differential thermal expansion. Rigid quality-control in steelmaking and fabrication is therefore highly important, as is frequent and thorough maintenance inspection during operation, for pressure containing equipment in elevated temperature hydrogen services.

Literature Cited

1. Skei and Wachttr: "Hydrogen Blistering of Steel in Hydrogen Sulfide Solutions," *Corrosion,* May 1953.
2. Bonner, W. A., et al: "Prevention of Hydrogen Attack on Steel in Refinery Equipment," paper presented at API Div. of Refining 18th Mid-year Meeting, May 1953.
3. NACE Publication 1F166: "Sulfide Cracking-Resistant Metallic Materials for Valves for Production and Pipeline Service," *Materials Protection,* Sept. 1966.
4. Walter, R. J., and Chandler, W. T.: "Effects of High Pressure Hydrogen on Storage Vessel Materials," paper presented at meeting of American Society for Metals, Los Angeles, March 1968.
5. API Div. of Refining Publication 1941: "Steels for Hydrogen Service at Elevated Temperatures and Pressures in Petroleum Refineries and Petrochemical Plants," July 1970.

Pitting Corrosion

Pits occur as small areas of localized corrosion and vary in size, frequency of occurrence, and depth. Rapid penetration of the metal may occur, leading to metal perforation. Pits are often initiated because of inhomogeneity of the metal surface, deposits on the surface, or breaks in a passive film. The intensity of attack is related to the ratio of cathode area to anode area (pit site), as well as the effect of the environment. Halide ions such as chlorides often stimulate pitting corrosion. Once a pit starts, a concentration-cell is developed since the base of the pit is less accessible to oxygen.

Pits often occur beneath adhering substances where the oxidizing capacity is not replenished sufficiently within the pores or cavities to maintain passivity there. Once the pit is activated, the surface surrounding the point becomes cathodic and penetration within the pore is rapid.

Pitting environments. In ordinary sea water the dissolved oxygen in the water is sufficient to maintain passivity, whereas beneath a barnacle or other adhering substance, metal becomes active since the rate of oxygen replenishment is too slow to maintain passivity, activation and pitting result.

Exposure in a solution that is passivating and yet not far removed from the passive-activity boundary may lead to corrosion if the exposure involves a strong abrasive condition as well. Pitting of pump shafts under packing handling sea water is an example of this abrasive effect.

The 18-8 stainless steels pit severely in fatty acids, salt brines, and salt solutions. Often the solution for such chronic behavior is to switch to plastics or glass fibers that do not pit because they are made of more inert material.

Copper additions seem to be helpful to avoid pitting. Submerged specimens in New York Harbor produced excessive pitting on a 28 percent Chromium stainless but no pitting on a 20 Cr-1 Cu alloy for the period of time tested.

Higher alloys. Some higher alloys pit to a greater degree than lower alloyed materials. Inconel 600 pits more than type 304 in some salt solutions. Just adding alloy is not necessarily the answer or a sure preventative.

In a few solutions such as distilled, tap, or other fresh waters, the stainless steels pit but it is of a superficial nature. In these same solutions carbon steels suffer severe attack.

Pitting in amine service. In the early 1950s the increased use of natural gas from sour gas areas multiplied the number of plants using amine systems for gas sweetening. Here severe pitting occurred in carbon steel tubed reboilers, in fact, failures occurred in two or three months. The problem was solved by lowering temperatures or by going to more alloyed materials.

In this same amine service it was found that carbon steel pits much more readily if it has not been stress relieved.

Aluminum was tried in the reboiler of an aqueous amine plant but it was found to pit very quickly.

Type 304 stainless steel does not seem to be a satisfactory material in all instances either. Monel and Type 316 appear most suitable in this service where pitting must be avoided.

Salt water cooling tower. A somewhat unusual occurrence where certainly the proper choice of materials was made but pitting still occurred involved a salt water cooling tower. The makeup water for this cooling tower was brackish and high in sulfur and from a highly industrialized area. By allowing the sulfur to concentrate in this closed circuit cooling system, 90-10 Cu-Ni tubes were pitted, with the pits going entirely through the tubes in three months.

A comparison unit with identical design used once through water cooling. This is the same high sulfur water that was used as makeup for the above cooling tower. Here the sulfur was not allowed to concentrate and the unit is working fine after several years.

Tests for pitting. A test which is in popular use to determine pitting characteristics is the 10 percent Ferric Chloride test which can be conducted at room temperature. This test is usually used for stainless or alloy steels.

Salt Spray (Fog) Testing such as set forth in ASTM B-117, B-287, and B-368 are useful methods of determining pitting characteristics for any given alloy. These are also useful for testing inorganic and organic coatings, etc., especially where such tests are the basis for material or product specifications.

Avoiding pitting. Often, this is a matter of design tied in with proper material selection. For example, a heat exchanger that uses cooling water on the shell side will cause tube pitting, regardless of the alloy used. If the cooling water is put through the tubes, no pitting will occur if proper alloy selection is made.

Creep and Creep-Rupture Life

Walter J. Lochmann, The Ralph M. Parsons Company, Los Angeles

Creep is that phenomenon associated with a material in which the material elongates with time under constant applied stress, usually at elevated temperatures. A material such as tar will creep on a hot day under its own weight. For steels, creep becomes evident at temperatures above 650°F. The term *creep* was derived because, at the time it was first recognized, the deformation which occurred at the design conditions occurred at a relatively slow rate.

Depending upon the stress load, time, and temperature, the extension of a metal associated with creep finally ends in failure. Creep-rupture or stress-rupture are the terms used to indicate the stress level to produce failure in a material at a given temperature for a particular period of time. For example, the stress to produce rupture for carbon steel in 10,000 hours (1.14 years) at a temperature of 900°F is substantially less than the ultimate tensile strength of the steel at the corresponding temperature. The tensile strength of carbon steel at 900°F is 54,000 psi, whereas the stress to cause rupture in 10,000 hours is only 11,500 psi.

To better understand creep, it is helpful to know something of the fracture characteristics of metals as a function of temperature for a given rate of testing. At room temperature, mechanical failures generally occur through the grains (transcrystalline) of a metal, while at more elevated temperature, failure occurs around the grains (intercrystalline). Such fracture characteristics indicate that at room temperature there is greater strength in the grain boundaries than in the grain itself. As temperature is increased, a point is reached called the "equicohesive temperature," at which the grain and the grain boundaries have the same strength. Below the equicohesive temperature, initial deformation is elastic, whereas above the equicohesive temperature deformation becomes more of a plastic rather than an elastic property.

The factors that influence creep are:

- For any given alloy, a coarse grain size possesses the greatest creep strength at the more elevated temperatures, while at the lower temperatures a fine grain size is superior.
- Creep becomes an important factor with different metals and alloys at different temperatures. For example, lead at room temperature behaves similarly to carbon steel at 1,000°F and to certain of the stainless steels and superalloys at 1,200°F and higher.
- Relatively slight changes in composition often alter the creep strength appreciably, with the carbide-forming elements being the most effective in improving the strength.

Creep-rupture life. Because failure will ultimately occur if creep is allowed to continue, the engineering designer must not only consider design stress values such that the creep deformation will not exceed a limiting amount for the contemplated service life, but also that fracture does not occur. In refinery practice, the design stress is usually based on the average stress to produce rupture in 100,000 hours or the average stress needed to produce a creep rate of 0.01 percent per 1,000 hours (considered to be equivalent to one percent per 100,000 hours) with appropriate factors of safety. A criterion based on rupture strength is preferable because rupture life is easier to determine than low creep rates.

Hydrogen reformers. The refinery unit in which the creep problem is more prevalent is the steam-methane, hydrogen producing reformer. In reforming furnace design, creep and creep-rupture life of the catalyst tubes, outlet pigtails and collection headers usually set the upper limit for the possible operating temperatures and pressures of the reforming process.

In a reforming furnace, the economically available materials for catalyst tubes will creep under stress. However, from creep-rupture curves, or stress-rupture curves as they are usually called, the designer can predict the expected service life of the catalyst tubes. The relative magnitude of the effects of stress and temperature on tube life are such that at operating temperatures of 1,650°F, a small increase in the tube-metal temperature drastically reduces service life. For example, a prediction that is 100°F too low will result in changing the expected life from 10 years into an actual life of 1½ years. Conversely, a prediction that is 100°F too high will increase the tube thickness unnecessarily by 40 percent with a corresponding increase in cost. The tube material cost for a hydrogen reformer heater is almost 30 percent of the total cost and is in the neighborhood of $350,000 for a 250 MM Btu/hr. heater; therefore, judicious use of the optimum tube design and selection of materials is essential for adequate service life and for significant cost savings.

ASTM A297 Gr. HK or A 351 HK-40, a 26 Cr-20 Ni alloy with a carbon range of 0.35 to 0.45 percent is the material almost always specified for catalyst tubes. A recent API Survey indicated that for most plants the tube wall was designed on the basis of stress to produce rupture in 100,000 hours. Other design bases were 50 percent of the stress to produce rupture in 10,000 hours or 40 to 50 percent of the stress to produce one percent creep in 10,000 hours.

Failures of catalyst tubes have occurred principally because of overheating and consequent creep-rupture cracking. Overheating may have been caused by local hot spots in the furnace as a result of faulty burners, inadequate control of furnace temperature, ineffective catalyst, or plugging of the catalyst tube inlet pigtail or catalyst tube. Headers and transfer line failures have resulted from inability of the materials to withstand strains from thermal gradients and internal loads during cyclic operation or from poor design and/or material selection.

Solution annealed Incoloy 800 is the material almost universally selected for the outlet pigtails. Pigtail failures have been particularly troublesome and have frequently been the result of creep failure associated with the stresses resulting from thermal expansion and bending moments transmitted from the catalyst tubes or collection headers. HK-40 material, although it has a higher creep strength, is not used for pigtails because it has insufficient high temperature ductility for this sensitive application.

Special consideration must also be given to selecting welding electrode materials with adequate creep strength and ductility for joining HK-40 to Incoloy 800. Inco weld "A" is the material most commonly used, with some refineries using Inco 82, 182, or 112. For joining HK-40, a high-carbon, Type 310 stainless steel filler material is generally used. For joining Inconel 800, Inco "A," Inco 82, or Inco 182 are the filler materials used.

The present trend of material selected for collection headers is toward Incoloy 800. The cast alloys used, HK and HT, have failed in most instances because of their inherently low ductility—especially after exposure to elevated temperature. It now appears that wrought alloys should be used in preference to cast alloys unless the higher creep strength of the cast alloy is required and the inherently low ductility of the aged cast alloy is considered in the design.

For hydrogen reformer transfer lines, materials used are Incoloy 800, HK, and HT cast stainless steels, Wrought 300 series stainless steels and internally insulated carbon, carbon-½, Mo, and 1¼ Cr-½ Mo steels. Reported failures of transfer lines indicate that failures are associated with design that did not account for all the stresses from thermal expansion, or failure or cracking of the insulation in internally insulated lines. The general trend is toward headers of low alloy steel internally lined with refractory.

Piping, exchanger tubing, and furnace hardware. Above temperatures of 900°F, the austenitic stainless steel and other high alloy materials demonstrate increasingly superior creep and stress-rupture properties over the chromium-molybdenum steels. For furnace hangers, tube supports, and other hardware exposed to firebox temperatures, cast alloys of 25 Cr-20 Ni and 25 Cr-12 Ni are frequently used. These materials are also generally needed because of their resistance to oxidation and other high temperature corrodents.

Furnace tubes, piping, and exchanger tubing with metal temperatures above 800°F now tend to be an austenitic stainless steel, e.g., Type 304, 321, and 347, although the chromium-molybdenum steels are still used extensively. The stainless steels are favored because not only are their creep and stress-rupture properties superior at temperatures over 900°F, but more importantly because of their vastly superior resistance to high-temperature sulfide corrosion and oxidation. Where corrosion is not a significant factor, e.g., steam generation, the low alloys, and in some applications, carbon steel may be used.

Pressure vessels. Refineries have many pressure vessels, e.g., hydrocracker reactors, cokers, and catalytic cracking regenerators, that operate within the creep range, i.e., above 650°F. However, the phenomenon of creep does not become an important factor until temperatures are over 800°F. Below this temperature, the design stresses are usually based on the short-time, elevated temperature, tensile test.

For desulfurizers, coking, and catalytic reforming units, many of the pressure vessels operating with metal temperatures of 700 to 900°F are constructed from carbon-½ Mo or 1¼ Cr-½ Mo alloy steels. These steels have marked creep-rupture strength properties over carbon steel. For example, compare the design stress value of 15,000 psi for 1¼ Cr-½ Mo steel at 900°F with the 6,500 psi value for carbon steel at the corresponding temperature.

For hot wall vessels, the increased strength may be such that the use of chromium and molybdenum alloy steels will be cheaper. Also, these steels may be required to prevent hydrogen attack and to reduce oxidation and sulfidation.

For refinery units such as hydrocrackers in which the hydrogen partial pressure is much higher, e.g., above 1,350 psi and the operating temperatures are above 800°F, the 2¼ Cr-1 Mo steel is commonly used. The higher alloy content is necessary to prevent hydrogen attack in these applications. The 2¼ Cr-1 Mo steel also has better creep and stress-rupture properties than the 1¼ Cr-½ Mo and carbon-

½ molybdenum steels, but is about 20 percent higher in cost. It is therefore selected in most of its refinery applications over the lower alloyed steel for its better resistance to corrosion and hydrogen attack. Finally, this alloy can be obtained in plate thicknesses of 10 to 12 inches, whereas there is now about a 6-inch thickness limitation for the 1¼ Cr-½ Mo alloy.

Carbon steel, where not limited by sulfur corrosion or hydrogen attack, can be the most economical material for elevated temperature service. Where the creep and stress-rupture temperature conditions are so high as to severely limit the service life or require too low design stress values, it is often advantageous to refractory line carbon steel and thus reduce the metal temperature rather than use materials with higher creep strength. Furnace stacks; sulfur plant combustion chamber, reaction furnaces, converters and incinerators; catalytic cracking regenerators; and some cold-shell catalytic reformer reactors are all examples of refinery equipment operating in the creep range that use refractory lined carbon steel.

Metal Dusting

Robert C. Schueler, Phillips Petroleum Company, Bartlesville, Oklahoma

Metal dusting is a form of metal deterioration that occurs in carbonaceous gas streams containing carbon monoxide and/or hydrocarbons at elevated temperatures. Until the early 1940s, carbon monoxide was the only recognized gaseous phase associated with this phenomenon. Few processes were operating at conditions conducive to this type of deterioration. With the development of butane dehydrogenation, coal conversion, and gas cracking processes in the 1940s, additional cases of metal dusting were reported. Specific problems were solved by changing temperatures and steam-hydrocarbon ratio, by changing alloy or by adding a sulfur compound inhibitor.

Figure 1. Pitting form of metal dusting.

The term, "metal dusting," was first used about this time to describe the phenomenon associated with hydrocarbon processing. Butane dehydrogenation plant personnel noted how iron oxide and coke radiated outward through catalyst particles from a metal contaminant which acted as a nucleating point. The metal had deteriorated and appeared to have turned to dust. The phenomenon has been called "catastrophic carburization" and "metal deterioration in a high temperature carbonaceous environment," but the term most commonly used today is metal dusting.

Variations. Metal dusting may appear in the form of pits, uniform thinning, or a combination of both in iron, nickel, and cobalt base alloys. The most common form is pitting. Pits (Figure 1) may vary in size, may be smooth or rough, may be undercut below the original metal surface, and may be filled with coke or carbonaceous deposits. Magnetic particles are always found in the coke deposits. Microscopic examination will usually show carburization, intergranular attack, and loss of grains at the surface. Reaction products (iron oxide, graphite, iron carbides) observed at high magnification may be in the form of filaments.

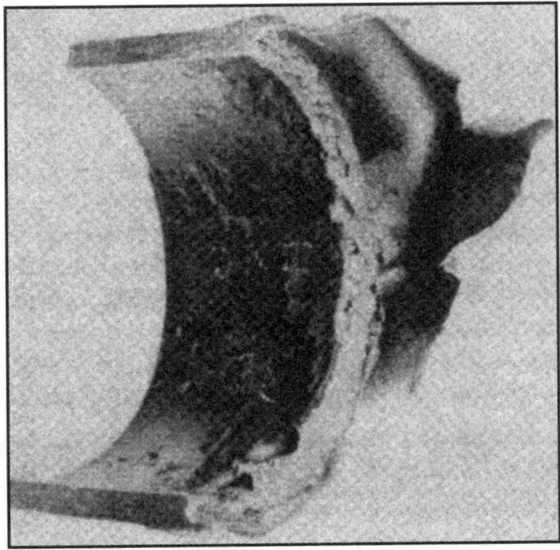

Figure 2. Section of tube at right shows uniform thinning by metal dusting.

Uniform thinning (Figure 2) is sometimes associated with high gas velocity where the surface reactants are carried

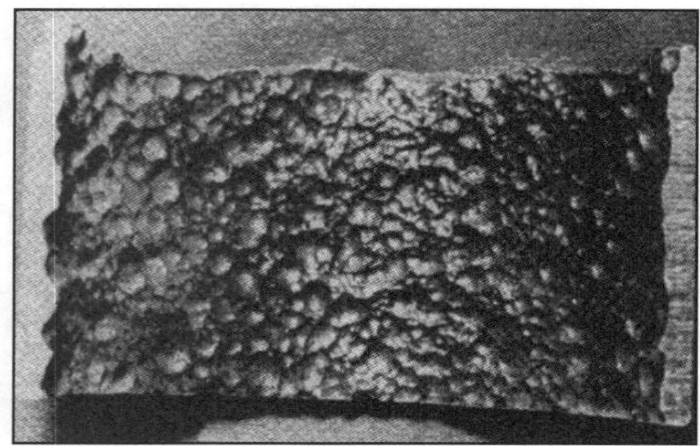

Figure 3. Metal dusting as a combination of pitting and thinning.

away. Magnetic reaction products are produced and are found downstream. They contain carbides and metal particles which are similar to those formed in the pitting form of dusting. There is very little or no carburization of the surface, but some slight intergranular attack may develop at the surface. Resistance of austenitic stainless steels to thinning is reduced and reaction rate is increased as the nickel content is increased while the Cr remains constant.

A combination of pitting and uniform thinning has the appearance of a corrosion-erosion (Figure 3). The tube wall is thinned, the surface is rippled, and some scattered pits may be visible. Reaction products are essentially the same as those formed in pitting and uniform thinning.

Occurrence. Metal dusting can occur at varying rates, but most reported cases indicate a rapid deterioration. Furnace tubes ¼-in. thick have been penetrated in less than 500 hours. Reaction temperature range in hydrocarbon atmosphere is about 800–1,700°F with very high activity about 1,350–1,450°F. In carbon monoxide only, the activity is limited to a range of 800–1,400°F. Metal dusting does not occur above the scaling temperature of the metal if oxygen or an oxygen compound is available in the process stream. It also does not occur in the *complete* absence of oxygen.

Prediction and control. Although research has been conducted intermittently since the 1940s, it is still difficult to predict the exact environments in which metal dusting will develop. The effects of other gases (H_2, steam, O_2) and impurities and the effects of temperature and pressure have not been clearly defined.

An effective way to determine if dusting will be a problem in a new high temperature hydrocarbon process is to conduct tests in small scale equipment in which the proposed materials and environment of the full scale unit are duplicated. These tests should be continued until metal deterioration is detected, but not over 500 hours. In the event that dusting occurs, additional testing can then be conducted to develop the most desirable preventive method. If thermal decoking will be involved in operations, the effect of the decoking should also be determined using the small scale equipment. Decoking has been found to definitely initiate or increase the severity of dusting. Carburization with accompanying reduction of available chromium will also initiate the attack.

Among the measures which have successfully prevented metal dusting are the use of additives (steam, and compounds of S, As, Sb, and P) in the feed, reduction of pressure, reduction of temperature, and material change. The most common additives are sulfur compounds and steam. Susceptibility can be reduced by using a material in which the total percent of Cr plus two times the percent of Si is in excess of 22 percent. In some environments, a small amount of a sulfur compound will stop the dusting. When sulfur compounds cannot be tolerated in the process stream, a combination of steam and an alloy with a Cr equivalent of over 22 percent may be most desirable.

Because of the inexactly defined conditions causing metal dusting, mitigation methods will not be the same for each occurrence. Each problem must be carefully studied to determine the most effective and economic measures that will be compatible with the process stream.

References

Camp, E. Q., Phillips, Cecil, and Gross, Lewis, "Stop Tube Failure in Superheater by Adding Corrosion Inhibitors," *National Petroleum News,* 38 (10) R-192-99, March 6, 1946.

Prange, F. A., "Corrosion In a Hydrocarbon Conversion System," *Corrosion,* 15 (12) 619t-21t (1959).

Hoyt, W. B., and Caughey, R. H., "High Temperature Metal Deterioration in Atmospheres Containing Carbon Monoxide and Hydrogen," *Corrosion,* 15 (12) 627t-30t (1959).

Eberle, F., and Wylie, R. D., "Attack on Metals by Synthesis Gas from Methane-Oxygen Combustion," *Corrosion,* 15 (12) 622t-26t (1959).

Hochman, R. F., and Burson, J. H., "The Fundamentals of Metal Dusting," API Division of Refining, Vol. 46, 1966.

Koszman, I., "Antifoulant Additive for Steam-Cracking Process," U.S. Patent 3,531,394, Sept. 29, 1970.

Hochman, R. F., "Fundamentals of the Metal Dusting Reaction," Proceedings, Fourth International Congress on Metallic Corrosion, NACE (1971).

Naphthenic Acid Corrosion

Cecil M. Cooper, Los Angeles

Naphthenic acid is a collective name for organic acids present in some but not all crude oils.[1] In addition to true naphthenic acids (naphthenic carboxylic acids represented by the formula X-COOH in which X is a cyclo-paraffin radical), the total acidity of a crude may include various amounts of other organic acids and sometimes mineral acids. Thus the *total* neutralization number of a stock, which is a measure of its total acidity, includes (but does not necessarily represent) the level of naphthenic acids present. The neutralization number is the number of milligrams of potassium hydroxide required to neutralize one gram of stock as determined by titration using phenolphthalein as an indicator, or as determined by potentiometric titration. It may be as high as 10 mg. KOH/gr. for some crudes. The neutralization number does not usually become important as a corrosion factor, however, unless it is at least 0.5 mg. KOH/gm.

Theoretically, corrosion rates from naphthenic acids are proportional to the level of the neutralization number of feed stocks; but investigators have been unable to find a precise correlation between these factors. Predicting corrosion rates based on the neutralization number remains uncertain. Published data, however, indicate a scattered trend toward increasing corrosion with increasing neutralization number.[2]

Temperatures required for corrosion by naphthenic acids range from 450 to 750°F, with maximum rates often occurring between 520 and 535°F.[1] Whenever rates again show an increase with a rise in temperature above 650°F, such increase is believed to be caused by the influence of sulfur compounds which become corrosive to carbon and low alloy steels at that temperature.

Where is naphthenic acid corrosion found? Naphthenic acid corrosion occurs primarily in crude and vacuum distillation units, and less frequently in thermal and catalytic cracking operations. It usually occurs in furnace coils, transfer lines, vacuum columns and their overhead condensers, sidestream coolers, and pumps. This corrosion is most pronounced in locations of high velocity, turbulence and impingement, such as at elbows, weld reinforcements, pump impellers, steam injection nozzles, and locations where freshly condensed fractions drip upon or run down metal surfaces.

What does the corrosion look like? Metal surfaces corroded by naphthenic acids are characterized by sharp-edged, streamlined grooves or ripples resembling erosion effects, in which all corrosion products have been swept away, leaving very clean, rough surfaces.

Which alloy to use. Unalloyed mild steel parts have been known to corrode at rates as high as 800 mils per year. The low-chrome steels, through 9-Cr, are sometimes much more resistant than mild steel. No corrosion has been reported, with both 2¼-Cr and 5-Cr furnace tubes, whereas carbon steel tubes in the same service suffered severe corrosion. The 12-Cr stainless steels are scarcely, if any, better than the low-chromes. But the 18-8 Cr-Ni steels, without molybdenum, are often quite resistant under conditions of low velocity although they are sometimes subject to severe pitting.

Type 316 (18-8-3 Cr-Ni-Mo) has, by far, the highest resistance to naphthenic acids of any of the 18-8 Cr-Ni alloys and provides adequate protection under most circumstances. It provides excellent protection against both high temperature sulfur corrosion and naphthenic acids whereas the 18-8 Cr-Ni alloys without molybdenum are not adequate for both. The high-nickel alloys, except those containing copper (such as monel) are also highly resistant but have little advantage in this respect over Type 316. Copper and all copper alloys, including aluminum-copper alloys such as Duralumin (5-Cu), are unsuitable.

Aluminum and aluminum-clad steels are highly resistant to naphthenic acids under most conditions. Aluminum-coated steels give good service until coatings fail at coating imperfections, cracks, welds, or other voids. In general, the use of aluminum and aluminized steels for the control of corrosion by naphthenic acids, as well as by other elevated temperatures corrodents, such as hydrogen sulfide, is somewhat unpredictable and less reliable than Type 316 stainless steel. Several large refiners have discontinued the use of aluminum materials for such services after thorough field trials.

Some of the restrictions on the use of aluminum are caused by manufacturing and fabrication problems and by its low mechanical strength. However, aluminum is widely used and is competitive with Type 316 stainless steel in many instances. The explosion-bonding process has made the aluminum cladding of steel practical, and improvements in the diffusion coating process are producing more reliable aluminum coatings for such parts as furnace tubes and piping.

Literature Cited

1. Derungs, W. A.: "Naphthenic Acid Corrosion—an Old Enemy of the Petroleum Industry," *Corrosion,* Dec. 1956, p. 41.

2. Heller, J. J.: "Corrosion of Refinery Equipment by Naphthenic Acid," Report of Technical Committee T-8, NACE, *Materials Protection,* Sept. 1963, p. 90.

Fuel Ash Corrosion

John A. Bonar, Esso Research and Engineering Co., Florham Park, N.J.

Fuel ash corrosion of components in process furnaces, utility boilers, and other equipment where high metals content fuels are fired is due to the effects of vanadium and sodium contained in the fuel. Crudes from Venezuela, other Caribbean sources, the western United States, and Canada have high concentrations of vanadium and other metals. These metals are found in the form of organometallic complexes called porphyrins which are concentrated during the refining cycle in heavy residual fuels. In addition to these organometallic complexes, sources of additional metallic salts such as spent caustic and entrained salt in the crude are also concentrated in the heavy fuels. These salts act synergistically with the vanadium compounds present and can cause accelerated corrosion of metals and refractories in utility boiler and process furnaces when firing these types of fuels.

Sodium plus vanadium in the fuel in excess of 100 ppm can be expected to form fuel ashes corrosive to metals by fuel oil ashes.

Corrosive compounds formed during combustion. Combustion of fuels containing significant amounts of porphyrins and other metallic salts results in formation of low melting point vanadium pentoxide (V_2O_5), vanadates such as $Na_2O{\bullet}6V_2O_5$ or alkali sulfates which are corrosive to metals at combustion temperatures. At high combustion temperatures, $T \simeq 2{,}800°F$, and especially when low amounts of excess O_2 are present, the sub-oxides of vanadium, V_2O_3 or VO_2, are formed, and these being refractory, do not cause corrosion. At lower temperatures, $T \leq 2{,}400°F$, and with higher excess air levels, the low melting point V_2O_5 is formed. Unfortunately, the latter condition is most prevalent during combustion, and thus corrosion of furnace components takes place.

Vanadium-sodium compounds most corrosive. Physical property data for vanadates, phase diagrams, laboratory experiments, and numerous field investigations have shown that the sodium vanadates are the lowest melting compounds and are the most corrosive to metals and refractories. These compounds are thought to form by either the vapor phase reaction of NaCl and V_2O_5 or by the combination of fine droplets of these materials upon the cooler parts of combustion equipment.

Excess sodium hydroxide present can also be troublesome as the alkali reacts with the SO_3 present in the gas stream to form a range of alkali sulfates which in themselves are highly corrosive to metallic components. In addition, the combination of alkali sulfate + V_2O_5 can result in compounds having melting points as low as 600°F. This situation is only encountered when alkali is present in amounts in excess of that which can react stoichiometrically with V_2O_5, since the formation of alkali vanadates is favored over that of alkali sulfates.

Maximum corrosion rates have been found to occur with alkali vanadates approximating the $Na_2O{\bullet}6V_2O_5$ compound (6–10 weight percent Na_2O). This corrosion rate maximum can be influenced by the presence of other oxides and sulfur. The sodium vanadates are most corrosive over the temperature range of 1,100–1,500°F. At temperatures below 1,100°F little or no liquid phase would be found, and corrosion of metals is low. At temperatures above 1,550°F, the maximum corrosion rate shifts toward melts having high proportions of V_2O_5.

Corrosion influenced by variables. Corrosion of metals by V_2O_5 or alkali vanadates is influenced by several variables, some of which are independent. Those found to affect the corrosion rate are:

- Conduction mechanism of the molten ash
- Oxygen transport capability of the ash
- Partial pressure of oxygen
- Characteristics of the metal vanadates formed
- Dissolution rate of the oxide films present
- Composition of the fuel ash.

Molten vanadate ashes (melts) can exhibit both semi-conducting and ionic conduction and experiments have shown that semiconducting melts are more corrosive than those exhibiting ionic conduction. Application of this knowledge as a corrosion control technique is not yet feasible, and a more complete discussion will not be attempted in this article.

Oxygen transport through molten vanadate ashes can also control metallic corrosion. Semiconducting melts readily corrode metals, for O_2 transport through the melt is by oxygen "vacancies" and no mass transport need be involved. In ionic melts, however, oxygen most likely diffuses as vanadate complexes. Total oxygen transfer is thus much slower than in semiconducting molten ashes. Experiments have shown that metals corrode much faster in semiconducting V_2O_5 or $Na_2O \bullet 6V_2O_5$ than in ionic melts approximating $Na_2O \bullet V_2O_5$.

Corrosion of metals by fuel ashes only occurs where the fuel ash contains a liquid phase. Temperatures at which the first liquid will form are inversely proportional to the oxygen partial pressure. Thus, when firing fuels at high excess air ratios, fuel ash corrosion occurs at lower temperatures than when firing fuels with low excess air ratios.

Metals forming oxide films high in Cr_2O_3, NiO, and CoO are found to corrode at much lower rates than those forming films predominantly of Fe_2O_3. This is thought to result for a number of reasons. Oxides of Cr_2O_3, NiO, or CoO tend to change the nature of molten ash conduction to ionic and thus lower the relative corrosiveness. Films of Cr_2O_3, NiO, and CoO are also tightly adherent and may act as barriers to O_2 diffusion. Dissolution rates for these oxides are also low and retard the overall corrosion rate. Furthermore, the vanadates that do form are refractory and tend to form skins which act as barriers to further transport of O_2 and molten ash to the metal surface. Iron vanadates show unique behavior. Fe_2O_3 is readily dissolved in vanadates over the range $Na_2O \bullet 6V_2O_5$ to $5Na_2O \bullet V_2O_5$ and can be dissolved in these melts in amounts equal to the weight of the melt before the melting point is raised above 1,300°F. Since these sodium vanadates cover the range of the compositions found in most fuel ashes, the corrosion of iron-bearing alloys is extensive when in contact with these molten ashes.

Fuel ash corrosion control. A variety of methods have been used to control fuel ash corrosion of metallic members in process furnaces, utility boilers, and other combustion equipment. Among these are:

- Fuel oil additives
- Excess air control
- Refractory coatings
- Choice of alloys.

Oil additives have been used for a number of years. They are normally introduced as suspensions of metallic oxides or other salts such as $Mg(OH)_2$, $Ca(OH)_2$, Al_2O_3, etc., in fuel oil or water. Use of these compounds has been shown to be uneconomic, ineffective and to cause problems, i.e, tube fouling, increased soot blower usage, solids disposal, etc., rather than cure them.

Use of excess air levels of 5 percent or less has been shown to reduce fuel ash corrosion in furnaces, most likely by stabilizing the vanadium as a refractory suboxide, VO_2 or V_2O_3. Utility plants have had some success using this method to control vanadium ash corrosion. However, practical application of excess air control in refinery and chemical plant operations is difficult, and has not been particularly successful. Problems with particulates, smoke, pollution, and flame control are encountered unless the necessary expensive control systems and operator attention are constantly available.

Monolithic refractory coatings have been applied to metallic components in furnaces for fuel ash corrosion control. Results have been less than satisfactory because of the large thermal expansion mismatch between the metal and refractory. Failure usually occurs upon thermal cycling which causes cracking, eventual spalling of the refractory, and direct exposure of the metal to the effects of the fuel ash.

High chromium alloys. Field experience and laboratory data indicate that alloys high in chromium offer the best fuel ash corrosion resistance. The table below shows laboratory corrosion rates for engineering alloys which have been exposed to several types of vanadium-sodium fuel ash melts.

Material	Melt	Corrosion rate 1,292°F	(mils/year) 1,472°F
Carbon steel	V:Na:S	690	—
5Cr-1Mo	V:Na:S	330	560
AISI Type 304	Na_2SO_4 - $6V_2O_5$	210	—
AISI Type 310	Na_2SO_4 - $6V_2O_5$	260	—
AISI Type 410	V:Na:S	550	—
AISI Type 430	V:Na:S	350	430
AISI Type 446	V: Na:S	190	—
Incoloy	V:Na:S	220	220
50Cr/50Ni	V:Na:S	—	120

Alloys having relatively high chromium contents, Type 446 stainless steel, and 50 Cr/50 Ni display improved fuel ash corrosion resistance. Type 310 stainless steel offers little or no more corrosion resistance than Type 304 stainless steel.

Extensive field experience has shown the 50 Cr/50 Ni and 60 Cr/40 Ni alloys to offer the best answer to controlling fuel oil ash corrosion. Type 446 stainless steel also shows acceptable corrosion rates but must be used judiciously due to its low strength at elevated temperatures and weldability. Since components of 50 Cr/50 Ni in contact with vanadium-sodium fuel ash melts still suffer high corrosion rates, they should be designed to minimize the amount of surface area available where ash may accumulate.

Thermal Fatigue

Howard S. Avery, Abex Corp., Mahwah, N.J.

Thermal fatigue characteristically results from temperature cycles in service. Even if an alloy is correctly selected and operated within normal design limits for creep strength and hot-gas corrosion resistance, it can fail from thermal fatigue.

Thermal fatigue damage is not confined to complex structures or assemblies. It can occur at the surface of quite simple shapes and appear as a network of cracks.

Thermal shock. As a simplified assumption, thermal fatigue can be expected whenever the stresses from the expansion and contraction of temperature changes exceed the elastic limit or yield strength of a material that is not quite brittle. If the material is brittle as glass, and the elastic limit is exceeded, prompt failure by cracking can be expected (as when boiling water is poured into a glass tumbler). The term *thermal shock* is logically applied to the treatment that induces prompt failure of a brittle material.

A metal tumbler would not crack because its elastic limit must be exceeded considerably before it fails. However, repetitions of thermal stress, when some plastic flow occurs on both heating and cooling cycles, can result in either cracking or so much deformation that a part becomes unserviceable.

Thermal stress. The formula for thermal stress can be rearranged to calculate the tolerable temperature gradient for keeping deformation within arbitrary limits.

$$S = aMTK\,(1 - v)$$

where S = Stress in psi, or elastic limit in psi, or yield strength in psi.

a = Coefficient of thermal expansion in microinches per inch per °F.

M = Modulus of elasticity (Young's Modulus) in psi, or modulus of plasticity.
Note: When the elastic modulus is used, the elastic limit or proportional limit should be used with it in the formula. When the plastic modulus or Secant Modulus is used, it should be used with the corresponding yield strength.

T = Temperature difference in °F.

v = Poisson's ratio.

K = Restraint coefficient.

The usefulness of this formula is restricted by the difficulty of obtaining good values to substitute in it. They must apply to the alloy selected and be derived from carefully controlled tests on it. The stress value, S, reflects an engineer's judgment in the selection of elastic limit or some arbitrary yield strength. The modulus value must match this. The restraint coefficent, K, is seldom known with any precision.

Fatigue life. The formulas for estimating fatigue life are more complex and usually require several assumptions. Without attempting to evaluate the merit of such formulas, it is suggested here that alloy selection or metallurgical variables operating within an alloy may invalidate the assumptions. One of the critical factors that enters such calculations is the amount of plastic strain (usually first in compression and then in tension as the temperature reverses) that occurs. Another is the tolerable flow before cracking develops.

Thermal gradients may be measured or calculated by means of heat flow formulas, etc. After they are established it is likely to be found from the formula that for most cyclic heating conditions the tolerable temperature gradient is exceeded. This means that some plastic flow will result (for a ductile alloy) or that fracture will occur. Fortunately, most engineering alloys have some ductibility. However, if the cycles are repeated and flow occurs on each cycle, the ductility can become exhausted and cracking will then result. At this point it should be recognized that conventional room temperature tensile properties may have little or no relation to the properties that control behavior at the higher temperatures.

Ductility. As a warning against the quick conclusion that high ductility is the premium quality desired, it should be pointed out that the amount of deformation is closely related to the hot strength with which deformation is resisted. A very strong alloy may avoid yielding, or yield so little that there is no need for much ductility.

Design approach. The engineering solution of thermal fatigue problems becomes chiefly a matter of design to minimize the adverse factors in the stress formula and alloy selection aimed carefully at the operating conditions. Much effort should be devoted to reducing the thermal gradients or the temperature differential to the practical minimum. This, and avoiding restraint, are rewarding areas where the designer can demonstrate his skill.

The restraint coefficient, K, in the thermal stress formula is very potent. It can be varied over a wider range than any of the other parameters. If a designer can build in flexibility, and thus substitute elastic deflection for plastic flow, he can achieve a major safety factor.

Thermal expansion. Alloys differ in their thermal expansion, but the differences are modest. Coefficients for the ferritic grades of steel are perhaps 30 percent below those of the austenitic steels at best, while expansion of the nickel-base austenitic types may be no more than 12 to 15 percent less than those of the less expensive, iron-base, austenitic, heat-resistant alloys. Unfortunately, the most economical grades for use in the 1,200–2,000°F range tend to have the highest expansion.

Poisson's ratio can vary somewhat but at present does not provide much choice. There is evidence that it is affected by crystal orientation but specifying and providing control is hardly practical. The value of 0.3 is frequently used for convenience in calculations.

Alloy selection. The designer must make two choices before using the stress formula. The first is a tentative selection of the alloy, because this determines the strength modulus and coefficient values. The second is the decision about strength and modulus limits, as typified by the term *elastic limit* vs. yield strength for some defined permanent set (e.g., 0.1 percent or 0.2 percent). After preliminary calculations for several candidate alloys, the range of choice may be narrowed. It is then advisable to seek out and examine results from comparative thermal fatigue tests using temperature conditions close to those of the intended operation. If such data can be found, or if arrangements for the tests can be made, the resultant ranking is more likely to integrate the various factors (particularly the role of strength and ductility) than any other procedure.

If some other criterion such as creep-rupture strength is of primary importance, the alloy choice may be restricted. Here it would be necessary to have thermal fatigue comparisons only for the alloys that pass the primary screening. When alloy selection reaches this stage some further cautions are in order.

For maximum temperatures below 800°F, suitable ferritic steels are usually good selections. Above 800°F their loss of strength must be considered carefully and balanced against their lower thermal expansion. It should be recognized that if they are heated through the ferrite to austenite transformation temperature their behavior will become more complex and the results probably adverse.

Environments. Among the environmental factors that can shorten life under thermal fatigue conditions are surface decarburization, oxidation, and carburization. The last can be detrimental because it is likely to reduce both hot strength and ductility at the same time. The usual failure mechanism of heat-resistant alloy fixtures in carburizing furnaces is by thermal fatigue damage, evidenced by a prominent network of deep cracks.

Aging (extended exposure to an elevated temperature) of an austenitic alloy can strengthen the material and increase its resistance to damage. Thus such alloys can be shown to improve after a heat treatment. However, there is little point in imposing such a treatment before use. The alloy will age anyway during its first few days of service. The sophisticated use of this information leads to aging of the thermal fatigue specimens before they are evaluated in a laboratory test rather than heat treating parts before service. (This remark does not apply to the "quench anneal" or solution heat treatment given to some wrought stainless steels to alter their metallurgical condition before service.)

The range in properties of a high-temperature alloy, within the compass of a nominal designation or a simplified specification, is much greater than most engineers recognize. Some of this is due to the influence of minor elements within the specification, some is due to production variables, and some may be due to subsequent environmental factors. Among the last, the cycling temperatures that lead to thermal fatigue cal also have an important effect on creep behavior, usually operating to shorten life expectancy. Variations in room temperature properties may not be especially serious but scatter can be extremely significant at high temperature. Unfortunately, the statistical probability limits have not been worked out for many of the available alloys. Moreover, the hot strength values offered to an engineer sometimes have a very sketchy basis.

Abrasive Wear

Howard S. Avery, Abex Corp., Mahwah, N.J.

Abrasive wear can be classified into three types. *Gouging abrasion* is a high stress phenomenon that is likely to be accomplished by high comprehensive stress and impact. *Grinding abrasion* is a high stress abrasion that pulverizes fragments of the abrasive that becomes sandwiched between metal faces. And *erosion* is a low-stress scratching abrasion.

Gouging abrasion. Most of the wearing parts for gouging abrasion service are made of some grade of austenitic manganese steel because of its outstanding toughness coupled with good wear resistance.

Grinding abrasion. The suitable alloys range from austenitic manganese steel (which once dominated the field) through hardenable carbon and medium alloy steels to the abrasion-resistant cast irons.

Erosion. The abrasive is likely to be gas borne (as in catalytic cracking units), liquid borne (as in abrasive slurries), or gravity pulled (as in catalyst transfer lines). Because of the association of velocity and kinetic energy, the severity of erosion may increase as some power (usually up to the 3d) of the velocity. The angle of impingement also influences severity. At supersonic speeds, even water droplets can be seriously erosive. There is some evidence that the response of resisting metals is influenced by whether they are ductile or brittle. Probably most abrasion involved with hydrocarbon processing is of the erosive type.

There is a large assortment of alloys available for abrasive service in the forms of wrought alloys, sintered metal compacts, castings and hard-surfacing alloys. They can be roughly classified as follows, with the most abrasion-resistant materials toward the top and the toughest at the bottom:

- Tungsten carbide and sintered carbide compacts
- High-chromium cast irons and hardfacing alloys
- Martensitic irons and hardfacing alloys
- Austenitic cast irons and hardfacing alloys
- Martensitic steels
- Pearlitic steels
- Ferritic steels
- Austenitic steels, especially the 13 percent manganese type

Since toughness and abrasion resistance are likely to be opposing properties, considerable judgment is required in deciding where, in this series, the best prospect lies, especially if economic considerations are important. The choice is easiest at the extremes.

Cobalt-base and nickel-base alloys may also be considered. Their chief merits are heat-resistance and corrosion-resistance, respectively. For simple abrasion resistance, the iron-base alloys are generally more economical. The basis of classification—total alloy content—sometimes put forward in the hardfacing field is misleading, especially as it implies that merit increases with alloy content. This can be disproved for abrasion. Carbon has a dominant role in determining abrasion resistance; moreover, it adds almost nothing to the cost of an alloy in the form of a casting or a surfacing electrode for welding. Unfortunately, many proprietary alloys do not reveal the amount of carbon used or its role.

Hardness. Though carbon is important, it must be used with proper insight to be most effective. In the form of graphite it usually is detrimental. As hard carbides, the form, distribution, and crystallographic character are important. Even hardness must be used with discretion for evaluating wear resistance; it should be considered as an unvalidated wear test until its relation to a given service has been proven. Simple and widely used tests (e.g., for Brinell or Rockwell hardness) tell almost nothing about the hardness of microscopic constituents.

For erosive wear, Rockwell or Brinell hardness is likely to show an inverse relation with carbon and low alloy steels. If they contain over about 0.55 percent carbon, they can be hardened to a high level. However, at the same or even at lower hardness, certain martensitic cast irons (HC 250 and Ni-Hard) can out perform carbon and low alloy steel considerably. For simplification, each of these alloys can be considered a mixture of hard carbide and hardened steel. The usual hardness tests tend to reflect chiefly the steel portion, indicating perhaps from 500 to 650 BHN. Even the Rockwell diamond cone indenter is too large to measure the hardness of the carbides; a sharp diamond point with a light load must be used. The Vickers diamond pyramid indenter provides this, giving values around 1,100 for the iron carbide in Ni-Hard and 1,700 for the chromium carbide in HC 250. (These numbers have the same mathematical basis as the more common Brinell hardness numbers). The microscopically revealed difference in carbide hardness accounts for the superior erosion resistance of these cast irons versus the hardened steels.

There is another interesting difference between the two irons. Ni-Hard (nominally 1½ Cr, 4½ Ni, 3C) has a matrix of the iron carbide that surrounds the areas of the steel constituent. This brittle matrix provides a continuous path if a crack should start; thus the alloy is vulnerable to impact and is weak in tension. In contrast, HC 250 (nominally 25 Cr, 2½C) has the steel portion as the matrix that contains island crystals of chromium carbide. As the matrix is tougher, HC 250 has more resistance to impact and the tensile strength is about twice as high as that of Ni-Hard. Moreover, by a suitable annealing treatment the matrix can be softened enough to permit certain machining operations and then rehardened for service.

Overlay. Design should also consider whether a tough substrate covered with a hard overlay is suitable or desirable. The process is quite versatile and contributes the advantages of protection to a depth of from 1/32 to about 3/8 inch, incorporation of a tough base with the best of the abrasion resistant alloys, economical use of the expensive materials like tungsten carbide, ready application in the field as well as in manufacturing plants, and usually, only simple welding equipment is needed. Sometimes high labor costs are a disadvantage.

There are size limitations. If large areas are surfaced by automatic welding, only tough alloys can be applied without cracking. The cracks tend to stop at the tougher base, but there is no simple answer to the question about the erosion resistance of a surface containing fine cracks.

Erosion and corrosion combined require special consideration. Most of the stainless steels and related corrosion-resistant alloys owe their surface stability and low rate of corrosion to passive films that develop on the surface either prior to or during exposure to reactive fluids. If conditions change from passive to active, or if the passive film is removed and not promptly reinstated, much higher rates of corrosion may be expected.

Where erosion by a liquid-borne abrasive is involved, the behavior of a corrosion-resistant alloy will depend largely on the rate at which erosion removes the passive film and the rate at which it reforms.

If the amount of metal removal by erosion is significant the surface will probably be continually active. Metal loss will be the additive effect of erosion and active corrosion. Sometimes the erosion rate is higher than that of active corrosion. The material selection judgment can then disregard corrosion and proceed on the basis of erosion resistance provided the corrosion rates of active surfaces of the alloys considered are not much different. As an example of magnitudes, a good high-chromium iron may lose metal from erosion only a tenth as fast as do the usual stainless steels.

Pipeline Toughness

There has been renewed interest in pipeline toughness in recent years. Pipeline flaws have caused failures, and increased toughness makes pipelines more tolerant of flaws thus helping to prevent or mitigate ruptures.

Predicting the appropriate level of ductile fracture resistance involves an analysis of fluid properties, operating conditions, and material properties. For natural gas pipelines containing mostly methane with very few heavy hydrocarbons, however, a simple equation applies:

$$R_{max} = 0.0072 S_o^2 (Rt)^{1/3}$$

where R_{max} = Required level of absorbed energy for a ⅔-size Charpy V-notch specimen to assure arrest of a ductile fracture, ft-lb

S_o = Maximum operating hoop stress level as determined by the Barlow formula using the maximum operating pressure, Ksi

R = ½ O.D. of the pipe, in.

t = Pipe W.T., in.

The source article discusses use of API Specification 5L-Supplementary Requirement 5 to obtain adequate fracture toughness. Certain small diameter, thin-wall pipe cannot use API 5L SR5, so the article discusses alternates for obtaining adequate toughness. If the purchaser is limited to only chemical analyses, the following guideline is given: "Materials with very high carbon contents (C > 0.25), very high sulfur (S > 0.015), and very high phosphorus (P > 0.01) should be avoided."

Source

Kiefner, J. F. and W. A. Maxey, "Specifying Fracture Toughness Ranks High In Line Pipe Selection," *Oil and Gas Journal,* Oct. 9, 1995.

Common Corrosion Mistakes

Kirby[1] gives sound advice for designers who are not experts in metallurgy. He gives seven pitfalls to avoid:

1. Not understanding details of the corrosion service. Stating only the predominant acid without the other details (such as presence of chloride ion) is an example. The author explains how to use the acronym SPORTS-FAN:

S = Solvent
P = pH
O = Oxidizing potential
R = Reducing potential
T = Temperature
S = Salts in solution
F = Fluid flow conditions
A = Agitation
N = New aspects or changes to a chemical process

2. Concentrating on overall or general corrosion. Ignoring pitting, crevice corrosion, stress corrosion, cracking, etc.
3. Ignoring alkaline service. Just because strong alkalies do not cause severe overall corrosion in carbon steel or stainless steel, don't overlook stress corrosion, cracking, or effects on other materials.
4. Not considering water or dilute aqueous solutions. This can be overlooked if the other side of the tube or coil has strong chemicals, such as sulfuric acid.
5. Confusion about L-grade of stainless steels. The L-grade such as 304L have lower carbon (0.03% vs 0.08%) than the standard grades, (e.g. 304). The L-grades are used to prevent "sensitization" from carbide precipitation during welding. This minimizes strong acid attack of the chromium-depleted areas along the welds. Don't forget to specify the L-grade for the filler metal as well as the base plate. Some confusion exists that the purpose of the L-grade is to handle chloride stress corrosion cracking at percent or multiple ppm levels.
6. Not accounting for the oxidizing or reducing potential of acidic solutions. For non-chromium-containing alloys that are capable of withstanding reducing acids, a small amount (ppm level) of oxidizing chemical can have devastating effects.
7. Neglecting trace chemicals. Watch for ppm levels of chloride with stainless steels or ppm levels of ammonia with copper-base alloys for example.

Source

Kirby, G. N., "Avoid Common Corrosion Mistakes for Better Performance," *Chemical Engineering Progress,* April 1997.

19
Safety

Estimating LEL and Flash

The lower explosive limit (LEL) is the minimum concentration of a vapor in air that will support a flame when ignited. The flash point is the lowest temperature of a liquid that produces sufficient vapor for an open flame to ignite in air.

Gooding provides ways to estimate these two important safety-related properties. The methods make use of the following observed rules:

1. The LEL occurs at about 50% of the stoichiometric oxidation concentration at ambient temperature and pressure.
2. The flash point occurs at about the temperature at which the liquid has a vapor pressure equal to the LEL partial pressure.
3. It follows then, that knowing the stoichiometry and having a vapor pressure chart, one can determine the LEL and flash point. Also if either the LEL or flash point is known, a vapor pressure chart can be used to estimate the other.

Example:
Estimate the LEL and flash point for ethanol.

The oxidation (combustion) equation is:

$$C_2H_5OH + 3O_2 = 2CO_2 + 3H_2O$$

For 1 mol of ethanol we need:
3 mols of O_2
or 3/0.21 = 14.28 mols of air

The stoichiometric concentration of ethanol in air is thus 1/15.28 = 0.0654 mol fraction. The LEL is 50% of this or 0.0327 mol fraction. This matches the reported value of 3.3% by volume.

The partial pressure of LEL ethanol is 0.0327 atm. The temperature that produces a vapor pressure of 0.0327 atm is 11°C, which is our predicted flash point. This is close to the reported 13°C.

The Gooding article presents graphs that show high accuracy for these methods.

Source

Gooding, Charles H., "Estimating Flash Point and Lower Explosive Limit," Chemical Engineering, December 12, 1983.

Tank Blanketing

Inert gas is used to blanket certain fixed-roof tanks for safety. Here is how to determine the inert gas requirements. Inert gas is lost in two ways: breathing losses from day/night temperature differential, and working losses to displace changes in active level.

Breathing losses

1. Determine the vapor volume, V_0

$$V_o = \pi D^2/4 \text{ (avg. outage)}$$

where

D = Tank diameter, ft
avg. outage = Average vapor space, ft
V_o = Vapor volume, scf

2. Calculate daily breathing loss (DBL)

$$DBL = V_o\{[(460 + T_s + T_{dc}/2)/(460 + T_s + \Delta - T_{dc}/2)] - 1.0\}$$

where

T_s = Storage temperature, °F
T_{dc} = Daily temperature change, °F
DBL = Daily breathing loss, scf
Δ = Adjustment for the differential between blanketing and pressure-relief settings (normally 2–4 °F)

3. See Figures 1 and 2 for T_s and T_{dc}

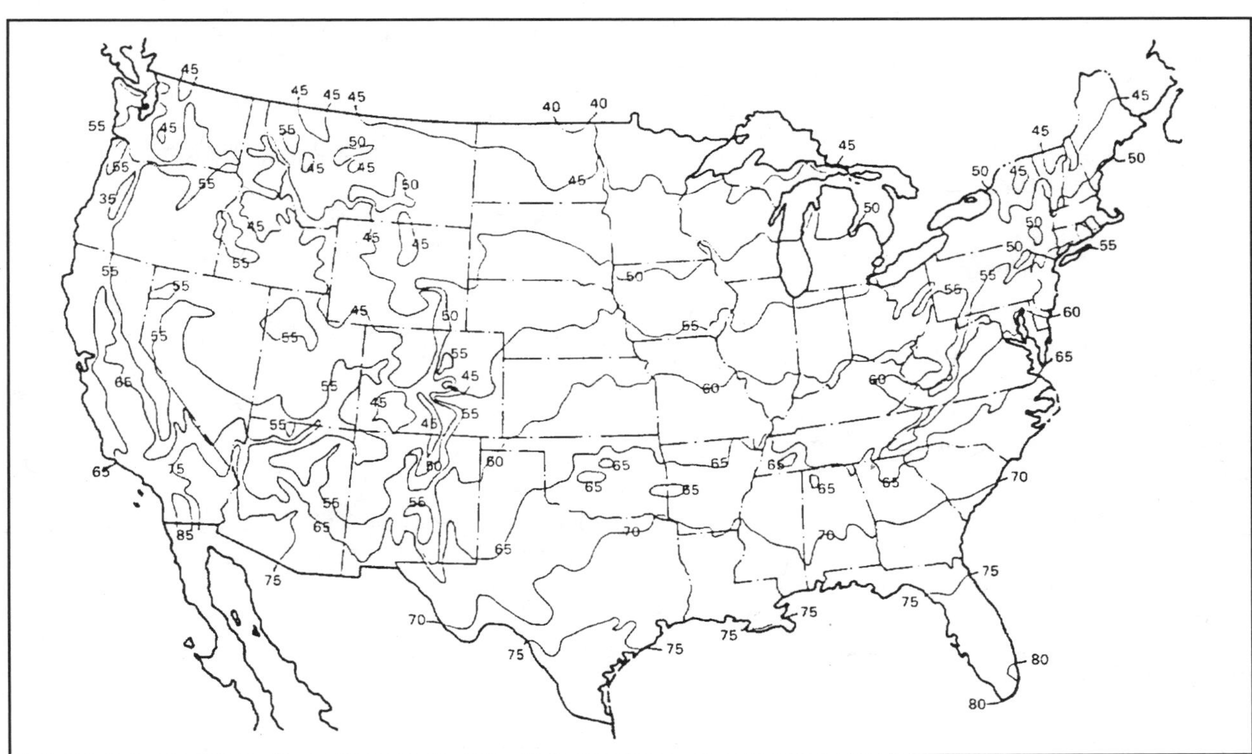

Figure 1. Average storage temperatures for the U.S. used to estimate breathing losses.

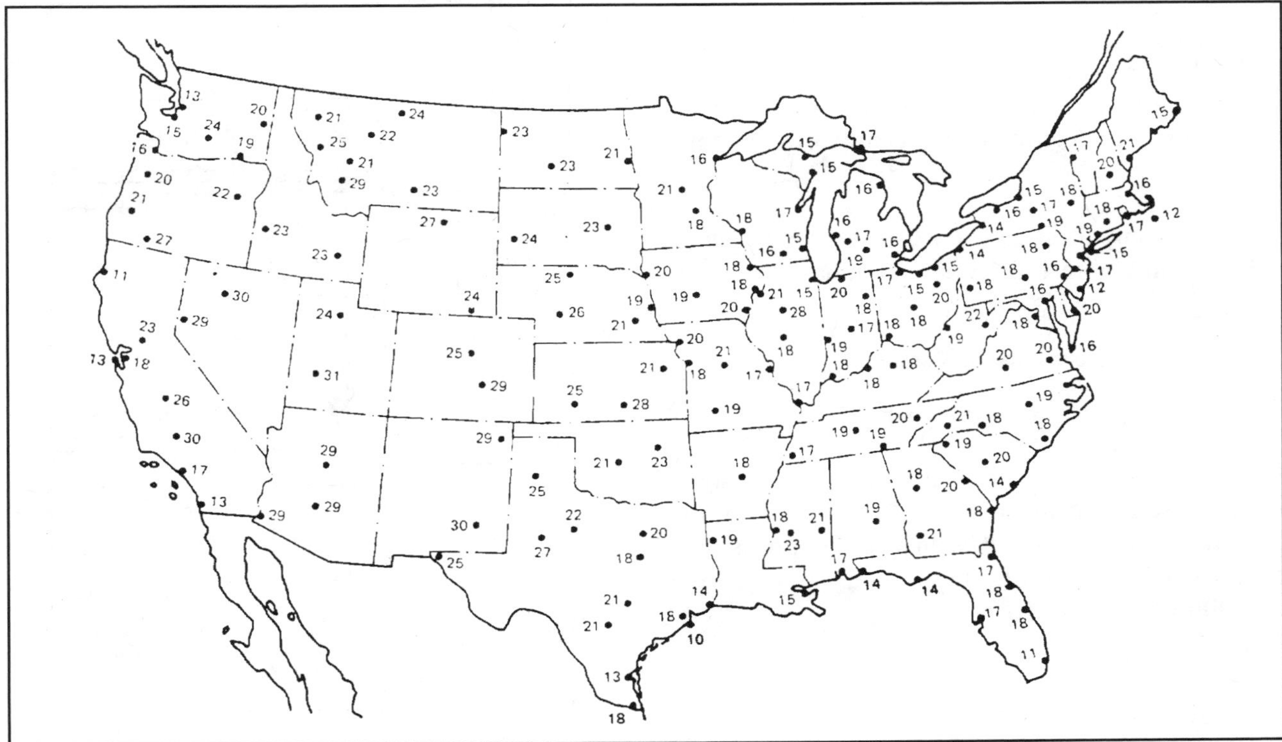

Figure 2. Average daily temperature changes for the U.S. used to estimate breathing losses.

Working Losses

Use the following displacement equivalents of inert gas to tank liquid:

1 gal × 0.1337 = 1 scf inert gas
1 bbl × 5.615 = 1 scf inert gas

Example: A fixed-roof tank

D = 128 ft
Height = 36 ft
Avg. outage = 12 ft
Annual throughput = 300,000 bbl
Location = New Orleans

Determine monthly inert gas usage

Solution:

$T_s = 75°F$ (Figure 1)
$T_{dc} = 15°F$ (Figure 2)
$\Delta = 2°F$ (Assumed)

Vapor volume $= \pi(128)^2/4(12) = 155{,}000$ ft^3

DBL $= 155{,}000\{[(460+75+15/2)/(460 + 75 + 2 - 15/2)] - 1.0\}$ scf/d

= 3805 scf/d or 3805(30) = 114,000 scf/mo

Monthly working loss = 300,000/12(5.615) = 140,000 scf/mo

Total inert gas usage 254,000 scf/mo

Source

Blakey, P. and Orlando, G., "Using Inert Gases For Purging, Blanketing and Transfer," *Chemical Engineering,* May 28, 1984.

Equipment Purging

Blakey and Orlando give useful methods for determining inert gas purging requirements.

Dilution Purging

- The inert gas simply flows through the vessel and reduces the concentration of unwanted component.
- It is used for tanks, reactors and other vessels.
- Use Figure 1 to determine the quantity of inert purge gas required.

Example: A tank full of air (21% O_2) needs to be purged to 1% O_2. From Figure 1, 3.1 vessel volumes of inert gas are required.

Pressure-Cycle Purging

- The vessel at 1 atm is alternately pressured with inert gas and vented.
- It is used for vessels that can withstand 30 psig or more, vessels with only one port, or vessels with coils or baffles inside. It is useful when pressurization is needed anyway, such as for testing.
- The dilution ratio is $(1/P)^n$
  where
  P = pressure, atm
  n = number of cycles
- The quantity of inert gas required for each cycle is P-1 vessel volumes.

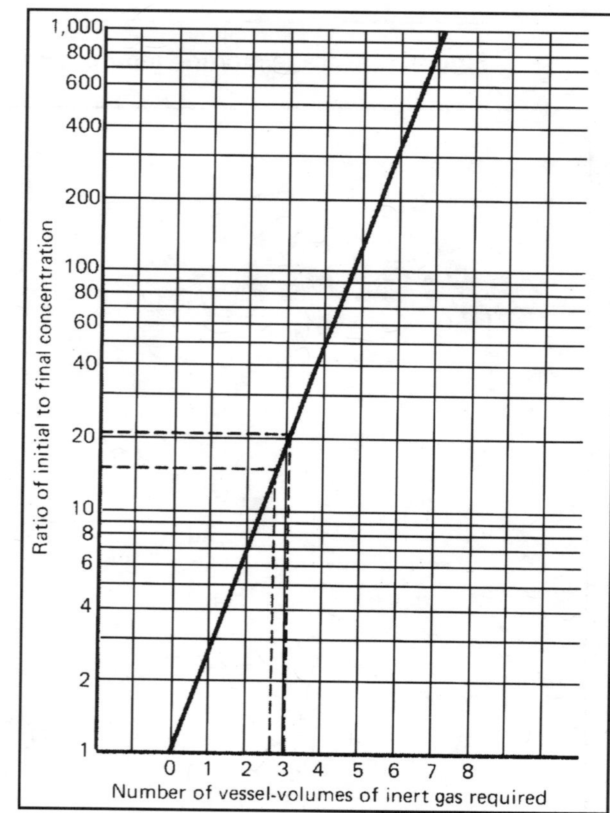

Figure 1. Vessel-volume of inert gas is determined by ratio of final to initial gas.

Example: Purge O_2 from 21% to 1%. Use pressure-cycle purging to 5 atm.

Two cycles will do the job since $(1/5)^2 = 0.04$ and $0.04(21) = 0.84\%$ O_2. For each cycle, purge volume is $5 - 1 = 4$ vessel volumes. So total purge volume is 8 vessel volumes.

Vacuum-Cycle Purging

- The vessel is alternately evacuated and fed with inert gas to 1 atm.
- It can only be used for vessels capable of withstanding a vacuum.
- Concentration of unwanted component is reduced from C to CP^n

where

C = concentration, %
P = pressure, atm
n = number of cycles

The quantity of inert gas required for each cycle is 1-P vessel volumes.

Example: Purge O_2 from 21% to 1%. Use vacuum-cycle purging to 0.5 atm.

$C = 21\%$; $CP^n = 1\%$ or $21(0.5)^n = 1$; $(0.5)^n = 0.048$
$n = 4.4$ so use 5 cycles
Purge gas required $= 5(1 - 0.5) = 2.5$ vessel volumes

Source

Blakey, P. and Orlando, G., "Using Inert Gases for Purging, Blanketing and Transfer," *Chemical Engineering,* May 28, 1984.

Static Charge from Fluid Flow

The following article written by Adam Zanker for *Hydrocarbon Processing* (March 1976) is reprinted here in its entirety.

Development of electrostatic charges in tanks and vessels in which hydrocarbons are pumped or stirred is widely recognized as a serious hazard. This electricity, which is generated in tanks during these normal operations, occasionally causes a spark in a tank vapor space.

As statistics show, during 10 years in one state and one oil company only, 18 fires have been attributed to static electricity, causing damage and product losses of millions of dollars.[1]

The order of magnitude of currents and voltages related to static electricity are of different orders of magnitude from those common to us from everyday life.

Voltages are extremely high; hundreds of thousands of volts are common. Currents, however, are usually less than one-millionth of an ampere. It is due to the extremely high resistivity of hydrocarbon products. This resistivity varies from 10^{11} to 10^{15} ohm-cm.

Voltage drop may be calculated from Ohms Law:

$$V = I R$$

When the resistivity equals 10^{13} ohm-cm and the current 10^{-8}A, the voltage drop equals

$$V = 10^{-8} \times 10^{13} = 10^5 \text{ volt-cm}$$

Some technical sources[1] state that a potential difference of 30,000 volts per centimeter is enough to cause a spark; other sources[2] say 18,000 volts per centimeter.

Information about the current generated and the resistivity of the particular hydrocarbon pumped is essential in order to calculate the voltage difference. This subsequently shows us whether the danger exists of self-ignition or explosion in any particular case.

A hydrocarbon's resistivity may be either directly measured or taken from typical average values listed in technical sources.

The value of current generated by pumping may be easily calculated or graphically found from the formula or nomograph which is attached to this article.

When the value of voltage difference exceeds the 30,000 volts per cm, the danger exists and one or some of the following precautions have to be taken:

- The velocity of pumping has to be reduced. It may be done by exchanging the pipes for bigger diameters, or by inserting at least an additional enlarged sector of pipe which acts as a "relaxation tank" and allows "self-relaxation" of the generated current.
- Roofs of tanks have to be exchanged for floating types to minimize the dangerous atmosphere of hydrocarbon vapor-air mixture which may explode by sparking due to the static electricity.

It is especially important for the JP-4 type products since they produce a typical explosive atmosphere at ambient temperatures above the liquid level.

The atmosphere in gasoline tanks is too rich in hydrocarbons, and in kerosene tanks is too lean in hydrocarbons, so that theoretically, both mixtures of air and vapor are outside the explosivity range.

- When the change of roof is impossible, it is advisable to exchange the air in the tank free space to carbon dioxide which will prevent ignition in the case of sparking.
- A widely used practice is the addition of antistatic additives which strongly reduces the resistivity of hydrocarbons, thus, decreasing the voltage difference.

An enlarged list of less common methods used to prevent the formation of static electricity or to reduce its dangerous influence is listed in the excellent work of W. M. Bustin & Co.[1] As has been told, a knowledge of current generated by pumping and stirring of hydrocarbons is of essential importance.

The order of magnitude of pumping currents is in the range of -100 to $-1{,}000 \times 10^{-10}$ A. When the hydrocarbon is pumped through filters or discharged in jet-form, these values may be even larger.

As experience shows (which, however, is not quite in agreement with the theory), the strongest current generation occurs when the hydrocarbon resistivity lies between 10^{12} to 10^{14} ohm-cm and decreases beyond this range.

Pipe line charge generation. The problem of electrical charges produced by flowing hydrocarbon products in pipe lines was studied by many investigators.

Bustin, Culbertson and Schleckser[1] designed a research unit consisting of:

(1) Source drum and pump section	(Section 1)
(2) Test line	(Section 2)
(3) Receiving drum	(Section 3)

Hydrocarbons were pumped through this unit and the currents-to-ground from each section were measured.

On the basis of these measurements, they have developed a semi-empirical and semi-theoretical formula which allows prediction of the current generated in pipe line during hydrocarbons flow.

This formula is as follows:

$$I_L = -[(T \times K \times V^{1.75}) + I_s]\,(1 - e^{L/TV})$$

where

I_L = Current from Section 2 (pipe line) to ground (in amperes $\times 10^{-10}$)
I_8 = Current from Section 1 (source drums and pump) to ground (in amperes $\times 10^{-10}$)
T = Time constant (found experimentally as equal 3.4 sec.)
V = Linear velocity of hydrocarbon in line, fps
L = Length of pipe, feet
K = Proportionality factor, determined by pipe diameter and charging tendency of hydrocarbon pipe interface.

A smooth stainless steel pipe with a diameter of ¼ inch was used for these tests.

The K value found for this particular case was:

$$K = 0.632$$

This K factor is approximately proportional to the roughness of pipe (friction factor) and inversely proportional to the pipe diameter:

$$K \propto \frac{f}{D}$$

The tested hydrocarbon was aviation fuel JP-4.

When taking into account the actual conditions of their tests (more than 40 runs performed), the authors computed the following equation:

$$I_L = -[(2.15)\,(V^{1.75}) + I_s]\,(1 - e^{-L/3.4V})$$

which shows excellent agreement with their test results (the error below ±5 percent). The (–) shows that current in the pipe line has negative polarity while current measured in the drum always had positive polarity.[1]

The algebraic sum of the three currents measured from all three sections have to be theoretically always zero; however, minute differences have been found in the separate tests.

The nomograph. The calculation of current generated in pipe during hydrocarbon flow, according to formula (3) is troublesome and time-consuming.

Therefore, a nomograph (Figure 1) is presented which allows the calculation of the I_L value within a couple of seconds using one movement of a ruler only.

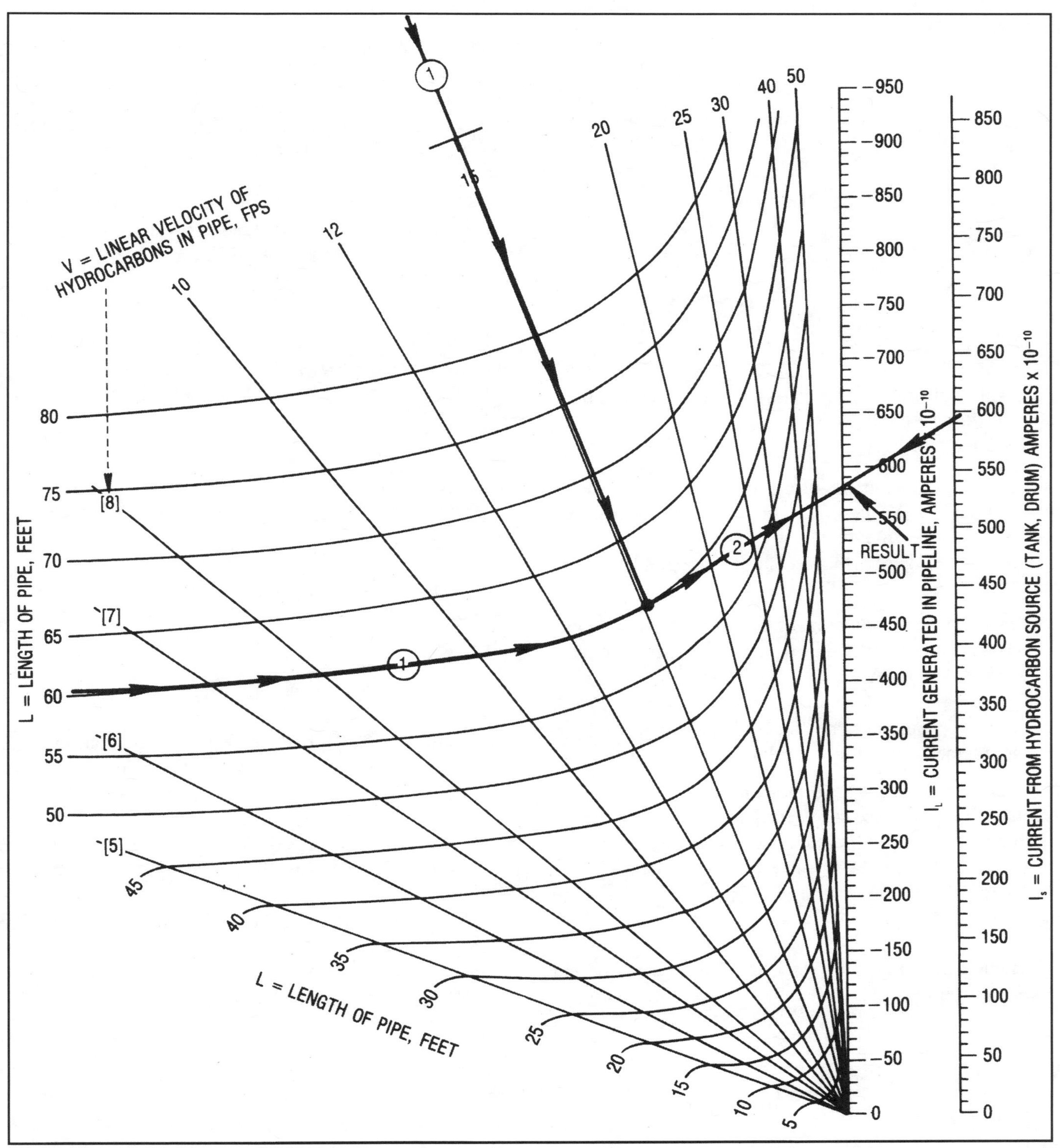

Figure 1. Nomograph for determining current generated in pipe lines from flowing hydrocarbons.

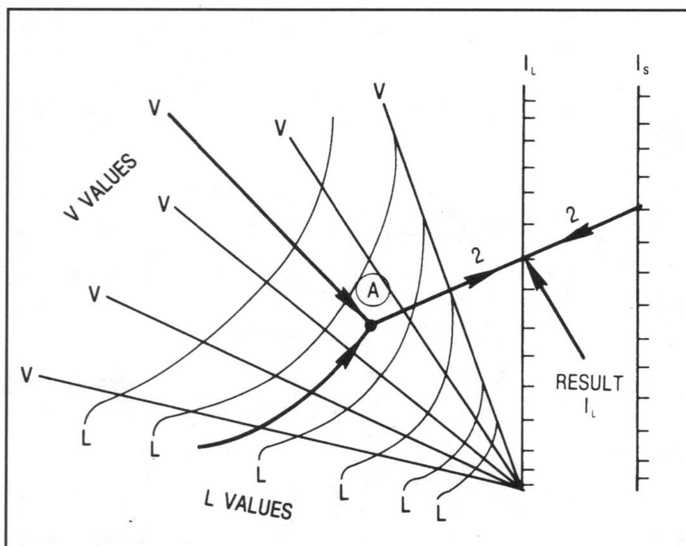

Figure 2. Example of how to use nomograph for finding current generated in pipe lines.

The nomograph consists of a grid where two families of curves (for L and V) are placed and from two scales (vertical) for I_s and I_L (in amperes $\times 10^{-10}$).

Referring to Figure 2, the use of this nomograph is as follows:

- Find the intersection point of the curves for known values of L and V on the grid. Interpolate, if necessary. Mark this point A.
- Connect point A with a ruler to the known value of I_s on the appropriate scale and read the final result I_L on the intersection point of a ruler with the I_L scale.

Typical example. Given: a drum containing JP-4. The current in this drum (source current) equals $+600 \times 10^{-10}$ amp.

From this drum, the liquid is pumped through a 60-foot line (¼-inch ID) with the linear velocity V = 15 fps.

It is to calculate the line current generated during pumping.

Solution. According to the "procedure" described, we find that this current equals:

$I_L = -585$ amp. $\times 10^{-10}$

Note: This example is shown in Figure 1.

Discussion. The designers of this formula[3] do not present a universal formula for all possible cases of pumping hydrocarbons.

However, from the results and graphs attached and from other sources cited,[2,3] the following approximate rules may be stated.

The result obtained by means of the nomograph may be roughly adapted to other conditions by means of multiplying by the following **approximate** factors:

New Condition	Approximate Multiplication Factor
Rough steel tube (stainless)	1.3
Clean carbon-steel tube	0.7
Rusted carbon-steel tube	2
Neoprene tube	1 to 1.5
Tube with any diameter, D (inch)	$\frac{0.25}{D}$
Flowing fluid: kerosine	2 to 3
Flowing fluid: gasoline	0.2 to 0.5

This formula may not be adapted for extremely low (resistivity < 10^{10} Ω cm) or extremely high resistive (resistivity > 10^{15} Ω cm) petroleum or related products.

Literature Cited

1. Bustin, W. M., Culbertson, T. L. and Schleckser, C. E., "General consideration of static electricity in petroleum products," Proceedings of American Petroleum Institute, Section III, Refining, Vol. 37 (III), p. 24–63, 1957.
2. Mackeown, S. S. and Wouk, V., "Electrical charges produced by flowing gasoline," *Ind. Eng. Chem.,* 34 (6), 659–64, June 1942.
3. Klinkenberg, A., "Production of static electricity by movement of fluids within electrically grounded equipment," Proceedings Fourth World Petroleum Congress, Sect. VII-C, Paper No. 5, 253–64, Carlo Colombo, Rome, 1955.

Mixture Flammability

Flammability limits for pure components and selected mixtures have been used to generate mixing rules.[1] These apply to mixtures of methane, ethane, propane, butane, carbon monoxide, hydrogen, nitrogen, and carbon dioxide. Work by Coward and Jones[2] can be consulted for other mixtures.

Case 1: A Combustible Mixture in Air

Use the following table.

Table 1
Flammability Limits

Gas	Lower limit	Upper limit
C_1	5.0	15.0
C_2	3.0	12.4
C_3	2.1	9.5
C_4	1.8	8.4
CO	12.5	74.0
H_2	4.0	75.0

Use LeChatelier's law to blend mixtures as in the following example (volume percentages are used):

Given a mixture of 70% C_1, 20% C_2, and 10% C_3

Lower limit = 100/(70/5.0 + 20/3.0 + 10/2.1) = 3.9%
Upper limit = 100/(70/15.0 + 20/12.4 + 10/9.5) = 13.6%

Case 2: A Combustible Mixture Containing the Inert Gases N_2 and CO_2—Complicated by Also Containing O_2.

The O_2/N_2 ratio must be less than air (< 20.9/79.1) to use this method.

Such a mixture is: 8% C_1, 4% C_2, 1% C_3, 1% C_4, 23% CO_2, 61% N_2, and 2% O_2.

First, of course, we must get rid of the air. The equivalent air is 2/0.209 = 9.6%. This leaves 90.4%. Normalizing to 100% we have: 8.8% C_1, 4.4% C_2, 1.1% C_3, 1.1% C_4, 25.4% CO_2, and 59.2% N_2.

Next we will make use of mixture data from the literature in Figures 1 and 2.

One or two component mixtures can be generated to fit the ranges in Figures 1 and 2, as shown in Table 2.

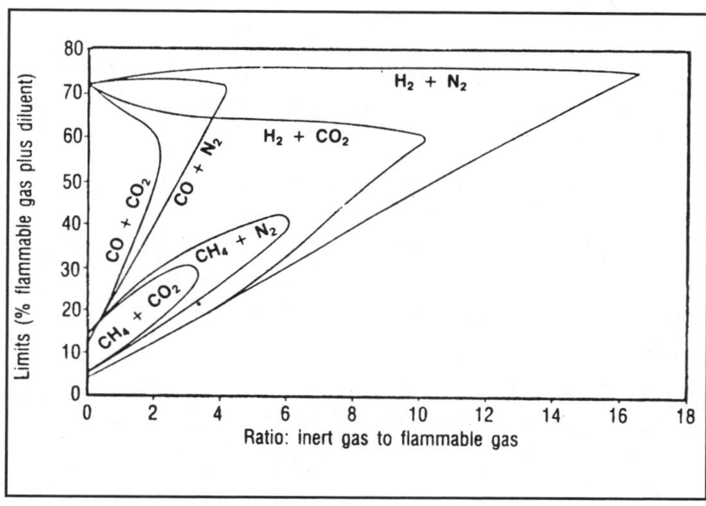

Figure 1. Flammability limits (from Coward and Jones).

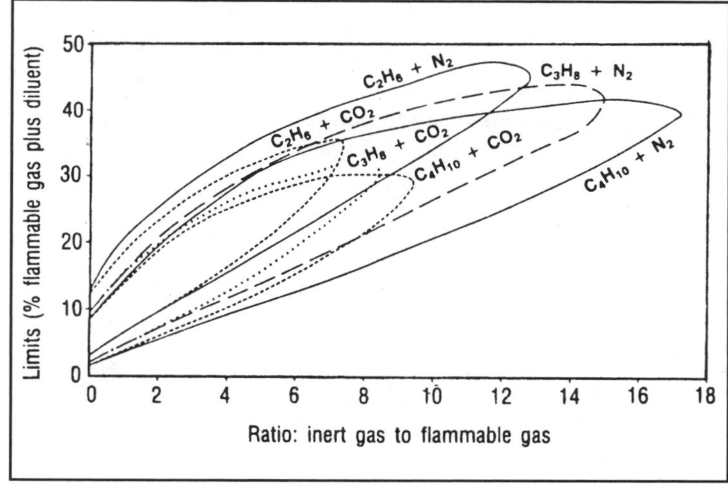

Figure 2. Flammability limits (from Jones and Kennedy).

Table 2
Example of Data Required*

Dissected Number Mixture	Combustible Components Species	%	Inert Components CO_2,%	N_2,%	Total, %	Ratio of Inert to Combustible	Lower Limit %	Upper Limit %	Source of Limit
1	C_1	8.8	22.0	0	30.8	2.5	20.5	29.0	Figure 1
2	C_2	4.4{0.68	3.4	0	4.08	5.0	20.4	32.4	Figure 2
3	C_2	3.72	0	44.6	48.32	12.0	41.2	47.5	Figure 2
4	C_3	1.1	0	14.6	15.7	13.3	35.0	44.1	Figure 2
5	C_4	1.1	0	0	1.1	0	1.8	8.4	Table 1
		15.4	25.4	59.2	100.0				

* To find the flammability limits of the example air-free mixture of 8.8% CH_4, 4.4% C_2H_6, 1.1% C_3H_8, 1.1% C_4H_{10}, 25.4% CO_2, and 59.2% N_2. After Coward and Jones.

LeChatelier's law is applied similarly to Case 1:

Lower limit = 100/(30.8/20.5 + 4.08/20.4 + 48.32/41.2 + 15.7/35.0 + 1.1/1.8) = 25.4%
Upper limit, calculated similarly = 37.1%

These are, of course, the limits for the air-free mixture. But remember, the original mixture, before we removed the air, had 2% O_2. So the lower limit for the original mixture (with the air added back) is 25.4% + 2/0.209 = 35.0 (the increased lower limit means *greater* percentage of the combustible mixture in air or *lower* percentage of air required with the combustible mixture to cause combustion). It is logical that *less* air is needed since the original mixture (with the air added back) has a head start on combustion (with the contained 2% O_2).

Similarly, the upper limit of the original mixture is 37.1% + 2/0.209 = 46.7%. The reasoning is the same as for the lower limit.

Case 3: Inerting a Flammable Mixture

It may be desirable to add enough N_2 to the Case 2 air-free mixture to make it nonflammable (or inert). The methods in Case 2 can be used to construct a "flammability envelope" as shown in Figure 3. N_2 is blended into the mixture until a final iteration at 72% original mixture plus 28% N_2 gives the required inert mixture.

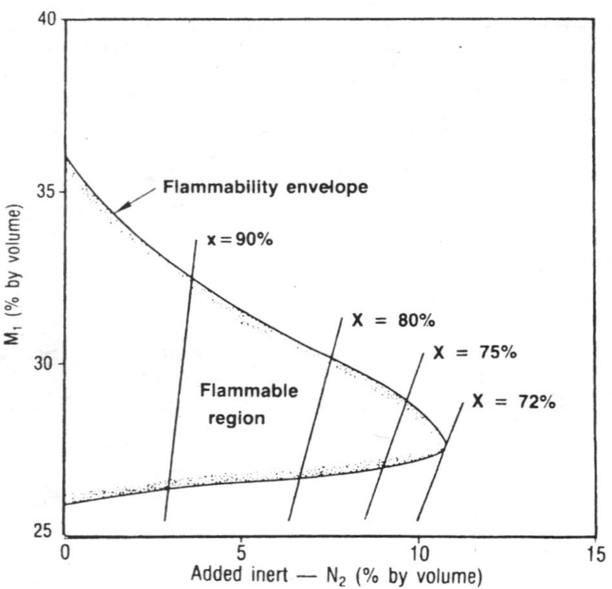

Figure 3. Flammability envelope, mixture M_1 and nitrogen.

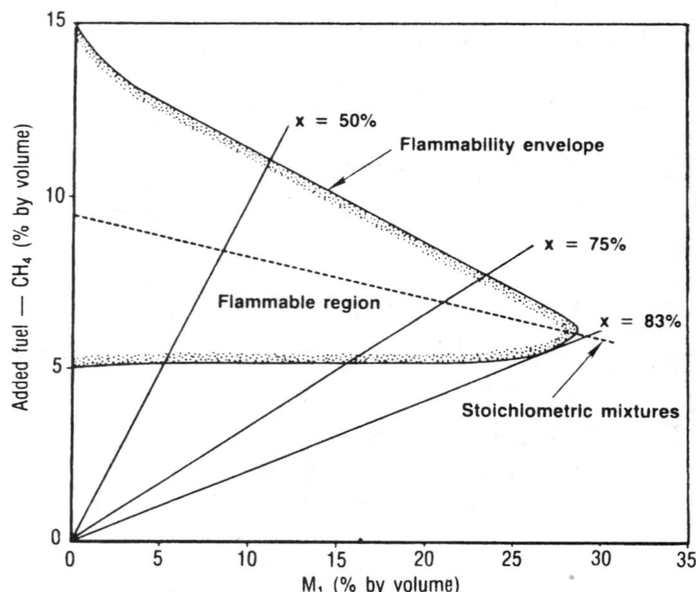

Figure 4. Flammability envelope, mixture M_1 and methane.

Case 4: Making Waste Gas Flammable

The converse of Case 3 is a desire to make an inert mixture (such as a waste gas stream) flammable. Figure 4 shows a similar flammability envelope for all concentrations of interest.

Sources

1. Heffington, W. M. and Gaines, W. R., "Flammability Calculations for Gas Mixtures," *Oil and Gas Journal,* November 16, 1981.
2. Coward, H. F. and Jones, G. W., "Limits of Flammability For Gases and Vapors," Bureau of Mines Bulletin 503, 1952.
3. Zabetakis, M. G., "Inflammability Characteristics of Combustible Gases and Vapors," Bureau of Mines Bulletin 627, 1965.
4. Jones, G. W., "Inflammability of Mixed Gases," Bureau of Mines Technical Paper 450, 1929.
5. Jones, G. W. and Kennedy, R. E., "Limits of Inflammability of Natural Gases Containing High Percentages of Carbon Dioxide and Nitrogen," Bureau of Mines Report of Investigation, 3216, 1933.

Relief Manifolds

Mak[1] has developed an improved method of relief valve manifold design. The API[2] has adopted this method, which starts at the flare tip (atmospheric pressure) and calculates backwards to the relief valves, thus avoiding the trial and error of other methods. This is especially helpful when a large number of relief valves may discharge simultaneously to the same manifold.

Mak's developed isothermal equation (based on the manifold outlet pressure rather than the inlet) is:

$$fL/D = (1/M_2^2)(P_1/P_2)^2[1 - (P_2/P_1)^2] - \ln(P_1/P_2)^2$$

where

D = Header diameter, ft
f = Moody friction factor
L = Header equivalent length, ft
M_2 = Mach number at the header outlet
P_1, P_2 = Inlet and outlet header pressures, psia

The equation for M_2 is as follows:

$$M_2 = 1.702 \times 10^{-5} (W/(P_2D^2))(ZT/M_w)^{1/2}$$

where W = Gas flow rate, lbs/hr
Z = Gas compressibility factor
T = Absolute temperature, °R
M_w = Gas molecular weight

To simplify and speed calculations, Mak provides Figure 1. The method is applied as follows:

1. Assume diameters of all pipes in the network.
2. Starting at the flare tip calculate logical segments using Figure 1 until all relief valve outlet pressures are found.
3. Check all relief valves against their MABP.
   Case A. The calculated back pressure at the lowest set relief valve on a header is much smaller than its MABP. Reduce header size.
   Case B. The calculated back pressure at the lowest set relief valve on a header is close to and below its MABP. The header size is correct.

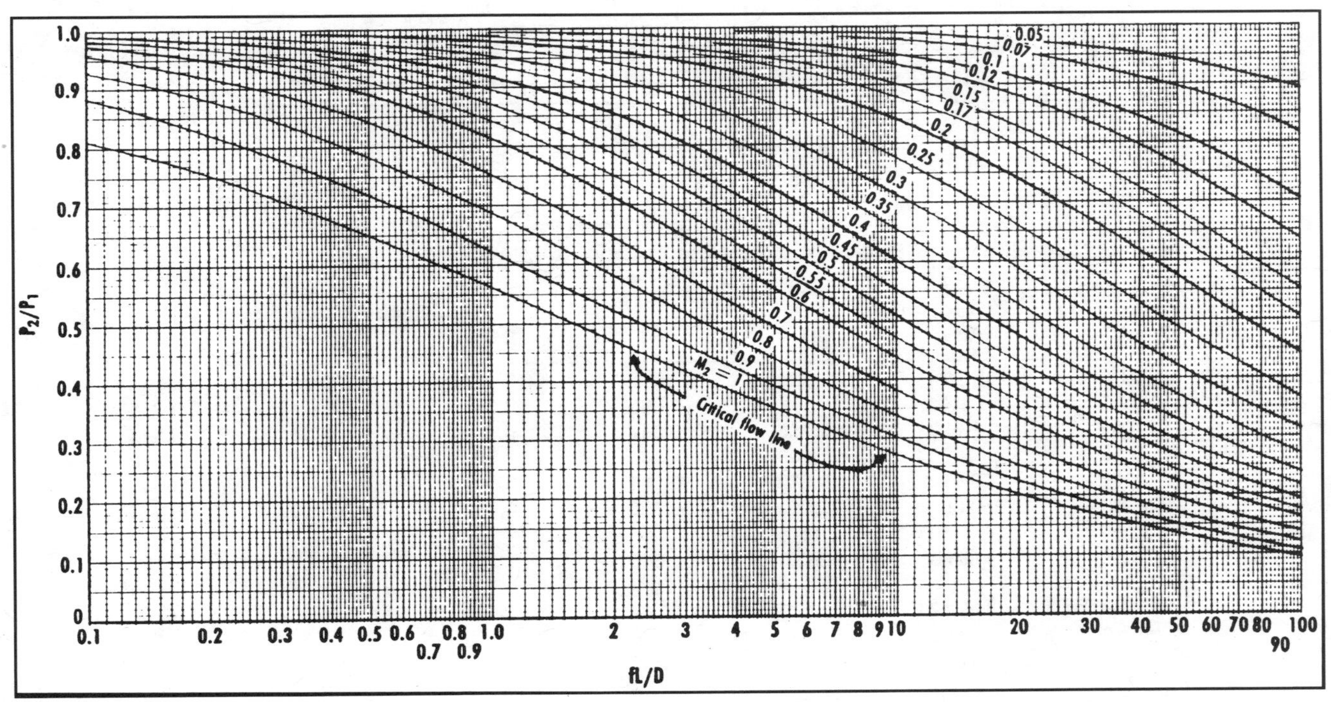

Figure 1. Isothermal flow chart.

Case C. The calculated back pressure at the lowest set relief valve on a header is above its MABP. Increase header size.

4. Use judgment in attempting to optimize. Try to preferentially reduce the sizes of the longest runs or those having the most fittings.

Figure 2 and Table 1 illustrate a sample problem (Z, the compressibility factor, is assumed to be 1.0).

Often, if both high and low pressure relief valves need to relieve simultaneously, parallel high and low pressure headers terminating at the flare knockout drum are the economical choice. Be sure to check for critical flow at key points in the high pressure header.

$$P_{crit} = (W/408d^2)(ZT/M_w)^{0.5}$$

where

P_{crit} = Critical pressure, psia
W = Gas flow rate, lb/hr
d = Pipe id, in.
Z = Gas compressibility factor
T = Gas temperature, °R
M_w = Gas molecular weight

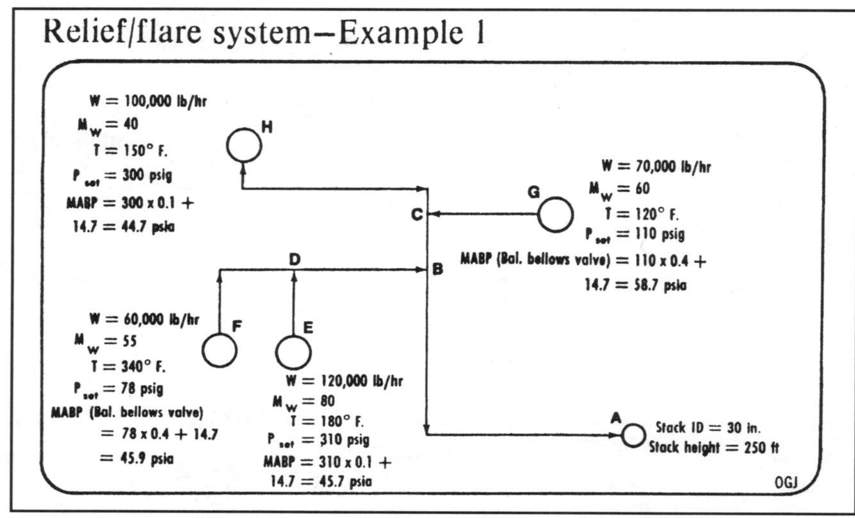

Figure 2. Relief/flare system—Example 1.

Table 1
Back-Pressure Calculations for Example 1

Line	Stack	AB	BD	DF	DE	BC	CH	CG
					Line Segment			
Size	30″ (std)	18″ (std)	12″ (std)	8″ (std)	8″ (std)	12″ (std)	10″ (std)	6″ (std)
D, ft	2.5	1.44	1	0.665	0.665	1	0.835	0.505
A, ft²	4.91	1.623	0.785	0.347	0.347	0.785	0.548	0.201
L, ft	250	1000	200	180	100	115	300	150
f	0.011	0.012	0.013	0.014	0.014	0.013	0.0135	0.015
W, lb/hr	350,000	350,000	180,000	60,000	120,000	170,000	100,000	70,000
M_W	56	56	69.5	55	80	46.4	40	60
T, °F.	186.6	186.8	233.3	340	180	137.6	150	120
P_2, psia	P_{atm} = 14.7	P_A = 15.12	P_B = 34.43	P_D = 38.04	P_D = 38.04	P_B = 34.43	P_C = 37.02	P_C = 37.02
fL/D	1.10	8.35	2.60	3.79	2.10	1.50	4.85	4.45
M_2	0.220	0.646	0.281	0.232	0.344	0.302	0.257	0.391
From Fig. 1	P_{atm}/P_A = 0.972	P_A/P_B = 0.440	P_B/P_D = 0.905	P_D/P_F = 0.910	P_D/P_E = 0.890	P_B/P_C = 0.930	P_C/P_H = 0.867	P_C/P_G = 0.755
P_1, psia	P_A = 15.12	P_B = 34.43	P_D = 38.04	P_F = 41.80	P_E = 42.74	P_C = 37.02	P_H = 42.70	P_G = 49.03

The check for critical pressure has two purposes:

1. If, for a segment, the P_2 calculated is less than P_{crit} then the flow is critical and P_{crit} is used in place of P_2.
2. The main flare header should not be designed for critical flow at the entrance to the flare stack, or else noise and vibration will result.

A reading below $M_2 = 1$ on Figure 1 also indicates critical flow. In such a case, read the graph at $M_2 = 1$.

Mak's article shows how the isothermal assumption is slightly conservative (higher relief valve outlet pressures) when compared to adiabatic.

Values for f and Z can be found using Figures 3–6.

Sources

1. Mak, "New Method Speeds Pressure-Relief Manifold Design," *Oil and Gas Journal,* Nov. 20, 1978.
2. API Recommended Practice 520:, "Sizing, Selection, and Installation of Pressure Relieving Devices in Refineries," 1993.
3. "Flow of Fluids through Valves, Fittings, and Pipe," Crane Co. Technical Paper 410.
4. Crocker, Sabin, *Piping Handbook,* McGraw-Hill, Inc., 1945.
5. Standing M. B. and D. L. Katz, Trans. AIME 146, 159 (1942).

(*text continued on page 285*)

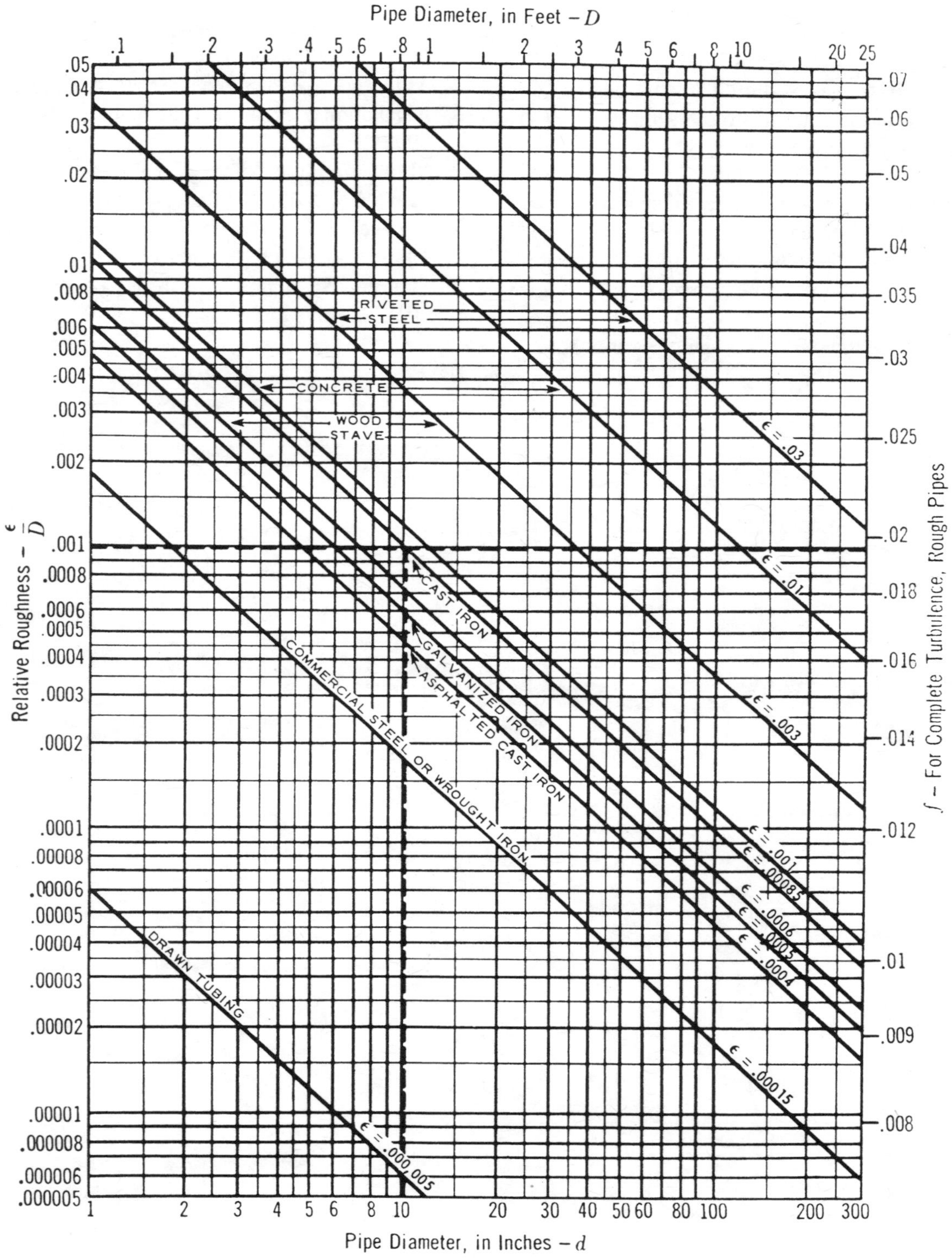

Figure 3. Relative roughness of pipe materials and friction factors for complete turbulence[3]

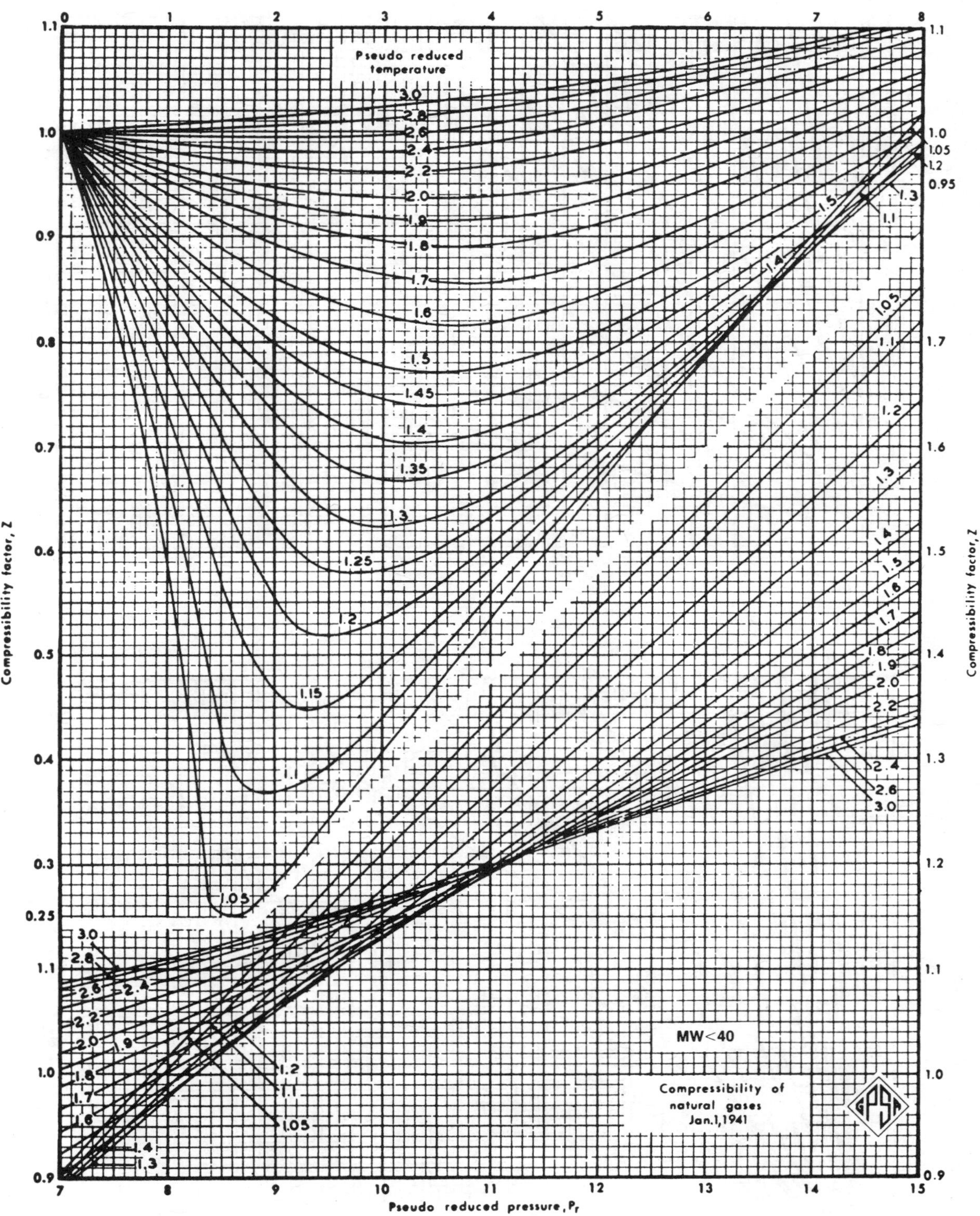

Figure 4. Compressibility factors for natural gas[5]

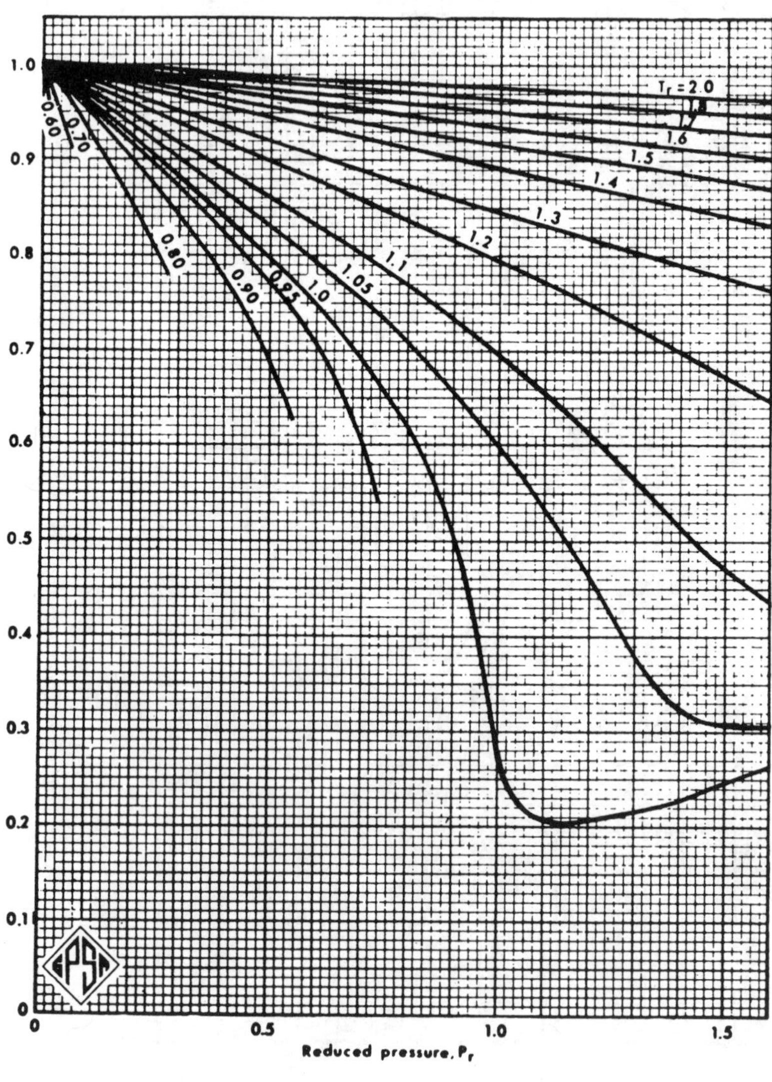

Figure 5. Compressibility factors at low reduced pressures[4]

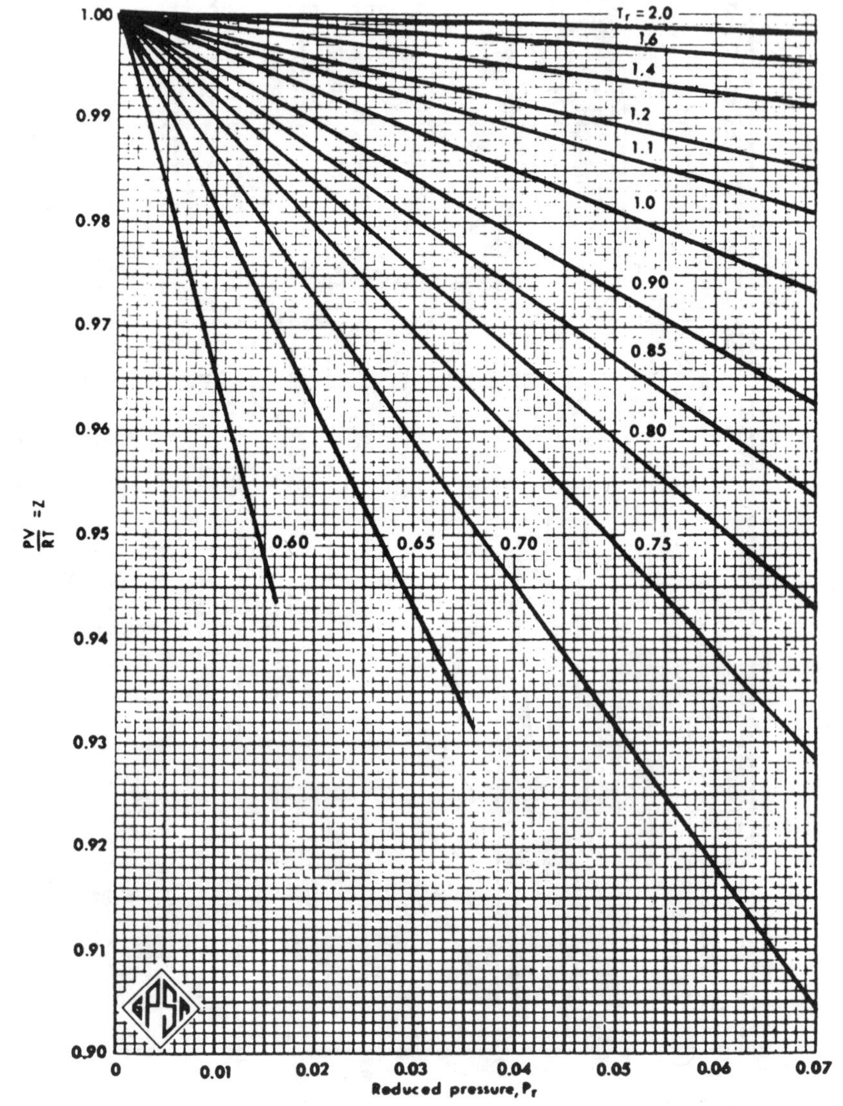

Figure 6. Compressibility factors for gases near atmospheric pressure[4]

(*text continued from page 281*)

Natural Ventilation

As part of a Process Hazards Analysis (PHA), I was required to check a naturally ventilated building containing electrical equipment and a fuel gas supply, for adequate air flow due to thermal forces (stack effect). API RP 500[1] has a method that they recommend for buildings of 1,000 ft^3 or less. The building in question was much larger. I used the RP 500 method for investigation because:

1. The building had exhaust fans so the calculation was only to check for natural ventilation when the fans were off.
2. Perhaps the calculation would show far more natural ventilation than required.

First, it was verified that the building ventilation geometry was to API Standards (no significant internal resistance and inlet and outlet openings vertically separated and on opposite walls). The building had many windows on both sides and 3-ft wide louvers at the roof peak. So the inlets were on both sides and the outlet at the top middle. This arrangement was judged adequate.

The calculations involve first finding:

$$h = H/(1 + [(A_1/A_2)^2 (T_i/T_o)])$$

where h = Height from the center of the lower opening to the Neutral Pressure Level (NPL), in feet. The NPL is the point on the vertical surface of a building where the interior and exterior pressures are equal.

H = Vertical distance (center to center) between A_1 and A_2, ft. This was weight averaged for various windows and open doors.

A_1 = Free area of lower opening, ft^2

A_2 = Free area of upper opening, ft^2

Then the required opening area is determined to provide 12 changes per hour:

$$A = V/[1200 (h (T_i - T_o)/T_i)^{0.5}]$$

where A = Free area of inlet (or outlet) openings, ft^2, to give 12 air changes per hour (includes a 50% effectiveness factor)

V = Volume of building to be ventilated, ft^3

T_i = Temperature of indoor air, °R

T_o = Temperature of outdoor air, °R

$(T_i - T_o)$ = Absolute temperature difference, so will always be positive

Fortunately, our calculations indicated that we had more than twice the free area required.

Source

API Recommended Practice 500 (RP 500), "Recommended Practice for Classification of Locations for Electrical Installations at Petroleum Facilities," 1991, American Petroleum Institute.

20
Controls

Introduction

Whether designing a new plant, modifying an existing one, or troubleshooting, it helps to understand the interface between process engineering and control engineering. For any process equipment, the controls are more an integral part of the design than many people realize. In fact, the process equipment itself is part of the controls.

For the purpose of this chapter, instrumentation will be considered everything from the primary element monitoring the controlled variable through the control valve performing the throttling action. The process equipment interacts with the instrumentation to provide the process control. The process equipment must often be designed with more capacity and sometimes with more utilities usage than is necessary to perform its basic job of fractionation, heat transfer, etc. This extra capacity, which is in addition to any extra *production* capacity, is needed for process control.

Source

Branan, C. R., *The Process Engineer's Pocket Handbook, Vol. 2,* Gulf Publishing Co., 1983.

Extra Capacity for Process Control

To illustrate how the control function requires extra capacity of process equipment, let us use a typical fractionation system, as shown in Figure 1. This example illustrates the point being made rather than recommending any particular fractionation control scheme. The best control scheme depends upon the situation, and a number of control schemes can be used. (For example, Reference 1 shows 21 ways to control fractionator pressure, giving advantages and disadvantages for each.)

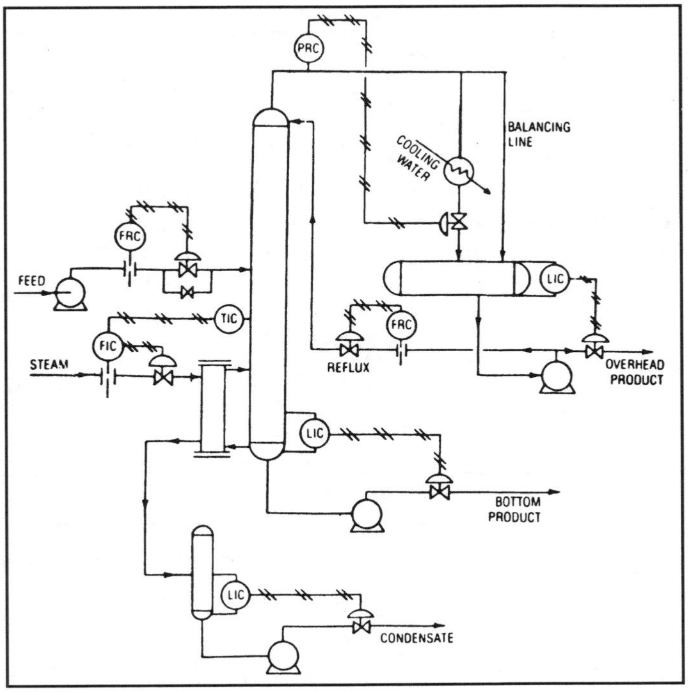

Figure 1. Typical fractionation system, which illustrates how process control requires extra equipment capacity.

Our example system has a flow-controlled feed, and the reboiler heat is controlled by cascade from a stripping section tray temperature. Steam is the heating medium, with the condensate pumped to condensate recovery. Bottom product is pumped to storage on column level control; overhead pressure is controlled by varying level in the overhead condenser; the balancing line assures sufficient receiver pressure at all times; overhead product is pumped to storage on receiver level control; and reflux is on flow control.

Now, let us see why extra capacity is needed. First, the feed pump must be sized large enough to overcome the pressure drops in the orifice and control valve while delivering feed at highest projected flow and pressure. The driver will often be sized large enough for the pumps' maximum impeller, even if a smaller impeller is initially installed. The driver, especially if it is an electric motor, should have enough power to be "non-overloading." The motor should therefore have enough power for the situation where the control valve is fully open or an operator fully opens the bypass. In either case, the pump will "run out on its curve" requiring more power. Similar considerations apply to the product and condensate pumps.

The reboiler must have enough surface area to provide a margin for control. The control valve introduces a varying pressure drop. The reboiler shell side (chest) pressure can vary from the steam line supply pressure to a high vacuum depending upon the situation. This pressure will balance itself at the necessary level to deliver the required heat duty.

An alternate means of reboiler control is to remove the control valve from the steam line and provide a condensate level controller for the chest cascaded from the tray temperature. The alternate method uses Δ tube surface for control, with the condensate covering more or less tube surface to vary the area exposed to condensing stream. Condensing area is many times more effective for heat transfer than area covered by relatively stagnant condensate. The reboiler must have extra surface to allow part of its surface to be derated for control purposes.

The top pressure controller varies the level of liquid in the condenser, so it, like the reboiler, must have extra surface for the derating required for control. Many other control methods also require some control surface. If noncondensibles are present, a vent should be provided. Otherwise, they collect at the liquid seal. With large amounts of noncondensibles, another type of system should be considered.

A vacuum condenser has vacuum equipment (such as steam jets) pulling the noncondensibles out of the cold end of the unit. A system handling flammable substances has a control valve between the condenser and jets (an air bleed is used to control nonflammable systems). The control method involves derating part of the tube surface by blanketing it with noncondensibles that exhibit poor heat transfer, similar to the steam condensate level in the reboiler or process condensate level in the pressure condenser.

The control valve allows the jets to pull noncondensibles out of the condenser as needed for system pressure control. In addition to requiring extra surface area for control, the vacuum condenser also needs enough surface area for subcooling to ensure that the jets do not pull valuable hydrocarbons or other materials out with the noncondensibles. To allow proper control and subcooling, some designers add approximately 50% to the calculated length.

For best control, a 2:1 pressure drop should be taken across the control valve, and the jets must be designed accordingly. This is still another good example of part of the equipment's capacity being used for control.

The fractionator shell itself should often have some extra trays. Conventional instrumentation alone cannot always be expected to handle all the things that can happen to a fractionation system, such as changes in feed composition, reboiler steam pressure, or coolant temperature (especially for an air condenser during a sudden cold front). Experience for a given service is the best guide for extra trays.

Sources

1. Chin, T. G., "Guide to Distillation Pressure Control Methods," *Hydrocarbon Processing,* October 1979.
2. Branan, C. R., *The Process Engineer's Pocket Handbook, Vol. 2,* Gulf Publishing Co., 1983.

Controller Limitations

A controller cannot do the impossible. We have already seen how the process equipment is part of the control system and has to be designed to accommodate its control function. Without the process equipment doing its part, the controller cannot adequately do its job.

There is also another limitation of controllers that is often overlooked. A controller must be given only its one job to do, not several. For example, a level controller might be designed to deliver feed to a process having a varying demand. A boiler level controller is such an example. This controller can be designed to accommodate changes in steam demand, but do not expect this same controller to also accommodate swings in upstream or downstream pressure. Make sure that other means are available to control the additional variations.

Source

Branan, C. R., *The Process Engineer's Pocket Handbook, Vol. 2,* Gulf Publishing Co., 1983.

False Economy

Avoid the use of instrumentation that may have low first cost, but is very expensive to operate or maintain. Blind controllers, for example, are completely unsatisfactory for most applications. The author has seen examples of temperature controllers set at the factory, but with no method of readout or calibration. These almost always require retrofitting of additional instrumentation later. Internal level floats on process vessels that require plant shutdown for maintenance and whose condition is hidden from view are frequently unsatisfactory.

Source

Branan, C. R., *The Process Engineer's Pocket Handbook, Vol. 2,* Gulf Publishing Co., 1983.

Definitions of Control Modes

Proportional Control

This is a mode of control that causes the output of a controller to change in a linear fashion to the error signal.

Integral (Reset) Control

This is a control algorithm that attempts to eliminate the offset (caused by proportional control) between the measurement and the setpoint of the controlled process variable. This control mode "remembers" how long the measurement has been off the setpoint.

Derivative Control

This is a mode of control that anticipates when a process variable will reach its desired control point by sensing its rate of change. This allows a control change to take place before the process variable overshoots the desired control point. You might say that derivative control gives you a little "kick" ahead.

Source

GPSA Engineering Data Book, Vol. I, Gas Processors Suppliers Association, 10th Ed., 1987.

Control Mode Comparisons

The following handy tabulation from the GPSA Data Book compares the various available control modes.

Source

GPSA Engineering Data Book, Vol. I, Gas Processors Suppliers Association, 10th Ed., 1987.

Control Mode Comparisons

Mode	Advantages	Disadvantages
On-Off	Simple, inexpensive	Constant cycling
Proportional	Does not add lag	Almost always has offset
Integral	Eliminates offset	Adds time lag to system
Derivative	Speeds up response	Responds to noise

Control Mode vs Application

The following handy tabulation from the GPSA Data Book compares the applicability of various control mode combinations.

Source

GPSA Engineering Data Book, Vol. I, Gas Processors Suppliers Association, 10th Ed., 1987.

Control Mode	Process Reaction Rate	Load Changes		Applications
		Size	Speed	
On-off: two-position with differential gap	Slow	Any	Any	Large-capacity temperature and level installations. Storage tanks, hot-water supply tanks, room heating, compressor suction scrubber
Floating, single-speed with adjustable neutral zone	Fast	Any	Small	Processes with small dead time. Industrial furnaces and air conditioning
Proportional	Slow to moderate	Small	Moderate	Pressure, temperature, and level where offset is not objectionable. Kettle reboiler level, drying-oven temperature, pressure-reducing stations
Proportional-plus-derivative (rate)	Moderate	Small	Any	Where increased stability with minimum offset and lack of reset wind-up is required. Compressor discharge pressure
Proportional-plus-integral (reset)	Any	Large	Slow to moderate	Most applications, including flow. Not suitable for batch operations unless overpeaking is allowed
Proportional-plus-integral-plus-derivative	Any	Large	Fast	Batch control; processes with sudden upsets; temperature control

Pneumatic vs Electronic Controls

The following handy tabulation from the GPSA Data Book compares pneumatic and electronic instrumentation.

Source

GPSA Engineering Data Book, Vol. I, Gas Processors Suppliers Association, 10th Ed., 1987.

Instrument Type Features

Pneumatic	Electronic
ADVANTAGES	
1. Intrinsically safe, no electrical circuits. 2. Compatible with valves. 3. Reliable during power outage for short period of time, dependent on size of air surge vessel.	1. Greater accuracy. 2. More compatible with computers. 3. Fast signal transit time. 4. No signal integrity loss if current loop is used and signal is segregated from A.C. current.
DISADVANTAGES	
1. Subject to air system contaminants. 2. Subject to air leaks. 3. Mechanical parts may fail due to dirt, sand, water, etc. 4. Signal boosters often needed on transmission lines of over 300 feet. 5. Subject to freezing with moisture present. 6. Control speed is limited to velocity of sound.	1. Contacts subject to corrosion. 2. Must be air purged, explosion proof, or intrinsically safe to be used in hazardous areas. 3. Subject to electrical interference. 4. More difficult to provide for positive fail-safe operation. 5. Requires consideration of installation details to minimize points 1, 2, 3, and 4.

Process Chromatographs

Process analyzers are used frequently, and gas chromatographs are one of the most common types. Here are some tips that will help improve the utilization of these valuable tools.

1. Assign a special team of instrument technicians to process analyzer maintenance. Having this specialized maintenance done by every instrument technician in the plant is simply not satisfactory, as demonstrated over the years at many plants.
2. The sampling system is as important as the analyzer itself. No stream is 100% free of water, extraneous liquids, scale, etc. In the early days, process chromatographs were little more than lab instruments mounted in the field. Each year increases the respect for the importance of the sampling system. Give this design careful consideration. Process engineers can have valuable input in this area, so do not hesitate to sharpshoot the vendor's design. You know more about the process streams in your plant than he does.
3. Provide for calibration *without using calibration gas.* It is surprising how few people realize that a process chromatograph can be calibrated using only the process stream itself. In fact, this method can even prove to be

more accurate than using a calibration gas. The chromatograph must have two features to use the method:

a) All components must be analyzed, but not necessarily routinely recorded, and none discarded or backflushed out without separation.

b) The instrumentation must be able to produce a recorded full spectrum. This will require attenuation controls to keep all component curves on the chart. Do not buy a chromatograph without this ability.

The calibration is done by observing enough bar graph analysis groups just prior to and after the spectrum generation to ensure that the process stream composition is unchanging. Next, simply integrate the spectrum peaks, calculate the component percentages, and compare to the bar graph recording. Then, dial in recalibration settings as required.

This method avoids errors inherent in handling, sampling, analyzing, storing, introducing into the system, and batch purging with calibration gas.

4. Avoid chromatographs requiring mixed carrier gas. Mixed carrier gas can introduce as many inherent accuracy problems as calibration gas, maybe more. There is almost always a way to avoid using mixed carrier gas. Find a way even if it is more expensive.
5. For new process chromatograph purchases, investigate for modern features that adjust for changes in sample size and flow. Some of these can be a large help in bridging the gaps between visits by the maintenance people.
6. Limit the number of streams sampled and the components per stream routinely recorded by a given chromatograph. Choose only those really necessary for control. This simplifies maintenance and readout.

Source

Branan, C. R., *The Process Engineer's Pocket Handbook, Vol. 2,* Gulf Publishing Co., 1983.

SECTION FOUR
Operations

21 Troubleshooting

Introduction

Much has been written on troubleshooting process equipment in recent years, especially on the topic of fractionation. Excellent material can be found in Lieberman[1], the GPSA Engineering Data Book two-volume set[2], various articles including the gas treating papers by Don Ballard and others at Coastal Chemical Co.[3], and Gulf Publishing Company's pocket handbook series[4,5]. See "Fractionation: Mechanical Problems" for recent troubleshooting articles.

Design and operations are separated in this book for clarity, and so troubleshooting for all types of equipment is included in this section.

Sources

1. Lieberman, N. P., *Process Design for Reliable Operations,* Gulf Publishing Co., 1983.
2. *GPSA Engineering Data Book, Vol. 1 and 2,* Gas Processors Suppliers Association, 10th Ed., 1987.
3. Ballard, Don, various gas treating articles.
4. *The Process Engineer's Pocket Handbook, Vol. 1,2,3,* Gulf Publishing Co.
5. *The Fractionator Analysis Pocket Handbook,* Gulf Publishing Co., 1978.

Fractionation: Checklists

Shah[1] and Drew[2] have provided checklists for troubleshooting fractionation systems. Shah's checklist is intended for use in design of equipment to insure good operations. It can also serve as a troubleshooting checklist.

Drew's Troubleshooter's Guide was included in a start-up article.[2] It includes a diagram to pinpoint instrumentation being discussed.

(text continued on page 299)

Table 1
Tower Checklist

Packed Towers	Tray Towers
1. Distributors, gas spargers: Misorientation can lead to channeling or blowing.	1. Type and number of trays.
2. Hold-down plates, demisters, and packing supports. Free area of the packing support must be equal to or greater than free area of packing to prevent premature flood.	2. Tray spacing, levelness, orientation (manways, seal pans, etc.)
3. Type, size, and volume of packing for different zones.	3. Downcomer sizing. Inlet and outlet weirs.
4. Instrumentation—bottoms level transmitter, thermocouples, delta p cell, etc.	4. Multipass trays—check liquid split.
5. Feed, reflux, bottoms, side-draw nozzle orientation.	5. For chimney trays—weir height and levelness.
6. General cleanliness.	6. Nozzle orientation—including feed, reflux, side draw, bottoms, etc.
	7. Clearance between bottom downcomer and column floor, distance between bottom tray and reboiler vapor return, bottom baffle.
	8. Instrumentation— interface level, bottoms level, thermocouples, delta p cell, etc.
	9. General cleanliness.

Table 2
Sample Troubleshooter's Guide

Condition Observed	As Shown by Instrument No.	Probable Cause	Check Instrument No.	Immediate Action Required (Until Repairs Can Be Made)
Feed				
Feed flow too low	FIC-1	Feed pump faulty or inlet plugged	PI-1 (low)	Stop feed pump.
		Filter or outlet line plugged	PI-1 (high)	
Feed temp. too low	TI-2	Economizer fouled	TI-5 (high)	Clean on shutdown
		Feed flow too high	FIC-1	Correct feed control
		Bottoms flow too low	LIC-4	Check column level
Feed tank full	LAH-O (alarm)	Stripper shut down	GG-0	Start operation of stripper
Column operation (bottom)				
Bottoms temp. too low	TR-3-1 (alarm)	Insufficient boilup	PDRC-6 (low)	Send bottoms to feed tank (TCV-3 shuts, automatically)
		Loss of steam	FIC-7, PI-7 (low)	Check boiler
		Trap plugged	FIC-7, PI-7 (normal)	Use trap bypass
		Feedrate too high, feed too rich	FIC-7 (high)	Reduce feedrate
		Feed distributor fouled	FIC-7 (normal)	Clean out, check filter
Bottoms temp. too high	TR-3-1	Column pressure too high	PI-6 (high)	
		Too much boilup	PDRC-6 & FIC-7 (high)	Reduce PDRC-6 setpoint
		Vent plugged	PDRC-6 (normal) TR-3-2 (high)	Shut down
		Insufficient condensing	TSH-9 (alarm)	See "High vent temp."
Column level too low	LIC-4	Column flooding	GG-4	
			PDRC-6 (high), PI-6 (high)	Reduce boilup
			PI-4 (low)	Stop bottoms pump
Column level too high	LIC-4, GG-4	Bottoms pump failure	PI-4 (low)	
		Economizer or line plugged or hand valve closed	PI-4 (high)	
		Column full after shutdown	PDRC-6, PI-15	Condition temporary
Column pressure too high	PI-6	Boilup too high	PDRC-6 (high)	See "Bottoms temp. too high"
		Vent plugged	TR3-3 (high)	See "Bottoms temp. too high"
		Column level high	LIC-4, GG4 (high)	See "Column level too high"
Column operation (top)				
Top temp. too high	TR-3-2	Rectifying temp. too high	TR3-3	Set TIC-12 lower
			Product assay (low)	Set TIC-12 lower
		Reflux flow too low	LIC-11, GG-9	Correct reflux level control
			PI-11	Correct reflux pump
		Distillate flow too high	FR-15	
		Column pressure high	PI-6, PDRC-6	See "Bottoms temp. too high"
		Column flooding	PDRC-6	See "Column level too low"
Top temp. too low	TR-3-2	Control temp. too low	TR-3-3 (low) FR-15 (low)	Set TIC-12 higher
		Insufficient boilup	FIC-7 (low) TR-3-1 (low)	Increase steam flow
High vent temperature	TSH-9 (alarm)	Cooling water failure	TI-10	Shut steam via SV-9 (automatic)
Distillate flow too low	FR-15	Recycle to feed tank too high	LIC-14	
			GG-14	Check interface level change
		Hand valve closed, line plugged	GG-16	Check rate of level rise
		Feed too low	FIC-1	Check feed supply
		Control temp. too low	TR-3-3	Check TIC-10

Table 2
Sample Troubleshooter's Guide

Condition Observed	As Shown by Instrument No.	Probable Cause	Check Instrument No.	Immediate Action Required (Until Repairs Can Be Made)
Column operation (top)				
		Boilup rate low	TR-3-1 (low)	Check steam supply, PDRC-6
		Reflux too high	TR-3-3	Check TIC-10, TCV-10
Distillate flow too high	FR-15	Reflux ratio too low	TR-3-2, 3 (high)	Check TIC-12
		Feed too high or too rich	TR-3-1 (low)	Reduce feedrate
		Reflux or recycle flow low	LIC-11, 14 (low)	Check these level controls
High distillate temp.	TI-13	Cooling water failure	TI-18	Check water supply
		Distillate flow too high	FI-15	See "Distillate flow too high"
Water in distillate tank	LAH-17 (alarm)	Distillate too hot	TI-13	Check distillate cooler
		Decanter water overflow	LIC-14, GG-14	Increase recycle from decanter
Temperature recorder	TR-3	Temperatures cycling sequentially	TIC-12	Controller cycling, check manual setting, tune controller
		Temperatures rising simultaneously	PI-6	Pressure rising, check condenser and vent
		Temperatures falling simultaneously	FIC-7	Loss of boilup
Composition problems (confirm by a second sample and assay before taking action)				
Solvent in feed too high or too low		Process problems		Reduce feedrate
Solvent in bottoms too high		Bottoms temp. too low	TSH-3	Raise TSH-3 setpoint, increase PDRC-6 setpoint or reduce feedrate
Water in distillate too high		Distillate cloudy		
		Insufficient cooling	TI-13	Check cooler
		Poor decanting	LIC-14	Check decanter
		Distillate clear, cool		Analysis error, soluble impurity present
Utility problems				
Indicated readings low, controllers sluggish		Air pressure low		
Differential pressure low or erratic		Steam pressure low		
Column top temperature high, distillate hot, column pressure high		Water pressure low		

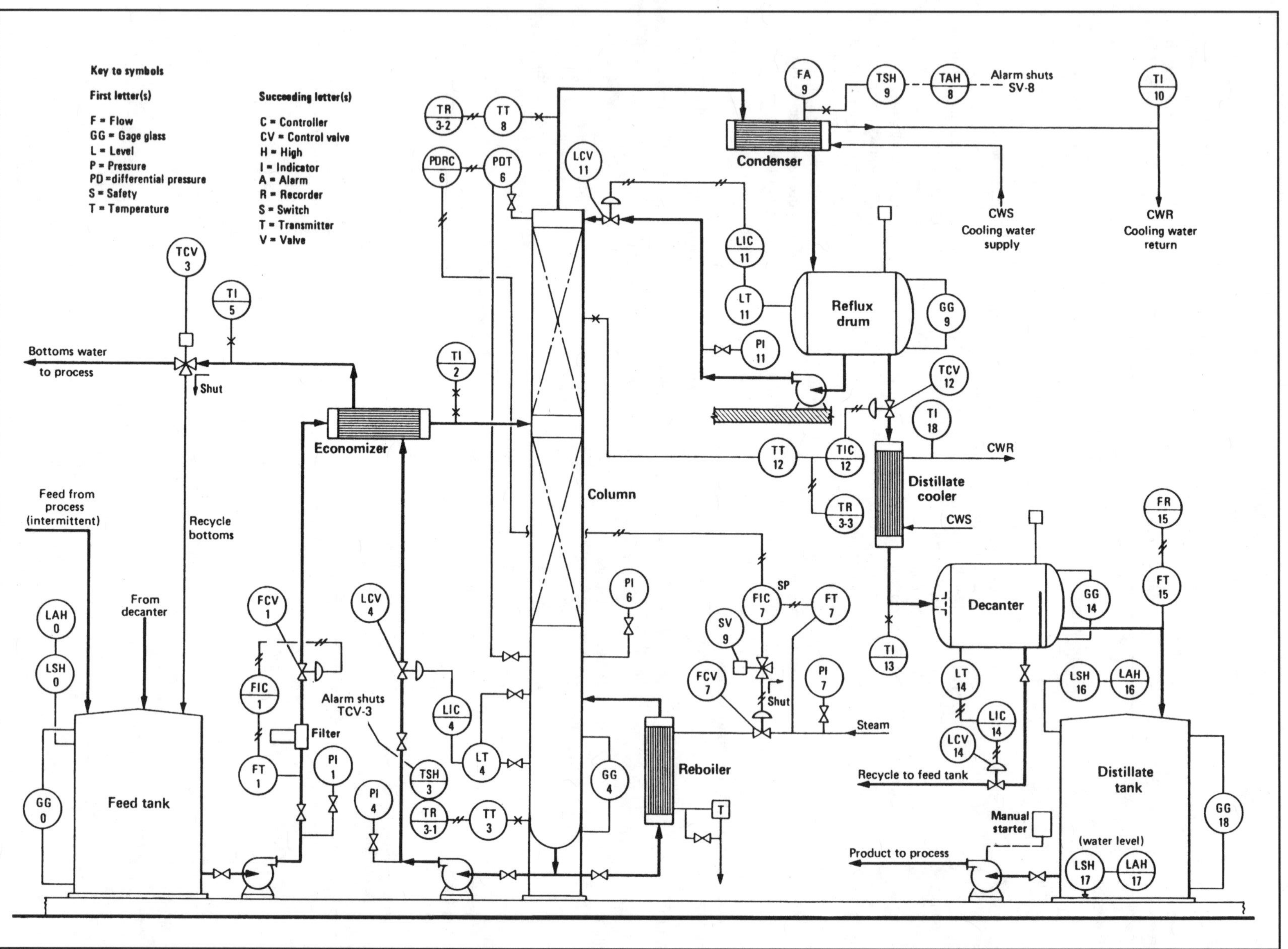

Figure 1. Solvent stripper—simplified piping and instrument diagram for use in conjunction with the Troubleshooter's Guide (Table 2).

(text continued from page 295)

Sources

1. Shah, G. C., "Guidelines Can Help Improve Distillation Operations", *Oil and Gas Journal,* September 25, 1978, p. 102.
2. Drew, J. W., "Distillation Column Startup," *Chemical Engineering,* November 14, 1983, p. 221.

Fractionation: Operating Problems

Some of the more common problems are briefly discussed here.[1] More detailed discussions, including unusual operating histories can be found in the references.[7-20]

Flooding

Design. Flooding is a common operating problem. Companies naturally wish to obtain maximum capacity out of fractionation equipment and thus often run routinely close to flooding conditions. New columns are typically designed for around 80% of flood. Clearly, the column needs some flexibility for varying operating conditions. Vendors state that their modern methods for determining percentage of flood represent very closely the true 100% flood point. The designer, therefore, shouldn't expect to design for 100% of flood and be able to accommodate variations in operating conditions. Reference 2 recommends designing for a percentage of flood of not more than 77% for vacuum towers or 82% for other services, except that for columns under 36″ diameter, 65–75% is recommended. Reference 2 warns against using its standard methods for systems that are frothy or have some other peculiar characteristics. Glycol dehydrators and amine absorbers are cited as examples of such systems.

Jet Flood. Flooding generally occurs by jet flood or downcomer backup. Reference 2 gives Equations 1, 2, and 3 for jet flood, using Ballast trays.

$$\%\ \text{Flood}/100 = \frac{V_{load} + (\text{GPM} \times \text{FPL}/13000)}{\text{AA} \times \text{CAF}} \quad (1)$$

$$\%\ \text{Flood}/100 = \frac{V_{load}}{\text{AT} \times \text{CAF}\ (0.78)} \quad (2)$$

$$\%\ \text{Flood}/100 = \quad (3)$$

$$\left[\frac{V_{load}}{\left(\text{AA} \times \text{CAF} - \dfrac{\text{AD} \times \text{VD}_{dsg} \times \text{FPL}}{13000}\right)(\text{DLF})}\right]^{0.625}$$

where

$$\text{DLF} = \left[\frac{\text{AD} \times \text{VD}_{dsg}}{\text{GPM}}\right]^{0.6}$$

The larger value from Equations 1, 2, and 3 applies. Equation 2 applies only for liquid rates less than 0.50 GPM/L_{wi}, where L_{wi} is the weir length in inches. Equation 3 applies where the downcomers are unusually small relative to the required downcomer area.

Downcomer Backup Flood. For downcomer backup, Equation 4 can be used. Reference 2 states that if the downcomer backup for valve trays exceeds 40% of tray spacing for high vapor density systems (3.0 lbs/ft^3), 50% for medium vapor densities, and 60% for vapor densities under 1.0 lb/ft^3, flooding may occur prior to the rate calculated by jet flood Equations 1, 2, and 3. Another good rule of thumb is that the downcomer area should not be less than 10% of the column area, except at unusually low liquid rates. If a downcomer area of less than 10% of column area is used at low liquid rate, it should still be at least double the calculated minimum downcomer area.

$$H_{dc} = H_w + 0.4(\text{GPM}/L_{wi})^{2/3} + [\Delta P_{tray} + H_{ud}]\ [\rho_L/(\rho_L - \rho_V)] \quad (4)$$

To find H_{ud}, use Equation 5.

$$H_{ud} = 0.65(U_{ud})^2 \text{ or}$$
$$H_{ud} = 0.06(\text{GPM}/\text{DCE}/\text{DCCL})^2 \quad (5)$$

Tower Operations. The tower operator can quickly determine which type of flooding will tend to be the limiting one for a particular system. If a rigorous computer run is available for the anticipated or actual operation, the operator can quickly calculate the expected limiting column section. The operator can then provide DP cell recording for the entire column and limiting section(s). As mentioned previously, a DP cell is the best measure of internal traffic and flooding tendency.

Many plant nontechnical operators do not understand that high vapor rates as well as high liquid rates can cause downcomer backup flooding. It is well to explain to the plant operators the mechanism of downcomer backup flooding and show them with a diagram how the head of liquid in the downcomer must balance the tray pressure drop. Then it can be explained how vapor flow is a major contributor to this pressure drop. Finally, showing them Equation 4 will demonstrate the significant effect of tray pressure drop on downcomer backup. Enlightened operators having the necessary recorded information represent a good investment.

Flooding across a column section reflects itself in an increase in pressure drop and a decrease in temperature difference across the affected section. Product quality is also impaired, but it is hoped that the other indicators will allow correction of the situation before major change in product quality. When a column floods, the levels in the accumulator and bottom often change. It can occur that the accumulator fills with liquid carried over while the reboiler runs dry.

Also, in a flooded column, the pressure will often tend to fluctuate. This may help to differentiate between flooding and a high column bottom level, if the bottom level indicator reading is suspect. The high bottom level will give higher than normal pressure drop, but often not the magnitude of pressure fluctuations associated with flooding.

Here is a tip for possible capacity increase for towers with sloped downcomers. Usually, the tray vendor doesn't use the dead area next to the bottom part of the sloped downcomer as active area if the trays are multipass, since he would require a different design for alternate trays. This area could be used for additional vapor capacity in an existing column.

Dry Trays

This problem, as with flooding, also impairs product quality. No fractionation occurs in the dry section, so the temperature difference decreases. However, unlike flooding, the pressure drop decreases and stays very steady at the ultimate minimum value. This problem is usually easier to handle than flooding. The problem is caused by either insufficient liquid entering the section or too much liquid boiling away. The problem is solved by reversing the action that caused the dry trays.

Since the changes usually occur close to the source of the problem, the source can usually be quickly found with proper instrumentation. Too little reflux or too much sidestream withdrawal are two examples of insufficient liquid entering a section. Too hot a feed or too much reboiling are examples of excessive liquid boiloff.

Damaged Trays

Effects. Trays can become damaged several ways. A pressure surge can cause damage. A slug of water entering a heavy hydrocarbon fractionator will produce copious amounts of vapor. The author is aware of one example where all the trays were blown out of a crude distillation column. If the bottom liquid level is allowed to reach the reboiler outlet line, the wave action can damage some bottom trays.

Whatever the cause of the tray damage, however, it is often hard to prove tray damage without column shutdown and inspection, especially if damage is slight. Besides poorer fractionation, a damaged tray section will experience a decrease in temperature difference because of the poorer fractionation. An increase in pressure difference may also result, since the damage is often to downcomers or other liquid handling parts. However, a decrease in pressure difference could also occur.

Bottoms Level. Trays are particularly vulnerable to damage during shutdown and startup operations. Glitsch, Inc., (Reference 21) provides several good tips to minimize the possibility of tray damage during such periods.

First, it is important to avoid high bottoms liquid level. Initial design should provide sufficient spacing above and below the reboiler return vapor line. A distance equal to at least tray spacing above the line to the bottom tray, or better, tray spacing plus 12″; and a distance of at least tray spacing below the line to the high liquid level is absolutely necessary. Probably more tray problems occur in this area of the column than any other. In spite of good initial design, however, the bottoms liquid level needs to be watched closely during startup.

If a tower does become flooded in the bottom section, a common operator error is to try to pump the level out too quickly. This can easily damage trays by imposing a downward acting differential pressure produced by a large weight of liquid on top of the tray and a vapor space immediately below the tray. To eliminate the flooding, it is better to lower feed rate and heat to the reboiler. It is important to be *patient* and avoid sudden changes.

Steam/Water Operations. Steam/water operations during shutdown have high potential for tray damage if not handled correctly. If a high level of water is built up in the tower and then quickly drained, as by pulling off a bottom manway, extensive tray damage can result, similar to pumping out hydrocarbons too fast during operation.

Adding steam and water to a tower together can be a risky operation. If the water is added first at the top, for instance, and is raining down from the trays when steam is introduced, the steam can condense and impose a downward acting differential pressure. This can result in considerable damage. If steam and water must be added together, start the steam first. Then *slowly* add water, not to the point of condensing all the steam. When finished, the water is removed first. One vendor estimates that he sees about six instances of tray failure per year resulting from mishandled steam/water operations.

Depressuring. Depressuring a tower too fast can also damage the trays by putting excessive vapor flow through them. A bottom relief valve on a pressure tower would make matters worse since the vapor flow would be forced across the trays in the wrong direction. The equivalent for a vacuum tower would be a top vent valve which would suck in inerts or air and again induce flow in the wrong direction. Such a top vent should not be designed too large.

Phase Change. Overlooking change of phase during the design stage can also cause tray damage. An example is absorber liquid going to a lower pressure stripper and producing a two-phase mixture. In one case, the absorber stream entered the stripper in a line that was elled down onto the stripper tray. The two-phase mixture beat out a section of trays. A ¼″ protection plate was provided and this had a hole cut in it in two years.

Water in Hydrocarbon Column

Here, we refer to small amounts of water rather than large slugs that could damage the trays. Often the water will boil overhead and be drawn off in the overhead accumulator bootleg (water drawoff pot). However, if the column top temperature is too low, the water is prevented from coming overhead. This plus too hot a bottom temperature for water to remain a liquid will trap and accumulate water within the column. The water can often make the tower appear to be in flood.

Many columns have water removal trays designed into the column. Top or bottom temperatures may have to be changed to expel the water if the column isn't provided with water removal trays. In some instances, the water can be expelled by venting the column through the safety relief system.

It should be remembered that water present in a hydrocarbon system, being immiscible, will add its full vapor pressure to that of the hydrocarbons. The author once wondered why the pressure was so high on a certain overhead accumulator until he noticed the installed bootleg.

A small steady supply of water entering a column through solubility or entrainment can, in some cases, cause severe cycling at constant intervals during which time the water is expelled. After expulsion of water, the column lines out until enough is built up for another cycle.

Besides water, other extraneous substances can leak into a column with varied effects. One example was a column separating two components and using the light component as seal flush for the reboiler pump. When excessive lights leaked into the tower system, the bottoms product went off specification. It took a long time to solve the problem, because at first the operators suspected loss of tray efficiency.

Foaming

The mechanism of foaming is not well understood. During the design phase, foaming is provided for in both the tray downcomer and active areas. "System factors" are applied that derate the trays for foaming. Table 1 contains some typical system factors. In Table 1, oil absorbers are listed as moderate foamers. A heavy oil mixed with light gases often tends to foam. The higher the pressure, the more foaming ten-

Table 1
System Factors (Reference 2)

Service	System Factor
Non-foaming, regular systems	1.00
Fluorine systems, e.g., BF_3, Freon	.90
Moderate foaming, e.g., oil absorbers, amine and glycol regenerators	.85
Heavy foaming, e.g., amine and glycol absorbers	.73
Severe foaming, e.g., MEK units	.60
Foam-stable systems, e.g., caustic regenerators	.30–.60

dency, since the heavy oil will contain more dissolved gases at higher pressures. Liquids with low surface tension foam easily. Also, suspended solids will stabilize foam. Foaming is often not a problem when a stabilizer is not present.

To the author's knowledge, no laboratory test has been developed to adequately predict foaming. An oil that doesn't foam in the laboratory or at low column pressure might well foam heavily at high column pressure. In general, aside from adding antifoam, there seems to be no better solution to foaming than providing adequate tray spacing, and column downcomer area. One designer solved a downcomer foaming problem by filling the downcomer with Raschig rings to provide coalescing area.

For troubleshooting suspected foam problems, vaporize samples of feed and bottoms to look for suspended solids. Also, one can look for the Tyndall effect as described in the section on condenser fogging.

In investigating foaming problems, it is helpful to have an estimate of foam density. Reference 3 contains a relationship for distillation and absorption column trays which is stated to agree well with published data. This relationship is shown as Equation 6.

$$\rho_F = \rho_L[1 - 0.46 \log (f/0.0073)] \qquad (6)$$

where

$$f = u[\rho_V/(\rho_L - \rho_V)]^{1/2}$$

Reboiler Problems

High Pressures. Thermosyphon reboilers present design problems at the two extremes of the pressure scale. Near the critical pressure, the maximum allowable flux drops.

Whereas a flux of about 12,000–16,000 Btu/hr ft^2 is a reasonable range for hydrocarbons at "normal" pressure, a flux of only, say, 7,000 Btu/hr ft^2 is reasonable near the critical pressure.

Low Pressures. At low pressures, the problem is typically not being able to achieve proper thermosyphoning. At low pressure, a longer sensible heat zone can be experienced which can inhibit proper circulation rate. This is related to the large effect of liquid head on boiling point at low pressure. For proper performance, a low percentage of vaporization per pass is required. A 4:1 liquid to vapor ratio is about minimum. For high vacuum service, forced circulation reboilers are sometimes used. It is not uncommon to see a circulation pump added to an existing low pressure reboiler originally designed for natural circulation.

Intermediate Pressures. At intermediate pressures, typical problems are fouling, noncondensibles on the steam (or other heating medium) side, improper withdrawal of condensate, and unstable circulation. Fouling is largely dependent upon the system properties, but often a more tightly designed unit will experience less fouling than one designed with a lot of fat. This may be due to the greater turbulence in the unit that is working harder. A once-through reboiler can be used if recirculation tends to increase fouling.

For steam side noncondensibles, a proper vent is required. A small amount of noncondensibles can greatly lower the steam side heat transfer coefficient. The improper removal of condensate is another way to reduce the steam side coefficient. A pumped system is preferable to steam traps for a large system.

Inlet Line. Unstable circulation can result if the inlet line to a vertical thermosyphon reboiler is too large. The tubes of a vertical thermosyphon reboiler "fire" individually. The tubes can backfire excessively if the liquid inlet line is too large. They don't have to backfire all the way into the tower to cause problems, just to the inlet tubesheet. It is common to put flanges in the inlet liquid line so an orifice can be added later, if required, to provide proper dampening effect.

Condenser Fogging

Fogging occurs in a condenser when the mass transfer doesn't keep up with the heat transfer. The design must provide sufficient time for the mass transfer to occur. A high temperature differential (ΔT) with noncondensibles present or a wide range of molecular weights can produce a fog. The high ΔT gives a high driving force for heat transfer. The driving force for mass transfer, however, is limited to the concentration driving force (ΔY) between the composition of the condensible component in the gas phase and the composition in equilibrium with the liquid at the tube wall temperature. The mass transfer driving force (ΔY) thus has a limit. The ΔT driving force can, under certain conditions, increase to the point where heat transfer completely outstrips mass transfer, which produces fogging.

Nature of a Fog. Fog, like smoke, is a colloid. Once a fog is formed, it is very difficult to knock down. It will go right through packed columns, mist eliminators, or other such devices. Special devices are required to overcome a fog, such as an electric precipitator with charged plates. This can overcome the zeta potential of the charged particles and make them coalesce.

A colloid fog will scatter a beam of light. This is called the "Tyndall Effect" and can be used as a troubleshooting tool.

Cures. Eliminate the source of fogging by using a smaller ΔT and thus more surface for mass transfer. Try to minimize $\Delta T/\Delta Y$.

Calculations. To check a design for possible fogging, a procedure is presented that rightly considers mass transfer and heat transfer as two separate processes.

Heat Transfer

$$Q = UA\,\Delta T_M \text{ or } Q = h_i\,A\Delta T_i \tag{7}$$

where i refers to the condensing side only.

Mass Transfer

$$W = K_m\,A\Delta Y = K_m A\,(Y - Y_i) \tag{8}$$

where

h_i = Condensing side film coefficient, Btu/hr ft^2 °F
W = Condensate rate, lbs/hr
K_m = Mass transfer coefficient, lb/hr ft$^2\Delta Y$
Y = Composition of the condensible component in the gas phase
Y_i = Composition of the condensible component in equilibrium with liquid at tube wall temperature.

Mass and heat transfer are related as follows:

$$h_i/K_m c = (K_T/c\rho_V K_D)^{2/3} \tag{9}$$

where

c = heat capacity (gas phase), Btu/lb °F
ρ_V = vapor density, lb/ft^3
K_T = thermal conductivity, Btu/hr ft^2 (°F/ft)
K_D = diffusivity, ft^2/hr

The ratio of heat transfer to mass transfer is:

$$Q/W = h_i\Delta T_i/K_m\Delta Y \tag{10}$$

since

$$h_i/K_m = c(K_T/c\rho_V K_D)^{2/3}$$

then

$$Q/W = c(K_T/c\rho_V K_D)^{2/3}\Delta T_i/\Delta Y$$

Note that Q refers only to sensible heat transfer. All latent heat is transferred via mass transfer. Likewise, h_i refers only to a dry gas coefficient (no condensation considered).

The calculations are made as follows. The exchanger is divided into small increments to allow numerical integrations. A tube wall temperature is first calculated and then Q/W. The gas temperature and composition from an increment can then be calculated. If the gas composition is above saturation for the temperature, any excess condensation can occur as a fog. This allows the degree of fogging tendency to be quantified. Whenever possible, experimental data should be used to determine the ratio of heat transfer to mass transfer coefficients. This can be done with a simple wet and dry bulb temperature measurement using the components involved.

Suspect Laboratory Analyses

Bad laboratory analyses are not always the fault of the laboratory. Sampling plays a big role. One plant superintendent investigated every instance of suspect analyses in his plant using elaborate around-the-clock methods over a considerable period. His results revealed that over one half of the bad analyses were not the fault of the laboratory. We are all human and bad analyses will result from time to time. Rather than resubmit samples, it may be well to spend a few minutes using the following methods as referees to evaluate the reasonableness of the results.

First, the old standby methods of checking the overall individual component balances and checking dew and bubble points will help verify distillate and bottoms concentrations. The total overhead (distillate plus reflux) calculated dew point is compared to the column overhead observed temperature and the bottoms calculated bubble point is compared to the column bottom observed temperature. If the analyses are not felt to be grossly in error, the following method will also prove very helpful.

The method proposed is the Hengstebeck Method described in Reference 4. The method is quite simple. First, a heavy key component is selected and the relative volatility (α) of all column components to this heavy key are determined. Once this is done for a column, minor operating changes won't affect α much and the same α can therefore often be used for subsequent sets of data. One can use α_{feed} or perhaps more accurately, $\alpha = \sqrt{\alpha_{Top}\,\alpha_{Bottom}}$. Then a plot is made of ln D/B versus α. A straight line should result. If a fairly straight line does not result, the analyses are probably faulty. Here, D is the mols/hr of each component in the distillate and B is the mols/hr of each component in the bottoms. Actually, from Fenske's equation (Equation 11 and Reference 5), at total reflux it is shown that ln D/B versus ln α is a straight line. However, at finite reflux and using α instead of ln α a line close enough to perfectly straight is obtained.

Armed with K and α values, the plant operators may be induced to do some of these calculations to occasionally sharpshoot laboratory results.

Time permitting, the more lengthy Smith-Brinkley Method (Reference 6) could also be used as an analysis troubleshooting tool.

$$N_m + 1 = \frac{\ln[(X_{LK}/X_{HK})_D(X_{HK}/X_{LK})_B]}{\ln(\alpha_{LK/HK})_{AVG}} \qquad (11)$$

Nomenclature

A = Heat transfer area, ft^2
AA = Active area, ft^2
AD = Downcomer area, ft^2
AT = Column cross-sectional area, ft^2
B = Bottoms molar rate or subscript for bottoms
c = Heat capacity (gas phase), Btu/lb °F
CAF = Vapor capacity factor
D = Distillate molar rate or subscript for distillate
DCCL = Downcomer clearance, ins.
DCE = Length of downcomer exit, ins.
DFL = Column flooding correlating factor
FPL = Tray flow path length, ins.
GPM = Column liquid loading, gal/min
H_{dc} = Downcomer backup, inches of liquid
h_i = Condensing side film coefficient, Btu/hr ft^2 °F
H_{ud} = Head loss under downcomer, inches of liquid
H_W = Weir height, ins.
K = Equilibrium constant equals y/x
K_D = Diffusivity, ft^2/hr
K_m = Mass transfer coefficient, lb/hr $ft^2 \Delta y$
K_T = Thermal conductivity, Btu/hr ft^2(°F/ft)
L_{Wi} = Weir length, ins.
N_m = Minimum theoretical stages at total reflux
Q = Heat transferred, Btu/hr
U = Overall heat transfer coefficient, Btu/hr ft^2 °F
u = Vapor velocity, ft/sec
U_{ud} = Velocity under downcomer, ft/sec
VD_{dsg} = Downcomer design velocity, GPM/ft^2
V_{load} = Column vapor load factor
W = Condensate rate, lbs/hr
X_{HK} = Mol fraction of heavy key component
X_{LK} = Mol fraction of the light key component
α_i = Relative volatility of component i versus the heavy key component
ΔP_{tray} = Tray pressure drop, inches of liquid
ΔT_i = Condensing side temperature difference, °F
ΔT_m = Log mean temperature difference, °F
ρ_F = Foam density, lbs/ft^3
ρ_L = Liquid density, lbs/ft^3
ρ_V = Vapor density, lbs/ft^3

References

1. Branan, C. R., *The Fractionator Analysis Pocket Handbook,* Gulf Publishing Co., Houston.
2. Glitsch Ballast Tray Design Manual, 5th Ed., Bulletin No. 4900, Copyright 1974, Printing 1989, Glitsch, Inc.
3. Zanker, Adam, "Quick Calculation for Foam Densities," *Chemical Engineering,* February 17, 1975.
4. Hengstebeck, R. J., *Trans. A.I.Ch.E., 42,* 309, 1946.
5. Fenske, M., *Ind. Eng. Chem., 24,* 482, 1932.
6. Smith, B. D., *Design of Equilibrium Stage Processes,* McGraw-Hill, 1963.
7. Shah, G. C., "Guidelines Can Help Improve Distillation Operations," *Oil and Gas Journal,* September 25, 1978, p. 102.
8. Drew, J. W., "Distillation Column Startup," *Chemical Engineering,* November 14, 1983, p. 221.
9. Hower, T. C., and Kister, H. Z., "Solve Process Column Problems," parts 1 and 2, *Hydrocarbon Processing,* May and June 1991.
10. Harrison, M. E. and France, J. J., "Trouble-Shooting Distillation Columns," parts 1–4, *Chemical Engineering,* March–June, 1989.
11. Hower, T. C., and Kister H. Z., "Unusual Operating Histories of Gas Processing and Olefins Plant Columns," paper presented at A.I.Ch.E. Annual Meeting, November 2–7, 1986, Miami Beach, Florida.
12. Yanagi, Tak, "Inside A Trayed Distillation Column," *Chemical Engineering,* November 1990, p. 120.
13. Biddulph, M. W., "When Distillation Can Be Unstable," *Hydrocarbon Processing,* September 1975, p. 123.
14. Shah, G. C., "Troubleshooting Distillation Columns," *Chemical Engineering,* July 31, 1978, p. 70.
15. Kister, H. Z., "When Tower Startup Has Problems," *Hydrocarbon Processing,* February 1979, p. 89.
16. Shah, G. C., "Troubleshooting Reboiler Systems," *Chemical Engineering Progress,* July 1979, p. 53.
17. Lieberman, N. P., "Basic Field Observations Reveal Tower Flooding," *Oil and Gas Journal,* May 16, 1988, p. 39.
18. Tammami, Ben, "Simplifying Reboiler Entrainment Calculations," *Oil and Gas Journal,* July 15, 1985, p. 134.
19. Pattinson, Scott, "Changes In Demethanizer Reboiler Solve Efficiency Problems," *Oil and Gas Journal,* May 1, 1989, p. 102.
20. Lieberman, N. P. and Lieberman, E. T., "Design, Installation Pitfalls Appear In Vac Tower Retrofit," *Oil and Gas Journal,* August 26, 1991, p. 57.
21. Kitterman, Layton, "Tower Internals and Accessories," Glitsch, Inc. Paper presented at Congresso Brasileiro de Petro Quimica, Rio de Janeiro, November 8–12, 1976.

Fractionation: Mechanical Problems

Many articles have been published recently on troubleshooting the fractionation problems of a larger magnitude than those under "Operating Problems." We have titled these larger-magnitude problems "Mechanical Problems," because they usually involve faulty equipment or needed revamps in contrast to operating "upsets." The following "Field Examples," summarize the literature. It is hoped you will find general solutions to some of your practical problems and can address the source articles for more details. These troubleshooting examples also provide several useful general principles.

Field Examples

Pinch

Problem:[1] Fractionator with 140 trays was a bottleneck. A revamp was proposed that would not have worked.

Cause: Computer simulation indicated design was okay, but didn't consider pinch point.

Solution: Constructed McCabe-Thiele[2] diagram. Pinch found.

Moral: Check computer designs with a graphical method. McCabe-Thiele recommended. Can use computer simulation to help construct McCabe-Thiele diagram.

Key Components

Problem:[1] Stripper computer simulation for a base case did not match test data. Proper simulation was needed for revamp design checking.

Cause: VLE data inaccuracies explained part of the problem. The computer simulation didn't allow understanding of the column operation. Different components were the effective key components in different parts of the column.

Solution: Henstebeck[3] diagrams were produced based upon using different key component pairs. These showed where pinch points were occurring and the resulting understanding allowed a proper computer simulation to be selected.

Moral: Check computer designs with a graphical method. In this case the Henstebeck method was good for the breaking out of key components from the multicomponent mix. The Henstebeck method uses a McCabe-Thiele type diagram.

Profiles

Problem:[1] A structured packing vacuum tower had too much heavy key in a vapor side-draw between the feed and the column bottom. The side-draw heavy key concentration was several times the design value.

Cause: The packing height was set by an optimistic HETP. It was delivering 6–7 stages versus 8 design.

Solution: Even though packing was only shorted by 1–2 stages this was enough to cause the problem during off-design conditions. Plotting the column heavy key profiles showed a very steep and unforgiving slope. The heavy key was designed to drop from 55% to 1% in only 5 stages. Loss of 1–2 stages, out of 8 total required, was devastating.

Moral: Examine column profiles before investing money.

Troubleshooting Technique

Problem:[4] A 100-tray vacuum distillation column was run in blocked operation mode. After a run on a previous product the column would not run properly for a new product.

Cause: The operators thought the cause was plugged trays, but a careful engineer looked deeper and found the water supply valve to the condenser only one-fourth open. The previous product run didn't require any more condenser cooling than this.

Solution: Opened the valve

Moral: Conduct a proper troubleshooting technique as described in the article:[4]

"1. The current and past operating data were compared and the timing of the operating problems was defined.

"2. The probable causes were compared to the data available and the physical conditions of equipment were checked. Probable causes were identified.

"3. Insufficient data were available to confirm the exact source of the problem. So, a program to collect the additional data was conducted.

"4. The new data directed the investigation to the correct part of the system, and the trouble was quickly identified."

Cutting Corners

Problem:[5] A deisobutanizer would not produce the required isobutane removal from the bottoms. The bottoms product ran 17% instead of 5%.

Cause: Several errors were made in the original design that are described as "cutting-corners."

Solution: After thorough analysis, including heat and material balances and hydraulic calculations, the initial design flaws were corrected including:

1. Installing the standard antijump baffles for certain inboard downcomer trays to keep droplets (produced by tray "blowing") from entering the downcomer.
2. "Picket fence" weirs were installed on other inboard downcomers also for shielding any blowing.

3. Added a feed distributor (omitted originally) to allow the proper lower feed point to be used for one feed stream.
4. Added a feed preheat exchanger so that feed was not subcooled. This unloaded the critical bottom tower section.

Moral: Cutting-corners for a design saves pennies and costs big dollars over the years.

Trapped Water

Problem:[6] A deethanizer overhead chiller experienced tubeside plugging due to freezing.
Cause: A recent process change had inadvertently trapped water in an endless loop. This allowed the water to build whereas before the change the water had a way out.
Solution: Added a small package TEG dehydrator to the stream ahead of the chiller.
Moral: The designer must look beyond any modification itself to see how it interacts with the existing system.

Cocurrent

Problem:[6] A gas plant absorber/deethanizer train was not achieving design separation. The absorber and the deethanizer seemed to be operating at poor efficiencies.
Cause: A cross exchanger designed to cool absorber lean oil while heating deethanizer rich oil feed was not doing its job. It was found to be fabricated for cocurrent flow instead of countercurrent.
Solution: Repiped one side of the exchanger to convert it to countercurrent.
Moral: Check design details carefully.

Foaming

Problem:[6] An amine absorber was carrying over due to foaming.
Cause: The amine had strong foaming tendencies and antifoam had not been added. When antifoam was added batchwise, the foaming became worse because too high an antifoam concentration actually causes foaming.
Solution: A dilution method was employed to achieve correct antifoam concentration. An injection pump was installed for slow, rather than batch, antifoam addition.
Moral: Add antifoam slowly with an injection pump.

Stacking Packing

Problem:[6] A packed, direct contact, water spray tower cooled acetylene furnace effluent. The bottom one foot, or so, of the bed would plug with polymer material. This is the hottest part of the bed.
Cause: Polymer deposits stuck in the random packing interstices. When the bed was hand stacked in a staggered arrangement, rather than random packed, the vapor and liquid channeled and the gas was not cooled.
Solution: Only the bottom one-foot was hand stacked and the rest of the bed was randomly packed using a wet-packing technique.
Moral: Sometimes conventional methods have to be modified to fit the conditions.

Tray Supports

Problem:[6] Absorber bubble caps were replaced with sieve trays. A severe flow upset dislodged the new trays.
Cause: The sieve trays had weaker connections at the support ring than the old bubble cap trays. No welding was allowed (heavy-metal acetylides present).
Solution: A unique support connection was designed and installed as shown in the article.
Moral: Check modifications for abnormal conditions to the extent possible.

Boilup Control

Problem:[6] An ethane/ethylene splitter exhibited poor control of overhead and bottoms concentrations.
Cause: Because the top product is very pure, the top tray temperature is insensitive to concentration. Therefore, the temperature difference between the top tray and tray 50 was the design control signal to adjust reflux. This was very sluggish because a change in reflux is slowly reflected down the column in changed liquid overflows from tray to tray.

Solution: Because vapor rate changes are reflected up and down the column much faster than liquid rate changes, the temperature difference controller was disconnected and the tower was controlled instead by boilup. A temperature 10 trays from the bottom set reboiler heating medium and the reflux was put on flow control.

Moral: In large superfractionators, the fast response of boilup manipulation is advantageous for composition control.

Condenser Velocity

Problem:[6] A knockback condenser mounted on a C_3 splitter reflux drum exhibited liquid carryover (as evidenced by the vent line icing-up). This indicated product loss from liquid carrying over rather than dripping back into the reflux drum. Also the vent line metallurgy would not withstand the cold temperatures produced.

Cause: The velocity was too high in the vent condenser, thus causing the vapor to entrain liquid.

Solution: A valve limiter was installed on the vent control valve to limit opening.

Moral: Excessive vapor flows in knockback condensers lead to entrainment. (Reference 7 shows how to predict the maximum allowable velocity.)

Blind Blind

Problem:[6] A fired reboiler outlet temperature would not rise above 350°F no matter how hard the heater was fired.

Cause: A blind had been left in the heater outlet circuit. The blind had no handle, so it was difficult to find this error. Troubleshooting was difficult because certain instruments had not been commissioned and the available ones indicated everything was okay except for the 350°F maximum.

Solution: Removed the blind.

Morals: Blinds should have long handles and tags. Instrumentation must be operational before the system is commissioned.

In a troubleshooting investigation, the obvious interpretation of an observation may not be the correct one.

Temperature Control of Both Ends

Problem:[6] A lean oil still had unstable control and erratic operation.

Cause: The instrumentation was attempting to control the temperature at both ends of the still which prevented steady state operation.

Solution: Disconnected the top temperature controller and used the more important bottom controller alone.

Morals: Don't attempt to control the temperature of both ends of a column.

Sources

1. Kister, H. Z., "Troubleshoot Distillation Column Simulations," *Chemical Engineering Progress,* June, 1995.
2. McCabe, W. L. and E. W. Thiele, "Graphical Design of Fractionating Columns," *Ind. Eng. Chem.*, 17, p. 605 (1925).
3. Hengstebeck, R. J., "An Improved Shortcut for Calculating Difficult Multicomponent Distillations," *Chem. Eng.*, Jan. 13, 1969, p. 115.
4. Hasbrouck, J. F., J. G. Kinesh and V. C. Smith, "Successfully Troubleshoot Distillation Towers," *Chemical Engineering Progress,* March 1993.
5. Sloley, A. W. and S. W. Golden, "Analysis Key to Correcting Debutanizer Design Flaws," *Oil and Gas Journal,* Feb. 8, 1993.
6. Hower, T. C. and H. Z. Kister, "Solve Process Column Problems," *Hydrocarbon Processing,* Part 1-May 1991, Part 2-June 1991.
7. Diehl, J. E. and C. R. Koppany, *Chemical Engineering Progress Symposium Series,* 92, (65), p. 77, 1969.

Fluid Flow

Here are four fluid flow problems with suggestions for corrections.

Water Hammer

This troublesome problem is often found in plants, particularly in condensate systems. The common variety is caused by rapid condensation of steam, which occurs when high- and low-pressure condensates are mixed.

Figure 1 shows high-pressure condensate (small line) being added to low-pressure condensate without the usual troublesome hammer. The high-pressure condensate has a chance to cool before emerging into the low-pressure condensate line.

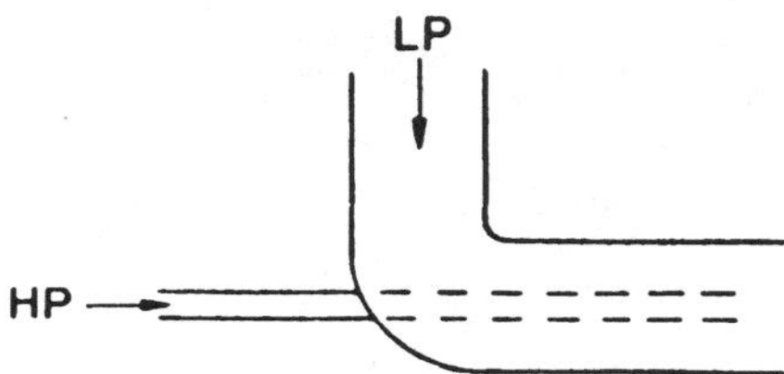

Figure 1. Method for adding high-pressure condensate to low-pressure condensated without troublesome water hammer.

Vertical Downflow Problems

Another problem can cause vibration and sounds that make the operators refer to it as water hammer. Without going into the theory of long bubbles getting trapped in vertical downflowing pipes, the avoidance of problems will be discussed. For theory, Reference 2 is recommended.

If a continuous source of vapor is available, Froude numbers between 0.31 and 1.0 will give trouble. The Froude number is defined as follows:

$$(N_{Fr})_L = V_L / \sqrt{g_L D}$$

where

V_L = Liquid velocity, ft/sec
$g_L = g_c(\rho_L - \rho_G)/\rho_L$
g_c = 32.2 ft/sec^2
D = Internal pipe diameter, ft
ρ_L = Liquid density, lb/ft^3
ρ_G = Vapor density, lb/ft^3

The source of vapor can be vapor already in the line ahead of the vertical downflow section or vapor that is produced if the downflow pulls a vacuum or sufficiently reduces pressure at the top of the downflow section.

Reference 2 gives a case history for a 30-in. line at 13,000 gpm, and the author has observed a plant example for a 6-in. line at 145 gpm. Both had Froude numbers in the troublesome range and both caused big problems.

Two-Phase Slug Flow

Slug flow must be avoided in all two-phase applications. The designer must be alert for two-phase flow developing in a system. In one case, an absorber liquid going to a lower pressure stripper produced a two-phase mixture. The absorber stream entered the stripper in a line that was elled down onto the stripper tray. The two-phase mixture beat out a section of trays. A ¼-in. protection plate was provided and this had a hole cut in it in two years.

Flashing/Cavitation

The most frequently encountered flashing problems are in control valves. Downstream from the control valve a point of lowest pressure is reached, followed by pressure "recovery." A liquid will flash if the low pressure point is below its vapor pressure. Subsequent pressure recovery can collapse the vapor bubbles or cavities, causing noise, vibration, and physical damage.

When there is a choice, design for no flashing. When there is no choice, locate the valve to flash into a vessel if possible. If flashing or cavitation cannot be avoided, select hardware that can withstand these severe conditions. The downstream line must be sized for two-phase flow, and a long conical adapter from the control valve to the downstream line must be used.

Sources

1. *The Process Engineer's Pocket Handbook, Vol. 3,* "Process Systems Development," Gulf Publishing Co., p. 22–23.
2. Simpson, L. L., "Sizing Piping For Process Plants," *Chemical Engineering,* June 17, 1968.

Refrigeration

Here is a troubleshooting checklist on refrigeration.

Source

GPSA Engineering Data Book, Gas Processors Suppliers Association, 10th Ed.

Refrigeration System Checklist

Indication	Causes
High Compressor Discharge Pressure	Check accumulator temperature. If the accumulator temperature is high, check: 1. Condenser operation for fouling. 2. High air or water temperature. 3. Low fan speed or pitch. 4. Low water circulation. If condensing temperature is normal, check for: 1. Noncondensables in referigerant. 2. Restriction in system which is creating pressure drop.
High Process Temperature	Check refrigerant temperature from chiller. If refrigerant temperature is high and approach temperature on chiller is normal, check: 1. Chiller pressure. 2. Refrigerant composition for heavy ends contamination. 3. Refrigerant circulation or kettle level (possible inadequate flow resulting in superheating of refrigerant). 4. Process overload of refrigerant system. If refrigerant temperature is normal, and approach to process temperature is high, check: 1. Fouling on refrigerant side (lube oil or moisture). 2. Fouling on process side (wax or hydrates). 3. Process overload of chiller capacity.
Inadequate Compressor Capacity	Check: 1. Process overload of refrigerant system. 2. Premature opening of hot gas bypass. 3. Compressor valve failure. 4. Compressor suction pressure restriction. 5. Low compressor speed.
Inadequate Refrigerant Flow to Economizer or Chiller	Check: 1. Low accumulator level. 2. Expansion valve capacity. 3. Chiller or economizer level control malfunction. 4. Restriction in refrigerant flow (hydrates or ice).

Firetube Heaters

The following firetube heater troubleshooting tips are provided by the GPSA data book.

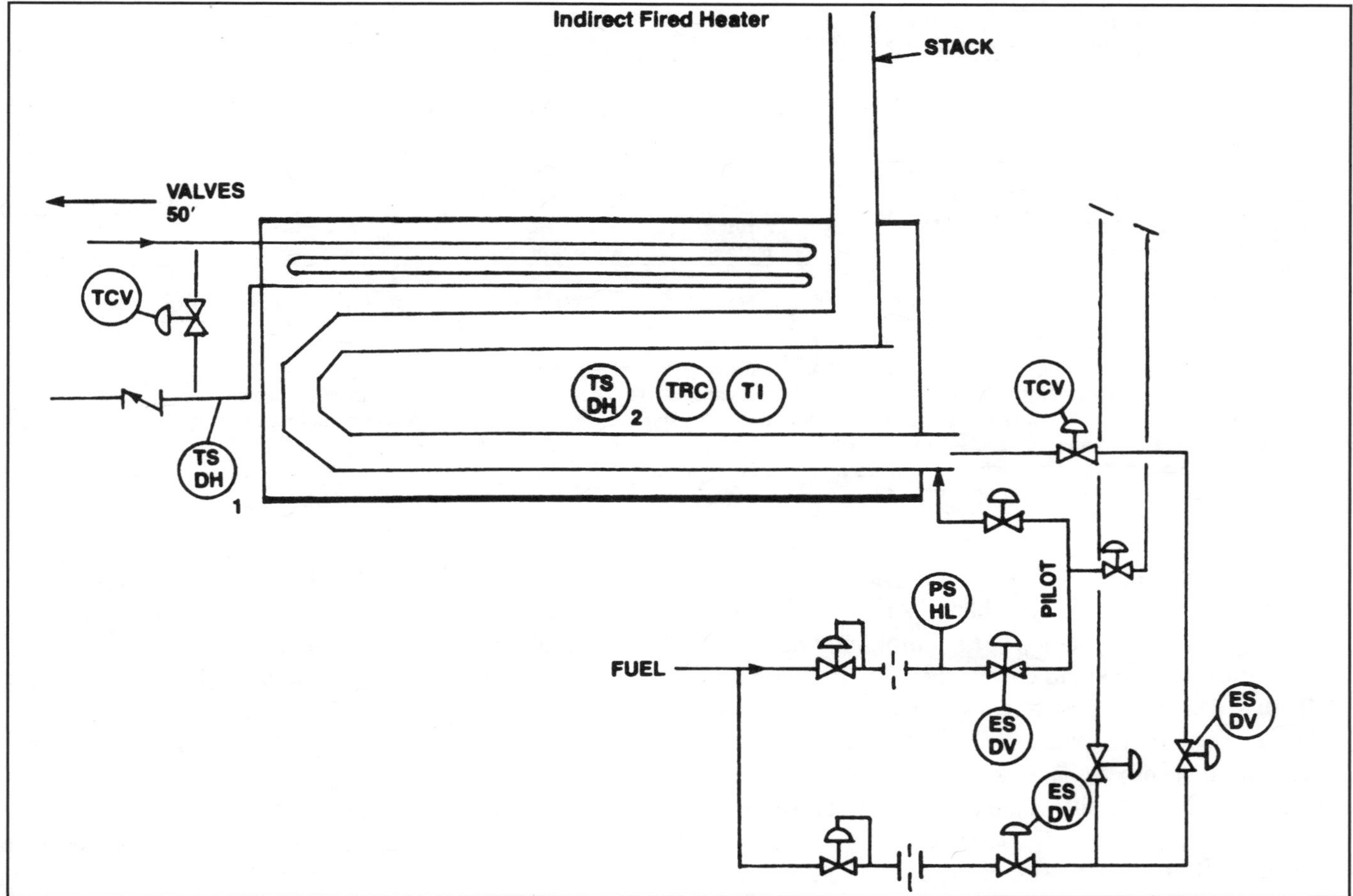

Figure 1. Indirect fired heater.

The following problems can occur with firetube heaters.

- **Bath level loss** can be the result of too high a bath temperature. This is often caused by the temperature controller on the process stream. Fouling of the process coil, internal and/or external, means a hotter bath is needed to accomplish the same heat transfer. The coil should be removed and cleaned.

If fouling of the coils is not the problem, water losses can be reduced with a vapor recovery exchanger mounted on top of the heater shell. It consists of thin tubes that condense the water vapor. Vapor losses can also be reduced by altering the composition of the heat medium or, in drastic cases, by changing the heat medium.

- **Shell side corrosion** is caused by decomposition of the bath. (The decomposition products of amines and glycols are corrosive.) Some decomposition and corrosion is inevitable; however, excessive decomposition is usually due to overheating near the firetube. Corrosion inhibitors are commonly added. There are numerous reasons for overheating the bath: localized ineffective heat transfer caused by fouling, excessive flame impingement, etc. An improper flame can sometimes be modified without system shutdown. Fouling, however, requires removal of the firetube.
- **Inadequate heat transfer** may result from improper flame, underfiring, firetube fouling, coil fouling, poor

Table 1
Bath Heater Alarm/Shutdown Description

Note: Alarms and shutdowns as shown are not to be considered as meeting any minimum safety requirement but are shown as representative of types used for control systems.

Schematic Label	Alarm/Shutdown Description	Line Heater	Hydrocarbon Reboiler	Low Pressure Steam Heater	Hot oil or Salt Heater	Glycol Reboiler	Amine Reboiler
TSDH-2	High Bath Temperature	No Note 1,2	Yes	No	Yes	Yes	Yes
LSL	Low Bath Level	No Note 2	No	Yes	Yes Note 3	No Note 2,3	Yes Note 2,3
PSL	Low Fuel Pressure	Yes Note 4	Yes	Yes	Yes Note 4	Yes	Yes
PSH	High Fuel Pressure	Yes Note 4	Yes	Yes	Yes Note 4	Yes	Yes
BSL	Flame Failure Detection	No Note 2,5	Yes Note 5	No Note 2	Yes Note 5	No Note 2	No Note 2
PSH	High Vessel Pressure	No	No	Yes Note 6	No	No	Yes Note 6

NOTES:

1. When the process stream is oil, a high bath temperature shutdown precludes the danger of coking.
2. This instrumentation is for heaters located in gas processing sections. Wellhead units have a minimum of controls.
3. Low bath level protects both the firetube and bath when it will coke (hot oils, glycol, amine) or decompose (molten salt).
4. Low/high fuel pressure is always used when the fuel gas is taken from the exit process stream.
5. Optical UV scanners or flame rods should be used because of the speed of response.
6. Code requirements, ASME Section IV or VIII.

shell fluid dynamics, too small a firetube or coil, etc. If it is not improper design, then it is most likely fouling or an improper flame. The solution may be a simple burner adjustment to correct the air to fuel mixture.

- **High stack temperature** can be the result of an improper air to fuel mixture. A leak of combustible material from the process side to the firetube is also a cause. It can also be the result of excessive soot deposition in the firetube.
- **Firetube failure** is most commonly caused by localized overheating and subsequent metallurgical failure. These "hot spots" are caused by hydrocarbon coking and deposition on the bath side. Firetube corrosion is caused by burning acid gases for fuel. The most damaging corrosion occurs in the burner assembly. There is little than can be done except to change the fuel and this may be impractical. Proper metallurgy is essential when burning acid gases.
- **High or low fuel gas pressure** can have a dramatic effect on the operation of a firetube heater. Burners are typically rated as heat output at a specified fuel pressure. A significantly lower pressure means inadequate heat release. Significantly higher pressure causes overfiring and over heating. The most common causes of a fuel gas pressure problem are the failure of a pressure regulator or an unacceptably low supply pressure.

Source

GPSA Engineering Data Book, Gas Processors Suppliers Association, 10th Ed.

Safety Relief Valves

The following problem discussions for safety relief valves from the GPSA data book will provide understanding of the mechanisms for the operating engineer.

Reactive Force

On high pressure valves, the reactive forces during relief are substantial and external bracing may be required. See equations in API RP 520 for computing this force.

Rapid Cycling

Rapid cycling can occur when the pressure at the valve inlet decreases at the start of relief valve flow because of excessive pressure loss in the piping to the valve. Under these conditions, the valve will cycle at a rapid rate which is referred to as "chattering". The valve responds to the pressure at its inlet. If the pressure decreases during flow below the valve reseat point, the valve will close; however, as soon as the flow stops, the inlet pipe pressure loss becomes zero and the pressure at the valve inlet rises to tank pressure once again. If the vessel pressure is still equal to or greater than the relief valve set pressure, the valve will open and close again.

An oversized relief valve may also chatter since the valve may quickly relieve enough contained fluid to allow the vessel pressure to momentarily fall back to below set pressure only to rapidly increase again. Rapid cycling reduces capacity and is destructive to the valve seat in addition to subjecting all the moving parts in the valve to excessive wear. Excessive back pressure can also cause rapid cycling as discussed above.

Resonant Chatter

Resonant chatter can occur with safety relief valves when the inlet piping produces excessive pressure losses at the valve inlet and the natural acoustical frequency of the inlet piping approaches the natural mechanical frequency of the valve's basic moving parts. The higher the set pressure, the larger the valve size, or the greater the inlet pipe pressure loss, the more likely resonant chatter will occur. Resonant chatter is uncontrollable; that is, once started it cannot be stopped unless the pressure is removed from the valve inlet. In actual application, however, the valve can self-destruct before a shutdown can take place because of the very large magnitude of the impact forces involved.

Seat Leakage of Relief Valves

Seat leakage is specified for conventional direct spring operated metal-to-metal seated valves by API RP 527. The important factor in understanding the allowable seat leak is that it is stated at 90% of set point. Therefore, unless special seat lapping is specified or soft seat designs used, a valve operating with a 10% differential between operating and set pressures may be expected to leak.

There are no industry standards for soft seated valves. Generally, leakage is not expected if the valve operates for one minute at 90% of set pressure and, in some cases, no leakage at 95%, or even higher. Longer time intervals could promote leakage. Application of soft seated valves is limited by operating temperatures. Manufacturer guidelines should be consulted.

Source

GPSA Engineering Data Book, Gas Processors Suppliers Association, 10th Ed.

API Standard 527—Commercial Seat Tightness of Safety Relief Valves with Metal-to-Metal Seats, 1978, American Petroleum Institute, 1220 L Street, NW., Washington, D.C., 20005.

API RP 520—Recommended Practice for the Design (Part I) and Installation (Part II) of Pressure Relieving Systems in Refineries, 1976(I) and 1973(II), American Petroleum Institute, 1220 L Street, NW., Washington, D.C., 20005.

Gas Treating

Here are two authoritative discussions[1,2] of troubleshooting in glycol dehydration plants.

Glycol plant troubleshooting information is presented from the article "Dehydration Using TEG" by Manning and Thompson.

Proper preventive maintenance minimizes upsets and shutdowns. Troubleshooting is greatly simplified if the glycol unit is appropriately instrumented and if operating records, glycol analyses, process flow sheets, and drawings of vessel internals are available (Ballard, 1979).[2] Interpreting the records requires knowledge of the design conditions for the plant. Major changes in key variables such as inlet gas flow rate, temperature or pressure require determination of a new set of optimum operating conditions.

Troubleshooting is described by suggesting possible causes of the more common problems and discussing corrective measures.

High Exit Gas Dew Point

- Change in gas flow rate, temperature, or pressure
- Insufficient glycol circulation (should be 1.5 to 3 gal TEG/lb water removed)

- Poor glycol reconcentration (exit gas dew point is 5–15 °F higher than dew point in equilibrium with lean glycol concentration)
- Current operating conditions different from design
- Inlet separator malfunctions

High Glycol Losses

First determine where loss is occurring: from contactor (i.e., down pipeline); out of still column; or leaking from pump.

- Loss from contactor
  - inlet separator passing liquids
  - carryover due to excessive foaming
  - mist extractor plugged or missing
  - lean glycol entering contactor too hot
  - plugged trays in contactor
  - excessive gas velocity in contactor
  - tray spacing less than 24 in.
  - mist pad too close to top tray
- Loss from separator
  - glycol dumped with hydrocarbons
- Loss from still
  - excessive stripping gas
  - flash separator passing condensate
  - packing in still column is broken, dirty, plugged
  - insufficient reflux cooling at top of still
  - too much cooling at top of still
  - temperature at top of still is too high
- Leaks, spills, etc.
  - check piping, fittings, valves, gaskets
  - check pumps, especially packings
  - pilferage

Many manufacturers will estimate maximum glycol losses to be 1 lb or .1 gal per MMscf. Improper operation can increase the actual glycol losses to 1, 10, or even 100 gallons per MMscf. One leaking pump can waste 35 gallons per day (Ballard, 1977).

Glycol Contamination

- Carryover of oils (e.g., compressor lube oils), brine, corrosion inhibitors, well treating chemicals, sand, corrosion scales, etc., from inlet separator
- Oxygen leaks into glycol storage tanks, etc.
- Inadequate pH control (low pH) increases corrosion
- Overheating of glycol in reboiler due to excessive temperature or hot spots on firetube
- Improper filtration; plugged filters; bypassing of filters
- Improper cleaning of glycol unit; use of soaps or acids (Ballard, 1979)[2]

Poor Glycol Reconcentration

- Low reboiler temperature
- Insufficient stripping gas
- Rich glycol is leaking into lean glycol in the glycol heat exchanger or in the glycol pump
- Overloading capacity of reboiler
- Fouling of fire tubes in reboiler
- Low fuel rate or low Btu content of fuel
- Glycol foaming in still column
- Dirty and/or broken packing in still column
- Flooding of still column
- High still pressure

Low Glycol Circulation—Glycol Pump

- Check pump operation (Caldwell, 1976). If it is a glycol powered pump, close lean glycol discharge valve; if pump continues running, it needs repair. If gas or electric, check circulation by stopping glycol discharge from contactor and timing fill rate of gauge column on chimney tray section.
- Check pump valves to see if worn or broken.
- Plugged strainer, lines, filters
- Vapor lock in lines or pump
- Low level in accumulator
- Excessive packing gland leakage
- Contactor pressure too high

High Pressure Drop Across Contactor

- Excessive gas flow rate
- Operating at pressures far below design
- Plugged trays
- Plugged demister pads
- Glycol foaming

High Stripping Still Temperature

- Inadequate reflux
- Still column flooded
- Glycol foaming
- Carryover of light HC in rich glycol
- Leaking reflux coil

High Reboiler Pressure

- Packing in still column is broken and/or plugged with tar, dirt, etc.
- Restricted vent line, not sloped
- Still column is flooded by excessive boil up rates and/or excessive reflux cooling.
- Slug of liquid HC enters stripper, vaporizes on reaching reboiler, and blows liquids out of still.

Firetube Fouling/Hot Spots/Burn Out

- Buildup of salt, dust, scales, etc., on firetube—check inlet separator.
- Deposition of coke, tar formed by glycol overheating and/or hydrocarbon decomposition.
- Liquid glycol level drops exposing firetubes. Low level shutdown.

Low Reboiler Temperature

- Inadequately sized firetube and/or burner
- Setting temperature controller too low
- More water in inlet gas because pressure is low or temperature high
- Temperature controller not operating correctly
- Carryover of water from inlet separator
- Inaccurate reboiler thermometer

Flash Separator Failure

- Check level controllers
- Check dump valves
- Excessive circulation

Here is another discussion of glycol plant troubleshooting from the article "How to Improve Glycol Dehydration" by Don Ballard.

The ability to quickly identify and eliminate costly operating problems can frequently save thousands of dollars. Here are some helpful troubleshooting tips.

High Gas Dew Points

The most obvious indication of a glycol dehydration malfunction is a high water content or dew point of the outgoing sales gas stream. In most cases, this is caused by an inadequate glycol circulation rate or by an insufficient reconcentration of the glycol. These two factors can be caused by a variety of contributing problems listed below.

Cause—Inadequate Glycol Circulation Rate

1. Glycol powered pump. Close the dry discharge valve and see if the pump continues to run; if so, the pump needs to be repaired.
2. Gas or electric driven pump. Verify adequate circulation by shutting off the glycol discharge from the absorber and timing the fill rate in the gauge glass.
3. Pump stroking but not pumping. Check the valves to see if they are seating properly.
4. Check pump suction strainer for stoppage.
5. Open bleeder valve to eliminate "air lock."
6. Be sure surge level is sufficiently high.

Cause—Insufficient Reconcentration of Glycol

1. Verify the reboiler temperature with a test thermometer and be sure the temperature is in the recommended range of 350 °F to 400°F for triethylene glycol. The temperature can be raised closer to 400°F, if needed, to remove more water from the glycol.
2. Check the glycol to glycol heat exchanger in the accumulator for leakage of wet, rich glycol into the dry, lean glycol.
3. Check the stripping gas, if applicable, to be sure there is a positive flow of gas. Be sure steam is not backflowing into the accumulator from the reboiler.

Cause—Operating Conditions Different From Design

1. Increase the absorber pressure, if needed. This may require the installation of a back pressure valve.
2. Reduce the gas temperature, if possible.
3. Increase the glycol circulation rate, if possible.

Cause—Low Gas Flow Rates

1. If the absorber has access manways, blank off a portion of the bubble caps or valve trays.
2. Add external cooling to the lean glycol and balance the glycol circulation rate for the low gas rate.
3. Change to a smaller absorber designed for the lower rate, if needed.

High Glycol Loss

Cause—Foaming

1. Foaming is usually caused by contamination of glycol with salt, hydrocarbons, dust, mud, and corrosion inhibitors. Remove the source of contamination with effective gas cleaning ahead of the absorber, improved solids filtration, and carbon purification.

Cause—Excessive Velocity in the Absorber

1. Decrease the gas rate.
2. Increase the pressure on the absorber, if possible.

Cause—Trays Plugged With Mud, Sludge, and Other Contaminants

1. If the pressure drop across the absorber exceeds about 15 psi, the trays may be dirty and/or plugged. Plugged trays and/or downcomers usually prevent the easy flow of gas and glycol through the absorber. If the absorber has access handholes, manual cleaning can be helpful. Chemical cleaning is recommended if handholes are not available.

Cause—Loss of Glycol Out of Still Column

1. Be sure the stripping gas valve is open and the accumulator is vented to the atmosphere.
2. Be sure the reboiler is not overloaded with free water entering with the gas stream.
3. Be sure excessive hydrocarbons are kept out of the reboiler.
4. Replace the tower packing in the still column, if fouled or powdered.

If the outlet gas temperature from the absorber exceeds the inlet gas temperature more than 20 to 30°F, the lean glycol entering the top of the absorber may be too hot. This could indicate a heat exchanger problem or an excessive glycol circulation rate.

If a glycol pump has been operating in a clean system, no major service will probably be needed for several years. Only a yearly replacement of packing is usually required. Normally the pump will not stop pumping unless some internal part has been bent, worn, or broken, a foreign object has fouled the pump, or the system has lost its glycol.

A pump which has been running without glycol for some time should be checked before returning to normal service. The pump will probably need at least new "O" rings. The cylinders and piston rods may also have been scored from the "dry run."

Here are some typical symptoms and causes for a Kimray pump operation. These are presented to assist in an accurate diagnosis of trouble.

Symptoms

1. The pump will not operate.
2. The pump will start and run until the glycol returns from the absorber. The pump then stops or slows appreciably and will not run at its rated speed.
3. The pump operates until the system temperature is normal and then the pump speeds up and cavitates.
4. The pump lopes or pumps on one side only.
5. Pump stops and leaks excessive gas from the wet glycol discharge.
6. Erratic pump speed. Pump changes speed every few minutes.
7. Broken pilot piston.

Causes

1. One or more of the flow lines to the pump are completely blocked or the system pressure is too low for standard pumps.
2. The wet glycol discharge line to the reboiler is restricted. A pressure gauge installed on the line will show the restriction immediately.
3. The suction line is too small and an increase in temperature and pumping rate cavitates the pump.
4. A leaky check valve, a foreign object lodged under a check valve, or a leaky piston.
5. Look for metal chips or shavings under the pump D-slides.
6. Traps in the wet glycol power piping send alternate slugs of glycol and gas to the pump.
7. Insufficient glycol to the Main Piston D-slide ports. Elevate the control valve end of the pump to correct.

The glycol pH should be controlled to avoid equipment corrosion. Some possible causes for a low, acidic pH are:

- Thermal decomposition, caused by an excessive reboiler temperature (over 404°F), deposits on the firetube, or a poor reboiler design.
- Glycol oxidation, caused by getting oxygen into the glycol with the incoming gas; it can be sucked in through a leaking pump or through unblanketed glycol storage tanks.
- Acid gas pickup from the incoming gas stream.

Salt accumulations and other deposits on the firetube can sometimes be detected by smelling the vapors from the still vent. A "burned" odor emitted from these vapors usually indicates this type of thermal degradation. Another detection method is to observe the glycol color. It will darken quickly if the glycol degrades. These detection methods may prevent a firetube failure.

Maintenance and production records, along with the used lean and rich glycol analyses, can be very helpful to the troubleshooter. A history of filter element, carbon, tower packing, and firetube changeouts can sometimes be very revealing. The frequency of pump repairs and chemical cleaning jobs is also beneficial. With this type of knowledge, the troubleshooter can quickly eliminate and prevent costly problems.

Sources

1. Francis S. Manning and Richard E. Thompson's *Oilfield Processing of Petroleum, Volume One: Natural Gas.* Copyright PennWell Books, 1991.
2. Ballard, Don, "How to Improve Glycol Dehydration," Coastal Chemical Co., Inc., 1979.

Compressors

Here are check lists for troubleshooting compressor problems in reciprocating compressors and centrifugal compressors.

Table 1
Probable Causes of Reciprocating Compressor Trouble

TROUBLE	PROBABLE CAUSE(S)
COMPRESSOR WILL NOT START	1. Power supply failure. 2. Switchgear or starting panel. 3. Low oil pressure shut down switch. 4. Control panel.
MOTOR WILL NOT SYNCHRONIZE	1. Low voltage. 2. Excessive starting torque. 3. Incorrect power factor. 4. Excitation voltage failure.
LOW OIL PRESSURE	1. Oil pump failure. 2. Oil foaming from counterweights striking oil surface. 3. Cold oil. 4. Dirty oil filter. 5. Interior frame oil leaks. 6. Excessive leakage at bearing shim tabs and/or bearings. 7. Improper low oil pressure switch setting. 8. Low gear oil pump bypass/relief valve setting. 9. Defective pressure gauge. 10. Plugged oil sump strainer. 11. Defective oil relief valve.
NOISE IN CYLINDER	1. Loose piston. 2. Piston hitting outer head or frame end of cylinder. 3. Loose crosshead lock nut. 4. Broken or leaking valve(s). 5. Worn or broken piston rings or expanders. 6. Valve improperly seated/damaged seat gasket. 7. Free air unloader plunger chattering.
EXCESSIVE PACKING LEAKAGE	1. Worn packing rings. 2. Improper lube oil and/or insufficient lube rate (blue rings). 3. Dirt in packing. 4. Excessive rate of pressure increase. 5. Packing rings assembled incorrectly. 6. Improper ring side or end gap clearance. 7. Plugged packing vent system. 8. Scored piston rod. 9. Excessive piston rod run-out.
PACKING OVERHEATING	1. Lubrication failure. 2. Improper lube oil and/or insufficient lube rate. 3. Insufficient cooling.

(table continued)

Table 1 (continued)
Probable Causes of Reciprocating Compressor Trouble

TROUBLE	PROBABLE CAUSE(S)
EXCESSIVE CARBON ON VALVES	1. Excessive lube oil. 2. Improper lube oil (too light, high carbon residue). 3. Oil carryover from inlet system or previous stage. 4. Broken or leaking valves causing high temperature. 5. Excessive temperature due to high pressure ratio across cylinders.
RELIEF VALVE POPPING	1. Faulty relief valve. 2. Leaking suction valves or rings on next higher stage. 3. Obstruction (foreign material, rags), blind or valve closed in discharge line.
HIGH DISCHARGE TEMPERATURE	1. Excessive ratio on cylinder due to leaking inlet valves or rings on next higher stage. 2. Fouled intercooler/piping. 3. Leaking discharge valves or piston rings.
	4. High inlet temperature. 5. Fouled water jackets on cylinder. 6. Improper lube oil and/or lube rate.
FRAME KNOCKS	1. Loose crosshead pin, pin caps or crosshead shoes. 2. Loose/worn main, crankpin or crosshead bearings. 3. Low oil pressure. 4. Cold oil. 5. Incorrect oil. 6. Knock is actually from cylinder end.
CRANKSHAFT OIL SEAL LEAKS	1. Faulty seal installation. 2. Clogged drain hole.
PISTON ROD OIL SCRAPER LEAKS	1. Worn scraper rings. 2. Scrapers incorrectly assembled. 3. Worn/scored rod. 4. Improper fit of rings to rod/side clearance.

Table 2
Probable Causes of Centrifugal Compressor Trouble

TROUBLE	PROBABLE CAUSE(S)
LOW DISCHARGE PRESSURE	1. Compressor not up to speed. 2. Excessive compressor inlet temperature. 3. Low inlet pressure. 4. Leak in discharge piping. 5. Excessive system demand from compressor.
COMPRESSOR SURGE	1. Inadequate flow through the compressor. 2. Change in system resistance due to obstruction in the discharge piping or improper valve position. 3. Deposit buildup on rotor or diffusers restricting gas flow.
LOW LUBE OIL PRESSURE	1. Faulty lube oil pressure gauge or switch. 2. Low level in oil reservoir. 3. Oil pump suction plugged. 4. Leak in oil pump suction piping. 5. Clogged oil strainers or filters. 6. Failure of both main and auxiliary oil pumps. 7. Operation at a low speed without the auxiliary oil pump running (if main oil pump is shaft-driven). 8. Relief valve improperly set or stuck open. 9. Leaks in the oil system. 10. Incorrect pressure control valve setting or operation. 11. Bearing lube oil orifices missing or plugged.
SHAFT MISALIGNMENT	1. Piping strain. 2. Warped bedplate, compressor, or driver. 3. Warped foundation. 4. Loose or broken foundation bolts. 5. Defective grouting.
HIGH BEARING OIL TEMPERATURE Note: Lube oil temperature leaving bearings should never be permitted to exceed 180°F.	1. Inadequate or restricted flow of lube oil to bearings. 2. Poor conditions of lube oil or dirt or gummy deposits in bearings. 3. Inadequate cooling water flow lube oil cooler. 4. Fouled lube oil cooler. 5. Wiped bearing. 6. High oil viscosity. 7. Excessive vibration. 8. Water in lube oil. 9. Rough journal surface.
EXCESSIVE VIBRATION Note: Vibration may be transmitted from the coupled machine. To localize vibration, disconnect coupling and operate driver alone. This should help to indicate whether driver or driven machine is causing vibration.	1. Improperly assembled parts. 2. Loose or broken bolting. 3. Piping strain. 4. Shaft misalignment. 5. Worn or damaged coupling. 6. Dry coupling (if continuously lubricated type is used). 7. Warped shaft caused by uneven heating or cooling. 8. Damaged rotor or bent shaft. 9. Unbalanced rotor or warped shaft due to severe rubbing. 10. Uneven buildup of deposits on rotor wheels, causing unbalance. 11. Excessive bearing clearance. 12. Loose wheel(s) (rare case). 13. Operating at or near critical speed. 14. Operating in surge region. 15. Liquid "slugs" striking wheels. 16. Excessive vibration of adjacent machinery (sympathetic vibration).
WATER IN LUBE OIL	1. Condensation in oil reservoir. 2. Leak in lube oil cooler tubes or tube-sheet.

Source

GPSA Engineering Data Book, Gas Processors Suppliers Association, 10th Ed.

Measurement

Here is a checklist for troubleshooting measurement problems:

Table 1
Common Measurement Problems

Variable	Symptom	Problem Source	Solution
Pressure	Zero shift, air leaks in signal lines.	Excessive vibration from positive displacement equipment.	Use independent transmitter mtg., flexible process connection lines. Use liquid filled gauge.
	Variable energy consumption under temperature control.	Change in atmospheric pressure.	Use absolute pressure transmitter.
	Unpredictable transmitter output.	Wet instrument air.	Mount local dryer. Use regulator with sump, slope air line away from transmitter.
	Permanent zero shift.	Overpressure.	Install pressure snubber for spikes.
Level (diff. press.)	Transmitter does not agree with level.	Liquid gravity change.	Gravity compensate measurement, or recalibrate.
	Zero shifts, high level indicated.	Water in process absorbed by glycol seal liquid.	Use transmitter with integral remote seals.
	Zero shift, low level indicated.	Condensable gas above liquid.	Heat trace vapor leg. Mount transmitter above connections and slope vapor line away from transmitter.
	Noisy measurements, high level indicated.	Liquid boils at ambient temp.	Insulate liquid leg.
Flow	Low mass flow indicated.	Liquid droplets in gas.	Install demister upstream; heat gas upstream of sensor.
	Mass flow error.	Static pressure change in gas.	Add pressure recording pen.
	Transmitter zero shift.	Free water in fluid.	Mount transmitter above taps.
	Measurement is high.	Pulsation in flow.	Add process pulsation damper.
	Measurement error.	Non-standard pipe runs.	Estimate limits of error.
Temperature	Measurement shift.	Ambient temperature change.	Increase immersion length. Insulate surface.
	Measurement not representative of process.	Fast changing process temperature.	Use quick response or low thermal time constant device.
	Indicator reading varies second to second.	Electrical power wires near thermocouple extension wires.	Use shielded, twisted pair thermocouple extension wire, and/or install in conduit.

Source

GPSA Engineering Data Book, Gas Processors Suppliers Association, 10th Ed.

22 Startup

Introduction

More has been written about plant startup than one might imagine. Since considerable *troubleshooting* is needed during *startup* the two subjects tend to overlap. Therefore, the reader is encouraged to seek startup techniques also within the troubleshooting chapter. For example, J. W. Drew's excellent article[1] "Distillation Column Startup" was included in the section on Fractionation Checklists due to its specific nature.

Sources

1. Drew, J. W., "Distillation Column Startup," *Chemical Engineering,* November 14, 1983, p. 221.
2. Gans, M., Kiorpes, S. A., and Fitzgerald, F. A., "Plant Startup-Step by Step," *Chemical Engineering,* October 3, 1983, p. 74.
3. Anderson, G. D., "Initial Controller Settings to Use at Plant Startup," *Chemical Engineering,* July 11, 1983, p. 113.
4. Talley, D. L., "Startup of A Sour Gas Plant," *Hydrocarbon Processing,* April 1976, p. 90.
5. Matley, J., "Keys to Successful Plant Startups," *Chemical Engineering,* September 8, 1969, p. 110.
6. Troyan, J. E., series "Hints for Plant Startup," *Chemical Engineering,* Part I—Troubleshooting New Processes, November 14, 1960, p. 223; Part II—Troubleshooting New Equipment, March 20, 1961, p. 147; Part III—Pumps, Compressors, and Agitators, May 1, 1961, p. 91.
7. Klehm, J. J. and Singletary, J. E., "Expander Plant Successful," *Hydrocarbon Processing,* August 1974, p. 89.
8. Godard, K. E., "Gas Plant Startup Problems," *Hydrocarbon Processing,* September 1973, p. 151.
9. Dyess, C. E., "Putting A 70,000 bbl/day Crude Still Onstream," *Chemical Engineering Progress,* August 1973, p. 98.
10. Gans, Manfred, "The A to Z of Plant Startup," *Chemical Engineering,* March 15, 1976, p. 72.

Settings for Controls

G. D. Anderson's article[1] recommends initial controller settings for those control loops set on automatic rather than manual for a plant startup. For liquid level, the settings depend upon whether the sensor is a displacer type or differential pressure type, or a surge tank (or other surge) is installed in the process:

	Liquid Level		
	Displacers	Differential Pressure	Surge
Proportional band, %	20	20	Equation 1
Reset, repeats/min	5	5	1
Rate, min	0	0.1	0

$$P.B. = (T)\,(100)\,(Q)/(A)\,(R.S.) \quad (1)$$

where

P.B. = Proportional band setting
T = Filter time constant provided by the surge, min
Q = Maximum throughput, ft^3/min
A = Surge tank cross-sectional area, ft^2
R.S. = Rated span of sensor, ft

For temperature, the settings depend upon whether the system is a basic thermal system with negligible dead time, or a mixing process, such as injection of steam into a stream of processed food to maintain its temperature.

	Temperature	
	Thermal	Mixing
Proportional band, %	Equation 2	Equation 2
Reset, repeats/min	0.2	0.2
Rate, min	0.5	0

$$P.B. = 200\ (\text{overshoot})/\text{span} \quad (2)$$

where

P.B. = proportional band setting
overshoot = amount +/- of setpoint, °F
span = transmitter span, °F

For pressure, "slow" and "fast" systems are compared. An example of a slow system is control of the pressure of compressible gas in a large process volume. A fast system could be the pressure of fuel oil (incompressible liquid) supply to a burner. For flow, a single set of settings is recommended.

	Pressure		Flow
	Slow	Fast	
Proportional band, %	Equation 2	200	200
Reset, repeats/min	2	20	20
Rate, min	0	0	0

Source

Anderson, G. D., "Initial Controller Settings To Use At Plant Startup," *Chemical Engineering,* July 11, 1983, p. 113.

Probable Causes of Trouble in Controls

J. E. Troyan's series of articles[1] on plant startup has a cause/effect table on instrumentation in Part II. This article also has troubleshooting hints for distillation, vacuum systems, heat transfer, and filtration. Here is the table on instrumentation.

50 Tips on Instrument Troubleshooting

Recorders and Recorder-Controllers

Trouble: Pen Dead at Zero.

Cause:

1. Check isolating valves at process impulse line; see that equalizing valve is shut.
2. Look for plugged impulse line.
3. In cases of pneumatic transmitters, check air supply to transmitter.
4. Look for broken or loose lines.
5. Primary sensory device not installed (i.e., orifice plate, temperature bulb).
6. Be sure process step is actually in operation (pump running, etc.).

Trouble: Recorders With Two or More Pens Record at the Same Point on the Chart, But Should be at Different Points.

Cause:

1. Pens may not pass owing to mechanical interference.
2. Make sure instrument leads are clear and properly connected.

Trouble: Chart Not Moving.

Cause:

1. Chart loose on hub.
2. Chart loose on chart rewind roll.
3. Clock not running; power not on drive.

Trouble: Chart Readings Apparently Not Correct or Not Consistent with Local Gauges.

Cause:

1. In case of chemical seal attachments, sealing fluid may have leaked out or there may be a hole in diaphragm seal.
2. Trouble in gauge movements.
3. Gauge may not have been read properly.
4. If dual-pointer instrument, be sure to read right pointer.
5. Check agreement of chart with transmitter range.
6. Check agreement of output pressure from transmitter with pen.

Manometers

Trouble: Manometer Does Not Indicate When Equipment is in Service.

Cause:

1. If manometer has sealing fluid, make sure it is the proper type fluid, and that the manometer is properly filled. (This is an instrument technician's job).

2. Check all valves and valve arrangements to be sure they are properly set.

Control Valves

Trouble: Valves Do Not Respond to Air Change.

Cause:

1. If valve has a valve positioner, check its air supply.
2. Check for broken air lines.
3. Check any interlocks on the control loop.
4. Valve action may be reversed.

Trouble: Valve Changes Position, but Brings About No Change In Process Flow.

Cause:

1. Valve may be plugged.
2. Block valves in process line may be closed.
3. Valve plug may have broken off from the valve stem.
4. Valve action may not be correct for application.

Level Controllers

General Suggestions:

1. If level controller has pneumatic transmitter, look for isolated or broken lines or leaks.
2. Excessive pressure during pressure test can collapse floats and wreck level-measuring device if unit is not suitably protected.

Rotameters

Trouble: Rotameter Shows No Indication of Flow.

Cause:

1. Check valving arrangement to be sure all valves are in proper position.
2. Be sure material being measured is being delivered to rotameter.
3. Check whether float is stuck with dirt or foreign matter.
4. Be sure that meter has proper tube and float installed.
5. In case of pneumatic transmitter, check air supply, loose or broken lines.

Purge-Fluid Units

Trouble: No Indication of Flow.

Cause:

1. Purge meter may be valved off.
2. Supply of fluid to purge meter may have been turned off.
3. Look for broken or loose lines from purge source.
4. Impulse lines may be plugged with process solids.

Weighing Scales

General Suggestions:

1. Once set, scales should never be moved or left exposed to contamination. Scale heads should be locked out when scale is not being used.
2. Check-weights can be used by Operations for spot checks on scales, but all repairs and adjustments should be sole responsibility of scale manufacturer or instrument department.

Electronic Recorders

Trouble (A): Analyzers (Inoperable or Questionable Readings).

Cause:

1. Power off or fuse blown.
2. No sample flow.
3. Sampling lines plugged.
4. Sampling streams contaminated; do a better purge job.
5. Pressure excessive in sample cell.

Trouble (B): Temperature Recorders (Inoperable or Questionable Readings).

Cause:

1. Power off or fuse blown. Check battery.
2. Loose connection on thermocouple terminal strip.
3. Read indicator scale plate; dimension of chart paper may be affected by humidity.

Reference

1. Troyan, J. E., series "Hints for Plant Startup," *Chemical Engineering,* Part I—Troubleshooting New Processes, November 14, 1960, p. 223; Part II—Troubleshooting New Equipment, March 20, 1961, p. 147; Part III—Pumps, Compressors, and Agitators, May 1, 1961, p. 91.

Checklists

Here is a set of startup checklists. First is a generalized list.[1]

General Startup Preparation Checklist

Maintenance

Organization staffed
Shop set up and equipped
Spare parts and materials in warehouse
Special tools and procedures
Equipment inspection procedures established
Proper packings and lubricants on hand
Vendor equipment instructions cataloged

Inspections

Vessel interiors
Vessel packings
Piping according to piping instrument diagrams
Equipment arrangement for access and operation
Cleanliness of critical piping
Insulation, stream tracing, etc.
Temporary strainers and blinds installed
Provisions for sampling

Pressure Testing, Cleaning, Flushing, Drying, and Purging

Pressure test piping and equipment
Flush and clean piping and equipment
Drain water to prevent freezing
Blow out piping
Continuity testing with air
Orifice plates (install after checking bore and location)
Dry out process
Purge
Vacuum test
Piping expansion and support (check free movement)

Utilities

Electric Power and Lighting
  Continuity check (megger)
  Trip settings at substations

Utilities (Cont.)

  Isolation and safety
  Sample and check transformer oil
Water Treating
  Load filter beds
  Load ion exchanger
  Make up injection systems
Cooling Water
  Flush inlet headers, laterals, and return lines
  Drain to prevent freezing
  Clean tower basin
  Adjust tower fans
Service Air
  Clean air header by blowing
  Keep water drained
  Load desiccants and dry out header
Underground Drains
  Cleanliness and tightness
  Seals established
Steam
  Line-warming procedures
  Blow main headers
  Blow laterals
Condensate
  Disposal to drains initially
  Check trap operation
Inert Gas
  Identify and provide warnings
  Blow lines with air
  Isolate and purge, if required
Fuel Oil
  Identify and provide warnings
  Air blow lines
  Isolate and purge
Fuel Gas
  Identify and provide warnings

Utilities (Cont.)

Blow lines with air
Isolate and purge

Control Laboratory

Staffed and equipped
Sample schedule published
Specifications for all products and raw materials
Sample-retention policy established

Equipment

Fired Heaters
- Instruments and controls checked
- Refractory dried out

Piping Strains on Equipment
Electric Motors
- Rotation
- Drying out
- No-load tests

Steam-Turbine Drivers
- Auxiliary lubrication and cooling systems checked
- Instrumentation and speed control checked
- No-load tests
- Light-load tests

Gas-Engine Drivers
- Auxiliary lubrication and cooling systems checked
- Instruments
- Idling tests

Gas-Turbine Drivers
- Lubricating-oil governor and seal system cleaned
- Instrumentation and speed control checked
- Auxiliary heat recovery system

Centrifugal Compressors
- Lubricating and seal oil systems cleaned
- Instrumentation and controls checked
- Preliminary operation of lubricating and seal oil systems
- Operation with air

Vacuum Equipment
- Alignment run-in testing

Pumps
- Alignment run-in adjustment

Hot-Alignment of Machinery
Vibration Measurements
Instruments
- Blow with clean air
- Dry out
- Check continuity zero and adjust
- Calibrate as required

Chemical Cleaning
- Activation
- Passivation
- Rinse

Operating Preparations

Log sheets
Hand tools, hose, and ladders on hand
Containers, bags, drums, and railroad cars available
Miscellaneous supplies (work-order pads, requisition forms, etc.) on hand

Safety

Protective clothing, goggles, face shields, hard hats, work gloves, rubber gloves, aprons, hoods, gas masks (with spare canisters), and self-contained breathing apparatus on hand
Safety procedures for lockout, tank entry, hot work permits, and excavation written
First aid and medical assistance available
First-aid kits, blankets, stretchers, antidotes, and resuscitators on hand
Safety valve settings and installations checked

Fire Protection

Asbestos suits, axes, ladders, hand extinguishers, hoses, wrenches, and nozzles on hand
Fire-fighting procedures planned
Foam chemicals on hand
Fire brigade organized

These three checklists concern utilities,[2] pressure testing/purging,[2] and mechanical equipment.[3]

Table 1
Checklist for Utilities and Service Systems Can Be Expanded to Suit Any Plant

Step	Description	Constraints	Safety Precautions	Prerequisite Steps	Remarks
1. Activate electrical systems	Handled by electrical maintenance.		Clear with electrical group before using any equipment.	Removal of lockouts, completion of pre-operational maintenance work.	
2. Activate fire, process, and potable water systems	Start water systems, venting and testing hydrants.	Control valves operated manually until air is available.		Ensure piping is ready.	Vent and test deluge systems, checking for leaks.
3. Fill cooling-water system and circulate		Same as in Step 2.		Same as in Step 2.	Vent exchanger high points; check for leaks; charge chemicals.
4. Activate instrument and plant air systems	After checkout, pressurize systems; with Instrument Dept., activate air to instruments.		Check control valves to see that valve action does not affect process equipment when air is activated.	Same as in Step 2.	Check for leaks; drain water from air headers and piping; check driers and filters, check dewpoint.
5. Activate steam and condensate-return systems	Introduce steam to high-pressure header; with it in service, pressurize low-pressure systems.	Boiler feed water must be available, and instruments activated.	Vent inerts from steam system; slowly pressurize to avoid hammer; do not over-pressure reduced-pressure systems.	Check all headers, traps, etc., before heating steam lines.	Check for leaks.
6. Activate nitrogen system	Purge and pressurize N_2 headers.	Vessel entry procedures must be enforced before introduction of N_2.	Purge to less than 1% O_2 by successive pressuring and venting; no entries can be permitted unless N_2 inlets are blanked.	Ensure piping is complete.	Check dewpoint.
7. Waste disposal	Water lines to battery limits must be functional.	Outside-battery-limit lines and facilities must be ready.			

Table 2
Procedural Checklist Ensures Safe, Complete Pressure-Testing and Purging of Process Equipment

Step	Description	Constraints	Safety Precautions	Prerequisite Steps	Remarks
1. Unit checkout	Check that required mechanical work has been completed, tags and blinds pulled, and temporary piping disconnected.	Plant supervision must certify completion of work.	Cancel all entry and work permits.	Utility system has been commissioned.	Check blind list and inspect lines; close bleed, drain and sample valves.
2. Flange taping	Apply masking tape to all pipe and vessel flanges.			Same as in Step 1.	Seal space between flanges and punch 1.5mm hole in tape for monitoring leaks.
3. Nitrogen pressuring	With N_2, pressurize all process systems to designated test pressures.	Test at pressures near to normal; avoid lifting relief valves.	Manual valves requiring it must be carsealed after being set.	Same as in Step 1.	Prepare lists for equipment groupings and test pressures; isolate leaking sections, if possible.
4. Locate leaks	Apply soap solution to tape bleed holes and threaded connections; depressure when tests are done.	Block in N_2 and check for loss of pressure.	Check all connections.	Same as in Step 1.	Tag flanges to identify leaks; tighten or repair leaks; each time a system is pressurized, drain condensate at low-point bleeds.
5. Purging (O_2 removal)	Alternately depressure and repressure system to test (or header) pressure, until O_2 by meter is less than 3.0%.		Regularly check O_2 meter to ensure its accuracy.	Passing leak check at test (or N_2 header) pressure.	

Checklist for Commissioning Mechanical Equipment

Before any mechanical equipment is put into service, there should be a:

1. Field disassembly and reassembly.
2. Clean-out of the lubrication system, including chemical cleaning and passivation, where required.
3. Circulation of lubrication to check the flow and temperature.
4. Clean-out and checkout of the cooling water system.
5. Checkout and commissioning of instruments.
6. Checkout of free and unhindered rotation.
7. Test-tightening of anchor bolts.
8. Disconnection and reconnection of piping to verify that it does not stress system.
9. Installation of temporary filters.
10. Preparation for running at a load.
11. Operation with the driver uncoupled.
12. Recoupling with driver and verification of alignment.
13. Checkout of vent system.
14. Checkout of seal system.
15. Operation empty to check for vibrations and heating of bearings.
16. Operation under load.

Sources

1. Matley, Jay, "Keys To Successful Plant Startups," *Chemical Engineering,* September 8, 1969, p. 110.
2. Gans, M., Kiorpes, S. A., and Fitzgerald, F. A., "Plant Startup—Step by Step," *Chemical Engineering,* October 3, 1983, p. 80 and 92.
3. Gans, Manfred, "The A to Z of Plant Startup," *Chemical Engineering,* March 15, 1976, p. 79.

23
Energy Conservation

Target Excess Oxygen

References 1 and 2 give target excess oxygen to shoot for as a guide in heater efficiency improvement. Table 1 summarizes the recommended targets.

Table 1
Target Excess Oxygen

Situation	% Excess Oxygen Gas Firing	Oil Firing
Portable analyzer weekly check	5.0	6.0
Permanently mounted oxygen recorder	4.5	5.0
Oxygen recorder with remote manual damper control	4.0	4.5
Automatic damper control	3.0	4.0

In an operating plant, the air rate can be adjusted at fixed heat output (constant steam rate for a boiler) until minimum fuel rate is achieved. This is the optimum so long as the warnings below are heeded.

Woodward, References 1 and 2, warns that one must not get so low on excess oxygen that combustibles can get in the flue gas. This can quickly lose the efficiency that one was trying to gain, plus pose a safety hazard. In oil firing, a fuel rich condition can be detected by smoking. In gas firing, however, substantial loss from unburned combustibles can occur before smoking is seen. Woodward believes that below 3 percent excess oxygen the percentage of combustibles in the flue gas should be monitored.

Reference 3 speaks of controlling the excess air at 10% for plants having highly variable fuel supplies by using gas density to control air/fuel ratio. Reference 3 gives the following equations relating theoretical air and density.

Fuel gas/air ratio = 1:14.78 $(1 + 0.0921/r_d)$ mass/mass (1)

Fuel oil/air ratio = 1:0.115 $(X + 3Y)$ mass/mass (2)

where

r_d = Density relative to air; M.W./M.W. air
X = % Carbon
Y = % Hydrogen

Equation (1), the fuel gas equation, does not hold for unsaturated hydrocarbons, but for small percentages of unsaturates the error is not serious.

High flame temperature and high excess air increase NO_x emissions.

"Percentage excess oxygen" refers to the oxygen left over from combustion and appearing in the flue gas. This is different from the "percentage excess air" which is combustion air used above that required for complete combustion with no oxygen left over. The relationship between excess oxygen (percentage in the flue gas) and excess air (percentage above minimum) is shown by Reed.[4]

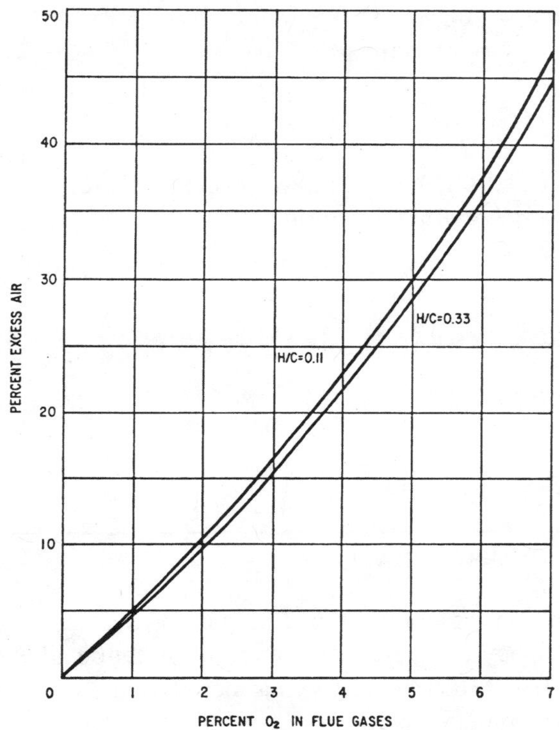

Figure 1. Excess air can be determined from flue gas analysis and hydrogen-to-carbon weight ratio of the fuel.

Sources

1. Woodward, A. M., "Control Flue Gas to Improve Heater Efficiency," *Hydrocarbon Processing,* May 1975.
2. Woodward, A. M. "Reduce Process Heater Fuel," *Hydrocarbon Processing,* July 1974.
3. Ferguson, B. C., "Monitor Boiler Fuel Density to Control Air/Fuel Ratio," *Hydrocarbon Processing,* February 1974.
4. Reed, R. D., "Save Energy At Your Heater," *Hydrocarbon Processing,* July 1973, p. 120.
5. Branan, C. R., *The Process Engineer's Pocket Handbook, Vol. 1,* Gulf Publishing Co., 1976, pp. 116, 117.

Stack Heat Loss

Woodward[1] gives a graph to estimate stack heat loss at various temperatures and percentage excess oxygens. To compare "percentage excess oxygen" and "percentage excess air" see the previous section, Target Excess Oxygen. Knowledge of the heat loss will allow determination of the economic incentive to lower the stack temperature or percentage excess oxygen.

When lowering stack temperature, be aware of the dew point as discussed in the next topic.

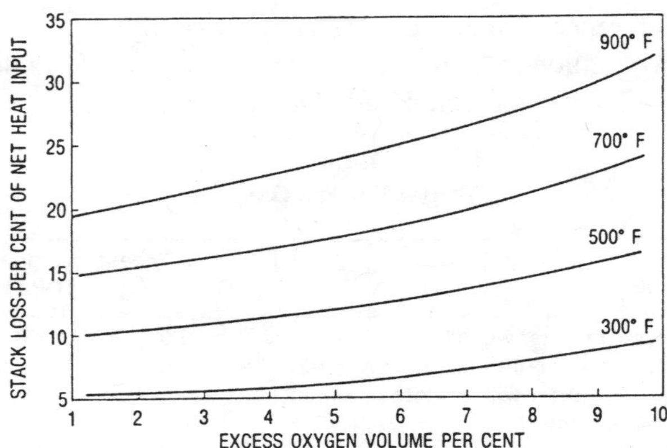

Figure 1. Stack loss vs. excess oxygen and stack temperature.

Source

Woodward, A. M., "Reduce Process Heater Fuel," *Hydrocarbon Processing,* July 1974, p. 106.

Stack Gas Dew Point

When cooling combustion flue gas for heat recovery and efficiency gain, the temperature must not be allowed to drop below the sulfur trioxide dew point. Below the SO_3 dew point, very corrosive sulfuric acid forms. The graph[1] in Figure 1 allows determination of the acid dew point as shown in Example 1.

Example 1. A pseudo single compound is generated for the natural gas composition in Table 1.

Table 1
Natural Gas Composition

Component	Mol frac	C	H	N	S
CH_4	0.750	0.750	3.000	—	—
C_2H_6	0.043	0.086	0.258	—	—
C_3H_8	0.013	0.039	0.104	—	—
C_4H_{10}	0.004	0.016	0.040	—	—
N_2	0.173	—	—	0.346	—
H_2S	0.017	—	0.034	—	0.017
Total	1.000	0.891	3.436	0.346	0.017

The elemental concentration of S, expressed as 1.7 vol.%, in Table 1, is used to enter Figure 1. One then proceeds to the elemental concentration of carbon as determined from Table 1, which is 0.891. Assuming 10% excess air and proceeding upward to the 0.891 carbon line and then to the left gives 152°C dew point.

Example 2. For a fuel oil, the wt.% sulfur is used instead of the vol.%. In this example the wt.% S is 2.5, the carbon elemental concentration is 2.4, and the excess air 25%. Follow the second dotted line to 147°C.

Alternate calculations can be made using equations to calculate flue gas compositions and partial pressures of H_2O and SO_3 to calculate the dew point. The alternate method is shown in the source document also.

Source

Okkes, A. G., "Get Acid Dew Point of Flue Gas," *Hydrocarbon Processing,* July 1987, p. 53.

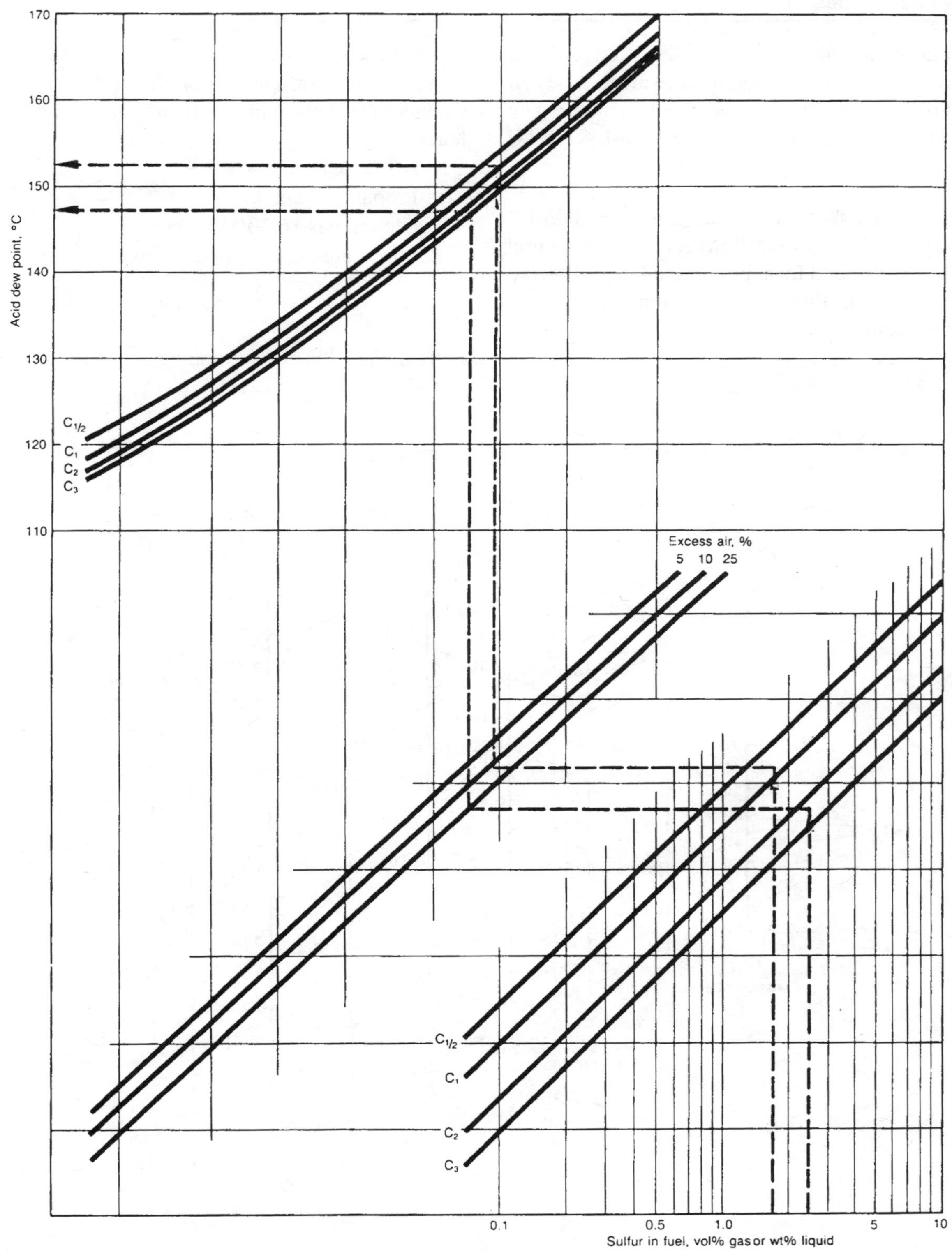

Figure 1. Graph accounts for fuel type, sulfur, and excess air.

Equivalent Fuel Values

In generating economics for fuel conservation projects, it is handy to have quick equivalent fuel values. The following nomograph (Figure 1) ranks fuels according to their gross heating value, thus providing cost per BTU equivalency.

Examples. Find the fuel values equivalent to \$5.00/MM BTU natural gas. Connect \$5.00/MM BTU on the natural gas line with the left-hand focal point. Read the following values along the connecting line, for example:

ethane 33 c/gal
butane 52 c/gal
heavy crude oil \$32/bbl

All of these values correspond to the \$5.00/MM BTU.

As another example start with \$40/bbl No. 6 Fuel Oil. Connect the point with the focal point. Example equivalencies are:

No. 2 fuel oil 88 c/gal
Propane 57 c/gal
Natural gas \$6.30/MM BTU

All the values correspond to \$6.30/MM BTU.

Source

Leffler, W. L., "Fuel value nomogram updated," *Oil and Gas Journal,* March 31, 1980, p. 150.

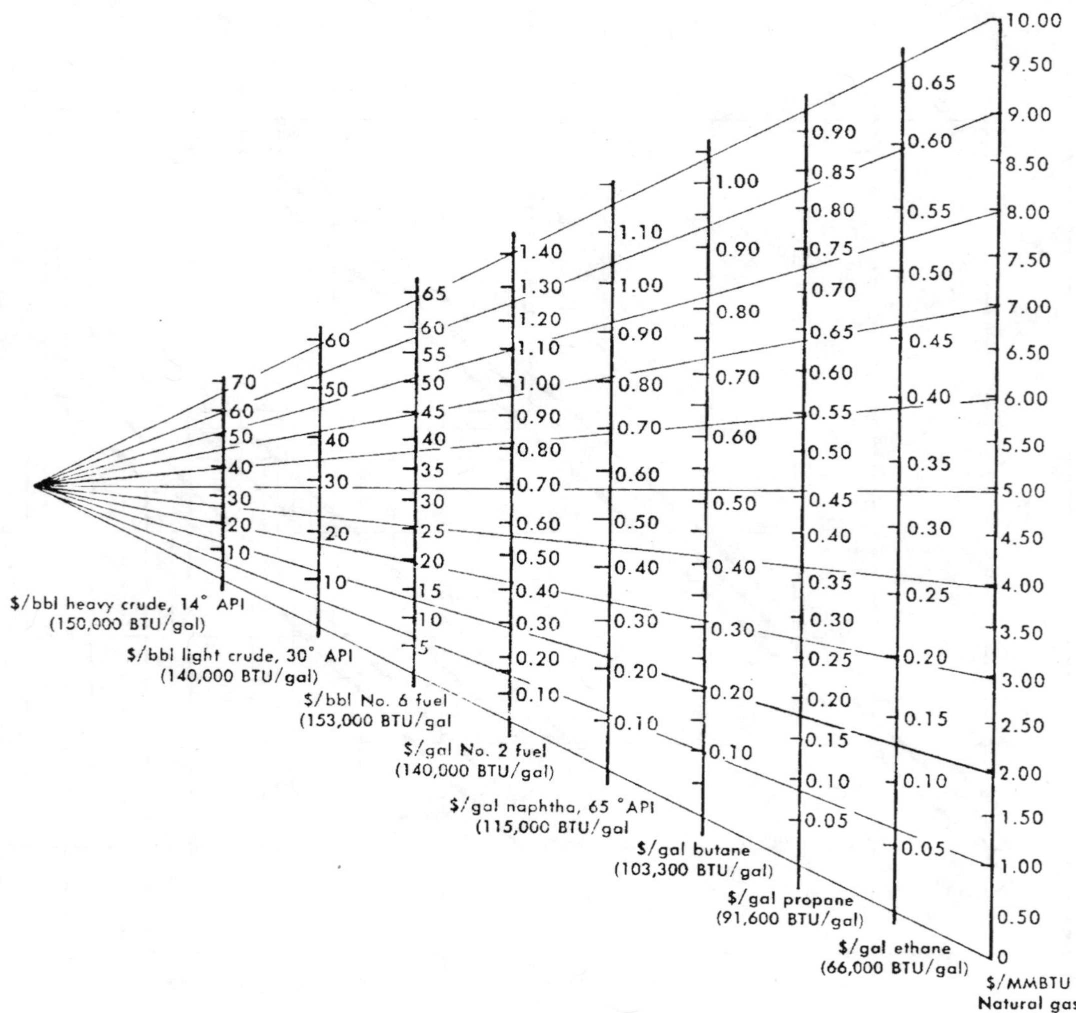

Figure 1. Equivalent values for selected hydrocarbons.

Heat Recovery Systems

Goossens[1] gives the following summary of heat recovery systems, subdivided by temperature range:

Table 1
Heat Recovery

Temperature Range	Range of η	Applications		
		Production of heat in furnaces and boilers	Recovery of furnace heat	Other Applications
500 to ⩾1000°C	0.65 to ⩾0.80	• Gas turbine cycle with furnace heat production in the turbine exhaust. • Steam turbine topping cycles for the production of electricity in back-pressure turbines with exhaust steam for process applications.	• Preheating cracking or reforming furnace feed streams. • Generating high-pressure steam (. . . 100 . . . bar). • Steam superheating. • Preheating off-gases for turbo-expander operation.	
300 to 500°C	0.53 to 0.65		• Boiler or furnace air preheating. • Generation of medium to high pressure steam.	• Generation of medium to high pressure steam from exothermic heat. • Organic Rankine cycle.
150 to 300°C	0.35 to 0.53		• Boiler feedwater preheating. • Air preheating.	• Generation of low pressure steam for heating, drying, distillation . . . • Organic Rankine cycle. • Absorption cooling.
⩾150°C	⩾0.35			• Absorption cooling. • Recovery of steam condensate and flash steam. • Heat pump for evaporation, drying, etc.
			• Recovery of low level waste heat for space heating, district heating system.	

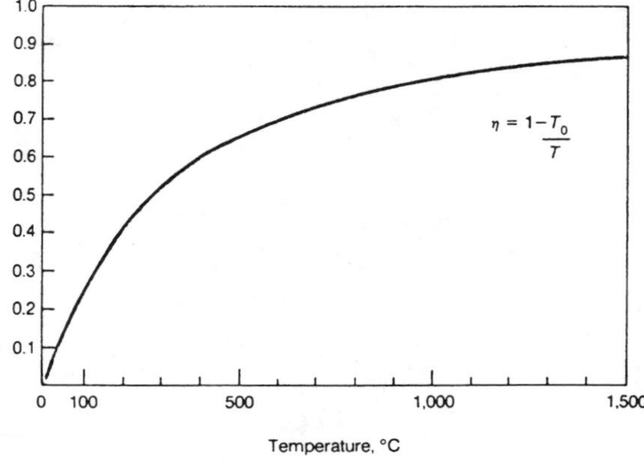

Figure 1. The exergetic efficiency of heat as a function of temperature (reference temperature: 0° C).

Source

Goossens, G., "Principles Govern Energy Recovery," *Hydrocarbon Processing,* August 1978, p. 133.

Process Efficiency

Because fuel costs are high, the search is on for processes with higher thermal efficiency and for ways to improve efficiencies of existing processes. One process being emphasized for its high efficiency is the gas turbine "combined cycle." The gas turbine exhaust heat makes steam in a waste heat boiler. The steam drives turbines, often used as helper turbines. References 6, 7, and 8 treat this subject and mention alternate equipment hookups, some in conjunction with coal gasification plants.

Arrangements that combine the gas turbine, steam helper turbine, and electric generator on a single shaft are gaining acceptance.[11] Already established overseas, these are gaining a foothold in the U.S. Advantges are the possibility of lower first cost, operating simplicity, a common lubricating oil system for the steam turbines and electric generator, and a smaller site footprint. In addition, the heat recovery steam generator bypass stack is eliminated. On multishaft units the bypass stack allows phased construction and simple-cycle operation. The single shaft units can employ a synchronous clutch between the generator and steam turbine(s). This permits startup in simple-cycle mode while warming the steam systems. The clutch engages when the steam turbine shaft rotating speed reaches that of the generator. Many plants that must shutdown and restart daily are using the synchronous clutch.

Reference 2 is a well written report which discusses power plant coal utilization in great detail. It gives a thermal efficiency of 80–83% for modern steam generation plants and 37–38% thermal efficiency for modern power generating plants at base load (about 70%). A modern base load plant designed for about 400 MW and up will run at steam pressures of 2400 or 3600 psi and 1000 °F with reheat to 1000 °F and regenerative heating of feedwater by steam extracted from the turbine. A thermal efficiency of 40% can be had from such a plant at full load, and of 38% at high annual load factor. The 3600 psi case is supercritical and is called a once-through-boiler since it has no steam drum. Plants designed for about 100–350 MW run around 1800 psi and 1000 °F with reheat to 1000 °F. Below 100 MW, a typical condition would be about 1350 psi and 950 °F with no reheat. Reference 2 states that below 60% load factor, efficiency falls off rapidly and that the average efficiency for all steam power plants on an annual basis is about 33%.

For any process converting heat energy to mechanical efficiency, the Carnot efficiency is the theoretical maximum. It is calculated as

$$\frac{T_1 - T_2}{T_1} \times 100 \qquad (1)$$

where

T_1 = Temperature of the heat source, °R
T_2 = Temperature of the receiver where heat is rejected, °R

Therefore, the efficiency is raised by increasing the source temperature and decreasing the receiver temperature.

The efficiency for a boiler or heater is improved by lowering its stack temperature. The stack minimum temperature is frequently limited by SO_3 gas dew point. References 3, 4, and 5 discuss this important subject. A stack as hot as 400°F (or perhaps higher) can have problems if the SO_3 concentration is high enough. Reference 10 states that SO_3 condensation will produce a blue-gray haze when viewed against a clear blue sky.

A very useful relationship for determining the maximum available energy in a working fluid is

$$\Delta B = \Delta H - T_o \Delta S \qquad (2)$$

where
ΔB = Maximum available energy in Btu/lb
ΔH = Enthalpy difference between the source and receiver, Btu/lb. For a typical condensing steam turbine it would be the difference between the inlet steam and the liquid condensate
T_o = Receiver temperature, °R
ΔS = Entropy difference between the source and receiver, Btu/lb °F

To obtain lb/hr-hp, make the following division:

$$\frac{2545}{\Delta B}$$

Equation 2 will yield the same result as the Theoretical Steam Rate Tables (Reference 9). Therefore, this is a handy way of getting theoretical steam rates when only a set of steam tables sans Mollier diagram are available.

Sources

1. Branan, C. R., *The Process Engineer's Pocket Handbook, Vol. 1,* Gulf Publishing Co., 1976.
2. Locklin, D. W., Hazard, H. R., Bloom, S. G., Nack, H., "Power Plant Utilization of Coal," A Battelle Energy

Program Report, Battelle Memorial Institute, Columbus, Ohio, September 1974.

3. Verhoff, F. H., and Banchero, J. T., "Predicting Dew Points of Flue Gases," *Chemical Engineering Progress,* Vol. 70, No. 8, August, 1974.
4. Martin, R. R., Manning, F. S., and Reed, E. D., "Watch for Elevated Dew Points in SO_3-Bearing Stack Gases," *Hydrocarbon Processing,* June, 1974.
5. "Fuel Additives Control Preburner and Fireside Combustion Problems," Betz Bulletin 713, Copyright © 1974, Betz Laboratories, Inc.
6. Moore, R., and Branan, C., "Status of Burnham Coal Gasification Project," *Proceedings, 54th Annual Convention, Gas Processors Association,* Houston, Texas, March 10–12, 1975.
7. Zanyk, J. P., "Power Plant Provides 86% Efficiency," *Oil and Gas Journal,* May 27, 1974.
8. Ahner, D. J., May, T. S., and Sheldon, R. C., "Low BTU Gasification Combined-Cycle Power Generation," Presented at Joint Power Generation Conference, Miami Beach, September 15–19, 1974
9. Keenan, J. H., and Keyes, F. G., "Theoretical Steam Rate Tables," *Trans. A.S.M.E.* (1938).
10. Reed, R. D., "Recover Energy from Furnace Stacks," *Hydrocarbon Processing,* January, 1976.
11. Swanekamp, R., "Single-Shaft Combined Cycle Packs Power at Lower Cost," *Power,* Jan. 1996.

Steam Traps

No section on energy conservation would be complete without rules of thumb on steam traps. Here are three handy correlations for steam traps:

1. How to estimate steam loss from Yarway Steam Conservation Service.
2. Understanding and design of inverted bucket traps.
3. Test results from a DuPont study.

How to Estimate Steam Loss

The C_v Concept. From a quantitative standpoint, steam loss can be estimated through the application of the C_v concept. A familiar term in control valve technology, C_v expresses the flow capability of a fluid-controlling device—in this case, a steam trap. A large C_v means a high flow rate; a low C_v means a low flow rate.

Figure 1 provides a means for estimating trap C_v from 0 to 600 psi, given the manufacturer's "rating" for the trap. The rating must define quantity (lb/hr), pressure and liquid or condensate temperature. The resulting C_v information can then be applied to Figure 2 to estimate the actual loss of steam. Multiply the steam flow values in Figure 2 by the previously determined trap C_v to obtain steam loss in pounds per hour of dry steam.

Example. A trap on a 150 psi steam line has been found to be blowing live steam on the basis of contact pyrometer measurements taken immediately upstream and downstream of the trap. The catalog rating of the trap is 5,000 lb/hr at saturation temperature (°F sub-cooled) at 150 psi.

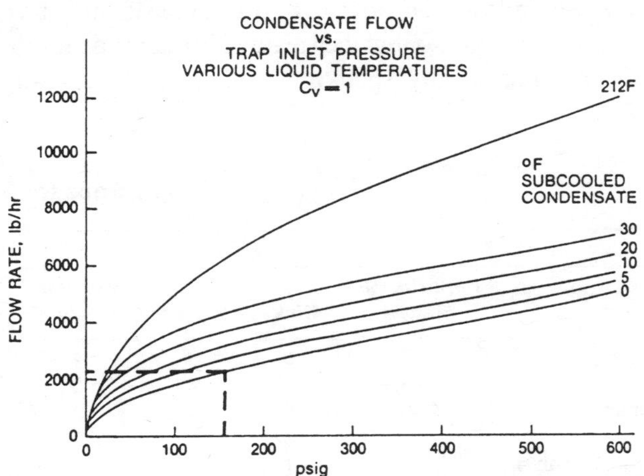

Figure 1. Estimating trap C_v.

Figure 1 indicates the value of 2,100 lb/hr at 150 psi for °F sub-cooled condensate. The flow factor is:

$$\frac{5,000}{2,100} = 2.38$$

Figure 2 shows 220 lb/hr steam flow at 150 psi for a flow factor of 1. For the trap in question:

$220 \times 2.38 = 525$ lb/hr steam loss

Assuming the rather modest cost of steam of $1.35 per thousand lb, the loss of steam is estimated to cost:

$$\frac{525}{1,000} \times 1.35 = \$0.71/\text{hr or } \$119/\text{wk}$$

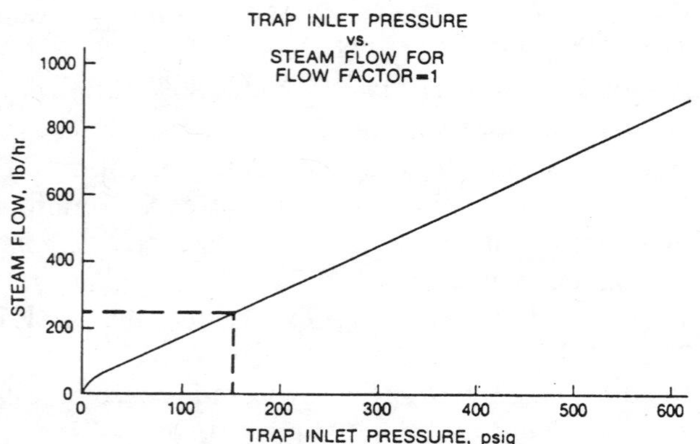

Figure 2. Estimating actual steam loss.

Inverted Bucket Steam Traps

It is easy to improperly design a steam trap. The design must work for two circumstances and often a designer will check only one of these. The circumstance often overlooked is as follows: On startup or upset the steam control valve can open wide so that the steam chest (assume for this discussion that we are speaking of a reboiler) pressure rises to full steam line pressure. At a time like this the steam trap downstream pressure can be atmospheric due to process variations or the operators opening the trap discharge to atmosphere in an attempt to get it working.

If the trap orifice has been designed too large, the trap valve cannot open to discharge condensate, creating or amplifying serious plant problems. For any steam trap, for a given trap pressure differential, there is a maximum orifice size above which the bucket can't exert sufficient opening power for the trap to operate. So, when designing a trap, check manufacturers' data to stay within the maximum sized orifice for full steam line pressure to atmospheric. If a larger orifice is required by the alternate circumstance discussed below, a larger trap body size must be specified whose bucket can service the larger orifice.

The required orifice continuous flow capacity is determined at steam chest pressure to condensate system pressure at a flow 6 to 8 times design. If designed for normal flow the trap would have to be open 100% of the time. Then, as stated above, a body size is selected that can contain the required orifice (not be above the stated maximum for that body size in the manufacturer's sizing tables) at the con-

Table 1
Steam Traps for Small Condensate Loads

		Trap Type								
Criterion	Importance, Range	Disk	Bucket	Impulse (piston)	Bellows	Float	Bimetallic	Orifice	Expansion	Instrumented
Steam loss	(0-10)	6	8	4	5	7	3	3	9	NA
Life	(0-8)	6	7	5	4	5	3	5	4	NA
Reliability	(0-6)	5	4	3	3	3	2	3	2	NA
Size	(0-3)	3	2	3	2	1	1	3	2	NA
Noncondensable venting	(0-2)	1	1	1	2	2	1	1	0	NA
Cost	(0-1)	1	1	1	0	0	0	1	0	NA
Small-trap types, total	**(30)**	**22**	**23**	**17**	**16**	**18**	**10**	**16**	**17**	**NA**

Table 2
Steam Traps for Medium Condensate Loads

		Trap Type								
Criterion	Importance, Range	Disk	Bucket	Impulse (piston)	Bellows	Float	Bimetallic	Orifice	Expansion	Instrumented
Steam loss	(0-10)	6	7	5	8	7	4	2	NA	NA
Life	(0-8)	3	7	5	5	4	2	6	NA	NA
Reliability	(0-6)	4	4	4	4	4	3	3	NA	NA
Size	(0-3)	3	2	3	3	3	2	3	NA	NA
Noncondensable venting	(0-2)	1	1	1	2	2	1	1	NA	NA
Cost	(0-1)	1	1	1	1	0	0	1	NA	NA
Medium-trap types, total	**(30)**	**18**	**22**	**19**	**23**	**20**	**12**	**16**	**NA**	**NA**

Table 3
Steam Traps for Large Condensate Loads

		Trap Type								
Criterion	Importance, Range	Disk	Bucket	Impulse (piston)	Bellows	Float	Bimetallic	Orifice	Expansion	Instrumented
Steam loss	(0-10)	4	5	6	NA	7	4	2	NA	9
Life	(0-8)	3	5	5	NA	4	3	4	NA	8
Reliability	(0-6)	3	4	4	NA	5	4	5	NA	5
Size	(0-3)	3	1	3	NA	2	2	3	NA	0
Noncondensable venting	(0-2)	1	1	1	NA	1	1	1	NA	1
Cost	(0-1)	1	1	1	NA	0	1	1	NA	0
Large-trap types, total	**(30)**	**15**	**17**	**20**	**NA**	**19**	**15**	**16**	**NA**	**23**

dition of full steam line pressure to atmospheric. The various vendor catalogs provide sizing charts and tables.

DuPont Test Results

Traps for small (test stand used), medium (in-plant testing) and large (in-plant testing) condensate loads were tested against a perfect rating of 30. The results are in Tables 1, 2, and 3.

Sources

1. Yarway Steam Conservation Service, "How To Estimate Steam Loss," *Chemical Engineering,* April 12, 1976, p. 63.
2. Branan, C., *The Process Engineer's Pocket Handbook, Vol. 1,* Gulf Publishing Co., 1976.
3. Vallery, S. J., "Setting Up A Steam-Trap Standard," *Chemical Engineering,* February 9, 1981.

Gas Expanders

When energy costs are high, expanders are used more than ever. A quickie rough estimate of actual expander available energy is

$$\Delta H = C_p T_1 \left[1 - \left(\frac{P_2}{P_1}\right)^{(K-1)/K}\right] 0.5 \qquad (1)$$

where

ΔH = Actual available energy, Btu/lb
C_p = Heat capacity (constant pressure), Btu/lb °F
T_1 = inlet temperature, °R
P_1, P_2 = Inlet, outlet pressures, psia
$K = C_p/C_v$

To get lb/hr-hp, divide as follows:

$$\frac{2545}{\Delta H}$$

A rough outlet temperature can be estimated by

$$T_2 = T_1\left(\frac{P_2}{P_1}\right)^{(K-1)/K} + \left(\frac{\Delta H}{C_p}\right) \qquad (2)$$

For large expanders, Equation 1 may be conservative. A full rating using vendor data is required for accurate results. Equation 1 can be used to see if a more accurate rating is worthwhile.

For comparison, the outlet temperature for gas at critical flow across an orifice is given by

$$T_2 = T_1\left(\frac{P_2}{P_1}\right)^{(K-1)/K} = T_1\left(\frac{2}{K+1}\right) \qquad (3)$$

The proposed expander may cool the working fluid below the dew point. Be sure to check for this.

Source

Branan, Carl R., *The Process Engineer's Pocket Handbook, Vol. 1,* Gulf Publishing Co., 1976.

Fractionation

Here is a checklist of methods for conserving energy in fractionation. The authors have divided the list into those methods that modify the distillation process (in contrast to modifying individual fractionators) and those that do not.

Do Not Require Process Modifications

1. More efficient separation
   a. Control retrofit
      i. Material balance
      ii. Feed forward
      iii. Floating pressure
   b. Tray internals retrofit
      i. More efficient internals
      ii. Lower pressure drop internals
2. More efficient heat use
   a. Insulation
   b. Feed/product heat exchange
   c. Heat exchanger retrofit
      i. High heat flux tubing
      ii. Augmented heat transfer surfaces
   d. Power or steam generation

Require Process Modifications

1. More efficient separation
   a. Additional draw point to eliminate a column
      i. Intermediate product
      ii. Intermediate impurity
      iii. Pasteurization
   b. Resequencing of separations
   c. Alternate separation techniques
      i. Azeotropic distillation
      ii. Extractive distillation
      iii. Liquid-liquid extraction
      iv. Thermal swing adsorption
      v. Pressure swing adsorption
      vi. Continuous countercurrent or cyclic adsorption with displacement regeneration
      vii. Crystallization
2. More efficient heat use
   a. Heat pumping
      i. External refrigerant cycle
      ii. Vapor recompression
      iii. Reboiler flashing
   b. Heat cascading
      i. Split column
      ii. Thermal coupling
      ii. Reboiler-condenser coupling
   c. Intermediate heat exchange
      i. Condensers
      ii. Reboilers

Source

"Energy Conservation in Distillation," by T. J. Mix, J. S. Dweck, M. Weinberg & R. C. Armstrong, *Chemical Engineering Progress,* Vol. 74, No. 4, pp. 49–55 (1978). "Reproduced by permission of the American Institute of Chemical Engineers. © 1978 AIChE."

Insulating Materials

Here is a comparison of insulating materials.

Table 1
Comparison of Insulating Materials

	Calcium Silicate	Cellular Glass	Fiber-glass	Mineral Wool	Poly-Urethane
Thermal conductivities (Btu/sq. ft., hr., in., °F)					
@100°F M.T.	0.33	0.32	—	0.25	0.16
@300°F M.T.	0.41	0.47	0.30	0.39	—
@500°F M.T.	0.51	0.64	—	0.54	—
Density (lbs./cu.ft.)	10–13	8.5	7	6	2
Max. temp. limit (°F)	1,200	800	450	1,000	250
Compressive strength (lbs./sq. in. @ % Deformation)	100 @ 5%	100	—	—	15 @5 %

Source

Duckham, Henry and Fleming, James, "Better Plant Design Saves Energy," *Hydrocarbon Processing,* July 1976, p. 78.

24 Process Modeling Using Linear Programming

Process Modeling Using Linear Programming

Many process engineers think of linear programming (L.P.) as a sophisticated mathematical tool, which is best applied by a few specialists extremely well grounded in theory. This is certainly true for your company's central linear program. The layman does not *write* a linear program, he only provides input that will model the process in which he is interested.

Still, even providing modeling input can be very involved for large linear programs, such as corporate models or plant models for refineries or olefin plants. However, there are many cases where linear programs can be used on a smaller scale by a person with minimal training. The purpose of this chapter is to encourage wider use of linear programming for the specific purpose of process design. Methods and philosophy are presented that will be easy to follow and particularly useful for process design applications.

Advantages for Process Design

Some process designs can be improved with linear programming. If a linear program model is available for a plant being modified or if a model can be generated, certain advantages are available to the designer. The more choices the operator has in running the plant, the greater the advantage of the optimizing power of a linear program. An olefin plant with a choice of feedstocks (for example, ethane, propane, and butane) and capable of achieving a reasonably broad range of conversions in its steam crackers is a good example of such a plant. The first advantage to the designer, then, even before starting the design calculations, is an optimized base case.

The linear program can be designed to deliver a wealth of base case information, such as:

1. Material balance (overall and by sections)
2. Energy balance (overall and by sections)
3. Refrigeration balance
4. Utility balances (fuel, steam of various levels, electricity, etc.)
5. Feedstock requirements
6. Production rates of major products and byproducts
7. Equipment bottlenecks and some insight into the value of debottlenecking
8. The value of every flowing component at every major point in the process
9. Insight into the economics of limited excursions from the optimized base case
10. Equipment utilization percentages

After the base case is digested and accepted by the designer as valid, various modification cases can be obtained. Because the base case and each modification case is presented in its best light (at the optimum plant operation for that case), bias between cases is eliminated. Therefore, the designer can compare cases on the same basis.

If plant expansion is being studied with the idea of determining the best practical upper limit, the linear program can be a great help. Each incremental expansion step can be evaluated and payouts determined off line. Also, each piece of proposed new equipment can be evaluated as to its effect on the entire plant.

It is important to note, however, that all this is not free. The designer must invest the time to set up the cases and evaluate the results. Only the designer can make the final decision as to whether the cases and the comparisons are valid (a true representation of the plant). The computer printout is simply the results of matrix manipulation and should be considered suspect until given the designer's stamp of approval.

Having a linear program available allows the designer to generate better designs, but does not necessarily make his job easier. In fact, it may put additional pressure on him to be sure his results are optimal rather than just a satisfactory design.

The designer will become better with time at using a model limited to linear relationships to approximate nonlinear parts of the plant. Although modern linear programs provide the means of side nonlinear programming, it is best to use only linear relationships whenever possible and feasible. There are many means available to approximate nonlinear with linear, such as breaking a curve into straight segments.

Process Design vs. Accounting Linear Program Models

The designations of "process design model" and "accounting model" are strictly the author's. Therefore, these designations will first be explained. Most linear program models that the reader is likely to encounter are the accounting type. In a large refinery, periodic balances are made with the aid of an accounting model for crude and other component purchases, optimizing runs on the units within the refinery, blending the pools of gasoline grades and other products, and determining optimum product slate, storage requirements, and economics. A polyolefin plant might also use an accounting model to determine optimum slate of the many product grades that are produced.

These models are usually overall plant or corporation oriented and are geared toward running the business. If utilities are included, they are probably keyed to production levels of operating units or, in the case of the polyolefin plant, a characteristic set of utilities is charged against each unit of each specialty product. Often, recycle streams within segments of the operation are not pertinent and not included.

It is not surprising that engineers accustomed to the accounting type of linear program model would not envision its application to process design. The process design linear program model looks at the process as a design engineer. Utilities are charged against the piece of equipment in the plant using the utility. The smaller the breakdown, the more accurate. Usually, the author will key utilities to individual fractionation systems containing column, condenser, reboiler, and pumps rather than to an individual pump.

Recycles are meticulously accounted for because they load equipment and draw utilities. An olefin plant sustaining relatively low conversion per pass often builds up large amounts of unreacted feed that is recycled to the steam crackers. With utilities charged to ultimate products, these recycles would seem to the model to be free. The model would likely opt for very low conversion, which usually gives high ultimate yield and saves feedstock. Assigning the utility costs to users causes the compressor to pay for the extra recycle and the model raises conversion to the true optimum value.

The process design linear program model is best written with flexibility in mind, such as extra matrix rows to provide flexibility in recycling, adding outside streams intermediate in the process, and determining component incremental values at each processing stage. This subject is discussed more fully later in this chapter.

Modeling the Process Design Linear Program

The method of approach given here works best if a diagram of the process model is first produced, such as shown in Figure 1. To simplify matrix development and look at the problem as a process design, do not think of the matrix in terms of equations. Rather, the rows are thought of as "tanks" and columns are thought of as "flows" to or from

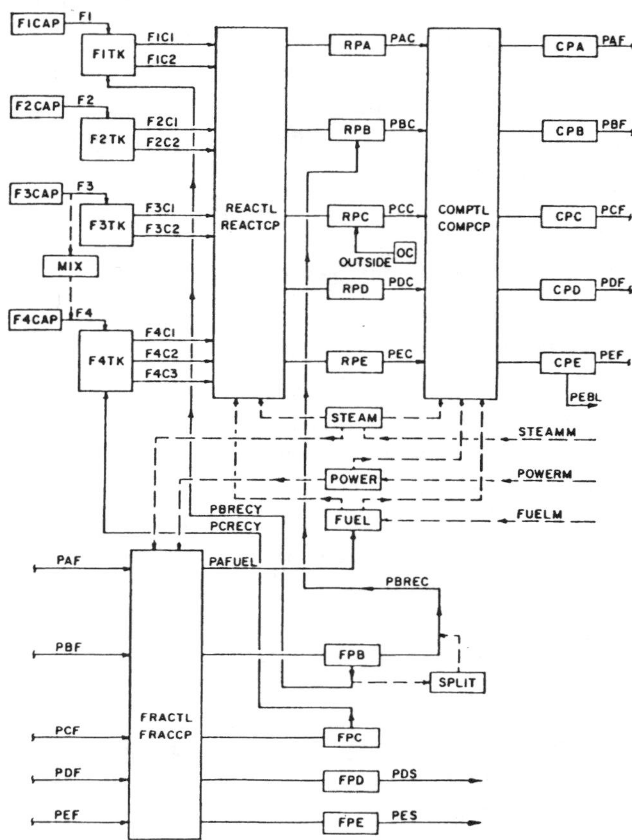

Figure 1. Process model for illustrating how process design linear programming can be achieved.

the tanks. The "tanks" may or may not have material in them at the beginning of the run (RHS) and are acted upon by the "flows," which either add material (minus sign) or subtract material (plus sign) from the "tanks".

All limits imposed on the model are done using the rows and price inhibitions where possible. Bounds that would provide hard limits using columns are therefore not used. The limits method is used as follows. For equipment capacity, the limit is usually keyed to total feed or some more accurate parameter. So each "flow" (column) to the equipment draws units of capacity from a "capacity tank" (row) until the capacity is used up.

Feed capacity is limited by putting a given amount of feed (RHS) in a "feed capacity tank." The other type of limit is that necessary to keep flows from accumulating or stacking in the middle of the plant in one of the intermediate "tanks" (rows). To inhibit this stacking, high prices are placed on any material accumulation. These are called slack prices (SPRICES).

The benefits of the method of handling limits just described are many. First, if the capacity of certain equipment is used up, the model generates a shadow price giving the value (example $/lb of increased capacity) to remove the capacity limit. Similarly, a shadow price for more feed would be generated.

ROW AND COLUMN NAMES

Rows (35 total)

F1CAP	The availability of feed 1
F2CAP	The availability of feed 2
F3CAP	The availability of feed 3
F4CAP	The availability of feed 4
F1TK	Mixer for feed and recycle
F2TK	Mixer for feed and recycle
F3TK	Mixer for feed and recycle
F4TK	Mixer for feed and recycle
MIX	Row to require feed 4 in given ratio to feed 3
REACTL	Totalizer for reactor feed
REACTCP	Reactor capacity limit
RPA	Reactor product A
RPB	Reactor product B
RPC	Reactor product C
RPD	Reactor product D
RPE	Reactor product E
OC	Capacity of product C from another plant
COMPTL	Totalizer for compressor feed
COMPCP	Compressor capacity limit
CPA	Compressor discharge of product A
CPB	Compressor discharge of product B
CPC	Compressor discharge of product C
CPD	Compressor discharge of product D
CPE	Compressor discharge of product E
FRACTL	Totalizer for fractionation section feed
FRACCP	Fractionation section capacity limit
FPB	Fractionation outlet of product B
FPC	Fractionation outlet of product C
FPD	Fractionation outlet of product D
FPE	Fractionation outlet of product E
SPLIT	Row to require fixed percentages of recycle to reactor and compressors for product B
STEAM	Steam supply reservoir
POWER	Power supply reservoir
FUEL	Fuel supply reservoir
$$$$	Objective function

Columns (34 total)

F1	Feed 1
F2	Feed 2
F3	Feed 3
F4	Feed 4
F1C1	Feed 1 at conversion level 1
F1C2	Feed 1 at conversion level 2
F2C1	Feed 2 at conversion level 1
F2C2	Feed 2 at conversion level 2
F3C1	Feed 3 at conversion level 1
F3C2	Feed 3 at conversion level 2
F4C1	Feed 4 at conversion level 1
F4C2	Feed 4 at conversion level 2
F4C3	Feed 4 at conversion level 3
PAC	Product A to compressor
PBC	Product B to compressor
PCC	Product C to compressor
PDC	Product D to compressor
PEC	Product E to compressor
OUTSIDE	Feed of product C from another plant
PAF	Product A to fractionation section
PBF	Product B to fractionation section
PCF	Product C to fractionation section
PDF	Product D to fractionation section
PEF	Product E to fractionation section
PEBL	Product E transferred to another plant
PBRECY	Product B recycle to reactor
PCRECY	Product C recycle to reactor
PAFUEL	Product A to fuel
PBREC	Product B recycle to compressor
PDS	Product D to sales
PES	Product E to sales
STEAMM	Steam supply
POWERM	Power supply
FUELM	Fuel supply in addition to product A going to fuel

Inhibition of stacking by pricing gives the model flexibility to stack if user errors have created an impossible situation. It is better for the model to be able to stack and yield an illogical case with large negative objective value than to get no run at all. This way, there is a better chance of obtaining a run that one can troubleshoot rather than just obtaining some diagnostic statements. This is particularly useful when trying to run a model for the first time.

Mixing or splitting of flows in the plant model is handled by having "flows" (columns) add or subtract material in given ratios from special "tanks" (rows) set up for this purpose. The method of modeling can now be seen in action with the example in Figure 1.

As was mentioned earlier, extra rows are provided as can be seen in Figure 1. This gives extra flexibility needed for process design case studies. Often, the set of case studies initially envisioned by the engineer or manager are not sufficient to answer all questions. The extra rows allow intermediate recycles or intermediate flows from or additions to the plant that might not have been envisioned initially. Another value of the intermediate rows is to provide component values at intermediate points in the process. The breakeven value of a proposed purchase stream from another plant would be one use of intermediate component values.

Example of Process Design Linear Program

The example used here does not represent any particular process and is simpler than most real life plant cases. It does, however, have the elements needed to talk our way through the development of a process design linear program model. All models should have the following:

1. Diagram
2. Nomenclature explanations
3. Matrix

As seen in Figure 1, the example plant has three major processing steps: Reaction, Compression, and Fractionation. There are four available feeds to the plant (F1, F2, F3, and F4). The desire is to use all of the available Feed 1. The model is inhibited from leaving any Feed 1 capacity unused by a large negative SPRICE as seen in the matrix. Feeds 2, 3, and 4 are given a choice of how much of the available material to use. However, Feeds 3 and 4 must be utilized in a 2:1 ratio as dictated by row MIX in the matrix. Feeds 1, 2, and 3 have a choice of 2 conversion levels, while Feed 4 has a choice of 3.

Flows (columns) F1C1 through F4C3 each create units of REACTL the total flow to the reactor and use up units REACTCP reactor capacity. Flows F1C1 through F4C3 create yields of reactor products RPA through RPE. Pure component "C" is added from an outside source into tank (row) RPC by flow (column) OUTSIDE.

Flows to the compressor and fractionation sections are handled similarly. Unreacted feed appearing in the fractionation section as components "B" and "C" are recycled to the reactor. These are taken from tanks (rows) FPB and FPC and flow to the feed tanks. Component "B" also has

(*text continued on page 344*)

MATRIX

Rows	Columns					SPRICES	RHS
	F1						
F1CAP	1					-L	A
	F2						
F2CAP	1						A
	F3						
F3CAP	1						A
	F4						
F4CAP	1						A
	F1C1	F1C2	F1	PBRECY			
F1TK	1	1	–1	–1		-L	
	F2C1	F2C2	F2				
F2TK	1	1	–1			-L	
	F3C1	F3C2	F3				
F3TK	1	1	–1			-L	
	F4C1	F4C2	F4C3	PCRECY	F4		
F4TK	1	1	1	–1	–1	-L	
	F3	F4					
MIX	–1	2				-L	
	(F1C1-through-F4C3)						
REACTL		–1					
	(F1C1-through-F4C3)						
REACTCP		E					R
	(F1C1-through-F4C3)			PAC			
RPA		-X		1		-L	
	(F1C1-through-F4C3)			PBC	PBREC		
RPB		-X		1	–1	-L	
	(F1C1-through-F4C3)			PCC	OUTSIDE		
RPC		-X		1	–1	-L	
	(F1C1-through-F4C3)			PDC			
RPD		-X		1		-L	
	(F1C1-through-F4C3)			PEC			
RPE		-X		1		-L	
	OUTSIDE						
OC	1						A
	(PAC-through-PEC)						
COMPTL		–1					
	(PAC-through-PEC)						
COMPCP		E					R
	PAC	PAF					
CPA	–1	1				-L	
	PBC	PBF					
CPB	–1	1				-L	
	PCC	PDF					
CPC	–1	1				-L	
	PDC	PCF					
CPD	–1	1				-L	
	PEC	PEF	PEBL				
CPE	–1	1	1			-L	

(table continued)

MATRIX (continued)

Rows			Columns				SPRICES	RHS
	(PAF-through-PEF)							
FRACTL		–1						
	(PAF-through-PEF)							
FRACCP		E						R
	PBF	PBREC	PBRECY					
FPB	–1	1	1				-L	
	PBREC	PBRECY						
SPLIT	Y	-Y					-L	
	PCF	PCRECY						
FPC	–1	1					-L	
	PDF	PDS						
FPD	–1	1					-L	
	PEF	PES						
FPE	–1	1					-L	
	(F1C1-through-F4C3)				(PAC-through-PEC)			
STEAM	U				U		-L	
	(PAF-through-PEF)			STEAMM				
STEAM (continued)	U			-1				
	(PAC-through-PEC)				(PAF-through-PEF)			
POWER	U				U		-L	
	POWERM							
POWER (continued)	-1							
	(F1C1-through-F4C3)				(PAC-through-PEC)			
FUEL	U				U		-L	
	FUELM	PAFUEL						
FUEL (continued)	-1	-1						
	(F1-through-F4)		OUTSIDE	STEAMM		POWERM	FUELM	
$$$$	C		C	C		C	C	
(The objective value to be maximized)								
	PEBL	PDS	PES					
$$$$ (continued)	-V	-V	-V					

Note: The L.P. is instructed to maximize the objective value and to apply all equations (each row in the matrix represents an equation) in the form: "LESS THAN," in contrast to "GREATER THAN" or "EQUAL TO," except for the objective value row equation, which is set at "FREE" (non-constraining, released from inhibition against negative slack value).

Nomenclature for Matrix Coefficients

A = Availability of feed or utility (example lb/hr)
C = Cost of purchased feed, stream from another plant, or utility (example $/lb)
E = Amount of equipment capacity used up by flowing component (units of R)
L = Selected large number
R = Capacity of equipment (example lb/hr feed)
U = Amount of utility used up by flowing component (example Kw/lb)
V = Value of product (example $/lb)
X = Fractional product yield
Y = Fractional stream split or fractional proportion in a mixture

Note: The above are changing quantities. For example, the two Y's that appear as intersections with row "SPLIT" and columns "PBREC" and "PBRECY" are not equal quantities.

(text continued from page 343)

a recycle to the compressor. Tank (row) SPLIT assures that for every unit of recycle to the reactor, a ratioed amount recycles to the compressor.

Utility usages are keyed to feed to the various sections of the plant. The objective row $$$$ receives credit for products sold and debits for feedstock and utility costs. Also stream OUTSIDE bought from another plant is debited.

Extras

Thus far, we have dealt with the major parts of a standard linear program package, but other parts are available and useful. Some of these are:

1. Range information
2. Report writer
3. Matrix generator

The range information is of two basic types. Shadow prices are available for items not in the solution basis, such as an equipment capacity that was completely used up with none remaining. The shadow price is the incremental value of additional capacity and ranges are provided to show the capacity range to which this value applies.

For items in the solution basis, such as feeds or conversion levels that were selected by the model for the run in question, range information is given. This shows what would happen if the basis item were removed from the basis.

Generally, it is good practice to print out full range information along with the solution. "What if's" from management may be already answered by the range information without having to make another computer run. Also, range information is valuable if a run library is being built.

Report writers are available to put the solution into layman's terms rather than matrix names.

Matrix generators are available to eliminate the task of preparing a handwritten matrix as we have done for the example.

25
Properties

Introduction

There are many sources of common properties of hydrocarbons and well-known chemicals. The more difficult to find properties are therefore provided in this handbook.

Approximate Physical Properties

Gravity of a boiling hydrocarbon liquid mixture at all pressures:

4 lb/gal or 30 lb/ft^3

Liquid volume contribution of dissolved gases in hydrocarbon mixtures:

Component	MW	gals/lb mol
O_2	32	3.37
CO	28	4.19
CO_2	44	6.38
SO_2	64	5.50
H_2S	34	5.17
N_2	28	4.16
H_2	2	3.38

Liquid volume contribution of light hydrocarbons in hydrocarbon mixtures:

Component	MW	gals/lb mol
C_1	16	6.4
$C_2^=$	30	10.12
C_2	28	9.64
$C_3^=$	44	10.42
C_3	42	9.67
iC_4	58	12.38
nC_4	58	11.93
iC_5	72	13.85
nC_5	72	13.71
iC_6	86	15.50
C_6	86	15.57
iC_7	100	17.2
C_7	100	17.46
C_8	114	19.39

Higher heating values of various fuels:

Solid Fuels	HHV, Btu/lb
Low volatile bituminous coal and high volatile anthracite coal	11,000–14,000
Subbituminous coal	9,000–11,000
Lignite	7,000–8,000
Coke and coke breeze	12,000–13,000
Wood	8,000–9,000
Bagasse (dry)	7,000–9,000

Gaseous Fuels	HHV, Btu/scf
Coke oven gas	500–700
Blast furnace gas	90–120
Natural gas	900–1,300
Refinery off gas	1,100–2,000

Liquid Fuels	HHV, Btu/gal
Heavy crude (14°API)	150,000
Light crude (30°API)	140,000
Fuel Oils No. 1	137,400
No. 2	139,600
No. 3	141,800
No. 4	145,100
No. 5	148,800
No. 6	152,400
15°API	178,710
Kerosine (45° API)	159,000
Gasoline (60° API)	150,000
Naphtha (65° API)	115,000
Butane	103,300
Propane	91,600
Ethane	66,000

Assumed gravities used commercially in determining rail tariffs:

Propane or butane	4.7 lb/gal
Natural gasoline	6.6 lb/gal

Volume of a "perfect gas":

379 ft^3/lb mol at 60°F and 14.7 psia

This volume is directly proportional to the absolute temperature = °F plus 460. It is inversely proportional to the absolute pressure.

Molal heat of vaporization at the normal boiling point (1 atm):

8,000 cal/gm mol or 14,400 Btu/lb mol

Compressibility factor (Z) for mixtures when using pseudo-critical mixture constants to determine:

Add 8°C to the critical temperature and 8 atmospheres to the critical pressure of Hydrogen and Helium.

The latent heat of water is about 5 times that for an organic liquid.

Heat Capacities, Btu/lb°F

Water	1.0
Steam/Hydrocarbon Vapors	0.5
Air	0.25

For structural strength:

Weight of concrete = weight of equipment supported

For desert landscaping (home or office)—contractors use: yards (yds^3) rock = ft^2 covered/100

Source

Branan, C., *The Process Engineer's Pocket Handbook, Vol. 3,* Gulf Publishing Co.

Liquid Viscosity

Uncommon Liquids

Here are viscosities of uncommon liquids. The table[1] below shows the point number to be used on the nomograph and the temperature scale to be used (both are °C). For example, for allyl iodide at 20°C, use point 8 and scale B to obtain 0.71 cp.

Table 1
Viscosities of Uncommon Liquids

No.	Chemical	Temperature Scale	No.	Chemical	Temperature Scale	No.	Chemical	Temperature Scale
1	Acetamide	B	30	Diethylene glycol	A	59	Methyl *n*-butyrate	B
2	Acetic anhydride	B	31	Diethyl ether	A	60	Methyl isobutyrate	B
3	Acetonitrile	A	32	Diethyl ketone	B	61	Methyl formate	B
4	Acetophenone	A	33	Dimethyl aniline	A	62	Methyl propionate	B
5	Acrylic acid	A	34	Diphenyl	A	63	Methyl propyl ketone	B
6	Allyl bromide	B	35	Diphenylamine	B	64	Methyl sulfide	A
7	Allyl chloride	A	36	Dipropyl ether	B	65	Methylene bromide	A
8	Allyl iodide	B	37	Dipropyl oxalate	A	66	Nitrogen dioxide	B
9	*n*-Amyl alcohol	B	38	Ethyl alcohol	A	67	Nitromethane	A
10	Benzophenone	A	39	Ethylaniline	B	68	Nitrotoluene	A
11	Benzotrichloride	B	40	Ethyl chloride	B	69	Phenetole	B
12	Benzyl ether	A	41	Ethyl iodide	B	70	Phenol	A
13	Benzylaniline	A	42	Ethyl propionate	B	71	Phosphorus	B
14	Bromine	B	43	Ethyl propyl ether	B	72	Propyl acetate	A
15	m-Bromo aniline	A	44	Ethyl sulfide	B	73	Propyl aldehyde	B
16	Bromobenzene	A	45	Ethylene bromide	B	74	*n*-Propyl bromide	A
17	Bromoform	A	46	Eugenol	A	75	Isopropyl bromide	B
18	Isobutyl acetate	B	47	Fluorobenzene	A	76	*n*-Propyl chloride	A
19	Isobutyl bromide	B	48	Formamide	A	77	Isopropyl chloride	B
20	Isobutyl chloride	B	49	Furfural	A	78	*n*-Propyl iodide	B
21	Isobutyl iodide	B	50	Hydrazine	A	79	Isopropyl iodide	B
22	*n*-Butyl formate	A	51	Iodobenzene	B	80	Succinonitrite	B
23	Carbon tetrachloride	B	52	Isoamyl acetate	A	81	Thiophene	B
24	*m*-cresol	A	53	Isoheptane	A	82	Toluene	B
25	Cyclohexane	B	54	Isohexane	A	83	Turpentine	B
26	Cyclohexene	A	55	Isopentane	A	84	Urethane	B
27	Diallyl	B	56	Isoprene	B	85	*o*-Xylene	B
28	Dibromomethane	A	57	Menthol, liquid	B	86	*m*-Xylene	B
29	Diethylaniline	A	58	Methyl alcohol	A	87	*p*-Xylene	B

Temperature Effect

For a quick estimate of viscosity change with temperature, simply place a point on the graph[2] shown as Figure 2, and move to a new temperature, °C represented by the 100°C increments on the X-axis.

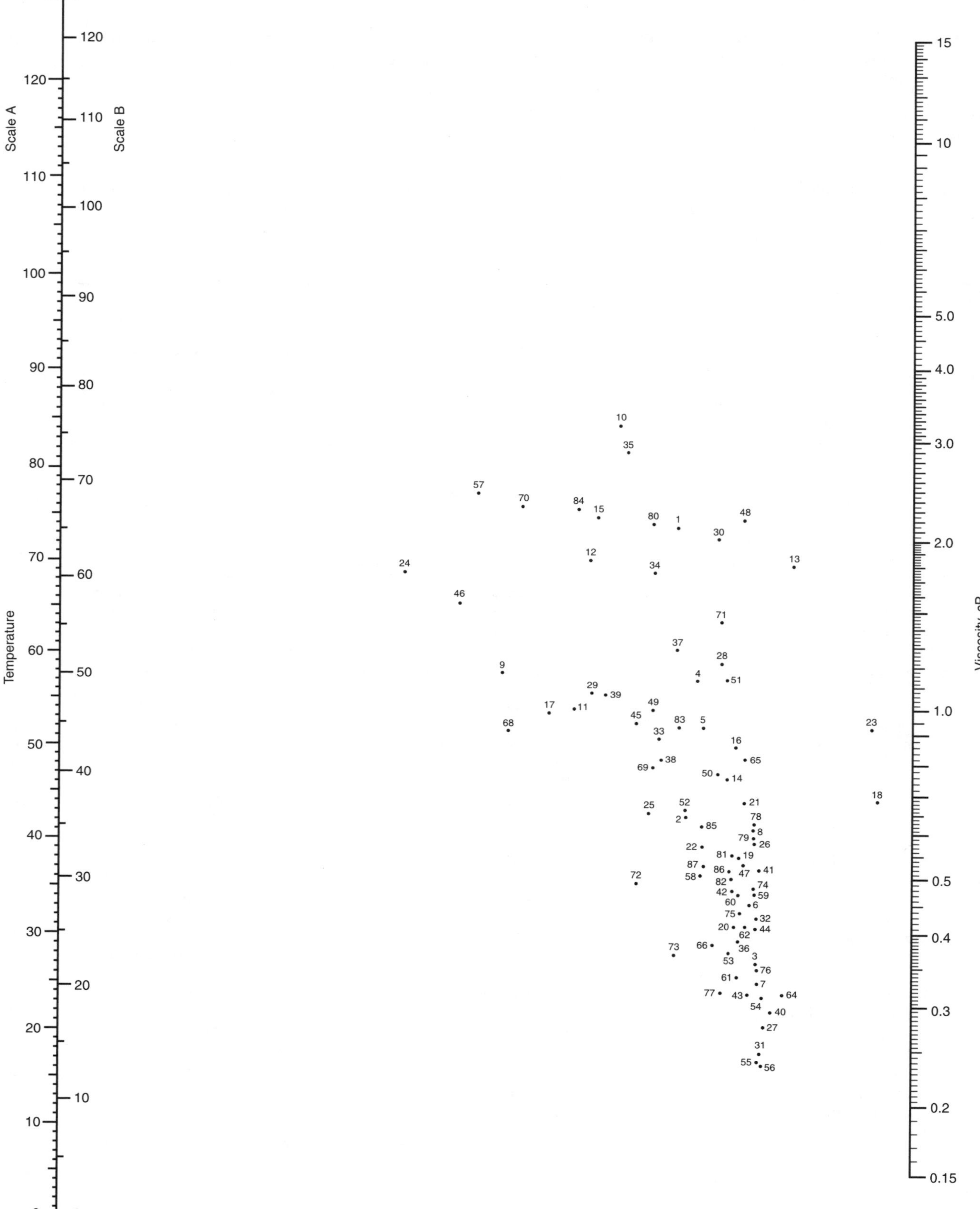

Figure 1. Nomograph for viscosity of uncommon liquids.

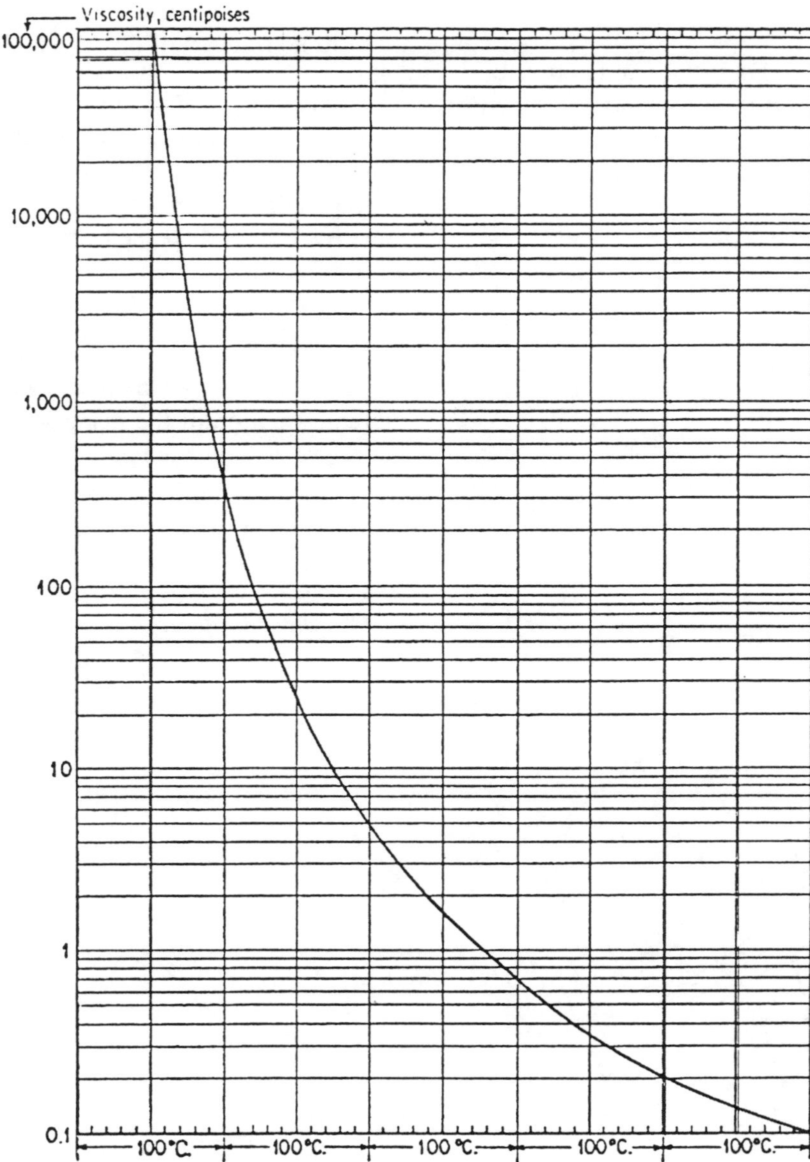

Figure 2. Easiest way to get from one temperature to another.

Viscosity Blending

For nonpolar mixtures, the following equation gives mixture viscosity usually within ±5 to 10%:

$$\ln \eta_{mix} = \Sigma W_i \ln \eta_i$$

where η_{mix} = Mixture viscosity
η_i = Viscosity of pure component i
W_i = Weight fraction of i

Do not use this equation for systems containing polar components including water.

Source

Perry, R. H. and Green, D. *Perry's Chemical Engineering Handbook,* 6th Ed., McGraw-Hill, Inc.

Emulsions

The nomograph (Figure 3) finds the emulsion viscosity of nonmiscible liquids and is based on the following equation:

$$\mu = \frac{\mu_c}{\delta_c}\left(1 + \frac{1.5\,\mu_D\,\delta_D}{\mu_D + \mu_c}\right)$$

Where

μ = Desired viscosity (any units)
δ = Volume fraction of liquid

Subscripts

$_D$ = Dispersed phase
$_c$ = Continuous phase

The range of variables was chosen as follows:

Viscosity of continuous phase = 0.5 to 50.0
Viscosity of dispersed phase = 0.5 to 100.0
Volume fraction of continuous phase = 0.1 to 1.0
Volume fraction of dispersed phase = 0.0 to 0.9

The use of the nomograph is as follows: Find the intersecting point of the curves of continuous phase and dispersed phase viscosities on the binary field (left side of nomograph). A line is drawn from this point to the common scale volume fraction of dispersed phase and continuous phase liquids. The intersection of this line with the Viscosity of Emulsion scale gives the result.

Example. The viscosity of the continuous phase liquid is 20. The viscosity of the dispersed phase liquid is 30. The volume fraction of the dispersed phase liquid is 0.3. The nomograph shows the emulsion viscosity to be 36.2.

Source

1. Kasipati, Rao, K. V., "Viscosities of Uncommon Liquids," *Chemical Engineering,* May 30, 1983, pp. 90–91.

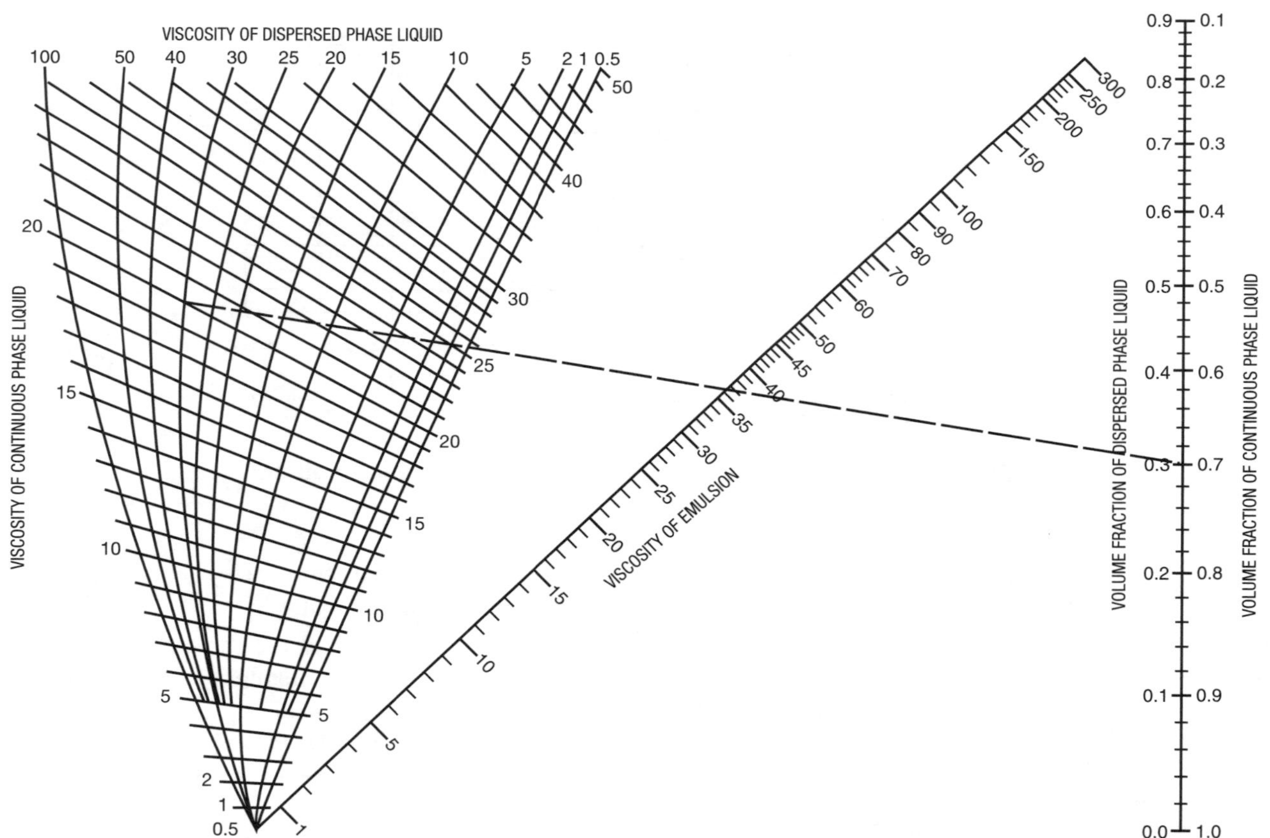

Figure 3. Nomograph for finding emulsion viscosity.

2. Gambill, W. R., "Estimate Engineering Properties—How P and T Change Liquid Viscosity," *Chemical Engineering,* February 9, 1959, p. 123.

3. Zanker, Adam, "Find Emulsion Viscosity Through Use of Nomograph," *Hydrocarbon Processing,* January 1969, p. 178.

Relative Humidity

Here is an equation for relative humidity (RH) and a short BASIC computer program that enhances its value. The average difference between the equation and actual values runs only 0.33 percentage points and the highest absolute difference seen in comparisons was 2 percentage points.

$$RH = 100\left(\frac{P_w - P_m}{P_d}\right) \tag{1}$$

where

$$P_d = P_c 10^{K_d(1 - T_c/T_d)} \tag{2}$$

$$P_w = P_c 10^{K_w(1 - T_c/T_w)} \tag{3}$$

$$P_m = 0.000367\left[1 + \left(\frac{t_w - 32}{1{,}571}\right)\right] \times P_b(t_d - t_w) \tag{4}$$

where K_d and K_w are functions of dry and wet bulb temperatures, T_d and T_w.

```
10 '% RH CALCULATION/Consultant Apalachin, NY
20 CLS:KEY OFF:SCREEN 2:SCREEN 0:COLOR
   15,1,0:CLS:PRINT
30 PRINT" PERCENT RELATIVE HUMIDITY CAL-
   CULATION":'SAVE"CE-RH.BAS
40 PRINT:PRINT:DEFDBL A-Z
50 PC=6567.6569#*25.4#:TC=1165.608#:KF=459.67#
60 PRINT" [1] Enter Dry-Bulb Temperature in Degrees..F
   ";:INPUT ": ",TD
70 TDR=TD+KF
80 PRINT" [2] Enter Wet-Bulb Temperature in Degrees..F
   ";:INPUT ":",TW
90 TWR=TW+KF
100 PRINT" [3] Enter Bar-Pres Read as Milimeters of
   Hg ";:INPUT ": ",BP
110 C0=4.39553#:C1=-.0034691#:C2=.0000030721#:
   C3=-.00000000088331#
120 PD1=C1*TDR:PW1=C1*TWR:PD2=C2*TDR*
   TDR:PW2=C2*TWR*TWR
130 PD3=C3*TDR*TDR*TDR:PW3=C3*TWR*
   TWR*TWR
140 KD=C0+PD1+PD2+PD3:KW=C0+PW1+PW2+PW3
150 PD=PC*10(KD*(1#-TC/TDR)):PW=PC*10(KW*
   (1#-TC/TWR))
160 K0=.000367#:K1=1571#:K2=32#
170 PM=K0*(1#+((TW-K2)/K1))*BP*(TD-TW):
   RH=100*((PW-PM)/PD):PRINT
180 IF RH <0 OR RH> 100 GOTO 200
190 GOTO 210
200 PRINT" Meaningless Result for % RH. ":PRINT
210 PRINT USING" Relative Humidity, RH = ###.# %
   ";RH
220 END
```

Nomenclature

- K — Temperature-dependent parameter, dimensionless; K_d = value at T_d; K_w = value at T_w
- P — Pressure of water vapor at any temperature T(°R), mm Hg
- P_b — Barometric pressure, mm Hg
- P_c — Critical pressure of water, 166,818 mm Hg
- P_d — Saturation pressure of water at the dry-bulb temperature, mm Hg
- P_m — Partial pressure of water vapor due to solution depression $(t_d - t_w)$, mm Hg
- P_w — Saturation pressure of water at the wet-bulb temperature, mm Hg
- RH — Relative humidity, %
- T_c — Critical temperature of water = 1,165.67°R
- t_d — Dry-bulb temperature, °F
- T_d — Dry-bulb temperature, °R
- t_w — Wet-bulb temperature, °F
- T_w — Dry-bulb temperature, °R

Source

Pallady, P. H., "Compute Relative Humidity Quickly," *Chemical Engineering,* November 1989, p. 255.

Surface Tension

For a liquid with known critical temperature and surface tension at one temperature, use the nomograph to estimate its surface tension at other temperatures.

Surface tension for most liquids can be estimated for temperatures other than the ones given in the literature (usually 15 or 20°C) by using the following equation:

$$\sigma_2 = \sigma_1\ [(T_c - T_2)/(T_G - T_1)]^{1.2}$$

where

σ_1 = surface tension at temp. T_1
σ_2 = surface tension at temp. T_2
T_c = critical temperature of the liquid

Figure 1 determines the foregoing temperature effect and is easier to use than the equation or a nomograph proposed by Kharbanda[1] for this relation. The results are fairly accurate, provided the temperatures for which the surface tensions are considered are not close to the critical temperature of the material in question. Best results are obtained for nonpolar compounds.

To assure correct use of the nomograph, let σ_L be the surface tension in dynes/cm for the lower temperature T_L. Then σ_H is the surface tension for the higher temperature T_H. All temperatures are given in Centigrade degree. The nomograph is suitable for estimating surface tension either at higher or lower temperatures from the one for which the known surface tension is given.

An example. Determine the surface tension of benzene at 100°C. Its surface tension is 29 dynes/cm at 20°C and its critical temperature is 288.5°C.

On the nomograph, find the point where the critical temperature line 1 for T_c = 288.5°C intersects the lower temperature line 2 for T_L = 20°C. Easy interpolation is possible by noticing the lines for critical temperatures T_c meet at a focal point on the upper end of the reference line and intersect the same temperatures on the scale for the higher temperature T_H. Likewise, the lines for the lower temperatures T_L meet at a focal point on the lower end of the reference line and also intersect the same temperatures on the higher temperature scale.

Next, connect a straight line 3 from the point on the higher temperature scale representing T_H = 100°C and extend this line 3 through the intersection of lines 1 and 2 to find the point on the reference line. From the reference line point, extend another straight line 4 through the point representing σ_L = 29 dynes/cm on the scale for the surface tension at the lower temperature in order to find the point of intersection on the scale for the surface tension at the higher temperature. The estimated surface tension at T_H = 100 °C is σ_H = 19 dynes/cm.

The measured surface tension for benzene at 100°C is 18.2 dynes/cm. Hence the error for the estimated value does not exceed 4.5% of the measured value.

LITERATURE CITED

1. Kharbanda, O.P., *Ind. Chemist,* April 1955, Vol. 31, p. 187 as cited in R. H. Perry and C. H. Chilton (Editors), "Chemical Engineers' Handbook," 5th ed., McGraw-Hill Book Co., New York, 1973, Ch. 3 p. 241.

Source

Zanker, Adam, "Estimating Surface Tension Changes For Liquids," *Hydrocarbon Processing,* April 1980, p. 207.

Gas Diffusion Coefficients

Bailey and Chen[1] have improved on previous correlations with the following equations:

$$DP = (186)(T/298)^{1.7}\ \frac{\left(\frac{1}{M_1} + \frac{1}{M_2}\right)^{0.5}}{(V_{c_1}{}^{0.4} + V_{c_2}{}^{0.4})^2}$$

An equation has been developed to relate the diffusion coefficients of two systems at the same temperature, as follows:

$$\frac{DP_{12}}{DP_{13}} = (0.99)\left[\frac{\alpha_{12}{}^{1.096}}{\alpha_{13}{}^{1.096}}\right]^{1.02}$$

where:

$$\alpha = \frac{\left(\frac{1}{M_1} + \frac{1}{M_2}\right)^{0.5}}{(V_{c_1}{}^{0.4} + V_{c_2}{}^{0.4})^2}$$

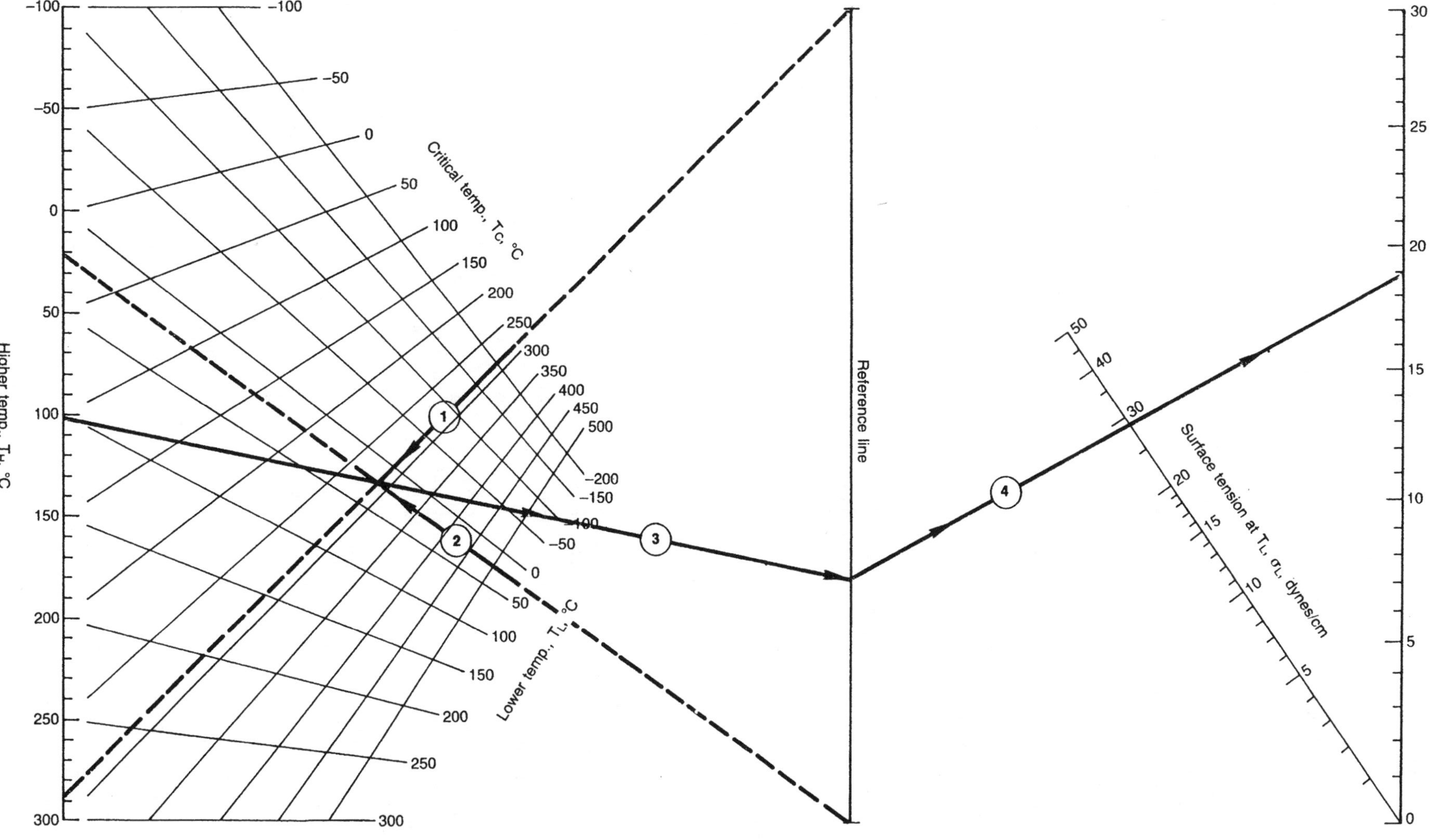

Figure 1. This nomograph helps estimate surface tension of liquids at various temperatures.

Finally, experimental data have been used to calculate the ratios DP/DP_{298}. If the coefficient is known at 298°K, it can be predicted at any temperature by the relation:

$$DP = (0.989)(DP)_{298}(T/298)^{1.7}$$

Nomenclature

D_{12} = diffusion coefficient, component 1 through component 2, sq cm/s
M = molecular weight
P = pressure, atm.
T = absolute temperature, °K
V_c = critical molal volume, cc/g mole
$_{1,2}$ = subscripts indicating components 1 and 2, respectively

Source

Bailey, R. G. and Chen, H. T., "Predicting Diffusion Coefficients In Binary Gas Systems," *Chemical Engineering,* March 17, 1975, p. 86.

Water and Hydrocarbons

Water in Natural Gas

Figure 1[1] shows water content of lean sweet natural gas. It can be used also for gases that have as much as 10% CO_2 and/or H_2S if the pressure is below 500 psia. Above 500 psia, acid gases must be accounted for by rigorous three-phase flash calculations[1] or approximation methods.[2]

Hydrocarbons in Water

An easy to use nomograph has been developed for the solubility of liquid hydrocarbons in water at ambient conditions (25°C). The accuracy of the nomograph has been checked against available solubility data. Performance of the nomograph has been compared with the predictions given by two available analytical correlations. The nomograph is much simpler to use and far more accurate than either of the analytical methods.

Development of the nomograph. Two main sources of data were used to develop the nomograph: McAuliffe[3] and Price.[4] The hydrocarbons were divided into 14 homologous series as listed in Table 1. Solubilities at 25°C were then regressed with the carbon numbers of the hydrocarbons in order to obtain the best fit for each homologous series. A second order polynomial equation fits the data very well:

$$-\log S = a + bC^2$$

where

S = the solubility in mole fraction
a,b = regression constants
C = the carbon number of the hydrocarbon

Table 1
Regression Constants and Nomograph Coordinates for the Solubilities of Liquid Hydrocarbons in Water at 25°C

Homologous Series	No. of Compounds	Regression Constants		Nomograph Coordinates	
		a	b	X	Y
Normal paraffins	7	3.81	0.0511	15.0	20.0
Isoparaffins-monosubstituted	8	3.91	0.0476	18.0	21.5
Isoparaffins-disubstituted	7	4.02	0.0409	23.2	25.1
Isoparaffins-trisubstituted	3	5.23	0.0186	47.1	20.8
Naphthenes (nonsubstituted)	4	3.45	0.0378	26.0	40.0
Naphthenes-monosubstituted	4	3.48	0.0441	20.7	33.1
Naphthenes-di and trisubstituted	4	4.75	0.0228	41.8	27.2
Alkenes-normal and monosubstituted	10	3.25	0.0479	18.0	34.0
Dienes	5	2.79	0.0460	19.1	43.9
Cycloalkenes	4	2.69	0.0460	19.1	46.1
Acetylenes	6	2.32	0.0469	18.9	52.5
Diynes	2	1.51	0.0391	25.0	76.8
Aromatics-single ring, mono and disubstituted by *n*-paraffins	6	1.91	0.0418	23.5	66.1
Aromatics-single ring, mono substituted by isoparaffins	2	2.99	0.0264	37.5	62.6

The constants a and b for the different homologous series are given in Table 1. The nomograph shown in Figure 2 was then built with the help of equations for the 14 homologous series. The X and Y coordinates for the different homologous series are listed in Table 1. The carbon number range of the nomograph is 4 to 10 while the mole fraction range is 10^{-9} to 10^{-4}. Its accuracy beyond these ranges is not known.

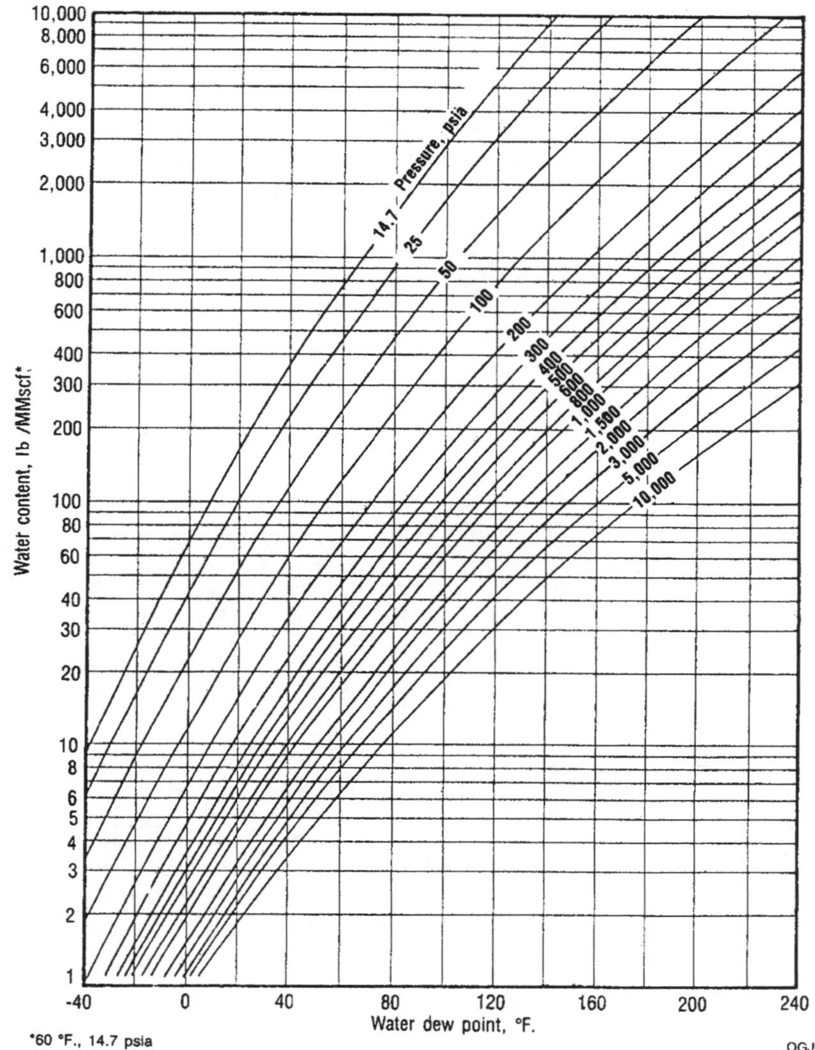

Figure 1. Water content of lean, sweet natural gas.

As an example of the use of the nomograph, the line is shown which would be drawn to determine the solubility of n-hexane in water at 25°C. The coordinates given in Table 1 for normal paraffins have been used: X = 15.0, Y = 20.0. The predicted solubility is 2×10^{-6} mole fraction: The experimental value is 1.98×10^{-6} as given by McAuliffe[3] and Price.[4]

Acknowledgment

Part of the revision to the *American Petroleum Institute's Technical Data Book—Petroleum Refining.* Funds were provided by the Refining Department of the American Petroleum Institute through the Subcommittee on Technical Data.

Literature Cited

1. Fredenslund, A., Jones, R. L., and Prausnitz, J. M., *AIChE Journal,* 21, 1086, 1975.
2. Leinonen, P. J., Mackay, D., Phillips, C. R., Canadian J. Chem. Engr., 19, 288, 1971.
3. McAuliffe, C., J. Phys. Chem., 70, 1267, 1966.

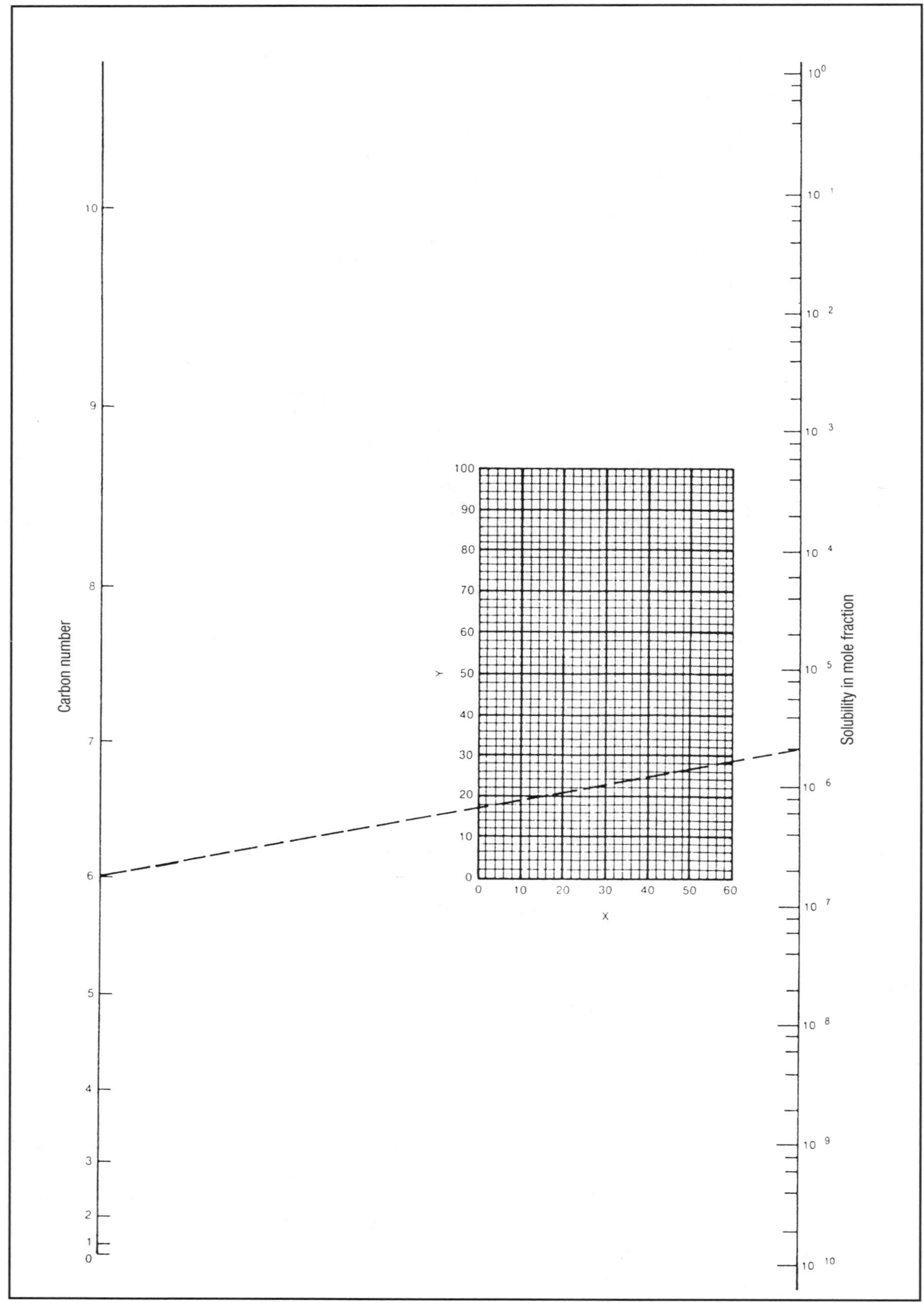

Figure 2. Nomograph for the solubility of liquid hydrocarbons in water at 25°C.

4. Price, L. C., Amer. Assoc. Petrol. Geol. Bull., 60, 213, 1976.
5. Fredenslund, A., Gmehling, J., and Rasmussen, P., Vapor-Liquid Equilibria using UNIFAC, Elsevier Scientific Publishing Co., 1977.

Methanol Injection

Methanol is frequently used to inhibit hydrate formation in natural gas so we have included information on the effects of methanol on liquid phase equilibria. Shariat, Moshfeghian, and Erbar[4] have used a relatively new equation of state and extensive calculations to produce interesting results on the effect of methanol. Their starting assumptions are the gas composition in Table 2, the pipeline pressure/temperature profile in Table 3 and methanol concentrations sufficient to produce a 24°F hydrate-formation-temperature depression. Resulting phase concentrations are shown in Tables 4, 5, 6. Methanol effects on CO_2 and hydrocarbon solubility in liquid water are shown in Figures 3 and 4.

Here are some conclusions of the study:

1. Water or water plus methanol has negligible effect on predicted liquid-hydrocarbon knockout.
2. Methanol causes substantial enhancement of CO_2 solubility in the water phase (Figure 3). Watch for increased corrosion.
3. Methanol concentration has essentially no effect on predicted water content of the liquid-hydrocarbon phase. The water content (not shown in the tables) was about 0.02 mol%.

Table 2
Gas Composition

Component	Mol %
N_2	2.85
C_1	79.33
CO_2	5.46
C_2	4.19
C_3	2.00
iC_4	0.44
nC_4	0.78
iC_5	0.40
nC_5	0.34
Lt. arom.	0.34
C_6^+	3.87
Total	100.00

Table 3
Pipeline Pressure Temperature Profile*

P, psia	T, °F.
†2,000	†150
1,800	100
1,600	80
1,400	70
1,200	60
‡1,000	‡60

**Primary separator operates at 150°F. and 2,000 psi. †Inlet from separator. ‡Discharge from pipeline.*

Table 4
Results Based on Dry Hydrocarbon Flash

P/T, psia/°F.	Hydro-carbons, bbl/MMscf	H_2O, lb/MMscf Liquid	Vapor	Methanol Water rich liq., wt %	Methanol Vapor phase, lb/MMscf[2]
2,000/150	0	—	131.0	—	86.1
1,800/100	3.36	95.0	36.0	—	—
1,600/80	4.57	107.0	24.0	—	—
1,400/70	5.19	114.5	16.5	—	—
1,200/60	5.75	118.8	12.2	—	—
1,000/60	5.59	117.8	13.2	24.75	47.03

Table 5
Results Based on Wet Hydrocarbon Flash

P/T, psia/°F.	Hydro-carbons, bbl/MMscf	H_2O, lb/MMscf Liquid	Vapor	Methanol Water rich liq., wt %	Methanol Vapor phase, lb/MMscf[2]
2,000/150	0	0	137.8	—	86.1
1,800/100	3.36	94.5	43.3	—	—
1,600/80	4.56	111.4	26.4	—	—
1,400/70	5.18	116.9	20.9	—	—
1,200/60	5.72	121.3	16.5	—	—
1,000/60	5.58	119.7	18.1	24.75	47.03

Table 6
Results for Wet Methanol-Hydrocarbon Flashes

P/T, psia/°F.	Hydro-carbons, bbl/MMscf	H_2O, lb/MMscf Liquid	Vapor	Methanol Water rich liq., wt %	Methanol Vapor phase, lb/MMscf[2]
2,000/150	0	0	137.8	—	85.15
1,800/100	3.36	98.8	39.0	16.22	64.43
1,600/80	4.69	114.9	22.9	20.40	52.07
1,400/70	5.33	120.1	17.7	22.68	45.24
1,200/60	5.90	124.1	13.7	24.78	38.20
1,000/60	5.74	122.8	15.0	24.77	38.56

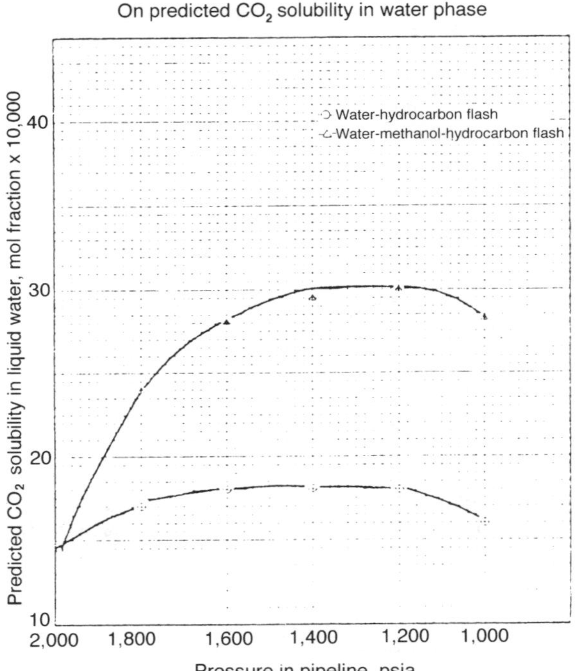

Figure 3. Effects of methanol on predicted CO_2 solubility in water phase.

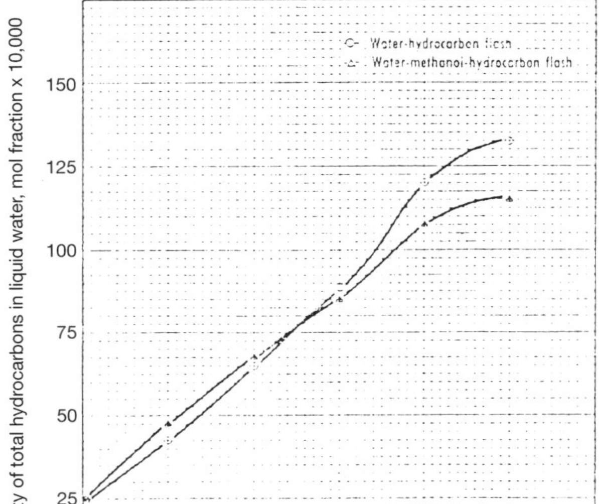

Figure 4. Effects of methanol on predicted hydrocarbon solubility in liquid water.

4. For this problem, methanol had no practical effect on the solubility of hydrocarbons in the water phase (Figure 4).
5. Methanol causes a 10–15% reduction in the predicted water content of the pipeline gas.

Sources

1. Erbar, J. H. and Maddox, R. N., "Water-Hydrocarbon System Behavior," *Oil and Gas Journal,* March 16, 1981, p. 75.
2. *GPSA Engineering Data Book,* Gas Processors Suppliers Association, 10th Ed., part 20—Dehydration.
3. Kabadi, V. N. and Danner, R. P., "Nomograph Solves For Solubilities of Hydrocarbons in Water," *Hydrocarbon Processing,* May 1979, p. 245.
4. Shariat, A., Moshfeghian, M., and Erbar, J. H., "Predicting Water Knockout In Gas Lines," *Oil and Gas Journal,* November 19, 1979, p. 126.

Natural Gas Hydrate Temperature

This graphical method was developed for field operations personnel. It is designed for sour natural gas at 100 to 4000 psia up to 50% H_2S or sweet gas with C_3 up to 10%.

Example. Calculate the hydrate temperature for the following conditions:

Pressure = 500 psia
H_2S = 10%
Gravity = 0.70 (rel. to air)
Propane = 0.5%

Enter at the left at 500 psia, go to the right to 10% H_2S, and drop down to 0.7 gravity. This hits just below the sloped line near 65°F. Parallel this line down to about 65.5°F.

Next, correct for C_3 content. Enter the small adjustment graph at the left at 10% H_2S, go to the right to 0.5% C_3, and

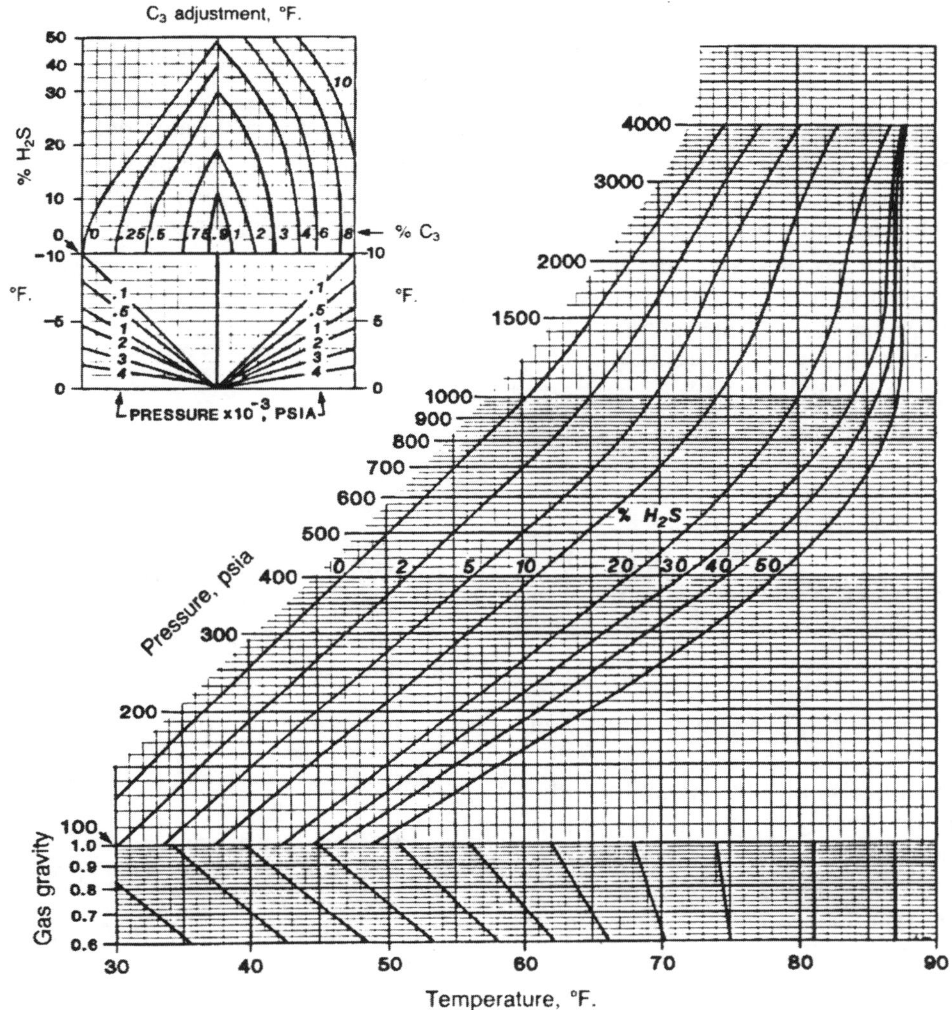

Figure 1. Imperial unit hydrate chart for sour natural gas.

drop down to 500 psia (0.5 line). Move left to a correction of −3.5°F.

The corrected hydrate temperature is 65.5 − 3.5 = 62°F.

Source

Baillie, C. and Wichert, E., "Chart Gives Hydrate Formation Temperature For Natural Gas," *Oil and Gas Journal,* April 6, 1987, p. 37.

Inorganic Gases in Petroleum

The Solubility of Inorganic Gases In Petroleum Liquids Can Be Estimated by This Nomograph

The equilibrium solubility of common inorganic gases in petroleum liquids can now be estimated by nomograph. The relationship is based on an earlier correlation[1] established by the U.S. National Aeronautics and Space Administration and the subject of a standard method[2] approved by the American Society of Testing and Materials.

The method applies to petroleum liquids having densities from 0.63 to 0.90 grams per milliliter at 15.5°C (approximately 93 to 25°API at 60°F). The results are given in Figure 1 as parts per million by weight for the specified partial pressure of the gas and for specified temperatures in the range from –45° to 150°C (–50° to 302°F).

Example. A gas oil having a density of 0.84 grams per milliliter (37°API) is stored at 60°C (140°F) under a gas blanket of carbon dioxide (CO_2) at a pressure of 1 atmosphere absolute. To estimate the amount of carbon dioxide dissolved in the gas oil at equilibrium, take the following steps:

Step 1. Find the intersection of 0.84 liquid density and 60°C on Figure 1.

Step 2. With a straight line, connect the point for carbon dioxide on the primary scale with the density-temperature intersection and extend the line to intersect the reference line 1.

Step 3. Extend a straight line from the point on the reference line 1 through the point on the secondary scale for carbon dioxide and intersect the reference line 2.

Step 4. A straight line between the point on the reference line 2 and the point on the pressure line for 1 atmosphere absolute will cross the line for gas dissolved at 1,500 parts per million by weight.

Nomograph defined. This method assumes the application of the Clausius-Clapeyron equation, Henry's law, and the perfect gas law. The method uses the Ostwald coefficient to define the solubility of a gas expressed as volume of gas dissolved per volume of liquid at equilibrium and 0°C (32°F). The reference liquid density is 0.85 grams per milliliter. These coefficients are given in Table 1.

For temperatures other than 0°C, the equation is as follows:

$$V = 0.300\, e^{0.639\,[(700 - T)/T]\,\ln(3.333\, V_o)} \qquad (1)$$

where

V = Ostwald coeff. for 0.85 liq. and temp. T
V_o = Ostwald coeff. for 0.85 liq. and 0°C (273° K)
T = Specified temp., °K
e = Natural base (2.7183)

For densities of petroleum liquid other than 0.85, the equation is as follows:

$$V_c = 7.70\, V\, (0.980 - d) \qquad (2)$$

where

V_c = Ostwald coeff. corrected for T and d
d = Liquid density at 15.5°C, g/ml.

For pressure effects, the Bunsen coefficient is used and is calculated from the equation as follows:

$$B = 273\, pV_c/T \qquad (3)$$

where

B = Bunsen coeff.
p = Gas pressure, atm. abs.

Then solubility of a gas in the liquid on a weight basis is given by the equation as follows:

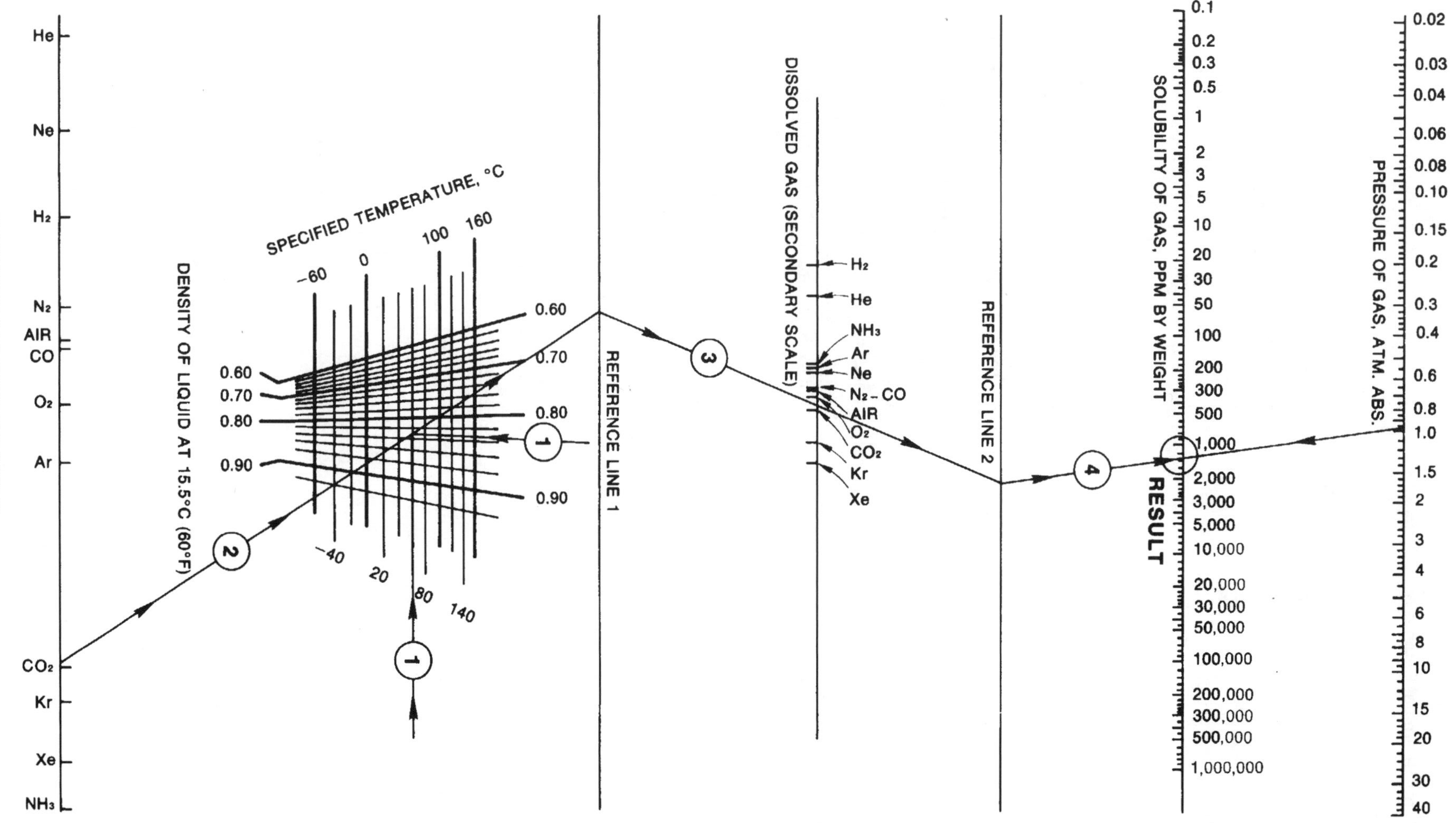

Figure 1. Physical properties of the gases are not needed to find their solubility in petroleum liquids with this nomograph.

Table 1
Ostwald Coefficient V_o for O °C

He	Ne	H_2	N_2	Air	CO	O_2	Ar	CO_2	Kr	Xe	NH_3
0.010	0.021	0.039	0.075	0.095	0.10	0.15	0.23	1.0	1.3	2.0	2.8

$$G = \left(\frac{BM_g}{0.0224}\right)\left\{d\left[1 - 0.000595\left(\frac{T - 288.6}{d^{1.21}}\right)\right]\right\}^{-1} \quad (4)$$

where

G = Gas solubility in liquid, ppm by wt.
M_g = Gas molecular wt.

Vapor pressure of the petroleum liquid is assumed to be less than 10 percent of the total pressure. For higher vapor pressure, the correction is as follows:

$$G' = [(P - p_v)/P]G \quad (5)$$

where

P = Total system pressure
p_v = Vapor pressure of liquid.

The nomograph combines all of these corrections except the one for higher liquid vapor pressures. Physical properties of each of the inorganic gases are incorporated in the primary and secondary scales identifying the dissolved gas.

Literature Cited

1. NACA Technical Note 3276, National Aeronautics and Space Administration (formerly National Advisory Committee on Aeronautics), 1956.
2. ASTM Standard D 2779-69 (reapproved 1974), 1974 Annual Book of ASTM Standards, Part 24, pp. 679–682.

Source

Zanker, Adam, "Inorganic Gases in Petroleum," *Hydrocarbon Processing,* May 1977, p. 255.

Foam Density

Zanker[1] gives this method for distillation and absorption column trays, which is stated to agree well with published data:

$$d_F = d_L[1 - 0.46\log(f/0.0073)]$$

where

$f = u[d_v/(d_L - d_v)]^{0.5}$
d_F = Foam density, lbs/ft^3
d_L = Liquid density, lbs/ft^3
d_V = Vapor density, lbs/ft^3
u = Superficial vapor velocity, ft/sec.

Source

Zanker, Adam, "Quick Calculation for Foam Densities," *Chemical Engineering,* February 17, 1975.

Equivalent Diameter

Zanker's article[1], reprinted here, gives a way to estimate the elusive equivalent diameter of solids.

Estimating Equivalent Diameters of Solids

Particle diameter is a primary variable important to many chemical engineering calculations, including settling, slurry flow, fluidized beds, packed reactors, and packed distillation towers. Unfortunately, this dimension is usually difficult or impossible to measure, because the particles are small or irregular. Consequently, chemical engineers have become familiar with the notion of "equivalent diameter" of a particle, which is the diameter of a sphere that has a volume equal to that of the particle.

The equivalent diameter can be calculated from the dimensions of regular particles, such as cubes, pyramids, cylinders, cones, etc. For irregular particles, the equivalent diameter is most conveniently determined from such data as the fractional free volume and the specific surface or area of the particle in a bed of the particles, as determined experimentally from measurements that determine particle volume as a displaced fluid.

These two nomographs provide a convenient means of estimating the equivalent diameter of almost any type of particle: Figure 1 of regular particles from their dimensions, and Figure 2 of irregular particles from fractional free volume, specific surface, and shape.

Also, in cases where the dimensions of a regular particle vary throughout a bed of such particles or are not known, but where the fractional free volume and specific surface can be measured or calculated, the shape factor can be calculated and the equivalent diameter of the regular particle determined from Figure 2.

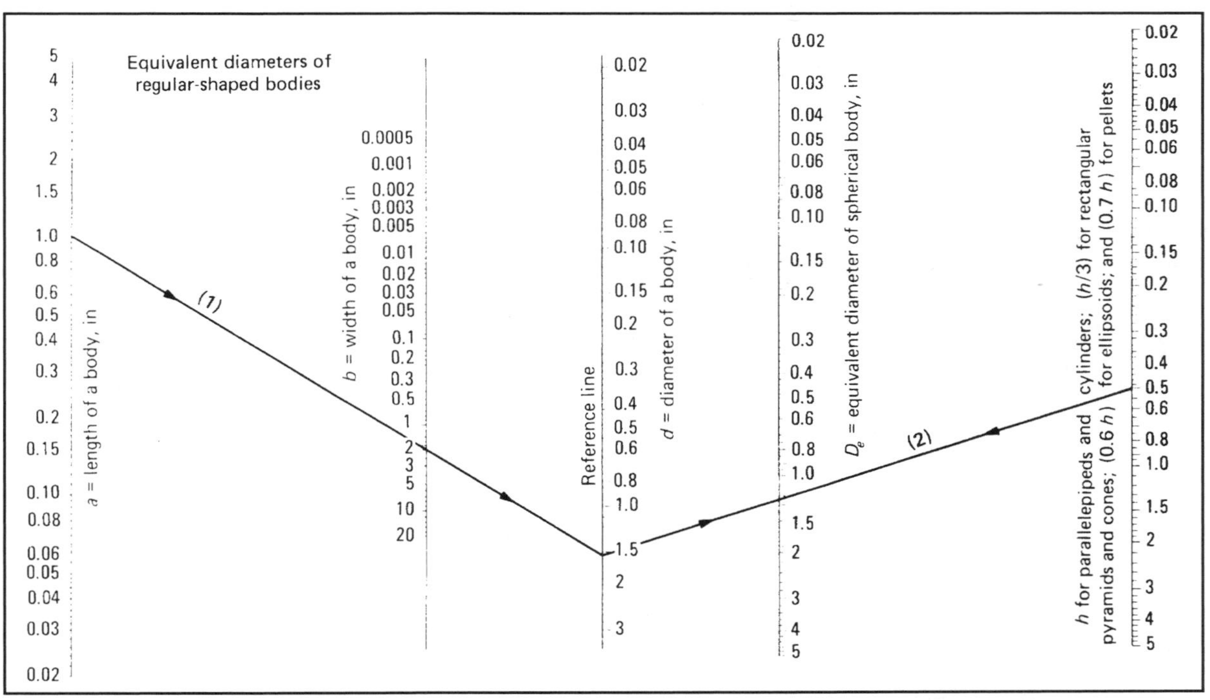

Figure 1. Equivalent diameters of regular-shaped bodies.

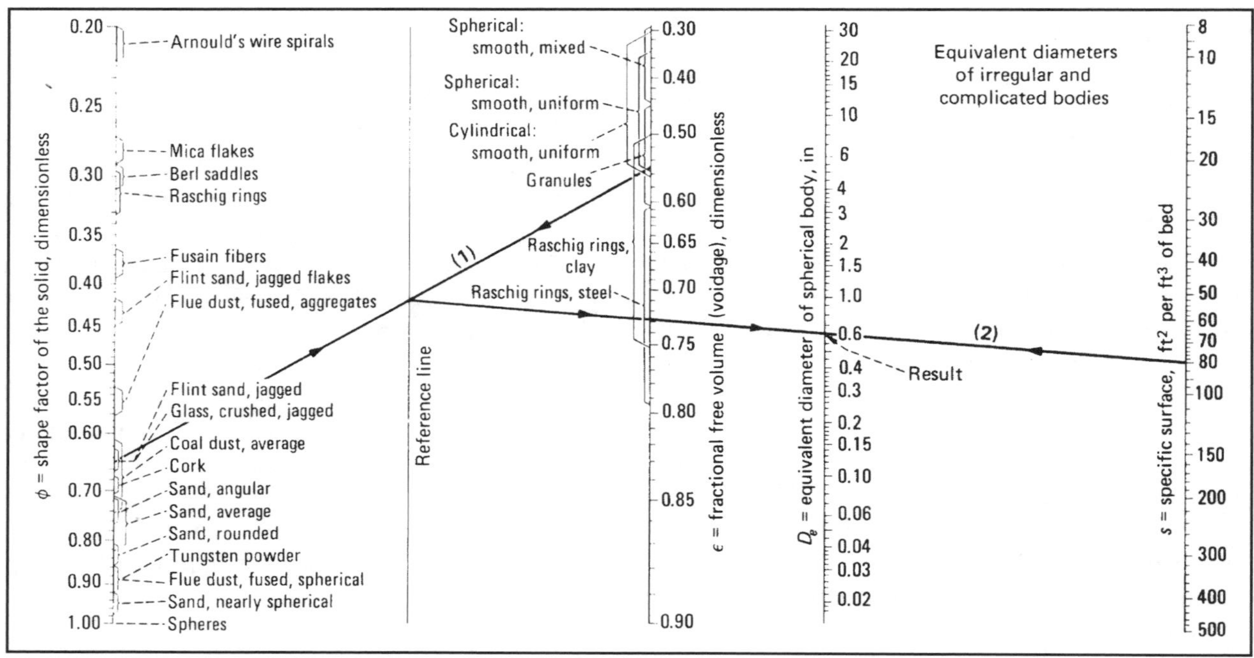

Figure 2. Equivalent diameters of irregular and complicated bodies.

In Figure 1, the equivalent diameter is related to regular shapes through equivalent volumes by the following formulas:

$$D_e = \sqrt[3]{6V/\pi} = \frac{6(1-\varepsilon)}{\phi s}$$

Shape	**Formula**
parallelepiped	$V = abh$
cube	$V = a^3$
rectangular pyramid	$V = (ab)h/3$
cylinder	$V = h(d^2\pi/4)$
cone	$V = (d^2\pi/4)(h/3)$
pellet	$V \cong 0.7h\,(d^2\pi/4)$
ellipsoid	$V \cong 0.6h(d^2\pi/4)$

where the terms, all in consistent units of length, are defined as:

V = volume
a = length of base edge
b = width of base measured perpendicular to a
h = height measured perpendicular to plane ab
d = diameter
ε = fractional free volume, dimensionless
s = specific surface = particle surface per unit volume of bed
φ = shape factor = area of sphere divided by the surface area of a particle of equal volume

Examples: What is the equivalent diameter of a rectangular pyramid with base sides 1 and 2 in, and a height of 1.5 in?

On Figure 1, align 1.0 (or 2.0) on the a scale with 2.0 (or 1.0) on the b scale and note the intersection of this line on the reference line. Align this intersection with h/3 = 0.5 and read $D_e = 1.24$ in.

What is the equivalent diameter of a pellet with a diameter of 0.5 in and a length of 0.5 in? On Figure 1, align 0.5 on the d scale with 0.35 (0.7h), and read $D_e = 0.51$.

What is the equivalent diameter of crushed glass ($\phi = 0.65$) with a fractional free volume of 0.55 and a specific surface of 80? On Figure 2, align $\phi = 0.65$ with $\varepsilon = 0.55$ and note the intersection on the reference line. Align this with s = 80, and read $D_e = 0.62$.

Source

Zanker, Adam, "Estimating Equivalent Diameters of Solids," *Chemical Engineering,* July 5, 1976, p. 101.

Autoignition Temperature

Hilado and Clark[1] report difficult-to-find autoignition temperatures while eliminating apparent discrepancies in literature values. Table 1 shows their compilation.

(text continued on page 370)

Table 1
Compilation of AIT Data

Material	Lowest Reported AIT in Glass			Lowest Reported AIT if Other Than Glass or Unspecified		
	°C.	°F.	Ref.	°C.	°F.	Ref.
1.1 Straight-Chain Paraffins						
Methane				537	999	17
Ethane	515	959	2	472	882	24
Propane	493	919	3	450	842	20
Butane	372	702	14	288	550	14
Pentane	258	496	5	243	470	20
Hexane	227	441	6	225	437	20
Heptane	213	415	6	204	399	14
Octane	206	403	6			
Nonane	205	401	6			
Decane	201	394	6			
Dodecane (dihexyl)	203	397	5			
Tetradecane	202	396	2	200	392	20
Hexadecane (cetane)	202	396	5			
Octadecane				227	441	8
Nonadecane				230	446	8
Eicosane				232	450	8
1.2 Single-Branched Paraffins						
2-Methylpropane (isobutane)	462	864	3	460	860	20
2-Methylbutane (isopentane)	420	788	3			
2-Methylpentane (isohexane)	264	507	6			
3-Methylpentane	278	532	6			
2-Methylhexane				280	536	20
3-Methylhexane				280	536	20
2-Methylheptane	420	788	5			
2-Methyloctane (isononane)	227	440	19	220	428	20
3-Methyloctane	228	442	19	220	428	20
4-Methyloctane	232	450	19	225	437	20
2-Methylnonane	214	418	19	210	410	20
2-Methyldecane				225	437	8
2-Methylundecane				500	932	12
2-Ethyloctane				231	448	8
3-Ethyloctane				230	446	20
4-Ethyloctane	237	458	19	229	445	20

**Union Carbide Corp. trademark.*

Table 1 (continued)
Compilation of AIT Data

Material	Lowest Reported AIT in Glass			Lowest Reported AIT if Other Than Glass or Unspecified		
	°C.	°F.	Ref.	°C.	°F.	Ref.
1.3 Two-Branched Paraffins						
2,2-Dimethylpropane (neopentane)	456	853	19	450	842	20
2,2-Dimethylbutane (neohexane)	405	761	6			
2,3-Dimethylbutane	396	745	6			
2,3-Dimethylpentane	338	640	19	335	635	20
2,3-Dimethylhexane				438	820	12
3,3-Dimethylheptane	330	626	19	325	617	20
2,3-Dimethyloctane	231	447	19	225	437	20
4,5-Dimethyloctane				290	554	8
2-Methyl-3-ethylpentane				460	860	20
2-Methyl-4-ethylhexane				280	536	20
3,3-Diethylpentane	290	554	2			
4-Isopropylheptane				255	491	20
1.4 Three-Branched Paraffins						
2,2,3-Trimethylbutane	412	774	6			
2,2,4-Trimethylbutane	407	765	6			
2,2,3-Trimethylpentane	396	745	6			
2,2,4-Trimethylpentane	418	784	2	415	779	20
2,3,3-Trimethylpentane	430	806	19	425	797	20
2,5,5-Trimethylheptane				275	527	8
2,4-Dimethyl-3-ethylpentane	390	734	19			
1.5 Four-Branched Paraffins						
2,2,3,3-Tetramethylpentane	430	806	2			
2,3,3,4-Tetramethylpentane	437	818	19			
2.1 Straight-Chain Olefins						
Ethene (ethylene)				490	914	20
Propene (propylene)				455	851	13
1-Butene (butylene)	455	851	22	385	725	20
2-Butene-*cis*				325	617	20
2-Butene-*trans*				325	617	20
1-Pentene (amylene)	298	569	19	275	527	20
1-Hexene	253	487	6			
2-Hexene	245	473	6			
1-Heptene	263	505	19			
1-Octene	230	446	6			
1-Decene	244	471	19	235	455	20
1-Dodecene (dodecylene)				255	491	8
1-Tetradecene	239	463	19	235	455	20
1-Hexadecene	240	464	19			
1-Octadecene (octadecylene)				250	482	20
2.2 Single-Branched Olefins						
2-Methyl-1-propene (isobutylene)	465	869	2			
3-Methyl-1-butene	374	706	19	365	689	20
2-Methyl-1-pentene	306	582	19	300	572	20
4-Methyl-1-pentene	304	580	19	300	572	20
2-Ethyl-1-butene	324	615	19	315	599	20
2.3 Two-Branched Olefins						
2,3-Dimethyl-1-butene	369	697	19	360	680	20
2,3-Dimethyl-2-butene	407	764	19	401	753	20
2.4 Three-Branched Olefins						
2,3,3-Trimethyl-1-butene	383	721	19			
2,2,4-Trimethyl-1-pentene	377	711	6			
2,4,4-Trimethyl-1-pentene	420	788	19	415	799	20
2,3,4-Trimethyl-1-pentene	257	495	19			
2,4,4-Trimethyl-2-pentene	308	587	19	305	581	20
3,4,4-Trimethyl-2-pentene	330	626	19	325	617	20
2.5 Dienes						
Dicyclopentadiene	505	941	22			
1,3-Butadiene (erythrene)				420	788	20
Dipentene				237	458	20
2-Methyl-1,3-butadiene (isoprene)				220	428	20
3.1 Single-Ring Aromatics						
Benzene	498	928	22			
Toluene	482	900	5	480	896	20
Ethylbenzene	432	810	2			
Propylbenzene	456	853	19			
Isopropylbenzene (cumene)	424	795	2			
Butylbenzene	412	774	2	410	770	20
Isobutylbenzene	428	802	2			
sec-Butylbenzene	418	784	2			
tert-Butylbenzene	450	842	2			
1,2-Dimethylbenzene (o-xylene)	464	867	2			
1,3-Dimethylbenzene (*m*-xylene)	528	982	2			
1,4-Dimethylbenzene (*p*-xylene)	529	984	2			
1,2,3-Trimethylbenzene (hemellitol)	479	895	19			
1,2,4-Trimethylbenzene	521	970	19			
1,3,5-Trimethylbenzene (mesitylene)	559	1,039	19			
1-Methyl-2-ethylbenzene	448	836	19	440	824	20
1-Methyl-3-ethylbenzene	485	905	19	480	896	20
1-Methyl-4-ethylbenzene	483	902	19	475	887	20
1,2-Diethylbenzene	404	759	19	395	743	20
1,3-Diethylbenzene	455	851	19	450	842	20
1,4-Diethylbenzene	430	806	2			
1-Methyl-3,5-diethylbenzene	461	861	19			
1-Methyl-4-isopropylbenzene (*p*-cymene)				436	817	20
Diisopropylbenzene				449	840	20
Styrene (vinylbenzene) (cinnamene)				490	914	20
α-Methylstyrene				574	1,066	20
3.2 Two-Ring Aromatics						
Biphenyl	566	1,051	2	540	1,004	20
2-Methylbiphenyl	502	936	19	495	923	20
2-Ethylbiphenyl	449	840	19			

(table continued)

Table 1 (continued)
Compilation of AIT Data

Material	Lowest Reported AIT in Glass °C.	°F.	Ref.	Lowest Reported AIT if Other Than Glass or Unspecified °C.	°F.	Ref.
2-Propylbiphenyl	452	845	19	445	833	20
2-Isopropylbiphenyl				435	815	20
2-Butylbiphenyl	433	811	19	430	806	20
Diphenylmethane	486	907	2	485	905	20
1,1-Diphenylethane,						
symmetrical	487	909	19	480	896	20
unsymmetrical				440	824	20
1,1-Diphenylpropane	466	870	19	460	860	20
1,1-Diphenylbutane	462	863	19	455	851	20
Naphthalene	526	979	2			
1-Methylnaphthalene	529	984	2			
1-Ethylnaphthalene	481	898	19	480	896	20
Tetrahydronaphthalene (tetralin)	423	794	19	384	723	21
Decahydronaphthalene (decalin)	288	521	19	250	482	20
trans-Decahydronaphthalene				255	491	20
Dimethyl decalin				235	455	20
3.3 Three-Ring Aromatics						
Anthracene	540	1,004	2			
4.1 Single-Ring Cycloparaffins						
Cyclopropane	498	928	3			
Ethylcyclobutane	212	414	2	210	410	20
Cyclopentane	361	682	6			
Methylcyclopentane	258	496	6			
Ethylcyclopentane	262	504	2	260	500	20
Propylcyclopentane	269	516	15			
Butylcyclopentane	250	482	15			
Hexylcyclopentane	228	442	15			
Cyclohexane	246	475	5	245	473	20
Methylcyclohexane	258	496	6			
Ethylcyclohexane	238	460	6			
Propylcyclohexane	248	478	15			
Isopropylcyclohexane	283	541	15			
Butylcyclohexane	246	475	15			
Isobutylcyclohexane	274	525	15			
sec-Butylcyclohexane	277	531	15			
tert-Butylcyclohexane	342	648	15			
Amylcyclohexane	239	462	15			
Octylcyclohexane	236	457	15			
1,2-Dimethylcyclohexane	304	579	15			
1,3-Dimethylcyclohexane	306	583	15			
1,4-Dimethylcyclohexane	304	579	15			
1,2,4-Trimethylcyclohexane	315	599	15			
1,3,5-Trimethylcyclohexane	314	597	15			
Diethylcyclohexane				240	464	20
Diisopropylcyclohexane	275	527	15			
1-Methyl-4-isopropylcyclohexane	306	583	15			
Hexamethylcyclohexane	349	660	15			
Cyclodecane	235	455	15			
4.2 Two-Ring Cycloparaffins						
Dicyclopentyl	252	486	15			
Cyclohexylcyclopentane	255	491	15			
Dicyclohexyl	256	493	15			
Dicyclohexylmethane	244	471	15			
1,1-Dicyclohexylethane	270	518	15			
1,2-Dicyclohexylethane	237	459	15			
Hydrindane	296	565	15			
Decahydronaphthalene (decalin)	268	514	15	250	482	20
Methyldecalin	264	507	15			
Dimethyldecalin	257	495	15	235	455	20
Bicyclo (2,2,1) heptane	314	597	15			
Pinane	273	523	15			
4.3 Three-Ring Cycloparaffins						
1-Cyclohexyldecalin	263	505	15			
1,2-Cyclopentanodecalin	261	502	15			
Perhydroacenaphthene	275	527	15			
Perhydrofluorene	266	511	15			
1,2,3,4-Dicyclopentanocyclohexane	272	522	15			
Perhydroanthracene	252	486	15			
Perhydrophenanthrene	246	475	15			
Perhydrophenalene	262	504	15			
endo-Tetrahydrodicyclopentadiene	273	523	15			
Adamantan	287	549	15			
1,3-Dimethyladamantan	288	550	15			
1-Ethyladamantan	271	520	15			
4.4 Four-Ring and Five-Ring Cycloparaffins						
1,4,5,8-Diendomethylenedecalin	284	543	15			
1,4,5,8-Dimethylene-2,3-trimethylenedecahydronaphthalene	232	450	15			
4.5 Cycloolefins						
Cyclopentene	395	743	15			
Cyclohexene	244	471	6			
1,2,3,4-Dicyclopentanocyclohexene	249	480	15			
Bicyclo (2,2,1) heptene-2	505	941	15			
Dihydrodicyclopentadiene	495	923	15			
Dicyclopentadiene	503	937	15			
1,4,5,8-Diendomethyleneoctahydronaphthalene	508	946	15			
trans,trans,cis-1,5,9-Cyclodecatriene	243	469	15			
5.1 Straight-Chain Alcohols						
Methanol (methyl alcohol)	386	727	5	385	725	20
Ethanol (ethyl alcohol)	363	685	5			
1-Propanol (propyl alcohol)	432	810	16			
2-Propanol (isopropyl alcohol)				399	750	20
1-Butanol (butyl alcohol)	359	678	19	343	650	21
2-Butanol	414	777	3	405	761	20
tert-Butanol	478	892	3			
1-Pentanol (amyl alcohol)	300	572	2			
2-Pentanol				343	650	20

Table 1 (continued)
Compilation of AIT Data

Material	Lowest Reported AIT in Glass			Lowest Reported AIT if Other Than Glass or Unspecified		
	°C.	°F.	Ref.	°C.	°F.	Ref.
5.1 Straight-Chain Alcohols (continued)						
3-Pentanol	437	819	2	435	815	20
Decanol (decyl alcohol)				288	550	8
1-Dodecanol (lauryl alcohol)				275	527	20
5.2 Branched-Chain Alcohols						
2-Methyl-1-propanol				427	800	20
2-Methyl-2-propanol				480	896	20
2-Methyl-1-butanol	385	725	22			
3-Methyl-1-butanol (fusel oil)				350	662	20
1-Butoxyethoxy-2-propanol	265	509	22			
5.3 Cyclic Alcohols						
Cyclohexanol (hexalin)	300	572	2			
o-Methylcyclohexanol	296	565	2			
p-Methylcyclohexanol	295	563	2			
5.4 Aromatic Alcohols						
Phenol				715	1,319	20
Benzyl alcohol	436	817	16			
o-Cresol (cresylic acid)				599	1,110	20
m-Cresol				599	1,038	20
p-Cresol				599	1,038	20
5.5 Dihydroxy Alcohols						
Tetramethylene glycol				390	734	20
Ethylene glycol	398	748	5			
Diethylene glycol				229	444	17
Triethylene glycol				371	700	20
Propylene glycol	421	790	3	371	700	20
Octylene glycol				335	635	20
1,3-Butanediol				395	743	20
2,3-Butanediol				402	756	20
1,5-Pentanediol				335	635	20
2-Ethyl-1,3-hexanediol	360	680	22			
Thiodiglycol	298	568	22			
Butoxytriglycol	317	603	22			
p-Dihydroxybenzene (hydroquinone)				515	959	20
5.6 Trihydroxy Alcohols						
Glycerol (glycerin)	371	700	5	370	698	20
5.7 Unsaturated Alcohols						
Allyl alcohol				378	713	20
6.1 Saturated Acids						
Formic acid	540	1,004	22			
Acetic acid	464	867	5			
Propionic acid	475	887	22			
Butyric acid	452	846	2	450	842	20
Isobutyric acid	490	914	22			
Pentanoic acid (valeric acid)	400	752	22			
Isopentanoic acid	416	781	22			
Hexanoic acid (caproic acid)	402	756	22	380	716	20
2-Methylpentanoic acid	378	712	22			
2-Ethylhexanoic acid	377	711	22			
Adipic acid	422	792	2	420	788	20
Isooctanoic acid	392	738	22			
2-Ethylbutyric acid	400	752	22			

Material	Lowest Reported AIT in Glass			Lowest Reported AIT if Other Than Glass or Unspecified		
	°C.	°F.	Ref.	°C.	°F.	Ref.
Stearic acid				395	743	20
6.2 Unsaturated Acids						
Acrylic acid	438	820	22			
Oleic acid				363	685	20
6.3 Aromatic Acids						
Benzoic acid				570	1,058	20
7.1 Formate Esters						
Methyl formate	467	873	5	449	840	17
Ethyl formate	455	851	2			
Propyl formate				455	851	20
Isopropyl formate				485	905	20
Butyl formate	322	612	3			
Isobutyl formate				320	608	20
7.2 Acetate Esters						
Methyl acetate	502	935	16	454	850	21
Ethyl acetate	484	904	16	427	800	20
Propyl acetate	450	842	3			
Isopropyl acetate	460	860	16			
Butyl acetate	422	792	16	421	790	17
Isobutyl acetate	423	793	2	421	790	20
Amyl acetate				360	680	20
Isoamyl acetate (banana oil)	379	714	17	360	680	20
Vinyl acetate				427	800	17
Methyl vinyl acetate	431	808	22			
Benzyl acetate	460	860	16			
Cyclohexyl acetate				335	635	20
7.3 Propionate Esters						
Methyl propionate				469	876	20
Ethyl propionate	476	889	3	440	824	20
Butyl propionate	426	799	16			
Amyl propionate	378	712	2			
7.4 Butyrate and Higher Esters						
Ethyl butyrate	463	866	3	463	865	20
Butyl stearate	355	671	2			
Ethyl lactate				400	752	20
Butyl lactate				382	720	20
7.5 Unsaturated Esters						
Methyl acrylate	468	875	22			
Ethyl acrylate				383	721	21
Isobutyl acrylate	427	800	22			
2-Ethylhexyl acrylate				252	485	21
Isodecyl acrylate	282	540	22			
Glycidyl acrylate	415	779	22			
7.6 Aromatic Esters						
Ethyl benzoate				490	914	20
Benzyl benzoate	481	898	2	480	896	20
Methyl salicylate (oil of wintergreen)				454	850	20
7.7 Diesters						
Dimethyl phthalate				556	1,032	20
Dibutyl phthalate				403	757	20
Diethyl maleate	350	662	22			

(table continued)

Table 1 (continued)
Compilation of AIT Data

Material	Lowest Reported AIT in Glass			Lowest Reported AIT if Other Than Glass or Unspecified		
	°C.	°F.	Ref.	°C.	°F.	Ref.
7.8 Triesters						
Glyceryl triacetate (triacetin)				433	812	20
8.1 Aliphatic Oxides						
Ethylene oxide (oxirane)	457	855	22	429	804	17
Propylene oxide	464	867	22			
Butylene oxide	439	822	22			
Diethylene oxide (tetrahydrofuran)				321	610	20
Diethylene dioxide (*p*-dioxane)				180	356	20
Mesityl oxide				344	652	20
8.2 Aromatic Oxides						
Styrene oxide	538	1,000	22	498	929	21
Diphenyl oxide	646	1,195	19	620	1,148	20
9.1 Saturated Ethers						
Dimethyl ether	350	662	3			
Diethyl ether	160	320	5			
Dipropyl ether	215	419	22			
Diisopropyl ether				443	830	17
Dibutyl ether	198	388	22	194	382	20
Diamyl ether	171	340	2	170	338	20
Dihexyl ether	187	369	2			
Dioctyl ether				205	401	20
Didecyl ether				215	419	20
Methyl ethyl ether				190	374	17
1,2-Diethoxyethane				205	401	20
9.2 Unsaturated Ethers						
Divinyl ether				360	680	17
Vinyl methyl ether	287	549	22			
Vinyl isopropyl ether				272	522	20
Vinyl butyl ether	255	437	22			
9.4 Aromatic Ethers						
Methyl phenyl ether (anisole)				475	887	20
Diphenyl ether	618	1,144	2			
10.1 Saturated Aldehydes						
Formaldehyde	424	795	2			
Acetaldehyde (ethanal)	204	399	22	175	347	20
Propionaldehyde	227	441	22	207	405	20
Butyraldehyde (butanal)	230	446	3			
Isobutyraldehyde	261	502	22	254	490	20
Valeraldehyde				222	432	21
2-Ethylhexanal				197	387	20
3-Hydroxybutanal				250	482	20
10.2 Unsaturated Aldehydes						
Acrolein (acrylic aldehyde)	300	572	22	235	455	20
Crotonaldehyde (2-butenal)	280	536	22	232	450	13
2-Ethylcrotonaldehyde	250	482	22			
2-Furaldehyde (furfural)				316	600	20
10.3 Aromatic Aldehydes						
Benzaldehyde (almond oil)				192	377	20
11.1 Straight-Chain Ketones						
Propanone (acetone)	467	873	5	465	869	20
Butanone (methyl ethyl ketone)	404	759	22			
2-Pentanone (methyl propyl ketone)	452	846	22			
3-Pentanone (diethyl ketone)	454	846	2	450	842	20
2-Hexanone (methyl butyl ketone)				533	992	13
2-Heptanone (methyl amyl ketone)				533	991	20
2,4-Pentanedione (acetylacetone)	340	644	2			
2,5-Hexanedione (acetonylacetone)				493	920	20
11.2 Cyclic Ketones						
Cyclohexanone (pimelic ketone)	420	788	2			
11.3 Aromatic Ketones						
Methyl phenyl ketone (acetophenone)	571	1,060	2	570	1,058	20
11.4 Branched-Chain Ketones						
4-Methyl-2-pentanone (MIBK)				457	854	21
Diisobuyl ketone (isovalerone)				418	785	21
Isobutyl heptyl ketone	410	770	22			
12.1 Amines						
Methylamine				430	806	20
Dimethylamine				400	752	20
Ethylamine				384	725	20
Diethylamine				312	594	20
Propylamine				318	604	20
Dipropylamine	299	570	22			
Isopropylamine				402	756	20
Diisopropylamine	316	600	22			
Butylamine				312	594	20
tert-Butylamine				380	716	20
Allyl amine				374	705	20
Cyclohexylamine				293	560	20
12.2 Diamines						
Propylenediamine	416	780	22			
12.3 Triamines						
Diethylenetriamine				399	750	20
12.4 Tetramines						
Triethylenetetramine				338	640	20
12.5 Pentamines						
Tetraethylenepentamine	321	610	22			
12.6 Alkanolamines						
Monoethanolamine	410	770	22			
Diethanolamine				662	1,224	20
Monoisopropanolamine	374	705	22			
Diisopropanolamine	374	705	22			
Triisopropanolamine	320	608	22			
Dimethylethanolamine	295	563	22			
Diethylethanolamine	320	608	22			
2-Aminoethylethanolamine				368	695	20
n-Acetylethanolamine				460	860	20

Table 1 (continued)
Compilation of AIT Data

Material	Lowest Reported AIT in Glass °C.	°F.	Ref.	Lowest Reported AIT if Other Than Glass or Unspecified °C.	°F.	Ref.
12.7 Aromatic Amines						
Aniline (phenylamine)				615	1,139	20
o-Toluidine				482	900	20
p-Toluidine				482	900	20
Phenyldimethylamine				371	700	20
Phenyldiethylamine				630	1,166	20
Diphenylamine	634	1,173	2			
2-Biphenylamine	452	846	2			
13.1 Cellosolve* Glycol Ethers						
Methyl Cellosolve (2-methoxy ethanol)				285	545	20
Ethyl Cellosolve (2-ethoxy ethanol)				235	455	20
Butyl Cellosolve (2-butoxy ethanol)				244	472	17
Hexyl Cellosolve (2-hexoxy ethanol)	280	536	22			
13.2 Carbitol* Glycol Ethers						
Butyl Carbitol (butoxy diethylene glycol)				228	442	17
Diethyl Carbitol (diethoxy diethylene glycol)	205	401	22			
Dibutyl Carbitol (dibutoxy diethylene glycol)	310	590	22			
14.1 Aliphatic Anhydrides						
Acetic anhydride				316	600	18
Propionic anhydride	335	635	22			
Butyric anhydride				310	590	23
Isobutyric anhydride				352	665	20
14.2 Aromatic Anhydrides						
Phthalic anhydride				570	1,058	20
20.1 Chlorides						
Methyl chloride				632	1,170	20
Ethyl chloride				519	966	20
Propyl chloride				520	968	20
Butyl chloride	250	482	14	240	464	14
1-Chloropentane				260	500	20
3-Chloropentane				345	653	20
Allyl chloride (3-chloropropene)				485	905	20
Acetyl chloride	390	734	2			
Chlorobenzene	638	1,180	2			
Benzyl chloride	585	1,085	2			
2-Chloroethanol (ethylene chlorohydrin)				425	797	20
20.2 Dichlorides						
Dichloromethane				662	1,224	17
1,2-Dichloroethane	438	820	14	413	775	17
1,2-Dichloropropane				557	1,035	17
1,2-Dichlorobutane				275	527	20
Dichloroethylene				460	860	20
o-Dichlorobenzene				648	1,198	20
20.3 Trichlorides and Higher Chlorides						
1,1,1-Trichloroethane	486	907	14	458	856	14
Hexachlorobutadiene	618	1,144	19	610	1,130	20
Hexachlorodiphenyloxide	628	1,163	19	620	1,148	20
20.4 Bromides						
Methyl bromide				537	999	20
Ethyl bromide				511	952	17
Propyl bromide				490	914	20
Butyl bromide	265	509	2			
Allyl bromide (3-bromopropene)				295	563	20
Bromobenzene	566	1,051	2	565	1,049	20
1,1,2,2-Tetrabromoethane				335	635	20
20.5 Fluorides						
Perfluorodimethylcyclohexane	651	1,204	19			
Trichloroethylene	420	788	14	416	781	14

**Union Carbide Corp. trademark.*

(text continued from page 365)

Source

Hilado, C. J. and Clark, S. W., "Autoignition Temperatures of Organic Chemicals," *Chemical Engineering,* September 4, 1972, p. 75.

Gibbs Free Energy of Formation

Find Favorable Reactions Faster

A recent article[1] reported equations to help calculate the heat of reaction for proposed organic chemical reactions. In that article, enthalpy equations were given for 700 major organic compounds.

Now, equations are given to identify whether the reactions are thermodynamically favorable. The method uses Gibbs free energy of formation for the reactants and products.

(text continued on page 379)

Table 1
Gibbs Free Energy of Formation of Gas, $\Delta G_f = A + BT + CT^2$

NO.	FORMULA	NAME	A	B	C	DEL GF @298K
1	CCL4	CARBON-TETRACHLORIDE	−100.838	1.4561E-01	−8.6766E-06	−58.24
2	CCL3F	TRICHLORO-FLUOROMETHANE	−284.764	1.3487E-01	−5.3707E-06	−245.18
3	CCL2F2	DICHLORODI FLUOROMETHANE	−481.826	1.4482E-01	4.3398E-06	−438.06
4	CCLF3	CHLOROTRI-FLUOROMETHANE	−694.327	1.3545E-01	6.8525E-07	−653.96
5	CF4	CARBON-TETRAFLUORIDE	−933.816	1.5206E-01	6.7334E-08	−888.43
6	CO	CARBON MONOXIDE	−109.885	−9.2218E-02	1.4547E-06	−137.28
7	CO2	CARBON DIOXIDE	−393.360	−3.8212E-03	1.3322E-06	−394.38
8	COS	CARBONYL-SULFIDE	−136.376	−1.0575E-01	2.5162E-05	−165.64
9	CS2	CARBON-DISULFIDE	121.242	−1.9750E-01	5.0587E-05	66.90
10	CHCL3	CHLOROFORM	−101.846	1.1137E-01	8.6302E-07	−68.53
11	CHCL2F	DICHLORO-FLUOROMETHANE	−283.827	1.0295E-01	3.0690E-06	−252.81
12	CHCLF2	CHLORODI-FLUOROMETHANE	−482.201	1.0493E-01	4.2461E-06	−450.47
13	CHF3	TRIFLUORO-METHANE	−698.169	1.1528E-01	7.1054E-06	−663.08
14	CHI3	TRIIODOMETHANE	116.349	1.0932E-01	−2.3734E-06	177.95
15	CHNS	ISOTHIOCYANIC-ACID	129.895	−6.6463E-02	3.1215E-05	112.88
16	CH2CL2	DICHLORO-METHANE	−95.965	8.8080E-02	8.5438E-06	-68.87
17	CH2CLF	CHLOROFLUORO-METHANE	−262.549	8.3336E-02	1.0838E-05	−236.64
18	CH2F2	DIFLUORO-METHANE	−453.610	9.0621E-02	1.2721E-05	−425.35
19	CH212	DIIODOMETHANE	53.345	9.2134E-02	3.1467E-06	101.09
20	CH20	FORMALDEHYDE	−115.972	1.6630E-02	1.1381E-05	−109.91
21	CH202	FORMIC-ACID	−379.192	9.1431E-02	9.5249E-06	−351.00
22	CH3BR	BROMOMETHANE	−55.241	8.0199E-02	1.1207E-05	−28.16
23	CH3CL	CHLOROMETHANE	−86.903	7.5722E-02	1.4823E-05	−62.89
24	CH3F	FLUOROMETHANE	−234.475	7.6613E-02	1.6807E-05	−209.99
25	CH3I	IODOMETHANE	−20.627	8.3254E-02	8.0701E-06	15.65
26	CH3NO2	NITROMETHANE	−76.173	2.2736E-01	1.4355E-05	−6.95
27	CH3NO2	METHYL-NITRITE	−65.640	2.2023E-01	9.2945E-06	1.00
28	CH3NO3	METHYL-NITRATE	−122.358	3.0626E-01	8.2197E-06	−30.17
29	CH4	METHANE	−75.262	7.5925E-02	1.8700E-05	−50.84
30	CH40	METHANOL	−201.860	1.2542E-01	2.0345E-05	−162.51
31	CH4S	METHANETHIOL	−19.483	1.6198E-02	5.2738E-05	−9.92
32	CH5N	METHYLAMINE	−24.115	1.8179E-01	2.2182E-05	32.26
33	C2CL4	TETRA CHLOROETHENE	−12.176	1.1919E-01	−7.6720E-06	22.64
34	C2CL6	HEXACHLORO-ETHANE	−142.168	2.9199E-01	−1.8349E-05	−56.82
35	C2CL3F3	112TRICHLORO-TRIFLUOROETHANE	−696.127	2.7082E-01	−1.6970E-05	−617.14
36	C2CL2F4	12DICHLORO-TETRAFLUORO-ETHANE	−888.420	2.8167E-01	−1.2046E-05	−805.42
37	C2CLF5	CHLOROPENTA-FLUOROETHANE	−1108.395	2.8235E-01	−6.2604E-06	−1025.08
38	C2F4	TETRAFLUORO-ETHENE	−658.653	1.1776E-01	−1.8295E-06	−623.71
39	C2F6	HEXAFLUORO-ETHANE	−1344.410	2.9411E-01	−7.6769E-06	−1257.38
40	C2N2	CYANOGEN	309.375	−3.9443E-02	−4.0226E-06	297.19
41	C2HCL3	TRICHLOROETHENE	−9.886	8.7331E-02	−1.0392E-06	16.07
42	C2HCL5	PENTACHLORO-ETHANE	−143.234	2.5953E-01	−9.0227E-06	−66.65
43	C2HF3	TRIFLUORO-ETHENE	−491.035	8.7927E-02	3.6243E-06	−464.47
44	C2H2	ACETYLENE (ETHYNE)	227.187	−6.1675E-02	4.3960E-06	209.20
45	C2H2CL2	1,1-DICHLORO-ETHENE	1.994	7.6745E-02	5.2089E-06	25.40
46	C2H2CL2	CIS-1,2-DI-CHLOROETHENE	−3.264	7.5010E-02	5.9732E-06	19.66
47	C2H2CL2	TRANS-1,2-DI-CHLOROETHENE	−0.772	7.4474E-02	5.8258E-06	22.01
48	C2H2CL4	1,1,2,2-TETRA-CHLOROETHANE	−153.859	2.2902E-01	−5.6905E-07	−85.56
49	C2H2F2	1,1-DIFLUORO-ETHENE	−337.271	7.8515E-02	7.9741E-06	−313.09
50	C2H2F2	CIS-1,2-DIFLU-OROETHENE	−322.591	7.4587E-02	9.0983E-06	−299.49
51	C2H2F2	TRANS-1,2-DIFLUOROETHENE	−322.543	7.5116E-02	8.3671E-06	−299.32
52	C2H2O	KETENE	−60.986	2.0204E-04	6.7660E-06	−60.29
53	C2H3BR	BROMOETHYLENE	61.115	5.6200E-02	8.8699E-06	80.75
54	C2H3CL	CHLOROETHENE	31.557	3.2864E-02	1.8788E-05	42.93
55	C2H3CL3	1,1,2-TRICHLORO-ETHANE	−139.616	2.0630E-01	5.9155E-06	−77.49
56	C2H3CLO	ACETYL-CHLORIDE	−244.471	1.2418E-01	1.2494E-05	−206.23
57	C2H3F	FLUOROETHENE	−139.376	4.3674E-02	1.3376E-05	−125.06
58	C2H3F3	1,1,1-TRIFLUORO-ETHANE	−746.790	2.2467E-01	9.9950E-06	−678.77
59	C2H3N	ACETONITRILE	87.546	5.6339E-02	1.3233E-05	105.60
60	C2H4	ETHYLENE	51.752	4.9338E-02	1.7284E-05	68.12
61	C2H4BR2	1,2-DIBROMO-ETHANE	−72.860	1.9515E-01	3.5902E-06	−10.59
62	C2H4CL2	1,1-DICHLORO-ETHANE	−131.060	1.9006E-01	1.2939E-05	−73.09
63	C2H4CL2	1,2-DICHLORO-ETHANE	−130.549	1.8562E-01	1.3775E-05	−73.85
64	C2H4F2	1,1-DIFLUORO-ETHANE	−494.858	1.9101E-01	1.6948E-05	−436.22
65	C2H4I2	1,2-DIIODOETHANE	0.000	1.9267E-01	5.9773E-06	78.49
66	C2H4O	ETHYLENE-OXIDE	−53.838	1.3048E-01	1.8741E-05	−13.10
67	C2H4O	ACETALDEHYDE	−167.052	1.0714E-01	1.8665E-05	−133.30
68	C2H4O2	ACETIC-ACID	−435.963	1.9346E-01	1.6362E-05	−376.69
69	C2H4O2	METHYL-FORMATE	−350.914	1.7474E-01	1.6322E-05	−297.19
70	C2H4OS	THIOACETIC-ACID	−178.876	6.9466E-02	4.6167E-05	−154.01
71	C2H4S	THIACYCLO-PROPANE	85.451	2.2403E−02	5.2946E-05	96.90
72	C2H5BR	BROMOETHANE	−83.274	1.7808E-01	1.4544E-05	−26.32
73	C2H5CL	CHLOROETHANE	−112.949	1.7104E-01	1.9992E-05	−60.00
74	C2H5F	FLUOROETHANE	−262.745	1.7117E-01	2.2082E-05	−209.53
75	C2H5I	IODOETHANE	−45.122	1.8148E-01	1.1291E-05	21.34
76	C2H5N	ETHYLENIMINE	121.662	1.8205E-01	2.0521E-05	177.99
77	C2H5NO2	NITROETHANE	−103.424	3.2377E-01	1.9797E-05	−4.90
78	C2H5NO3	ETHYL-NITRATE	−156.474	3.9637E-01	1.3852E-05	−36.86
79	C2H6	ETHANE	−85.787	1.6858E-01	2.6853E-05	−32.93
80	C2H6O	METHYL-ETHER	−185.257	2.3378E-01	2.7075E-05	−112.93
81	C2H6O	ETHYL-ALCOHOL	−236.102	2.1904E-01	2.5659E-05	−168.28
82	C2H6O2	ETHYLENE-GLYCOL	−390.502	2.8378E-01	1.4492E-05	−304.47
83	C2H6S	METHYL-SULFIDE	−33.596	1.1704E-01	6.2933E-05	6.95
84	C2H6S	ETHANETHIOL	−42.415	1.0784E-01	6.1979E-05	−4.69
85	C2H6S2	METHYL-DISULFIDE	−18.556	8.6388E-02	8.3731E-05	14.73
86	C2H7N	ETHYLAMINE	−47.732	2.7622E-01	2.6932E-05	37.28
87	C2H7N	DIMETHYLAMINE	−20.837	2.8875E-01	2.7736E-05	67.99
88	C3H3N	ACRYLONITRILE	184.585	3.2496E-02	1.0680E-05	195.31
89	C3H4	ALLENE(PROPA-DIENE)	191.839	3.0752E-02	1.4194E-05	202.38
90	C3H4	PROPYNE(METHYL-ACETYLENE)	185.351	2.5730E-02	1.4771E-05	194.43
91	C3H4O2	ACRYLIC-ACID	−337.271	1.6672E-01	1.5209E-05	−286.06
92	C3H5BR	3-BROMO-1-PROPENE	30.623	1.5305E-01	1.3603E-05	79.96
93	C3H5CL	3-CHLORO-1-PROPENE	−1.565	1.4516E-01	1.9444E-05	43.60
94	C3H5CL3	1,2,3-TRICHLORO-PROPANE	−187.124	2.9501E-01	1.3833E-05	−97.78
95	C3H5I	3-IODO-1-PROPENE	60.299	1.6149E-01	8.7329E-06	120.16
96	C3H5N	PROPIONITRILE	49.759	1.4902E-01	2.0199E-05	96.15
97	C3H6	PROPENE	19.412	1.3685E-01	2.5749E-05	62.72

NO.	FORMULA	NAME	A	B	C	DEL GF @298K
98	C3H6	CYCLOPROPANE	51.643	1.6873E-01	2.4956E-05	104.39
99	C3H6BR2	1,2-DIBROMO-PROPANE	−108.362	2.8677E-01	1.0766E-05	−17.66
100	C3H6CL2	1,2-DICHLORO-PROPANE	−167.410	2.7636E-01	1.9161E-05	−83.09
101	C3H6CL2	1,3-DICHLORO-PROPANE	−162.984	2.6293E-01	2.0036E-05	−82.59
102	C3H6CL2	2,2-DICHLORO-PROPANE	−177.370	3.0654E-01	1.3973E-05	−84.56
103	C3H6I2	1,2-DIIODO-PROPANE	−33.335	2.8806E-01	7.3960E-06	74.52
104	C3H6O	PROPYLENE-OXIDE	−94.384	2.2200E-01	2.4413E-05	−25.77
105	C3H6O	ALLYL-ALCOHOL	−133.300	2.0021E-01	2.4167E-05	−71.25
106	C3H6O	PROPIONALDE-HYDE	−193.326	2.0318E-01	2.3446E-05	−130.46
107	C3H6O	ACETONE	−218.777	2.1177E-01	2.6619E-05	−153.05
108	C3H6S	THIACYCLO-BUTANE	64.908	1.2220E-01	6.8322E-05	107.49
109	C3H7BR	1-BROMO-PROPANE	−108.651	2.7327E-01	1.9373E-05	−22.47
110	C3H7BR	2-BROMO-PROPANE	−117.884	2.8906E-01	1.7035E-05	−27.24
111	C3H7CL	1-CHLORO-PROPANE	−131.804	2.6332E-01	2.6786E-05	−50.67
112	C3H7CL	2-CHLORO-PROPANE	−148.229	2.7943E-01	2.4248E-05	−62.51
113	C3H7F	1-FLUORO-PROPANE	−282.862	2.6769E-01	2.8109E-05	−200.29
114	C3H7F	2-FLUORO-PROPANE	−290.498	2.8041E-01	2.7211E-05	−204.22
115	C3H7I	1-IODOPROPANE	−69.399	2.8216E-01	1.4073E-05	27.95
116	C3H7I	2-IODOPROPANE	−80.618	2.9402E-01	1.3004E-05	20.08
117	C3H7NO2	1-NITROPROPANE	−127.329	4.1957E-01	2.5482E-05	0.33
118	C3H7-NO2	2-NITROPROPANE	−142.923	4.2813E-01	2.4633E-05	−12.80
119	C3H7NO3	PROPYL-NITRATE	−177.035	4.9531E-01	1.9599E-05	−27.32
120	C3H7NO3	ISOPROPYL-NITRATE	−194.076	5.0799E-01	1.8793E-05	−40.67
121	C3H8	PROPANE	−105.603	2.6475E-01	3.2500E-05	−23.47
122	C3H8O	ETHYL-METHYL-ETHER	−218.142	3.2628E-01	3.2734E-05	−117.65
123	C3H8O	PROPYL-ALCOHOL	−259.317	3.1232E-01	3.3063E-05	−162.97
124	C3H8O	ISOPROPYL-ALCOHOL	−274.608	3.2915E-01	2.9243E-05	−173.59
125	C3H8S	ETHYL-METHYL-SULFIDE	−55.315	2.0160E-01	7.3615E-05	11.42
126	C3H8S	1-PROPANETHIOL	−63.596	1.9853E-01	7.3301E-05	2.18
127	C3H8S	2-PROPANETHIOL	−72.286	2.1257E-01	7.0456E-05	−2.55
128	C3H9N	PROPYLAMINE	−74.693	3.7313E-01	3.2876E-05	39.79
129	C3H9N	TRIMETHYLAMINE	−26.507	4.0989E-01	3.2215E-05	98.91
130	C4F8	OCTAFLUORO-CYCLOBUTANE	−1530.065	4.4414E-01	−1.3877E-05	−1398.84
131	C4H2	BUTADIYNE (BIACETYLENE)	473.689	−9.9298E-02	−6.6000E-07	443.96
132	C4H4	1BUTEN3YNE (VINYLACETYLENE)	304.509	1.1570E-03	1.1754E-05	305.98
133	C4H4O	FURAN	−35.958	1.1791E-01	1.7007E-05	0.88
134	C4H4S	THIOPHENE	118.623	1.2478E-02	4.9100E-05	126.78
135	C4H6	1,2-BUTADIENE	161.456	1.1686E-01	2.2211E-05	198.45
136	C4H6	1,3-BUTADIENE	109.172	1.3296E-01	1.9003E-05	150.67
137	C4H6	1-BUTYNE (ETHYL-ACETYLENE)	164.525	1.1892E-01	2.1726E-05	202.09
138	C4H6	2-BUTYNE (DI-METHYLACETY-LENE)	145.727	1.2497E-01	2.5536E-05	185.43
139	C4H6	CYCLOBUTENE	128.171	1.4815E-01	2.4315E-05	174.72
140	C4H6O3	ACETIC-ANHY-DRIDE	−578.076	3.3162E-01	2.5188E-05	−476.68
141	C4H7N	BUTYRONITRILE	32.613	2.4673E-01	2.5125E-05	108.66
142	C4H7N	ISOBUTYRO-NITRILE	23.812	2.5960E-01	2.4170E-05	103.60
143	C4H8	1-BUTENE	−1.692	2.3442E-01	3.1582E-05	71.30
144	C4H8	2-BUTENE, CIS	−8.619	2.3793E-01	3.6144E-05	65.86
145	C4H8	2-BUTENE, TRANS	−12.497	2.4252E-01	3.2484E-05	62.97
146	C4H8	2-METHYL-PROPENE	−18.295	2.4609E-01	3.0860E-05	58.07
147	C4H8	CYCLOBUTANE	24.216	2.7677E-01	3.3570E-05	110.04
148	C4H8BR2	1,2-DIBROMO-BUTANE	−135.879	3.9161E-01	1.6038E-05	−13.14
149	C4H8BR2	2,3-DIBROMO-BUTANE	−139.240	4.0686E-01	1.5194E-05	−11.92
150	C4H8I2	1,2-DIIODOBUTANE	−58.978	3.9654E-01	1.1910E-05	82.09
151	C4H8O	BUTYRALDEHYDE	−206.702	2.9866E-01	2.9382E-05	−114.77
152	C4H8O	2-BUTANONE	−239.887	3.0451E-01	3.0949E-05	−146.06
153	C4H8O2	P-DIOXANE	−318.550	4.5161E-01	3.0870E-05	−180.79
154	C4H8O2	ETHYL-ACETATE	−444.940	3.8444E-01	2.9614E-05	−327.40
155	C4H8S	THIACYCLO-PENTANE	−29.540	2.2963E-01	7.8684E-05	46.02
156	C4H9BR	1-BROMOBUTANE	−129.464	3.7268E-01	2.3816E-05	−12.89
157	C4H9BR	2-BROMOBUTANE	−142.489	3.7386E-01	2.1226E-05	−25.77
158	C4H9BR	2-BROMO-2-METHYLPROPANE	−155.863	4.1286E-01	1.6443E-05	−28.16
159	C4H9CL	1-CHLOROBUTANE	−149.464	3.6039E-01	3.2740E-05	−38.79
160	C4H9CL	2-CHLOROBUTANE	−163.844	3.5979E-01	3.1321E-05	−53.47
161	C4H9CL	1-CHLORO-2-METHYLPROPANE	−161.758	3.6555E-01	3.1330E-05	−49.66
162	C4H9CL	2-CHLORO-2-METHYLPROPANE	−185.492	3.9793E-01	2.7814E-05	−64.10
163	C4H9I	2-IODO-2-METHYLPROPANE	−113.912	4.1708E-01	1.1495E-05	23.64
164	C4H9N	PYRROLIDINE	−6.977	3.9482E-01	3.9541E-05	114.68
165	C4H9NO2	1-NITROBUTANE	−147.096	5.1673E-01	3.1532E-05	10.13
166	C4H9NO2	2-NITROBUTANE	−166.954	5.2869E-01	3.0809E-05	−6.23
167	C4H10	BUTANE	−128.375	3.6047E-01	3.8256E-05	−17.15
168	C4H10	2-METHYL-PROPANE (ISOBUTANE)	−136.800	3.7641E-01	3.7497E-05	−20.88
169	C4H10O	ETHYL-ETHER	−254.382	4.3005E-01	3.8901E-05	−122.34
170	C4H10O	METHYL-PROPYL-ETHER	−239.994	4.2359E-01	3.8666E-05	−109.91
171	C4H10O	METHYL-ISO-PROPYL-ETHER	−254.443	4.3538E-01	3.8072E-05	−120.88
172	C4H10O	BUTYL-ALCOHOL	−276.720	4.0989E-01	3.9100E-05	−150.67
173	C4H10O	SEC-BUTYL-ALCOHOL	−294.630	4.1501E-01	3.6367E-05	−167.32
174	C4H10O	TERT-BUTYL-ALCOHOL	−328.304	4.4891E-01	3.4424E-05	−191.04
175	C4H10S	ETHYLSULFIDE	−78.870	2.9895E-01	8.3578E-05	17.78
176	C4H10S	ISOPROPYL-METHYL-SULFIDE	−85.853	3.0788E-01	8.3375E-05	13.43
177	C4H10S	METHYL-PROPYL-SULFIDE	−77.076	2.9483E-01	8.4110E-05	18.41
178	C4H10S	1-BUTANETHIOL	−83.578	2.9249E-01	8.2267E-05	11.05
179	C4H10S	2-BUTANETHIOL	−91.727	3.0122E-01	8.1363E-05	5.40
180	C4H10S	2-METHYL-1-PROPANETHIOL	−92.734	3.0510E-01	8.1489E-05	5.56
181	C4H10S	2-METHYL-2-PROPANETHIOL	−105.410	3.3262E-01	7.7086E-05	0.71
182	C4H10S2	ETHYL-DISULFIDE	−68.436	2.7360E-01	1.0156E-04	22.26
183	C4H11N	BUTYLAMINE	−94.792	4.7000E-01	3.9129E-05	49.20
184	C4H11N	SEC-BUTYLAMINE	−107.194	4.8307E-01	3.8180E-05	40.63
185	C4H11N	TERT-BUTYLAMINE	−123.074	4.9817E-01	3.4076E-05	28.87
186	C4H11N	DIETHYLAMINE	−75.362	4.8136E-01	3.9617E-05	72.09
187	C5H5N	PYRIDINE	138.330	1.6730E-01	2.0124E-05	190.20
188	C5H6S	2-METHYLTHIO-PHENE	87.227	1.0101E-01	6.1872E-05	122.93
189	C5H6S	3-METHYLTHIO-PHENE	86.422	9.9868E-02	6.2512E-05	121.84
190	C5H8	1,2-PENTADIENE	144.351	2.1311E-01	2.5804E-05	210.41
191	C5H8	1,3-PENTADIENE, CIS	76.492	2.2279E-01	2.9198E-05	145.77
192	C5H8	1,3-PENTADIENE, TRANS	76.388	2.2756E-01	2.5564E-05	146.73
193	C5H8	1,4-PENTADIENE	104.148	2.1318E-01	2.5915E-05	170.25
194	C5H8	2,3-PENTADIENE	137.273	2.2032E-01	3.0252E-05	205.89
195	C5H8	3-METHYL-1,2-BUTADIENE	128.505	2.2625E-01	2.7245E-05	198.61
196	C5H8	2-METHYL-1,3-BUTADIENE	74.305	2.3206E-01	2.4048E-05	145.85

NO.	FORMULA	NAME	A	B	C	DEL GF @298K
197	C5H8	1-PENTYNE	143.273	2.1581E-01	2.7158E-05	210.25
198	C5H8	2-PENTYNE	127.758	2.1215E-01	3.2628E-05	194.18
199	C5H8	3-METHYL-1-BUTYNE	135.083	2.2757E-01	2.6485E-05	205.52
200	C5H8	CYCLOPENTENE	30.597	2.5615E-01	3.8896E-05	110.79
201	C5H8	SPIROPENTANE	182.924	2.6664E-01	2.9124E-05	265.31
202	C5H10	1-PENTENE	−22.938	3.3006E-01	3.7449E-05	79.12
203	C5H10	2-PENTENE, CIS	−30.293	3.2915E-01	4.0930E-05	71.84
204	C5H10	2-PENTENE, TRANS	−33.667	3.3463E-01	3.9095E-05	69.91
205	C5H10	2-METHYL-1-BUTENE	−38.406	3.3676E-01	3.6799E-05	65.61
206	C5H10	3-METHYL-1-BUTENE	−30.771	3.4337E-01	3.2419E-05	74.77
207	C5H10	2-METHYL-2-BUTENE	−44.552	3.3614E-01	4.0962E-05	59.66
208	C5H10	CYCLOPENTANE	−80.495	3.8452E-01	4.5171E-05	38.62
209	C5H10-BR2	2,3-DIBROMO-2-METHYLBUTANE	−177.701	5.2964E-01	1.6349E-05	−13.35
210	C5H10O	VALERALDEHYDE	−229.987	3.9652E-01	3.5518E-05	−108.28
211	C5H10O	2-PENTANONE	−260.842	4.0269E-01	3.7677E-05	−137.07
212	C5H10S	THIACYCLO-HEXANE	−59.519	3.5205E-01	8.4273E-05	53.05
213	C5H10S	CYCLOPEN-TANETHIOL	−43.019	3.0720E-01	9.3808E-05	57.03
214	C5H11BR	1-BROMO-PENTANE	−152.750	4.7242E-01	2.7996E-05	-5.73
215	C5H11CL	1-CHLORO-PENTANE	−177.625	4.5745E-01	3.8765E-05	−37.40
216	C5H11CL	1-CHLORO-3-METHYLBUTANE	−183.095	4.5557E-01	3.6144E-05	−43.68
217	C5H11CL	2-CHLORO-2-METHYLBUTANE	−205.548	4.8810E-01	3.5442E-05	−56.48
218	C5H12	PENTANE	−149.141	4.5748E-01	4.4417E-05	−8.37
219	C5H12	2-METHYLBUTANE (ISOPENTANE)	−157.445	4.6399E-01	4.3411E-05	−14.81
220	C5H12	2,2-DIMETHYL-PROPANE	−169.088	5.0284E-01	3.9605E-05	−15.23
221	C5H12O	METHYL-TERT-BUTYL-ETHER	−296.013	5.5806E-01	4.2094E-05	−125.44
222	C5H12O	PENTYL-ALCOHOL	−305.155	5.0637E-01	4.5232E-05	−149.75
223	C5H12O	TERT-PENTYL-ALCOHOL	−332.312	5.4443E-01	4.2371E-05	−165.77
224	C5H12S	BUTYL-METHYL-SULFIDE	−97.062	3.8661E-01	9.3888E-05	26.65
225	C5H12S	ETHYL-PROPYL-SULFIDE	−99.575	3.8453E-01	9.4240E-05	23.56
226	C5H12S	2-METHYL-2-BUTANETHIOL	−122.516	4.1515E-01	8.8234E-05	9.20
227	C5H12S	1-PENTANETHIOL	−103.448	3.8392E-01	9.2649E-05	19.37
228	C6CL6	HEXACHLORO-BENZENE	−34.385	2.6934E-01	−1.8686E-05	44.18
229	C6F6	HEXAFLUORO-BENZENE	−957.239	2.6465E-01	−1.1629E-05	-879.39
230	C6H4CL2	O-DICHLORO-BENZENE	28.788	1.7721E-01	1.0481E-05	82.68
231	C6H4CL2	M-DICHLORO-BENZENE	25.238	1.7541E-01	1.0191E-05	78.58
232	C6H4CL2	P-DICHLORO-BENZENE	21.845	1.8209E-01	1.0115E-05	77.15
233	C6H4F2	M-DIFLUORO-BENZENE	−311.294	1.7786E-01	1.2420E-05	−257.02
234	C6H4F2	O-DIFLUORO-BENZENE	−295.933	1.7694E-01	1.1637E-05	−242.00
235	C6H4F2	P-DIFLUORO-BENZENE	−308.558	1.8283E-01	1.2013E-05	−252.84
236	C6H5BR	BROMOBENZENE	85.584	1.6480E-01	1.3600E-05	138.53
237	C6H5CL	CHLOROBENZENE	50.479	1.5737E-01	1.7618E-05	99.16
238	C6H5F	FLUOROBENZENE	−118.073	1.5842E-01	1.8022E-05	-69.04
239	C6H5I	IODOBENZENE	125.947	1.6680E-01	9.7930E-06	187.78
240	C6H6	BENZENE	81.512	1.5282E-01	2.6522E-05	129.66
241	C6H6O	PHENOL	−97.896	2.1140E-01	1.9980E-05	−32.89
242	C6H6S	BENZENETHIOL	114.974	9.1284E-02	6.0204E-05	147.61
243	C6H7N	2-PICOLINE	96.687	2.6032E-01	2.7720E-05	177.07
244	C6H7N	3-PICOLINE	103.829	2.6047E-01	2.7853E-05	184.26
245	C6H7N	ANILINE	84.822	2.6707E-01	2.2598E-05	166.69
246	C6H10	1-HEXYNE	121.936	3.1308E-01	3.3426E-05	218.57
247	C6H10	CYCLOHEXENE	−8.701	3.7574E-01	3.5208E-05	106.86
248	C6H10	1-METHYLCYCLO-PENTENE	−8.313	3.5608E-01	4.3221E-05	102.13
249	C6H10	3-METHYLCYCLO-PENTENE	5.787	3.5193E-01	4.3095E-05	114.98
250	C6H10	4-METHYLCYCLO-PENTENE	11.922	3.5346E-01	4.3307E-05	121.59
251	C6H10O	CYCLOHEXANONE	−234.483	4.6963E-01	3.6411E-05	−90.75
252	C6H12	1-HEXENE	−44.217	4.2734E-01	4.3352E-05	87.45
253	C6H12	2-HEXENE, CIS	−55.001	4.2475E-01	4.6828E-05	76.23
254	C6H12	2-HEXENE, TRANS	−56.244	4.3026E-01	4.4958E-05	76.44
255	C6H12	3-HEXENE, CIS	−50.434	4.3223E-01	4.6533E-05	83.01
256	C6H12	3-HEXENE, TRANS	−56.854	4.3669E-01	4.3504E-05	77.61
257	C6H12	2-METHYL-1-PENTENE	−54.752	4.3022E-01	4.1468E-05	77.61
258	C6H12	3-METHYL-1-PENTENE	−47.336	4.3630E-01	3.7470E-05	86.44
259	C6H12	4-METHYL-1-PENTENE	−46.656	4.4313E-01	4.6863E-05	90.04
260	C6H12	2-METHYL-2-PENTENE	−62.375	4.3289E-01	4.6135E-05	71.21
261	C6H12	3-METHYL-2-PENTENE, CIS	−60.369	4.3290E-01	4.6127E-05	73.22
262	C6H12	3-METHYL-2-PENTENE, TRANS	−61.309	4.2960E-01	4.6053E-05	71.30
263	C6H12	4-METHYL-2-PENTENE, CIS	−52.910	4.3887E-01	4.2824E-05	82.13
264	C6H12	4-METHYL-2-PENTENE, TRANS	−56.547	4.4368E-01	3.9806E-05	79.62
265	C6H12	2-ETHYL-1-BUTENE	−54.167	4.3620E-01	4.1384E-05	79.96
266	C6H12	2,3-DIMETHYL-1-BUTENE	−57.956	4.4708E-01	3.7750E-05	79.04
267	C6H12	3,3-DIMETHYL-1-BUTENE	−45.896	4.6814E-01	4.5620E-05	98.16
268	C6H12	2,3-DIMETHYL-2-BUTENE	−61.872	4.4549E-01	5.0110E-05	75.86
269	C6H12	CYCLOHEXANE	−127.917	5.2032E-01	4.4706E-05	31.76
270	C6H12	METHYLCYCLO-PENTANE	−110.437	4.7401E-01	4.9123E-05	35.77
271	C6H12O	CYCLOHEXANOL	−299.464	5.9552E-01	3.9211E-05	−117.91
272	C6H12O	HEXANAL	−251.064	4.9255E-01	4.1576E-05	−100.12
273	C6H12S	THIACYCLO-HEPTANE	−58.172	4.4956E-01	9.0710E-05	84.06
274	C6H14	HEXANE	−170.447	5.5417E-01	5.0303E-05	−0.25
275	C6H14	2-METHYL-PENTANE	−177.675	5.6303E-01	4.8313E-05	−5.02
276	C6H14	3-METHYL-PENTANE	−174.861	5.6271E-01	5.0351E-05	−2.13
277	C6H14	2,2-DIMETHYL-BUTANE	−189.225	5.8649E-01	4.7623E-05	−9.62
278	C6H14	2,3-DIMETHYL-BUTANE	−181.310	5.7783E-01	4.9722E-05	−4.10
279	C6H14O	PROPYL-ETHER	−296.085	6.2222E-01	5.0914E-05	−105.56
280	C6H14O	ISOPROPYL-ETHER	−322.050	6.5458E-01	5.0837E-05	−121.88
281	C6H14O	HEXYL-ALCOHOL	−322.918	6.0346E-01	5.1175E-05	−137.95
282	C6H14S	BUTYL-ETHYL-SULFIDE	−119.839	4.7788E-01	1.0404E-04	32.01
283	C6H14S	ISOPROPYL-SULFIDE	−136.423	5.1958E-01	9.5795E-05	27.11
284	C6H14S	METHYL-PENTYL-SULFIDE	−117.395	4.8025E-01	1.0343E-04	35.10
285	C6H14S	PROPYL-SULFIDE	−119.954	4.8210E-01	1.0475E-04	33.22
286	C6H14S	1-HEXANETHIOL	−123.627	4.7685E-01	1.0281E-04	27.82
287	C6H14S2	PROPYL-DISULFIDE	−110.285	4.5705E-01	1.2224E-04	36.99
288	C6H15N	TRIETHYLAMINE	−103.700	7.0072E-01	5.1024E-05	110.29
289	C7H5F3	A,A,A-TRIFLUORO-TOLUENE	−602.334	3.0018E-01	1.5014E-05	−511.28
290	C7H5N	BENZONITRILE	217.638	1.3974E-01	1.5912E-05	260.87
291	C7H6O2	BENZOIC-ACID	−292.081	2.6249E-01	3.5028E-05	−210.41
292	C7H7F	P-FLUORO-TOLUENE	−150.014	2.5689E-01	2.5593E-05	−70.88
293	C7H8	TOLUENE	47.813	2.3831E-01	3.1916E-05	122.01

NO.	FORMULA	NAME	A	B	C	DEL GF @298K
294	C7H8	1,3,5-CYCLOHEP-TATRIENE	179.884	2.4584E-01	2.2285E-05	255.39
295	C7H8O	M-CRESOL	−134.426	3.0573E-01	2.7772E-05	−40.54
296	C7H8O	O-CRESOL	−130.472	3.0478E-01	2.5844E-05	−37.07
297	C7H8O	P-CRESOL	−127.446	3.1458E-01	2.8223E-05	−30.88
298	C7H12	1-HEPTYNE	100.830	4.0998E-01	3.9550E-05	226.94
299	C7H14	1-HEPTENE	−65.354	5.2430E-01	4.9439E-05	95.81
300	C7H14	CYCLOHEPTANE	−124.727	6.1171E-01	5.3325E-05	63.01
301	C7H14	ETHYLCYCLO-PENTANE	−131.223	5.7136E-01	5.4772E-05	44.56
302	C7H14	1,1-DIMETHYL-CYCLOPENTANE	−142.804	5.9241E-01	5.2208E-05	39.04
303	C7H14	C-1,2-DIMETHYL-CYCLOPENTANE	−133.964	5.8514E-01	5.2430E-05	45.73
304	C7H14	T-1,2-DIMETHYL-CYCLOPENTANE	−141.075	5.8436E-01	5.2455E-05	38.37
305	C7H14	C-1,3-DIMETHYL-CYCLOPENTANE	−140.241	5.8437E-01	5.2448E-05	39.20
306	C7H14	T-1,3-DIMETHYL-CYCLOPENTANE	−137.981	5.8437E-01	5.2451E-05	41.46
307	C7H14	METHYLCYCLO-HEXANE	−160.038	6.1255E-01	4.6303E-05	27.28
308	C7H14O	HEPTANAL	−267.231	5.8999E-01	4.7332E-05	−86.65
309	C7H16	HEPTANE	−191.520	6.5052E-01	5.6444E-05	7.99
310	C7H16	2-METHYLHEXANE	−198.645	6.5837E-01	5.6475E-05	3.22
311	C7H16	3-METHYLHEXANE	−196.032	6.5427E-01	5.6454E-05	4.60
312	C7H16	3-ETHYLPENTANE	−193.374	6.6687E-01	5.6450E-05	11.00
313	C7H16	2,2-DIMETHYL-PENTANE	−209.894	6.8563E-01	5.6352E-05	0.08
314	C7H16	2,3-DIMETHYL-PENTANE	−203.028	6.6456E-01	5.6286E-05	0.67
315	C7H16	2.4-DIMETHYL-PENTANE	−205.762	6.8189E-01	5.6332E-05	3.10
316	C7H16	3,3-DIMETHYL-PENTANE	−205.327	6.7887E-01	5.6327E-05	2.64
317	C7H16	2,2,3-TRIMETHYL-BUTANE	−209.101	6.9811E-01	5.2621E-05	4.27
318	C7H16O	ISOPROPYL-TERT-BUTYL-ETHER	−361.844	7.6268E-01	5.6918E-05	−128.83
319	C7H16O	HEPTYL-ALCOHOL	−338.701	7.0060E-01	5.7147E-05	−124.18
320	C7H16S	BUTYL-PROPYL-SULFIDE	−140.231	5.6947E-01	1.1450E-04	39.87
321	C7H16S	ETHYL-PENTYL-SULFIDE	−140.038	5.7058E-01	1.1447E-04	40.38
322	C7H16S	HEXYL-METHYL-SULFIDE	−137.588	5.7293E-01	1.1389E-04	43.47
323	C7H16S	1-HEPTANETHIOL	−143.858	5.6969E-01	1.1320E-04	36.19
324	C8H6	ETHYNYLBENZENE	326.263	1.1297E-01	1.8500E-05	361.75
325	C8H8	STYRENE	145.657	2.1917E-01	2.8490E-05	213.80
326	C8H8	1,3,5,7-CYCLO-OCTATETRAENE	296.084	2.3877E-01	2.6787E-05	369.91
327	C8H10	ETHYLBENZENE	27.421	3.3327E-01	3.8542E-05	130.58
328	C8H10	M-XYLENE	15.063	3.3452E-01	4.1387E-05	118.87
329	C8H10	O-XYLENE	17.048	3.3940E-01	3.9428E-05	122.09
330	C8H10	P-XYLENE	15.763	3.3952E-01	4.2301E-05	121.13
331	C8H14	1-OCTYNE	79.739	5.0705E-01	4.5503E-05	235.39
332	C8H16	1-OCTENE	−86.500	6.2135E-01	5.5457E-05	104.22
333	C8H16	CYCLOOCTANE	−131.710	7.2244E-01	6.2236E-05	89.91
334	C8H16	PROPYLCYCLO-PENTANE	−152.696	6.6824E-01	6.0934E-05	52.59
335	C8H16	ETHYLCYCLO-HEXANE	−177.580	7.0980E-01	5.1198E-05	39.25
336	C8H16	1,1-DIMETHYL-CYCLOHEXANE	−187.269	7.2867E-01	5.1482E-05	35.23
337	C8H16	C-1,2-DIMETHYL-CYCLOHEXANE	−178.140	7.1809E-01	5.1867E-05	41.21
338	C8H16	T-1,2-DIMETHYL-CYCLOHEXANE	−185.997	7.2249E-01	4.9658E-05	34.48
339	C8H16	C-1,3-DIMETHYL-CYCLOHEXANE	−190.817	7.2267E-01	5.1077E-05	29.83
340	C8H16	T-1,3-DIMETHYL-CYCLOHEXANE	−182.416	7.1581E-01	5.2435E-05	36.32
341	C8H16	C-1,4-DIMETHYL-CYCLOHEXANE	−182.489	7.2157E-01	5.2435E-05	37.95
342	C8H16	T-1,4-DIMETHYL-CYCLOHEXANE	−190.646	7.2867E-01	5.0049E-05	31.71
343	C8H16O	OCTANAL	−293.423	6.8704E-01	5.3338E-05	−83.30
344	C8H18	OCTANE	−212.692	7.4774E-01	6.2361E-05	16.40
345	C8H18	2-METHYL-HEPTANE	−219.792	7.5940E-01	6.2231E-05	12.76
346	C8H18	3-METHYL-HEPTANE	−216.883	7.5284E-01	6.2428E-05	13.72
347	C8H18	4-METHYL-HEPTANE	−216.353	7.6119E-01	6.2325E-05	16.74
348	C8H18	3-ETHYLHEXANE	−215.139	7.5633E-01	6.2330E-05	16.53
349	C8H18	2,2-DIMETHYL-HEXANE	−229.039	7.8345E-01	6.2253E-05	10.71
350	C8H18	2,3-DIMETHYL-HEXANE	−218.217	7.7062E-01	6.2282E-05	17.70
351	C8H18	2,4-DIMETHYL-HEXANE	−223.697	7.6894E-01	6.2289E-05	11.72
352	C8H18	2,5-DIMETHYL-HEXANE	−226.952	7.7568E-01	6.2197E-05	10.46
353	C8H18	3,3-DIMETHYL-HEXANE	−224.411	7.7653E-01	6.2269E-05	13.26
354	C8H18	3,4-DIMETHYL-HEXANE	−217.260	7.6617E-01	6.2351E-05	17.32
355	C8H18	3-ETHYL-2-METHYLPENTANE	−215.532	7.7359E-01	6.2197E-05	21.25
356	C8H18	3-ETHYL-3-METHYLPENTANE	−219.251	7.8158E-01	6.2315E-05	19.92
357	C8H18	2,2,3-TRIMETHYLPEN-TANE	−224.335	7.8913E-01	6.2502E-05	17.11
358	C8H18	2,2,4-TRIMETHYLPEN-TANE	−228.383	7.9125E-01	6.2364E-05	13.68
359	C8H18	2,3,3-TRIMETHYLPEN-TANE	−220.671	7.8284E-01	6.2446E-05	18.91
360	C8H18	2,3,4-TRIMETHYLPEN-TANE	−221.742	7.8655E-01	6.2266E-05	18.91
361	C8H18	2,2,3,3-TETRA-METHYLBUTANE	−231.148	8.3203E-01	5.0497E-05	22.01
362	C8H18O	BUTYL-ETHER	−337.796	8.1490E-01	6.4263E-05	−88.53
363	C8H18O	SEC-BUTYL-ETHER	−364.965	8.5433E-01	6.2752E-05	−104.06
364	C8H18O	TERT-BUTYL-ETHER	−369.106	8.8957E-01	6.2818E-05	−97.70
365	C8H18O	OCTYL-ALCOHOL	−361.386	7.9753E-01	6.3199E-05	−117.36
366	C8H18S	BUTYL-SULFIDE	−161.106	6.6757E-01	1.2527E-04	49.20
367	C8H18S	ETHYL-HEXYL-SULFIDE	−160.100	6.6283E-01	1.2547E-04	48.83
368	C8H18S	HEPTYL-METHYL-SULFIDE	−157.568	6.6496E-01	1.2493E-04	51.92
369	C8H18S	PENTYL-PROPYL-SULFIDE	−160.310	6.6161E-01	1.2550E-04	48.24
370	C8H18S	1-OCTANETHIOL	−163.961	6.6230E-01	1.2366E-04	44.64
371	C8H18S2	BUTY-DISULFIDE	−150.651	6.4275E-01	1.4278E-04	53.85
372	C9H10	ALPHA-METHYL-STYRENE	111.072	3.1472E-01	3.7121E-05	208.53
373	C9H10	PROPENYL-BENZENE, CIS	119.443	3.1471E-01	3.7129E-05	216.90
374	C9H10	PROPENYL-BENZENE, TRANS	115.154	3.1893E-01	3.5506E-05	213.72
375	C9H10	M-METHYL-STYRENE	113.573	3.0888E-01	3.7107E-05	209.28
376	C9H10	O-METHYL-STYRENE	116.520	3.1469E-01	3.7144E-05	213.97
377	C9H10	P-METHYL-STYRENE	112.745	3.1472E-01	3.7119E-05	210.20
378	C9H12	PROPYLBENZENE	4.889	4.2937E-01	4.4012E-05	137.24
379	C9H12	CUMENE	0.983	4.4175E-01	4.3406E-05	136.98
380	C9H12	M-ETHYLTOLUENE	−4.705	4.2485E-01	4.5613E-05	126.44
381	C9H12	O-ETHYLTOLUENE	−1.301	4.2958E-01	4.3769E-05	131.08
382	C9H12	P-ETHYLTOLUENE	−5.947	4.2950E-01	4.6751E-05	126.69
383	C9H12	1,2,3-TRIMETHYL-BENZENE	−11.858	4.4140E-01	4.9389E-05	124.56
384	C9H12	1,2,4-TRIMETHYL-BENZENE	−16.358	4.3119E-01	4.8564E-05	116.94

NO.	FORMULA	NAME	A	B	C	DEL GF @298K
385	C9H12	MESITYLENE	−18.595	4.4166E-01	4.9679E-05	117.95
386	C9H16	1-NONYNE	58.588	6.0406E-01	5.1578E-05	243.76
387	C9H18	1-NONENE	−107.582	7.1838E-01	6.1441E-05	112.68
388	C9H18	BUTYLCYCLO-PENTANE	−173.390	7.6514E-01	6.7065E-05	61.38
389	C9H18	PROPYLCYCLO-HEXANE	−199.501	8.0874E-01	5.6176E-05	47.32
390	C9H18	C-C-135-TRIMETHYL-CYCLOHEXANE	−222.259	8.4009E-01	5.5762E-05	33.89
391	C9H18	C-T-135 TRIMETHYL-CYCLOHEXANE	−212.955	8.2810E-01	5.8630E-05	39.87
392	C9H18O	NONANAL	−314.560	7.8403E-01	5.9438E-05	−74.94
393	C9H20	NONANE	−233.826	8.4477E-01	6.8451E-05	24.81
394	C9H20	2-METHYLOCTANE	−240.805	8.5474E-01	6.6289E-05	20.59
395	C9H20	3-METHYLOCTANE	−238.219	8.4955E-01	6.6996E-05	21.71
396	C9H20	4-METHYLOCTANE	−238.219	8.4955E-01	6.6996E-05	21.71
397	C9H20	3-ETHYLHEPTANE	−235.493	8.5556E-01	6.7981E-05	26.32
398	C9H20	4-ETHYLHEPTANE	−235.493	8.5556E-01	6.7981E-05	26.32
399	C9H20	2,2-DIMETHYL-HEPTANE	−251.912	8.8046E-01	6.1573E-05	16.74
400	C9H20	2,3-DIMETHYL-HEPTANE	−240.580	8.6274E-01	6.7176E-05	23.30
401	C9H20	2,4-DIMETHYL-HEPTANE	−245.667	8.6328E-01	6.7859E-05	18.45
402	C9H20	2,5-DIMETHYL-HEPTANE	−245.667	8.6328E-01	6.7859E-05	18.45
403	C9H20	2,6-DIMETHYL-HEPTANE	−248.402	8.7460E-01	6.6904E-05	19.00
404	C9H20	3,3-DIMETHYL-HEPTANE	−246.821	8.7164E-01	6.2519E-05	19.29
405	C9H20	3,4-DIMETHYL-HEPTANE	−237.951	8.6040E-01	6.7955E-05	25.31
406	C9H20	3,5-DIMETHYL-HEPTANE	−243.061	8.6678E-01	6.8579E-05	22.18
407	C9H20	4.4-DIMETHYL-HEPTANE	−246.853	8.7750E-01	6.2463E-05	21.00
408	C9H20	3-ETHYL-2-METHYLHEXANE	−237.903	8.6316E-01	6.8037E-05	26.19
409	C9H20	4-ETHYL-2-METHYLHEXANE	−243.011	8.6950E-01	6.8697E-05	23.05
410	C9H20	3-ETHYL-3-METHYLHEXANE	−241.791	8.7215E-01	6.3284E-05	24.56
411	C9H20	3-ETHYL-4-METHYLHEXANE	−235.257	8.6369E-01	6.8766E-05	29.08
412	C9H20	2,2,3-TRIMETHYL-HEXANE	−246.598	8.8845E-01	6.2515E-05	24.52
413	C9H20	2,2,4-TRIMETHYL-HEXANE	−248.840	8.8895E-01	6.3241E-05	22.51
414	C9H20	2,2,5-TRIMETHYL-HEXANE	−259.495	8.9459E-01	6.2167E-05	13.43
415	C9H20	2,3,3-TRIMETHYL-HEXANE	−244.225	8.8514E-01	6.2489E-05	25.90
416	C9H20	2,3,4-TRIMETHYL-HEXANE	−240.368	8.7376E-01	6.8017E-05	26.90
417	C9H20	2,3,5-TRIMETHYL-HEXANE	−248.145	8.8262E-01	6.7777E-05	21.76
418	C9H20	2,4,4-TRIMETHYL-HEXANE	−246.460	8.8561E-01	6.3235E-05	23.89
419	C9H20	3,3,4-TRIMETHYL-HEXANE	−241.552	8.7981E-01	6.3300E-05	27.07
420	C9H20	3,3-DIETHYLPEN-TANE	−237.693	8.9338E-01	6.4026E-05	35.06
421	C9H20	3-ETHYL-2,2-DIMETHYLPEN-TANE	−243.872	8.9430E-01	6.3760E-05	29.12
422	C9H20	3-ETHYL-2,3-DIMETHYLPEN-TANE	−239.241	8.8583E-01	6.3120E-05	31.17
423	C9H20	3-ETHYL-2,4-DIMETHYLPEN-TANE	−240.323	8.8226E-01	6.8132E-05	29.50
424	C9H20	2,2,3,3-TRIMETHYLPEN-TANE	−242.926	9.1022E-01	5.8406E-05	34.31
425	C9H20	2,2,3,4-TETRAMETHYL-PENTANE	−242.527	9.0198E-01	6.2433E-05	32.64
426	C9H20	2,2,4,4-TETRAMETHYL-PENTANE	−247.825	9.2587E-01	5.7635E-05	34.02
427	C9H20	2,3,3,4-TETRA-METHYLPENTANE	−241.799	9.0451E-01	6.2351E-05	34.10
428	C9H20O	NONYL-ALCOHOL	−391.755	8.9463E-01	6.9190E-05	−118.20
429	C9H20S	BUTYL-PENTYL-SULFIDE	−181.275	7.5451E-01	1.3571E-04	55.90
430	C9H20S	ETHYL-HEPTYL-SULFIDE	−180.361	7.5587E-01	1.3553E-04	57.20
431	C9H20S	HEXYL-PROPYL-SULFIDE	−180.488	7.5455E-01	1.3567E-04	56.69
432	C9H20S	METHYL-OCTYL-SULFIDE	−177.894	7.5828E-01	1.3475E-04	60.33
433	C9H20S	1-NONANETHIOL	−184.234	7.5533E-01	1.3380E-04	53.01
434	C10H8	NAPHTHALENE	148.988	2.4014E-01	3.0705E-05	223.59
435	C10H8	AZULENE	277.714	2.3826E-01	3.1796E-05	351.87
436	C10H14	BUTYLBENZENE	−17.190	5.2633E-01	5.0190E-05	144.68
437	C10H14	M-DIETHYL-BENZENE	−25.143	5.2637E-01	4.9852E-05	136.69
438	C10H14	O-DIETHYL-BENZENE	−21.957	5.3091E-01	4.8193E-05	141.08
439	C10H14	P-DIETHYL-BENZENE	−25.535	5.3121E-01	5.0922E-05	137.86
440	C10H14	1,2,3,4-TETRAMETHYL-BENZENE	−44.515	5.4762E-01	4.7618E-05	123.43
441	C10H14	1,2,3,5-TETRAMETHYL-BENZENE	−47.443	5.4079E-01	5.0474E-05	118.74
442	C10H14	1,2,4,5-TETRAMETHYL-BENZENE	−47.801	5.4403E-01	5.1676E-05	119.45
443	C10H18	1-DECYNE	37.488	7.0119E-01	5.7516E-05	252.21
444	C10H18	DECAHYDRO-NAPHTHALENE, CIS	−175.788	8.5412E-01	6.9065E-05	85.81
445	C10H18	DECAHYDRO-NAPHTHALENE, TRANS	−189.001	8.5677E-01	6.9466E-05	73.43
446	C10H20	1-DECENE	−128.709	8.1530E-01	6.7574E-05	121.04
447	C10H20	1-CYCLO-PENTYLPENTANE	−194.532	8.6222E-01	7.3010E-05	69.79
448	C10H20	BUTYLCYCLO-HEXANE	−219.884	9.0572E-01	6.2305E-05	56.44
449	C10H20O	DECANAL	−335.674	8.8103E-01	6.5416E-05	−66.53
450	C10H22	DECANE	−255.000	9.4201E-01	7.4254E-05	33.22
451	C10H22	2-METHYLNONANE	−261.869	9.5206E-01	7.1758E-05	29.08
452	C10H22	3-METHYLNONANE	−259.232	9.4674E-01	7.2527E-05	30.21
453	C10H22	4-METHYLNONANE	−259.232	9.4674E-01	7.2527E-05	30.21
454	C10H22	5-METHYLNONANE	−259.270	9.5261E-01	7.2471E-05	31.92
455	C10H22	3-ETHYLOCTANE	−256.591	9.5302E-01	7.3335E-05	34.81
456	C10H22	4-ETHYLOCTANE	−256.563	9.4713E-01	7.3410E-05	33.10
457	C10H22	2,2-DIMETHYL-OCTANE	−272.907	9.7753E-01	6.7222E-05	25.23
458	C10H22	2,3-DIMETHYL-OCTANE	−266.771	9.6074E-01	7.3210E-05	26.94
459	C10H22	2,4-DIMETHYL-OCTANE	−261.653	9.6011E-01	7.2592E-05	31.80
460	C10H22	2,5-DIMETHYL-OCTANE	−266.771	9.6074E-01	7.3210E-05	26.94
461	C10H22	2,6-DIMETHYL-OCTANE	−266.771	9.6074E-01	7.3210E-05	26.94
462	C10H22	2,7-DIMETHYL-OCTANE	−269.396	9.7167E-01	7.2527E-05	27.49
463	C10H22	3,3-DIMETHYL-OCTANE	−267.928	9.6911E-01	6.7859E-05	27.78
464	C10H22	3,4-DIMETHYL-OCTANE	−259.027	9.5779E-01	7.3332E-05	33.81
465	C10H22	3,5-DIMETHYL-OCTANE	−264.140	9.5839E-01	7.3992E-05	28.95
466	C10H22	3,6-DIMETHYL-OCTANE	−264.093	9.6400E-01	7.4100E-05	30.67

NO.	FORMULA	NAME	A	B	C	DEL GF @298K
467	C10H22	4,4-DIMETHYL-OCTANE	−267.928	9.6911E-01	6.7859E-05	27.78
468	C10H22	4,5-DIMETHYL-OCTANE	−258.993	9.6344E-01	7.3443E-05	35.52
469	C10H22	4-PROPYL-HEPTANE	−256.533	9.6193E-01	7.3483E-05	37.53
470	C10H22	4-ISOPROPYL-HEPTANE	−256.743	9.6635E-01	7.3427E-05	38.66
471	C10H22	3-ETHYL-2-METHYLHEPTANE	−258.970	9.6045E-01	7.3515e-05	34.69
472	C10H22	4-ETHYL-2-METHYLHEPTANE	−264.077	9.6109E-01	7.4107E-05	29.83
473	C10H22	5-ETHYL-2-METHYL-HEPTANE	−264.106	9.6694E-01	7.4061E-05	31.55
474	C10H22	3-ETHYL-3-METHYL-HEPTANE	−262.888	9.6960E-01	6.8641E-05	33.05
475	C10H22	4-ETHYL-3-METHYLHEPTANE	−256.382	9.5829E-01	7.4126E-05	36.69
476	C10H22	3-ETHYL-5-METHYLHEPTANE	−261.414	9.6136E-01	7.5079E-05	32.68
477	C10H22	3-ETHYL-4-METHYLHEPTANE	−256.320	9.6100E-01	7.4238E-05	37.57
478	C10H22	4-ETHYL-4-METHYLHEPTANE	−262.888	9.6960E-01	6.8641E-05	33.05
479	C10H22	2,2,3-TRIMETHYL-HEPTANE	−267.615	9.8561E-01	6.8083E-05	33.01
480	C10H22	2,2,4-TRIMETHYL-HEPTANE	−269.875	9.8622E-01	6.8723E-05	31.00
481	C10H22	2,2,5-TRIMETHYL-HEPTANE	−277.782	9.8622E-01	6.8723E-05	23.10
482	C10H22	2,2,6-TRIMETHYL-HEPTANE	−280.476	9.9171E-01	6.7751E-05	21.97
483	C10H22	2,3,3-TRIMETHYL-HEPTANE	−265.181	9.8212E-01	6.8178E-05	34.43
484	C10H22	2,3,4-TRIMETHYL-HEPTANE	−261.391	9.7099E-01	7.3509E-05	35.40
485	C10H22	2,3,5-TRIMETHYL-HEPTANE	−266.466	9.7144E-01	7.4281E-05	30.54
486	C10H22	2,3,6-TRIMETHYL-HEPTANE	−269.149	9.7393E-01	7.3404E-05	28.53
487	C10H22	2,4,4-TRIMETHYL-HEPTANE	−267.448	9.8275E-01	6.8799E-05	32.43
488	C10H22	2,4,5-TRIMETHYL-HEPTANE	−266.466	9.7144E-01	7.4281E-05	30.54
489	C10H22	2,4,6-TRIMETHYL-HEPTANE	−259.166	9.8025E-01	7.4100E-05	40.46
490	C10H22	2,5,5-TRIMETHYL-HEPTANE	−275.356	9.8275E-01	6.8799E-05	24.52
491	C10H22	3,3,4-TRIMETHYL-HEPTANE	−262.647	9.7725E-01	6.8661E-05	35.56
492	C10H22	3,3,5-TRIMETHYL-HEPTANE	−264.902	9.7782E-01	6.9318E-05	33.56
493	C10H22	3,4,4-TRIMETHYL-HEPTANE	−262.647	9.7725E-01	6.8661E-05	35.56
494	C10H22	2,3,5-TRIMETHYL-HEPTANE	−259.802	9.7141E-01	7.5693E-05	38.24
495	C10H22	3-ISOPROPYL-2-METHYL-HEXANE	−261.387	9.7967E-01	7.3499E-05	37.99
496	C10H22	3,3-DIETHYL-HEXANE	−257.797	9.7908E-01	6.9505E-05	41.05
497	C10H22	3,4-DIETHYL-HEXANE	−253.621	9.7301E-01	7.5062E-05	43.93
498	C10H22	3-ETHYL-2,2-DIMETHYLHEXANE	−264.982	9.8620E-01	6.8786E-05	35.90
499	C10H22	4-ETHYL-2,2-DIMETHYLHEXANE	−267.225	9.9254E-01	6.9429E-05	35.65
500	C10H22	3-ETHYL-2,3-DIMETHYLHEXANE	−260.267	9.7727E-01	6.8648E-05	37.95
551	C12H12	1,2-DIMETHYL-NAPHTHALENE	80.611	4.4220E-01	3.8339E-05	216.23
552	C12H12	1,3-DIMETHYL-NAPHTHALENE	78.976	4.3881E-01	3.9571E-05	213.72
553	C12H12	1,4-DIMETHYL-NAPHTHALENE	79.522	4.4807E-01	3.8242E-05	216.90
554	C12H12	1,5-DIMETHYL-NAPHTHALENE	78.821	4.4803E-01	3.8272E-05	216.19
555	C12H12	1,6-DIMETHYL-NAPHTHALENE	79.690	4.3879E-01	3.9579E-05	214.43
556	C12H12	1,7-DIMETHYL-NAPHTHALENE	78.976	4.3881E-01	3.9571E-05	213.72
557	C12H12	2,3-DIMETHYL-NAPHTHALENE	81.060	4.3524E-01	4.2654E-05	215.02
558	C12H12	2,6-DIMETHYL-NAPHTHALENE	79.922	4.3839E-01	4.0694E-05	214.64
559	C12H12	2,7-DIMETHYL-NAPHTHALENE	79.918	4.3842E-01	4.0642E-05	214.64
560	C12H18	HEXYLBENZENE	−59.463	7.1997E-01	6.2172E-05	161.34
561	C12H18	1,2,3-TRIETHYL-BENZENE	−71.837	7.2867E-01	6.2504E-05	151.54
562	C12H18	1,2,4-TRIETHYL-BENZENE	−72.938	7.1000E-01	6.8084E-05	145.23
563	C12H18	1,3,5-TRIETHYL-BENZENE	−78.848	7.2900E-01	6.2716E-05	144.68
564	C12H18	HEXAMETHYL-BENZENE	−109.218	7.8642E-01	5.0247E-05	130.21
565	C12H22	1-DODECYNE	−4.724	8.9500E-01	6.9758E-05	268.99
566	C12H24	1-DODECYNE	−170.907	1.0093E+00	7.9654E-05	137.90
567	C12H24	1-CYCLO-PENTYL-HEPTANE	−236.801	1.0564E+00	8.5011E-05	86.61
568	C12H24	1-CYCLO-HEXYL-HEXANE	−262.182	1.1000E+00	7.4071E-05	73.26
569	C12H26	DODECANE	−297.182	1.1359E+00	8.6457E-05	50.04
570	C12H26O	DODECYL-ALCOHOL	−449.239	1.1858E+00	8.7192E-05	-87.07
571	C12H26S	BUTYL-OCTYL-SULFIDE	−241.769	1.0327E+00	1.6697E-04	81.17
572	C12H26S	DECYL-ETHYL-SULFIDE	−242.383	1.0341E+00	1.6662E-04	80.92
573	C12H26S	HEXYL-SULFIDE	−241.877	1.0389E+00	1.6649E-04	82.89
574	C12H26S	METHYL-UNDECYL-SULFIDE	−238.536	1.0370E+00	1.6539E-04	85.52
575	C12H26S	NONYL-PROPYL-SULFIDE	−241.027	1.0327E+00	1.6692E-04	81.92
576	C12H26S	1-DODECANETHIOL	−244.797	1.0337E+00	1.6475E-04	78.24
577	C12H26S2	HEXYL-DISULFIDE	−231.339	1.0136E+00	1.8450E-04	87.49
578	C13H14	1-PROPYLNAPHTHALENE	71.181	5.2690E-01	4.3941E-05	232.63
579	C13H14	2-PROPYLNAPHTHALENE	70.476	5.2339E-01	4.5280E-05	231.00
580	C13H14	2ETHYL-3-METHYLNAPHTHALENE	62.759	5.2530E-01	4.7025E-05	224.01

NO.	FORMULA	NAME	A	B	C	DEL GF @298K
581	C13H14	2ETHYL-6-METHYLNAPHTHALENE	58.499	5.2712E-01	4.6332E-05	220.20
582	C13H14	2ETHYL-7-METHYLNAPHTHALENE	58.499	5.2712E-01	4.6332E-05	220.20
583	C13H20	1-PHENYL-HEPTANE	−80.556	8.1688E-01	6.8286E-05	169.74
584	C13H24	1-TRIDECYNE	−25.829	9.9206E-01	7.5746E-05	277.44
501	C10H22	4-ETHYL-2,3-DIMETHYLHEXANE	−258.811	9.7449E-01	7.4202E-05	39.12
502	C10H22	3-ETHYL-2,4-DIMETHYLHEXANE	−258.787	9.7149E-01	7.4304E-05	38.24
503	C10H22	4-ETHYL-2,4-DIMETHYLHEXANE	−262.418	9.8330E-01	6.9521E-05	37.70
504	C10H22	3-ETHYL-2,5-DIMETHYLHEXANE	−266.527	9.7459E-01	7.4067E-05	31.42
505	C10H22	4-ETHYL-3,3-DIMETHYLHEXANE	−259.929	9.8334E-01	6.9567E-05	40.21
506	C10H22	3-ETHYL-3,4-DIMETHYLHEXANE	−257.579	9.7763E-01	6.9541E-05	40.84
507	C10H22	2,2,3,3-TETRA-METHYLHEXANE	− 264.021	1.0077E+00	6.3741E-05	42.80
508	C10H22	2,2,3,4-TETRA-METHYLHEXANE	−259.481	9.9664E-01	6.8878E-05	44.56
509	C10H22	2,2,3,5-TETRA-METHYLHEXANE	−275.197	9.9984E-01	6.8575E-05	29.75
510	C10H22	2,2,4,4-TETRA-METHYLHEXANE	−263.859	1.0089E+00	6.3813E-05	43.35
511	C10H22	2,2,4,5-TETRA-METHYLHEXANE	−272.355	9.9985E-01	6.8569E-05	32.59
512	C10H22	2,2,5,5-TETRA-METHYLHEXANE	−291.593	1.0233E+00	6.2952E-05	19.83
513	C10H22	2,3,3,4-TETRA-METHYLHEXANE	−260.022	9.9071E-01	6.8648E-05	42.22
514	C10H22	2,3,3,5-TETRA-METHYLHEXANE	−264.912	9.9652E-01	6.8562E-05	39.04
515	C10H22	2,3,4,4-TETRA-METHYLHEXANE	−257.185	9.9074E-01	6.8625E-05	45.06
516	C10H22	2,3,4,5-TETRA-METHYLHEXANE	−263.908	9.9031E-01	7.3463E-05	38.66
517	C10H22	3,3,4,4-TETRA-METHYLHEXANE	−258.982	1.0048E+00	6.4558E-05	47.07
518	C10H22	24DIMETHYL-3ISOPROPYLPENTANE	−264.215	1.0048E+00	6.8707E-05	42.26
519	C10H22	33-DIETHYL-2-METHYLPENTANE	−255.118	9.9226E-01	6.9725E-05	47.70
520	C10H22	3ETHYL-223-TRIMETHYLPENTANE	−258.971	1.0024E+00	6.4555E-05	46.36
521	C10H22	3ETHYL-224-TRIMETHYLPENTANE	−259.533	1.0062E+00	6.8727E-05	47.32
522	C10H22	3ETHYL-234-TRIMETHYLPENTANE	−257.589	9.9632E-01	6.8759E-05	46.32
523	C10H22	22334-PENTAMETHYLPENTANE	−253.281	1.0201E+00	6.4778E-05	57.36
524	C10H22	22344-PENTAMETHYLPENTANE	−253.379	1.0306E+00	6.3360E-05	60.25
525	C10H22O	DECYL-ALCOHOL	−408.594	9.9155E-01	7.5283E-05	−105.52
526	C10H22S	BUTYL-HEXYL-SULFIDE	−201.497	8.4742E-01	1.4603E-04	64.31
527	C10H22S	ETHYL-OCTYL-SULFIDE	−200.446	8.4814E-01	1.4635E-04	65.61
528	C10H22S	HEPTYL-PROPYL-SULFIDE	−200.752	8.4746E-01	1.4599E-04	65.06
529	C10H22S	METHYL-NONYL-SULFIDE	−198.094	8.5098E-01	1.4518E-04	68.70
530	C10H22S	PENTYL-SULFIDE	−201.402	8.5291E-01	1.4621E-04	66.07
531	C10H22S	1-DECANETHIOL	−204.389	8.4798E-01	1.4421E-04	61.42
532	C10H22S2	PENTYL-DISULFIDE	−190.970	8.2801E-01	1.6386E-04	70.67

NO.	FORMULA	NAME	A	B	C	DEL GF @298K
533	C11H10	1-METHYLNAPHTHALENE	114.458	3.3490E-01	3.4180E-05	217.69
534	C11H10	2-METHYLNAPHTHALENE	113.834	3.3151E-01	3.5442E-05	216.15
535	C11H16	PENTYLBENZENE	−38.343	6.2290E-01	5.6186E-05	152.93
536	C11H16	PENTAMETHYL-BENZENE	−77.647	6.5761E-01	4.9942E-05	123.34
537	C11H20	1-UNDECYNE	16.390	7.9808E-01	6.3658E-05	260.62
538	C11H22	1-UNDECENE	−149.838	9.1236E-01	7.3538E-05	129.45
539	C11H22	1-CYCLOPENTYL-HEXANE	−215.718	9.5953E-01	7.8827E-05	78.20
540	C11H22	PENTYLCYCLO-HEXANE	−241.092	1.0031E+00	6.7958E-05	64.85
541	C11H24	UNDECANE	−276.105	1.0389E+00	8.0406E-05	41.59
542	C11H24O	UNDECYL-ALCOHOL	−428.070	1.0886E+00	8.1234E-05	-95.44
543	C11H24S	BUTYL-HEPTYL-SULFIDE	−221.656	9.4009E-01	1.5642E-04	72.72
544	C11H24S	DECYL-METHYL-SULFIDE	−218.271	9.4391E-01	1.5535E-04	77.15
545	C11H24S	ETHYL-NONYL-SULFIDE	−220.622	9.4107E-01	1.5652E-04	74.06
546	C11H24S	OCTYL-PROPYL-SULFIDE	−220.786	9.4013E-01	1.5638E-04	73.60
547	C11H24S	1-UNDECANETHIOL	−224.542	9.4068E-01	1.5465E-04	69.83
548	C12H10	BIPHENYL	179.415	3.2590E-01	3.5313E-05	280.08
549	C12H12	1-ETHYLNAPHTHALENE	93.648	4.3103E-01	3.8318E-05	225.98
550	C12H12	2-ETHYLNAPHTHALENE	93.053	4.2755E-01	3.9627E-05	224.43
585	C13H26	1-TRIDECENE	−192.059	1.1064E+00	8.5626E-05	146.27
586	C13H26	1-CYCLOPENTYL-OCTANE	−257.926	1.1536E+00	9.0881E-05	95.06
587	C13H26	1-CYCLOHEXYL-HEPTANE	−283.301	1.1970E+00	8.0042E-05	81.67
588	C13H28	TRIDECANE	−318.273	1.2328E+00	9.2592E-05	58.45
589	C13H28O	1-TRIDECANOL	−470.339	1.2840E+00	9.3318E-05	−78.28
590	C13H28S	BUTYL-NONYL-SULFIDE	−261.912	1.1252E+00	1.7759E-04	89.54
591	C13H28S	DECYL-PROPYL-SULFIDE	−261.150	1.1251E+00	1.7763E-04	90.29
592	C13H28S	DODECYL-METHYL-SULFIDE	−258.629	1.1295E+00	1.7609E-04	93.97
593	C13H28S	ETHYL-UNDECYL-SULFIDE	−260.976	1.1265E+00	1.7738E-04	90.88
594	C13H28S	1-TRIDECANETHIOL	−264.946	1.1264E+00	1.7521E-04	86.65
595	C14H16	1-BUTYLNAPHTHALENE	49.101	6.2380E-01	5.0256E-05	240.08
596	C14H16	2-BUTYLNAPHTHALENE	48.475	6.2035E-01	5.1496E-05	238.53
597	C14H22	1-PHENYL-OCTANE	−101.612	9.1381E-01	7.4333E-05	178.20
598	C14H22	1,2,3,4-TETRAETHYL-BENZENE	−128.011	9.3083E-01	6.4969E-05	155.94
599	C14H22	1,2,3,5-TETRAETHYL-BENZENE	−127.656	9.2414E-01	6.7695E-05	154.56
600	C14H22	1,2,4,5-TETRAETHYL-BENZENE	−127.872	9.2707E-01	6.9124E-05	155.35
601	C14H26	1-TETRADECYNE	−46.990	1.0891E+00	8.1802E-05	285.81
602	C14H28	1-TETRADECENE	−213.140	1.2035E+00	9.1571E-05	154.77
603	C14H28	1-CYCLOPENTYL-NONANE	−279.030	1.2505E+00	9.7036E-05	103.43
604	C14H28	1-CYCLOHEXYL-OCTANE	−304.390	1.2940E+00	8.6187E-05	90.08
605	C14H30	TETRADECANE	−339.495	1.3301E+00	9.8399E-05	66.82
606	C14H30O	1-TETRADECANOL	−491.525	1.3812E+00	9.9188E-05	−69.87
607	C14H30S	BUTYL-DECYL-SULFIDE	−282.126	1.2181E+00	1.8779E-04	97.95
608	C14H30S	DODECYL-ETHYL-SULFIDE	−281.237	1.2195E+00	1.8745E-04	99.24
609	C14H30S	HEPTYL-SULFIDE	−282.102	1.2238E+00	1.8786E-04	99.70

NO.	FORMULA	NAME	A	B	C	DEL GF @298K
610	C14H30S	METHYL-TRIDECYL-SULFIDE	-278.793	1.2219E+00	1.8683E-04	102.34
611	C14H30S	PROPYL-UNDECYL-SULFIDE	-281.344	1.2181E+00	1.8773E-04	98.74
612	C14H30S	1-TETRADECANETHIOL	-285.169	1.2193E+00	1.8537E-04	95.06
613	C14H30S2	HEPTYL-DISULFIDE	-271.597	1.1987E+00	2.0566E-04	104.31
614	C15H18	1-PENTYLNAPHTHALENE	27.930	7.2052E-01	5.6005E-05	248.32
615	C15H18	2-PENTYLNAPHTHALENE	27.301	7.1700E-01	5.7364E-05	246.77
616	C15H24	1-PHENYLNONANE	-122.814	1.0110E+00	8.0315E-05	186.56
617	C15H28	1-PENTADECYNE	-68.040	1.1860E+00	8.7866E-05	294.26
618	C15H30	1-PENTADECENE	-234.403	1.3006E+00	9.7536E-05	163.05
619	C15H30	1-CYCLOPENTYLDECANE	-300.184	1.3476E+00	1.0295E-04	111.84
620	C15H30	1-CYCLOHEXYL-NONANE	-325.560	1.3912E+00	9.2165E-05	98.49
621	C15H32	PENTADECANE	-360.562	1.4269E+00	1.0458E-04	75.23
622	C15H32O	1-PENTADECANOL	-512.580	1.4780E+00	1.0540E-04	-61.46
623	C15H32S	BUTYL-UNDECYL-SULFIDE	-302.378	1.3111E+00	1.9792E-04	106.36
624	C15H32S	DODECYL-PROPYL-SULFIDE	-301.620	1.3111E+00	1.9794E-04	107.11
625	C15H32S	ETHYL-TRIDECYL-SULFIDE	-301.464	1.3125E+00	1.9758E-04	107.65
626	C15H32S	METHYL-TETRADECYL-SULFIDE	-298.940	1.3145E+00	1.9728E-04	110.75
627	C15H32S	1-PENTADECANETHIOL	-305.323	1.3120E+00	1.9578E-04	103.47
628	C16H26	1-PHENYLDECANE	-143.980	1.1082E+00	8.6141E-05	194.97
629	C16H26	PENTAETHYL-BENZENE	-181.006	1.1365E+00	7.1696E-05	164.98
630	C16H30	1-HEXADECYNE	-89.177	1.2831E+00	9.3850E-05	302.67
631	C16H32	1-HEXADECENE	-255.460	1.3976E+00	1.0363E-04	171.50
632	C16H32	1-CYCLOPENTYL-UNDECANE	-321.252	1.4445E+00	1.0913E-04	120.25
633	C16H32	1-CYCLOHEXYLDECANE	-346.683	1.4883E+00	9.8015E-05	106.90
634	C16H34	HEXADECANE	-381.697	1.5241E+00	1.1046E-04	83.68
635	C16H34O	1-HEXADECANOL	-533.675	1.5751E+00	1.1135E-04	-53.01
636	C16H34S	BUTYL-DODECYL-SULFIDE	-322.605	1.4041E+00	2.0806E-04	114.77
637	C16H34S	ETHYL-TETRADECYL-SULFIDE	-321.538	1.4049E+00	2.0836E-04	116.11
638	C16H34S	METHYL-PENTADECYL-SULFIDE	-319.146	1.4075E+00	2.0743E-04	119.20
639	C16H34S	OCTYL-SULFIDE	-322.409	1.4092E+00	2.0873E-04	116.52
640	C16H34S	PROPYL-TRIDECYL-SULFIDE	-321.856	1.4041E+00	2.0804E-04	115.52
641	C16H34S	1-HEXADECANETHIOL	-325.520	1.4048E+00	2.0614E-04	111.88
642	C16H34S2	OCTYL-DISULFIDE	-312.192	1.3851E+00	2.2550E-04	121.08
643	C17H28	1-PHENYLUNDECANE	-165.020	1.2050E+00	9.2372E-05	203.43
644	C17H32	1-HEPTADECYNE	-110.335	1.3802E+00	9.9805E-05	311.04
645	C17H34	1-HEPTADECENE	-276.569	1.4947E+00	1.0958E-04	179.91
646	C17H34	1-CYCLOPENTYL-DODECANE	-345.061	1.5413E+00	1.1522E-04	125.94
647	C17H34	1-CYCLOHEXYL-UNDECANE	-367.817	1.5854E+00	1.0407E-04	115.31
648	C17H36	HEPTADECANE	-402.745	1.6209E+00	1.1669E-04	92.09
649	C17H36O	1-HEPTADECANOL	-554.820	1.6721E+00	1.1740E-04	-44.64
650	C17H36S	BUTYL-TRIDECYL-SULFIDE	-342.679	1.4965E+00	2.1883E-04	123.22
651	C17H36S	ETHYL-PENTADECYL-SULFIDE	-341.808	1.4980E+00	2.1838E-04	124.47
652	C17H36S	HEXADECYL-METHYL-SULFIDE	-339.362	1.5003E+00	2.1778E-04	127.57
653	C17H36S	PROPYL-TETRADECYL-SULFIDE	-341.939	1.4965E+00	2.1877E-04	123.97
654	C17H36S	1-HEPTADECANETHIOL	-345.747	1.4978E+00	2.1627E-04	120.29
655	C18H30	1-PHENYL-DODECANE	-186.163	1.3021E+00	9.8370E-05	211.79
656	C18H30	HEXAETHYL-BENZENE	-231.018	1.3613E+00	7.6183E-05	182.46
657	C18H34	1-OCTADECYNE	-131.478	1.4774E+00	1.0574E-04	319.49
658	C18H36	1-OCTADECENE	-297.655	1.5915E+00	1.1579E-04	188.32
659	C18H36	1-CYCLOPENTYL-TRIDECANE	-363.548	1.6385E+00	1.2095E-04	136.98
660	C18H36	1-CYCLOHEXYL-DODECANE	-388.881	1.6821E+00	1.1029E-04	123.72
661	C18H38	1-OCTADECANE	-423.911	1.7180E+00	1.2257E-04	100.50
662	C18H38O	1-OCTADECANOL	-575.861	1.7690E+00	1.2355E-04	-36.19
663	C18H38S	BUTYL-TETRADECYL-SULFIDE	-362.931	1.5895E+00	2.2894E-04	131.59
664	C18H38S	ETHYL-HEXADECYL-SULFIDE	-361.840	1.5902E+00	2.2925E-04	132.93
665	C18H38S	HEPTADECYL-METHYL-SULFIDE	-359.398	1.5926E+00	2.2863E-04	136.02
666	C18H38S	NONYL-SULFIDE	-362.874	1.5950E+00	2.2922E-04	133.30
667	C18H38S	PENTADECYL-PROPYL-SULFIDE	-362.194	1.5896E+00	2.2887E-04	132.34
668	C18H38S	1-OCTADECANETHIOL	-365.792	1.5899E+00	2.2735E-04	128.74
669	C18H38S2	NONYL-DISULFIDE	-352.424	1.5702E+00	2.4668E-04	137.95
670	C19H32	1-PHENYL-TRIDECANE	-207.235	1.3990E+00	1.0443E-04	220.25
671	C19H36	1-NONADECYNE	-152.616	1.5744E+00	1.1175E-04	327.86
672	C19H38	1-NONADECENE	-318.805	1.6887E+00	1.2163E-04	196.73
673	C19H38	1-CYCLOPENTYL-TETRADECANE	-384.618	1.7355E+00	1.2721E-04	145.48
674	C19H38	1-CYCLOHEXYL-TRIDECANE	-409.956	1.7791E+00	1.1633E-04	132.17
675	C19H40	NONADECANE	-444.994	1.8150E+00	1.2867E-04	108.91
676	C19H40O	1-NONADECANOL	-597.017	1.8661E+00	1.2948E-04	-27.78
677	C19H40S	BUTYL-PENTADECYL-SULFIDE	-380.286	1.6678E+00	2.5645E-04	140.00
678	C19H40S	ETHYL-HEPTADECYL-SULFIDE	-382.174	1.6836E+00	2.3903E-04	141.29
679	C19H40S	HEXADECYL-PROPYL-SULFIDE	-382.299	1.6822E+00	2.3931E-04	140.79
680	C19H40S	METHYL-OCTADECYL-SULFIDE	-379.714	1.6859E+00	2.3849E-04	144.39
681	C19H40S	1-NONADECANETHIOL	-386.116	1.6833E+00	2.3717E-04	137.11
682	C20H34	1-PHENYL-TETRADECANE	-228.416	1.4962E+00	1.1037E-04	228.61
683	C20H38	1-EICOSYNE	-173.613	1.6710E+00	1.1807E-04	336.31
684	C20H40	1-EICOSENE	-339.913	1.7856E+00	1.2774E-04	205.14
685	C20H40	1-CYCLOPENTYLPENTADECANE	-405.771	1.8327E+00	1.3312E-04	153.89
686	C20H40	1-CYCLOHEXYL-TETRADECANE	-431.135	1.8762E+00	1.2227E-04	140.54
687	C20H42	EICOSANE	-466.086	1.9119E+00	1.3473E-04	117.32
688	C20H42O	1-EICOSANOL	-618.212	1.9633E+00	1.3535E-04	-19.41
689	C20H42S	BUTYL-HEXADECYL-SULFIDE	-403.297	1.7751E+00	2.4959E-04	148.41
690	C20H42S	DECYL-SULFIDE	-403.294	1.7809E+00	2.4954E-04	150.16
691	C20H42S	ETHYL-OCTADECYL-SULFIDE	-402.324	1.7761E+00	2.4974E-04	149.70
692	C20H42S	HEPTADECYL-PROPYL-SULFIDE	-402.579	1.7752E+00	2.4943E-04	149.16
693	C20H42S	METHYL-NONADECYL-SULFIDE	-399.878	1.7786E+00	2.4886E-04	152.80
694	C20H42S	1-EICOSANETHIOL	-406.155	1.7754E+00	2.4814E-04	145.52
695	C20H42S2	DECYL-DISULFIDE	-392.688	1.7552E+00	2.6793E-04	154.77
696	C21H36	1-PHENYLPENTADECANE	-249.518	1.5931E+00	1.1646E-04	237.02
697	C21H42	1-CYCLOPENTYL-HEXADECANE	-426.839	1.9296E+00	1.3923E-04	162.30

NO.	FORMULA	NAME	A	B	C	DEL GF @298K
698	C21H42	1-CYCLOHEXYL-PENTADECANE	−452.259	1.9733E+00	1.2822E-04	148.95
699	C22H38	1-PHENYLHEXA-DECANE	−270.675	1.6903E+00	1.2232E-04	245.43
700	C22H44	1-CYCLOHEXYL-HEXADECANE	−473.340	2.0702E+00	1.3439E-04	157.36

NOTE: 1. Temperature range is 298–1000K except for S, Br and I compounds.
2. Temperature range for S, Br and I compounds is 298–717.2K, 332.6–1000K and 458.4–1000K.
3. Additional compounds are: H2O(g) (A=−2.4174E+02, B=4.1740E-02, C=7.4281E-06) AND HCL(g) (A=−9.2209E+01, B=−1.1226E-02, C=2.6966E-06).

(text continued from page 370)

The equation for any of 700 major organic compounds is given as temperature coefficients. Then the reaction can be tested at various temperature levels beyond the standard 298°K conditions imposed by many other data tabulations. Data for water and hydrogen chloride are also included.

Gibbs free energy of formation of ideal gas (ΔG_f, kjoule/g-mol) is calculated from the tabulated coefficients (A, B, C) and the temperature (T, °K) using the following equation:

$$\Delta G_f = A + BT + CT^2 \tag{1}$$

Chemical equilibrium for a reaction is associated with the change in Gibbs free energy (ΔG_r) calculated as follows:

$$\Delta G_r = \Delta G_{f,\ products} - \Delta G_{f,\ reactants} \tag{2}$$

If the change in Gibbs free energy is negative, the thermodynamics for the reaction are favorable. On the other hand, if the change in Gibbs free energy is highly positive, the thermodynamics for the reaction are not favorable and may be feasible only under special circumstances. Rough criteria for screening chemical reactions are as follows:

$\Delta G_r < 0$ kjoules/g-mol	favorable
$0 < dG_r < 50$ kjoules/g-mol	possibly favorable
$\Delta G_r > 50$ kjoules/g-mol	not favorable

Example. Calculate the change in Gibbs free energy for the reaction of methanol and oxygen to produce formaldehyde and water at reaction temperatures of 600, 700, 800, 900, and 1,000°K:

$$CH_4O(g) + 0.5\ O_2(g) \rightarrow CH_2O(g) + H_2O(g)$$

Using correlation constants from Table 1 and Equation 1 at temperature of 600°K, we obtain:

$$\begin{aligned}
\Delta G_{f,CH_2O} &= -115.972 + 1.663*10^{-2}(600) + 1.138*10^{-5}(600^2) = -101.9 \\
\Delta G_{f,H_2O} &= -241.74 + 4.174*10^{-2}(600) + 7.428*10^{-6}(600^2) = -214.02 \\
\Delta G_{f,CH_4O} &= -201.860 + 1.254*10^{-1}(600) + 2.035*10^{-5}(600^2) = -119.28 \\
\Delta G_{f,O_2} &= 0
\end{aligned}$$

The change in Gibbs free energy for the reaction is determined from Equation 2 and the Gibbs free energy of formation for the products and reactants:

$$\begin{aligned}
\Delta G_r &= \Delta G_{f,CH_2O} + \Delta G_{f,H_2O} - \Delta G_{f,CH_4O} - 0.5*\Delta G_{f,O_2} \\
&= -101.9 + (-214.02) - (-119.28) - 0 \\
&= -196.64 \text{ kjoule/g-mol}
\end{aligned}$$

Repeating the calculations at the other temperatures gives these results:

T, °K	600	700	800	900	1,000
G_r, kJ/g-mol	−197	−204	−210	−217	−224

Since the change in Gibbs free energy for the reaction is highly negative, the thermodynamics for the reaction at these temperatures are favorable (reaction promising).

Data source. The correlation constants were determined from a least-squares fit of data from the literature.[2-13] In most cases, average deviations between calculated and reported data were less than 0.6 kjoules/g-mol.

A copy (5¼ inch floppy disk) of a menu-driven computer program to calculate Gibbs free energy of formation and change in Gibbs free energy for reactions (including random access data file of compound coefficients) is available for a nominal fee. For details, contact: C. L. Yaws, Dept. of Chem. Eng. Lamar University, P.O. Box 10053, Beaumont, Texas 77710, USA.

Literature Cited

1. Yaws, C. L. and Chiang, P-Y., *Chemical Engineering,* Vol. 95, No. 13, Sept. 26, 1988, pp. 81–88.
2. Gallant, R. W., *Physical Properties of Hydrocarbons,* Vols. 1 and 2, Gulf Publishing Co., Houston, 1968 and 1970.
3. Gallant, R. W., and Railey, J. M., *Physical Properties of Hydrocarbons,* Vol. 2, 2nd ed., Gulf Publishing Co., Houston, 1984.

4. Braker, W., and Mossman, A. L., *Matheson Gas Data Book, 6th ed.,* Matheson, Lyndhurst, N.J., 1980.
5. Dean, John A., *Handbook of Organic Chemistry,* McGraw-Hill Book Co., New York, 1987.
6. Perry, R. H., ed., *Perry's Chemical Engineers' Handbook, 6th ed.,* McGraw-Hill Book Co., 1984.
7. *Selected Values of Properties of Hydrocarbons and Related Compounds, Vol. 7,* TRC Tables (u-y), Thermodynamic Research Center, Texas A&M Univ., College Station, Texas, 1977 and 1987.
8. *Selected Values of Properties of Chemical Compounds, Vol. 4,* TRC Tables (u-y), Thermodynamic Research Center, Texas A&M Univ., College Station, Texas, 1977 and 1987.
9. Stull, D. R., et al., *The Chemical Thermodynamics of Organic Compounds,* Wiley, New York, 1969.
10. Stull, D. R. and Prophet, H., Project Directors, *JANAF Thermochemical Tables, 2nd ed.,* NSRDS-NBS 37, U.S. Govt. Printing Office, Washington, D.C., 1971.
11. Weast, R. C., ed., *CRC Handbook of Chemistry and Physics,* CRC Press, Boca Raton, Fla., 1985.
12. Yaws, C. L., *Physical Properties,* McGraw-Hill Book Co., New York, 1977.
13. Reid, R. C., et al., *The Properties of Gases and Liquids, 4th ed.,* McGraw-Hill Book Co., New York, 1987.

Source

Yaws, C. L. and Chiang, P., "Find Favorable Reactions Faster," *Hydrocarbon Processing,* November 1988, p. 81.

New Refrigerants

New hydroflurocarbons (HFCs) are replacing the chloroflurocarbons (CFCs) and hydrochloroflurocarbons (HCFCs) phased out to lessen damage to the ozone layer. The DuPont website lists physical properties for a number of their refrigerants.
Go to:

www.dupont.com/suva/412.html

These physical property tables give three vapor pressure points that can be connected on your favorite Cox type vapor pressure chart for a quick approximation of the whole range. These three points are:

normal boiling point
critical properties
vapor pressures at 25°C

If more accurate vapor pressure data across the range are needed, DuPont can supply them. I received full-range vapor pressure data for many products on request from their very helpful SUVA® Refrigerants Division.

Source

E. I. DuPont deNemours and Company website, "Physical properties and ASHRAE standard 34 classification of SUVA® refrigerants."

Appendixes

Appendix 1
Computer Programming

The first edition discussed porting engineering programs from a mainframe computer to a minicomputer and development towards a final port to PC level. Our Central Engineering Department loves the minicomputer with its UNIX operating system and FORTRAN 77 programs. The company's new Windows 95 wide area network is also helpful and we believe that certain of our programs would be better applied there.

Reasonable language choices for the PC port in the first edition were:

FORTRAN 77
Turbo Pascal
C
C++

The C language was our recommended choice based on such factors as wide commercial usage and increased use in the universities. C++ was touted as the wave of the future as the best representative of Object Oriented Programming.

Since that time, we see increased use of the powerful FORTRAN 90 and increased use of C++ over C. A recent university contact said that they accept FORTRAN, C, or C++ in their problems classes. Companies are beginning to embrace Sun Microsystems' Java for business applications on the Internet and elsewhere. Are engineering applications far behind? Java can replace C++ as an object-oriented language.

I am quite a fan of Visual Basic, having produced a commercial Windows application with it. Visual C++ would be a step up, but for my use the extra power is not worth the learning curve. Give Visual Basic a try some time, just for the fun of it.

Here is a tip for those who travel with a laptop computer, but don't wish to carry a printer. Use the hotel's fax machine as a printer by faxing from your room to the lobby. Also use the hotel's fax machine as a scanner by faxing the other way.

Source

Munson-McGee, Stu, Dept. of Chemical Engineering, New Mexico State University.

Appendix 2
Geographic Information Systems

There is a good chance that you have had some contact with geographic information systems (GIS). A GIS is a computer representation of a system that uses all pertinent survey and database information to generate intelligent maps and queries. Much has been published on the subject. In addition there are many vendors of the technology and many seminars. GIS is a complicated subject. Several early GIS projects were not successful. Often GIS projects are very expensive.

To me, the easiest GIS demonstration to understand is that for a typical city. We can all relate to the various utilities involved and can see the usefulness of a GIS for maintenance and emergencies as we view the on-screen city GIS map. Watching the layers come in and out and change color is impressive.

A pipeline GIS is a close cousin to a city GIS. Our company needed to produce a GIS for a two-mainline system from New Mexico to California. A subsidiary had already begun a GIS for the gathering systems, their part of the business.

To bring ourselves up-to-speed we attended seminars, talked to companies and municipalities with GIS projects, and invited vendors for presentations. Here is what we finally decided upon.

1. We made the drafting department the lead agency to produce the GIS aided by a consultant.
2. Because we were already an AutoCAD shop, we built the GIS around AutoCAD instead of using software that was more specialized for GIS purposes, but more expensive. AutoCAD is getting more powerful for GIS usage as new versions are released.
3. We began with a program to digitize all our alignment sheets into a GIS format. This gave us immediate savings over having the sheets updated by an outside company. It also gave us material for progress reports understandable to management. We feel it is important to have impressive presentations for management during a pilot project and for the early stages of the main GIS

project. The alignment sheets facilitated understanding similar to the city map discussed earlier.

4. During the alignment sheet project, we incorporated the databases from the engineering and right-of-way departments. This allowed the early presentations to include demonstration of the intelligent features of the GIS on-screen alignment sheets. We could launch queries by clicking on points on the on-screen alignment sheets.
5. Finally, the many other company databases were added.

GIS will provide us many advantages such as:

1. Easier and faster alignment sheet updating and distribution.
2. Queries across multiple databases. GIS is a basic tool for tying all of a company's databases together. In our case everything relates to the pipeline.
3. Full knowledge and control for system modifications.
4. Compliance with governmental agencies. GIS is being mandated.
5. Improvement in the integration of a global positioning system (surveys by satellite) data into our system.

To keep the various company departments informed, we provided numerous inhouse seminars. We also published the Geographic Information System Monthly (GISMO) on the company e-mail system.

Appendix 3
Internet Ideas

The Internet's growth is amazing. *Investor's Business Daily* reported on a research study by International Data Corp. that showed commercial sites on the World Wide Web grew by more than 45,000 in 1996.

What's in it for us? Trade magazines show numerous sites for chemical engineers, some with monthly lists. No list of sites can ever be complete because of frequent additions and turnover. The following are some examples of chemical engineering sites. As you browse these on line, you will find others:

AIChE—www.aiche.org
Alloys
    AlloyTech—www.alloytech.com
    Carpenter Technology Corp—www.cartech.com
Automation
    Allen-Bradley—www.ab.com
    Foxboro—www.foxboro.com
Chemical Engineering Information
    Chemical Engineering—
        www.che.ufl.edu/WWW-CHE/index.html
    Chemical Engineering URL's Directory—
    www.ciw.uni-karlsruhe.de/chem-eng.html
    Engineering Information—www.ei.org
Chemical Information
    ChemFinder Search Engine—http://chemfinder.camsoft.com
    Chemistry at Center for Scientific Computing (CSC)—www.csc.fi/lul/csc_chem.html
    Chemists' Club Library—
    http://members.aol.com/chemists/ccl.htm
    Arco—www.arcochem.com
    Dow—www.dow.com
    Dupont—www.dupont.com
Coatings
    General Magnaplate—www.magnaplate.com
    Protective coatings, linings and related resources—
    www.corrosion.com

Consultants
- Pathfinder Inc.—www.pfinet.com
- Radian International—www.radian.com
- World Batch Forum—www.cyboard.com/wbf/index/html

Controls
- Johnson Yokogawa Corp.—www.jyc.com
- Watlow Electric Mfg.—www.watlow.com

Clearinghouse
- Industry Net—www.industry.net/*(*Fill in a name such as GE, Foxboro, Bettis, Mueller, etc.)

Environmental
- Blymer Engineers—www.blymer.com
- ENSR—www.ensr.com
- Environmental Resources Management—www.erm.com
- Pace University Global Environmental Law Network—www.law.pace.edu
- Roy F. Weston—www.rfweston.com
- R & R International—www.rrint.com

Filtration
- Pall—www.pall.com

Fluids Handling
- Moyno Industrial Products—www.tpusa.com/moyno

Instruments
- Sierra Instruments—www.sierrainst.com

Mass Transfer
- Nutter Engineering—www.nutterl.com

Measurement
- Omega Engineering—www.omega.com

Mixing
- Chemineer—www.chemineer.com

Patents
- Questel* Orbit—www.questel.orbit.com/patents

Piping
- Crane Resistoflex—www.resistoflex.com

Properties
- National Institute of Standards—www.nist.gov/srd

Publications
- Gulf Publishing Company—www.gulfpub.com
- Metric Guides—http://physics.nist.gov/SI
- National Academy Press—www.nap.edu

Research
- Argonne National Laboratory—www.anl.gov
- Western Research Institute—www.wri.uwyo.edu

Sensors

Gas Tech—www.gastech-inc.com
- MIL-RAM Technology—www.mil-ram.com

Software
- Batchmaster—www.batchmaster.com
- Cherwell Scientific—www.cherwell.com/cherwell
- COADE Inc—ww.coade.com
- ESM Software—www.esm-software.com
- ExperTune—www.expertune.com
- Gulf Publishing Company—www.gulfpub.com
- National Instruments—www.natinst.com/hiq
- Neutral Ware—www.neutralware.com
- Rebis—www.rebis.com
- Simulation Sciences—www.simsci.com

Technical Employment
- National Ad Search Weekly—www.nationaladsearch.com

Technology
- NASA's Commercial Technology Network—http://nctn.hq.nasa.gov

Testing
- Panametrics—www.panametrics.com
- Sensorex—www.sensorex.com/process.html

Water/Waste Handling
- Aqua-Chem—www.aqua-chem.com

The information is voluminous and varied. The chemical manufacturers have good product information. A good example of useful product information is on the DuPont website (www.dupont.com).

As you know, most countries are phasing out some refrigerants to lessen damage to the ozone layer. The chemicals being phased out are chloroflurocarbons (CFCs) and hydrochloroflurocarbons (HCFCs). Replacements are hydroflurocarbons (HFCs) and certain blends. The DuPont website gives a table of recommended replacement refrigerants for various applications. For each application DuPont lists the previously used refrigerant being phased out and the recommended new refrigerant for two cases: "Retrofit" or "New System." This table of refrigerants is in Chapter 11, "Refrigeration."

For my automobile's air conditioning system, for example, I need to change R-12 for DuPont SUVA® 134a (also referred to as R-134a and HFC-134a elsewhere). Other listed applications are such things as building A/C, light/medium/low temperature commercial refrigeration, low temperature transport, marine A/C, and food storage.

DuPont also provides a computer program to simulate performance of different refrigerants in your system. It is called the "Dupont Refrigeration Expert" (DUPREX).

If you have questions about your auto air conditioning system the site www.aircondition.com has a question-and-answer bulletin board for interactive responses.

If you use the Internet for investing, you are probably familiar with the following free sites:

www.dljdirect.com—I like their daily Dow graph.

www.nasdaq.com—This is a convenient site to enter the name of a company or mutual fund (on the Nasdaq, New York, or American Exchanges) and be provided with its trading symbol. Just click on the option "Security Look-Up." There is also a link to the SEC Edgar Database for annual reports.

www.sec.gov/edaux/searches.htm—This is the place to find such things as a company's quarterly and annual reports, but you have to go through a lot of work to get to the goodies:

1. Click CIK
2. Click Ticker Symbol Look-up
3. Type company name
4. Receive a long number
5. Highlight it and hit Ctrl-C to copy it to the clipboard
6. Backtrack to the "Search the EDGAR Database" page
7. Select "Search the EDGAR Archives"
8. In the "Search Keywords" field hit Ctrl-V to paste in the CIK of the company or fund you are researching
9. Hit Enter to get a document listing

Here is a tip: to avoid all this a second time just bookmark this page and you can skip to it immediately for subsequent searches.

Sources

1. *Investor's Business Daily,* March 11, 1997.
2. *Chemical Engineering,* McGraw-Hill, Inc.
3. *Chemical Engineering Progress,* AIChE.
4. *Hydrocarbon Processing,* Gulf Publishing Co.
5. *Luckman's World Wide Web Yellow Pages,* Barnes and Noble.
6. E. I. DuPont de Nemours and Company Website—www.dupont.com
7. *Chemicals on the Internet,* Vols 1 and 2, William Crowley, (ed.), Gulf Publishing Co.

Appendix 4
Process Safety Management

Process Safety Management (PSM)[5], under OSHA, attempts to protect employees exposed to toxicity, fire, or explosion. Many plants employing chemical engineers must do a PSM consisting of fourteen parts. Some of the parts are greatly facilitated if the team includes chemical engineers.

I was included on teams for the following parts: Process Safety Information, Process Hazards Analysis, and Mechanical Integrity. The experience was rewarding and you are encouraged to participate in PSM if given the opportunity. Our plants are safer because of hard work by dedicated teams.

Safety relief systems are verified as part of PSM. This includes the PSVs themselves and also flare system piping networks. Safety relief valves are covered in Section 1—Fluid Flow. A good procedure for sizing the flare system piping is found in Section 19—Safety-Relief Manifolds. This method, first published in the *Oil and Gas Journal*[3], has been adopted by API[1]. I have also used the procedure to do compressible flow analyses of process piping as part of PSM.

Working with a Process Hazards Analysis (PHA) team was particularly rewarding. The plant operators on the team pointed out items that they had considered safety problems for years, but that hadn't been fully addressed until the PHA team was formed. We used the "What if" analysis method. Many methods are available for PHA.

As part of PHA the author checked the natural ventilation system for an auxiliary electrical generator building that contained gas engine generators. The procedure to perform this API recommended practice[2] can be found in Section 19—Safety-Natural Ventilation.

More focus is being put on safety. OSHA and EPA will soon team up to investigate major chemical accidents or releases. The AIChE has established a safety center and it seems like a higher percentage of AIChE publications relate to safety. I can envision more chemical engineers being involved part or full time in safety.

Sources

1. American Petroleum Institute (API), RP520, "Sizing, Selection and Installation of Pressure-Relieving Devices in Refineries."
2. API, RP500, "Recommended Practice for Classification of Locations for Electrical Installations at Petroleum Facilities."
3. Mak, Henry Y., "New Method Speeds Pressure-Relief Manifold Design," *Oil and Gas Journal,* Nov. 20, 1978.
4. Chadd, Charles M., and J. K. Bowman, "OSHA and EPA Team Up," *Chemical Engineering,* April 1997, p. 135.
5. Process Safety Management of Highly Hazardous Chemicals, Title 29, Code of Federal Regulations, Section 1910.119.

Appendix 5
Do-It-Yourself Shortcut Methods

This paper, presented by the author at the 1985 ASEE Conference in Atlanta, Georgia, is intended to show how classic shortcut design methods can be developed, and to encourage new and better ones.

DEVELOPMENT OF SHORTCUT EQUIPMENT DESIGN METHODS

Introduction

Shortly after my first book was published, a draftsman in our company bought one for his son Bruce, a mechanical engineering student. The father told me the following story. Bruce's engineering class had been assigned a problem related to available energy in a working fluid. Bruce's classmates spent a substantial amount of time working with Mollier charts and such. Fortunately for Bruce, the problem statement was tailor-made for simply plugging into a formula in his new book.

When Bruce presented his ten-minute solution to the graduate student instructor the next morning, the instructor remarked, "You have the right answer, but I have never seen it done this way." I will never forget Bruce's reply, "Well, I guess that's how they do it in the field."

Like Bruce, I am here to share the perspective of the field. Shortcut equipment design methods play a major role in process equipment installations. Virtually all process designs employ these methods and shortcut techniques are developed from many sources.

What exactly do we mean by the term "shortcut equipment design method"? Here is an example. Two rules are given for determining the thickness in inches of copper boiler level float balls:

Rule 1. Thickness = diameter in inches x pressure in lbs per square inch/16,000

Rule 2. Thickness = diameter x sq. rt. pressure/1240

These two rules give the same answer for a pressure of about the level for a modern boiler: 166 lbs per sq. in.—modern, that is, when the rules were published in the

October 1891 issue of *The Locomotive*. We must be careful that our shortcut methods don't become dated!

In this appendix, shortcut design methods will be defined to include rules of thumb and safety or experience factors. The following discussions catalog shortcut design methods according to their origin. This breakdown is useful in developing the perspective of the field.

Often in the field, designs must be done quickly, for example, during an unexpected plant shutdown. For such cases, shortcut design methods are invaluable. The field invariably has the best perspective on how designs will perform for startup, shutdown, upset, or offload conditions. Experience factors come into play for field review of these design aspects.

Sources of Shortcut Design Methods

Seven categories of shortcut design methods are discussed. These seven differ according to the origin of the methods. Some of the shortcut methods might have characteristics of more than one category, so some overlapping will be observed. The purpose here is to show the fertile fields from which shortcut methods can grow and to encourage new and better ones.

From Simplifying Assumptions

Many of the better known shortcut equipment design methods have been derived by informed assumptions and mathematical analysis. Testing in the laboratory or field was classically used to validate these methods but computers now help by providing easy access to rigorous design calculations.

The assumptions that go into the analysis often are the outgrowth of a thorough understanding of perfect or near perfect thermodynamic behavior. Some assumptions are based upon the near constancy of certain parameters for wide ranges of commercial practice. Linear variation of parameters is also used at times for simplification.

Development of useful shortcut methods by intuitive and evolutionary mental analysis alone must surely be extremely fulfilling. We will see later how other developmentally oriented individuals are able to build upon the original shortcut method ideas to create variations on the theme for specific applications.

Examples chosen for this category include the operations of vapor/liquid separation, heat transfer, and fluid flow. The Underwood[1] method of estimating the minimum reflux ratio for a fractionation column multicomponent mixture is one of the most famous shortcut methods of the type under discussion. The simplifying assumptions required are constant molal overflow and component relative volatility throughout the column. Although both these quantities show nonlinear column profiles, the net effect of using the method is acceptable accuracy for reasonably well behaved systems. Many commercial columns were built using this shortcut calculation.

The Underwood, as well as other methods such as Brown-Martin[2] and Colburn,[3] is used primarily for new column design when overhead and bottoms compositions are specified. The Smith-Brinkley method[4] was developed to apply to all types of equilibrium stage equipment existing in the field. For a fractionation column, overhead and bottoms compositions are calculated from known operating parameters. This is handy in the field for estimating the effect of parameter changes on composition. It is also useful for setting up predictive computer control algorithms. This method was developed by dividing the column in half and assuming constant molal flow rates and distribution coefficients (given constant K values) within each section. The complete derivation of the method is shown in B. D. Smith's book[5] and is complicated enough to be outside the scope of this paper and my mathematical abilities.

The Edmister methods[6,7] for absorption use factors for overall stripping and absorption built up by combining column top and bottom factors.

In the field of heat transfer, a good example of this category of shortcut design method is the famous "F" correction factor[8] to correct the log mean temperature difference of shell and tube heat exchangers for deviations from true countercurrent flow. For multipass heat exchangers, the assumptions are:

1. The overall heat-transfer coefficient U is constant throughout the heat exchanger.
2. The rate of flow of each fluid is constant.
3. The specific heat of each fluid is constant.
4. There is no condensation of vapor or boiling of liquid in part of the exchanger.
5. Heat losses are negligible.
6. There is equal heat transfer surface in each pass.
7. The temperature of the shell-side fluid in any shell-side pass is uniform over any cross section.

The assumptions hold so well that the F factor charts are standardized by TEMA[9] for shell and tube heat exchanger design.

The Lapple charts[10] for compressible fluid flow are a good example for this operation. Assumptions of the gas obeying the ideal gas law, a horizontal pipe, and constant friction factor over the pipe length were used. Compressible flow analysis is normally used where pressure drop produces a change in density of more than 10%.

Finally, there is an interesting article[17] that shows how to use Taylor's series to generate shortcut methods from established theory. Examples are given for developing a criterion for replacing log mean temperature differences with average differences and for estimating the effect of temperature on reaction rate.

From Correlation of Test Data

This source of shortcut design methods, like the first one discussed, often requires a high level of ingenuity. A case in point is the Lockhart and Martinelli two-phase gas/liquid flowing pressure drop correlation.[11] Test data can be viewed and correlated in a number of ways. These investigators found that a significant generalization would be possible if a factor was generated based upon the pressure drops of the individual phases calculated as though each were alone in the pipe. This factor could then be correlated against the multipliers necessary to generate the total system pressure drop starting with the pressure drop of either phase.

Lockhart and Martinelli used pipes of one inch or less in diameter in their test work, achieving an accuracy of about +/–50%. Predictions are on the high side for certain two-phase flow regimes and low for others. The same +/–50% accuracy will hold up to about four inches in diameter. Other investigators have studied pipes to ten inches in diameter and specific systems; however, no better, generalized correlation has been found.[12] The way that the data was correlated made all the difference in the success of this widely used method.

Being persistent enough to look at data from alternate perspectives often pays off in a useful correlation where none existed previously. Two examples from my own experience illustrate this. I attempted to estimate the naphtha content of catalytic cracking gas oil feed using ASTM distillation data. Naphtha was defined as material boiling below 430°F. I had many sets of data giving naphtha percentage from TBP analyses, along with companion ASTM distillation results.

The naphtha content showed very poor correlation when plotted against ASTM 10%, 20%, 30%, etc., temperatures or mixtures thereof. However, when naphtha content was related to ASTM % off at fixed temperatures (430°F, 450°F, 470°F, etc.) extremely good correlations were obtained. Hindsight showed the reasonableness of this. The ASTM % off data resembled in magnitude and variability the sought after naphtha content. The initially tried temperatures bore little resemblance.

Another interesting case occurred when attempting to correlate adipic acid plant product nitrate content with production rate. A plot of nitrate content versus production rate produced the worst shotgun effect I have ever seen. Each production level produced a range of nitrate contents.

To establish the required causal effect of production rate on nitrates a year's history was investigated. For various production rate ranges the percentage of nitrate contents of various magnitudes was tabulated as follows:

A plot of accumulative percentages demonstrates the hypothesis (Figure 1).

Generalized single phase pressure drop correlations for pipes and packed beds have used complicated factors composed of groups (sometimes dimensionless) of physical property data and system geometry.[13, 14]

Considerable work on methods for pre-predicting fractionator tray efficiency continues to the present. Shortcut

Adipic Acid Production	Percentages in Nitrate Range				
Mlb/Day	20ppm or below	21–25	26–30	31–40	41 or above
90–150	73.7	10.5	3.5	8.8	3.5
150–180	63.3	14.3	10.2	6.1	6.1
180–200	40.9	30.3	10.6	10.6	7.6
200–220	33.0	35.4	13.9	11.4	6.3
220–240	29.0	20.3	15.9	17.4	17.4
240–260	10.7	21.4	7.2	28.6	32.1

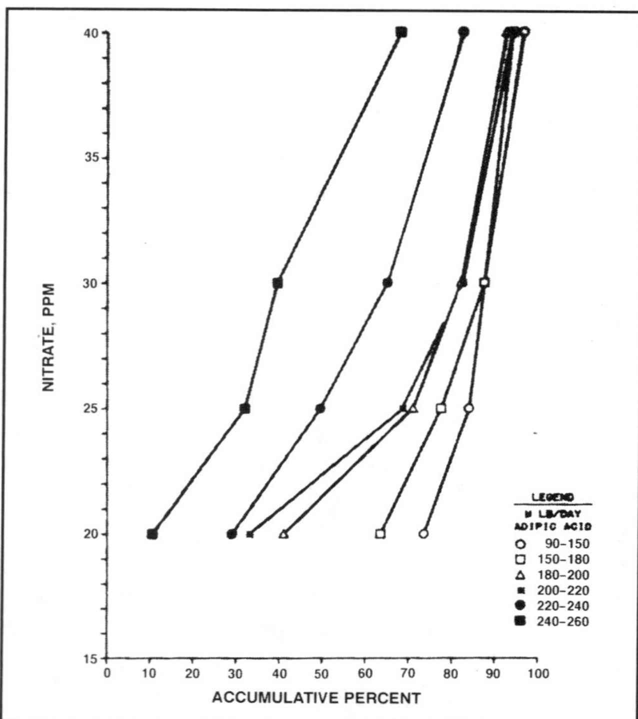

Figure 1. Adipic acid nitrate content.

methods from the past differed rather widely.[15, 16] The AIChE 1958 calculation method is 90% accurate,[16] but definitely cannot be termed shortcut.

From Correlation of Property Data

There are so many methods for generating properties or coefficients from simple property data that it is difficult to state where sophisticated methods end and shortcut methods begin. In the field of heat transfer, film coefficients are determined by sophisticated, though not rigorous, methods involving dimensionless groups composed largely of physical, chemical, or transport properties. Some of these are brief enough to border on what we could call shortcut.

Better examples of shortcut design methods developed from property data are fractionator tray efficiency, from viscosity[18, 19, 20] and the Clausius-Clapeyron equation[21] which is useful for approximating vapor pressure at a given temperature if the vapor pressure at a different temperature is known. The reference states that all vapor pressure equations can be traced back to this one.

From Operating and Maintenance Experience

One man's plant is another man's laboratory. The plant is indeed a fertile field for development of shortcut design methods.

Have you ever tried to decide which was the best article you ever read? For me, the answer comes quickly. It is "How to Operate an Amine Plant," by Don Ballard.[22] This article has helped me more than any other. I have gone back to the article time after time to troubleshoot and optimize amine plants. In addition, the article has design tips and shortcut rules of thumb developed by field experience.

If I ever have to design an amine plant I will know, for example, that the temperature of the lean amine solution entering the absorber should be about 10°F higher than the inlet gas temperature to prevent hydrocarbon condensation and subsequent foaming, that the reboiler tube bundle should be placed on a slide about six inches above the bottom of the shell to provide good circulation, that about two percent of the total circulating flow should pass through the carbon towers, and many other necessary requirements.

Such rules of thumb resulting from field experience give me not just isolated shortcut design parameters, but a system of understanding to do the design job. What a help!

A huge category of design rules of thumb born of field experience involves internal and external piping associated with fractionating columns.[23, 24, 25, 26, 27, 28] Critical dimensions, orientations, and angles for such things as feed and reboiler return headers, draw sumps, and downcomers mean the difference in the equipment working or not working. All the rules of thumb embrace sound thermodynamic and hydraulic principles but such principles would not always be so obvious within the time restraints of a design without the help of experience.

A good example of a disaster caused by overlooking seemingly small details is given in Reference 28. A series of reboiler related oversights caused a lengthy series of operating troubles for a new depropanizer. One by one, oversights were rectified, but before the final one could be fixed, local plant management lost patience, had the depropanizer torn down and sent the nameplate to the vice president over the engineering office.

Rules of thumb for piping are also particularly important for the following situations:

- Vacuum systems[29]
- Pump suction piping[30]
- Two-phase flow[31]
- Vertical downflow[32] (Froude number rules for avoiding trouble)

Maintenance records also supply valuable rules for things such as heat exchanger tube velocity limitations for various metals in seawater service and boiler tube heat density limitations.

Rules of thumb[15] are needed in design of process controls for the following (among others):

- Extra condenser or reboiler surface for control functions
- Extra fractionator trays for control margin
- Percentage of flowing pressure drop taken at the control valve.

From Rigorous Calculation Results

Gilliland[33] tells how his famous correlation was developed for relating actual and minimum reflux to actual and minimum theoretical stages for a fractionating column. Numerous plate-to-plate calculations were made and the results plotted using his well-known correlating parameters. The best curve was then drawn through points.

He observed that, theoretically, a single curve couldn't represent all cases exactly, when in fact the curve is actually a function of other things such as feed vapor fraction. However, he preferred to use a single curve since he envisioned the real value of his correlation to be rapid approximations for preliminary design and a guide for interpolating and extrapolating plate-to-plate calculations. He also eliminated slightly more accurate but more complicated correlating parameters for the same reason.

Gilliland's work should be an encouragement to us all in developing shortcut methods from rigorous results. The proliferation of computers makes our job relatively easy. I feel that when correlations of this type are developed, one shouldn't stop with the graph but continue until an equation is fitted to the data. Computers need equations.

Commercial computer computation services are being set up to provide shortcut calculation options. These provide considerable savings for preliminary design studies.

Even without a computer, or more likely, one with limited core, an engineer can build specialized fractionator cases by combining rigorous computer runs with shortcut techniques.[25] Usually a rigorous computer run is available for a plant case or a design case. Excursions for a plant fractionator can then be estimated using the Smith-Brinkley method.[4,5] The rigorous computer run gets the computational process started with accurate column profiles rather than an equal molal overflow approach.

Rigorous calculation results combined with plant data can be used to back calculate column tray efficiencies for plant revamp or expansion calculations. This is yet another specialized application. A similar approach was used by a friend to calculate "tuning factors" to adjust reactor calculations by comparing plant and rigorously calculated equilibrium yields.

From Other Shortcut Methods

Why not put new lyrics to an old tune? This is an excellent idea, and many have done this very thing. Rice[34] started with the Smith-Brinkley method[4, 5] used to calculate distillation, absorption, extraction, etc., overhead and bottoms compositions, and developed distillation equations for determining the liquid composition on any tray. This together with bubble point calculations yield a column temperature profile useful for column analysis.

Loeb[35] used Lapple's[10] compressible flow work, techniques, and reasoning to develop graphs useful for direct calculations between two points in a pipe. Lapple's graphs were designed for pressure drop estimations for flow from a large vessel into a length of pipe (having static velocity in the reservoir).

Mak[36] used the same Lapple article[10] to develop a method for designing flare headers. His method has the advantage of starting calculations from the flare tip (atmospheric) end, thus avoiding the trial and error calculations of methods starting at the inlet.

This author[25] developed a reboiled absorber alternate from the Smith-Brinkley[4, 5] generalized equation, and friends have developed plant fractionator control algorithms using Smith-Brinkley.

These examples should be sufficient to cause us all to go out and put our words to music.

From Industry Practice

Industry sets limits that bound our degrees of freedom and thus tend to shorten our design case study load. We are all aware of such limits and this last category is included primarily for completeness. Examples include minimum industrial thickness for carbon steel plate, and maximum baffle cut for shell and tube heat exchangers.

In addition, certain industry segments have ways of fabrication that seldom vary. An example is that for air-cooled heat exchangers the fanned section length divided by the bay width seldom exceeds 1.8.[37]

Summary and Conclusions

Shortcut equipment design methods are part of an engineering designer's life. In addition to using such methods frequently and in all of his designs, he should always be on the alert for chances to develop his own for future generations to use. The opportunities for doing this are great and the results can be very rewarding.

References

1. Underwood, A. J. V., Fractional Distillation of Multicomponent Mixtures, *Chem. Eng. Prog.*, 44, 603 (1948).
2. Martin, H. Z. and Brown, G. G., *Trans. A.I.Ch.E.*, 35, 679 (1939).
3. Colburn, A. P. *Trans. A.I.Ch.E.*, 37, 805 (1941).
4. Smith, B. D., and Brinkley, W. K., *A.I.Ch.E.J.*, 6, 446 (1960).
5. Smith, B. D., *Design of Equilibrium Stage Processes,* McGraw-Hill, 1963.
6. Edmister, W. C., *Petr. Engr.*, September 1947 Series to January 1948.
7. Edmister, W. C., "Absorption and Stripping-factor Functions for Distillation Calculation by Manual and Digital Computer Methods", *A.I.Ch.E.J.*, June 1957.
8. Bowman, Mueller, and Nagle, *Trans. Am. Soc. Mech. Engrs.*, 62, 283, 1940.
9. Standards of Tubular Exchanger Manufacturers Association (TEMA), Latest Edition.
10. Lapple, C. E., "Isothermal and Adiabatic Flow of Compressible Fluids", *Trans. A.I.Ch.E.*, 39, 1943.
11. Lockhart, R. W. and Martinelli, R. C., *Chem. Eng. Prog.*, January 1949, pp. 39–48.
12. Perry, R. H. and Green, D. W., *Perry's Chemical Engineers Handbook,* 6th Ed., McGraw-Hill, Inc., 1984.
13. Moody, L. F., "Friction Factors for Pipe Flow", *Trans. A.S.M.E.*, 66, 671–684 (1944).
14. Eckert, J. S., "How Tower Packings Behave", *Chem. Eng.* April 14, 1975.
15. Branan, C., *The Process Engineer's Pocket Handbook,* Vol. 2, Gulf Publishing Co., Houston, Texas 1983, p. 64–65.
16. Vital, Grossel, Olsen, "Estimating Separation Efficiency, Part 2—Plate Columns", *Hydro. Proc.*, November 1984, p. 147.
17. Douglas, J. M., "Derive Short-Cut Correlations", *Hydr. Proc.*, December 1978, p. 145.
18. Gunness, *Ind. Eng. Chem.*, 29, 1092, 1937.
19. Drickamer, H. G. and Bradford, J. B., "Overall Plate Efficiency of Commercial Hydrocarbon Fractionating Columns," *Trans. A.I.Ch.E.*, 39, 319, 1943.
20. Maxwell, J. B., *Data Book on Hydrocarbons,* Van Nostrand, 1965.
21. Reid, R. C. and Sherwood, T. K., *The Properties of Gases and Liquids, 2nd Ed.,* McGraw-Hill, 1966.
22. Ballard, Don, "How to Operate an Amine Plant," *Hydr. Proc.*, April 1966, pp. 137-144.
23. Kitterman, Layton, "Tower Internals and Accessories," Glitsch, Inc. Paper presented at Congresso Brasileiro de Petro Quimica, Rio de Janeiro, November 8–12, 1976.
24. Zister, H. Z., C. E. Refresher: Column Internals Parts 1–7, *Chem. Eng.*, Series beginning May 19, 1980.
25. Branan, C. R., *The Fractionator Analysis Pocket Handbook,* Gulf Publishing Co., Houston, Texas, 1978.
26. Kern, Robert, C. E. Refresher: Plant Layout, Part 3—Layout Arrangements for Distillation Columns; Part 9—How to Design Piping for Reboiler Systems; Part 10—How to Design Overhead Condensing Systems, *Chem. Eng.*, Series beginning August 15, 1977.
27. Lieberman, N. P., *Troubleshooting Refinery Processes,* Pennwell, 1981.
28. Lieberman, N. P., *Process Design for Reliable Operations,* Gulf Publishing Co., 1983.
29. Branan, C. R., *The Process Engineer's Pocket Handbook,* Gulf Publishing Co., Houston, Texas, 1976.
30. Kern, Robert, "C. E. Refresher—How to Design Piping for Pump-Suction Conditions," *Chem. Eng.*, April 28, 1975, p. 119.
31. Kern, Robert, "C. E. Refresher—Piping Design for Two-Phase Flow", *Chem. Engr.*, June 23, 1975, p. 145.
32. Simpson, L. L., "Sizing Piping for Process Plants", *Chem. Engr.*, June 17, 1968, p. 192 (Vertical downflow example p. 203).
33. Robinson, C. S., and Gilliland, E. R., *Elements of Fractional Distillation,* McGraw-Hill, 1950, p. 348.
34. Rice, V. L., "Way to Predict Tray Temperatures," *Hydr. Proc.*, August 1984, p. 83.
35. Loeb, M. B., "New Graphs for Solving Compressible Flow Problems," *Chem. Engr.*, May 19, 1969, p. 179.
36. Mak, H. Y., "New Method Speeds Pressure-Relief Manifold Design," *Oil and Gas J.*, November 20, 1978.
37. Lerner, J. E., "Simplified Air Cooler Estimating", *Hydr. Proc.*, February 1972.

Source

Branan, C. R., "Development of Short-Cut Equipment Design Methods," presented at the Proceedings 1985 Annual Conference (Computer Aided Engineering) of American Society For Engineering Education, Atlanta, Georgia, June 16–20, 1985, Vol. 1.

Appendix 6
Overview for Engineering Students

Engineering students can effectively use shortcut design methods and rules of thumb. This is especially true for seniors working on their plant design problems. Dr. Long[1], of New Mexico State University, shows how plausible designs are bounded by real world constraints in Figure 1. This helps the students narrow their focus (and their hard work) to a suitable range of commercial practice.

For a plant design problem, students will need guidance on the hierarchical levels of decision making to choose a process. Also checklists of information needed for the design of typical parts of commercial processes prove invaluable. These are provided in Dr. Long's student handout.

When the students are finally into the equipment design phase of their project, they need design rules of thumb. These "experience factors" provide the students ranges and bounds to save them time.

Process equipment has been optimized over the years. Initial cost has been balanced against performance and also against ease of transportation, construction, and maintenance. The students need to have normal design ranges as determined by this wealth of commercial experience. A good set of these handy rules of thumb follows[1]:

Handy Rules of Thumb

Distillation

1. Vapor velocity thru tower: 1.5–2.5 ft/sec (at top)
2. Reflux: R/R_{min} = 1.1–1.5
3. Height of column: 175 ft. max
4. Tray spacing: 18 in.–24 in.
5. Tray efficiency: ≈ 70% is usually conservative (it is usually a good idea to put a few extra trays—say 10% extra)
6. (a) Add 4 ft on top of tower for condenser
   (b) Add 6 ft on bottom for reboiler
7. If the tower diameter is < 3 ft use a packed tower instead of plates.
8. Pressure drop across trays: 4–5 in. H_2O

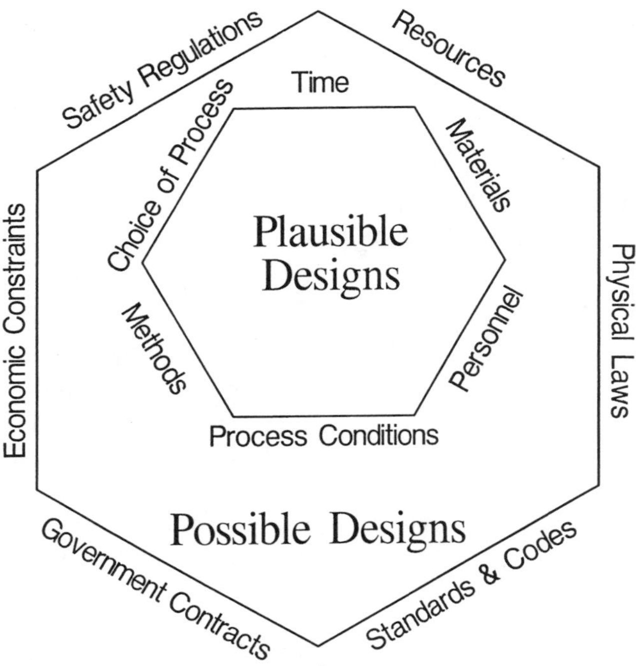

Figure 1.

9. Vacuum distillation velocity: 6 ft/sec @ top of tower.
10. Select a heavy key component and compute $\propto$ of all components to this key.
11. Compute and "average" $\propto$ for the column (say, $\propto_{feed}$ or $\propto = (\propto_{top} \propto_{bottom})^{1/2}$)
12. Plot ln (D/B) vs. ln $\propto$ and the points should fall on a straight line. (D = overhead molar rate, B = bottoms molar rate). If a fairly straight line does *not* result, the compositions are suspect.

Compressors:

T_{out} <250–300°F for multistage compressor (otherwise, special equipment is required).

Expanders

Avoid outlet temperatures in the dew point range.

Cooling Towers and Air-Cooled Heat Exchangers

1. Usually start with the 95% summer hours maximum wet bulb temperature as the ambient air temperature.
2. $T_{optimum} \approx$ 120–150°F.

Pumps

1. Usually a spare is made available for each major pump.
2. See Viking Pump books for data on sizing.
3. Also Perry's Handbook.

Reboilers

1. Max heat flux:
   (a) natural circulation: 12,000 Btu/hr ft^2
   (b) forced circulation: 30,000 Btu/hr ft^2
2. $\dfrac{\text{Liquid circulated in reboiler}}{\text{vapor generated in reboiler}} \; \dfrac{4}{1}$

Heat Exchangers

1. Max surface area = 5,000 ft^2
2. Liquid side velocity (in the tubes) = 1–2 m/sec (4m/s max)

Furnaces

ΔP = 25–40 psi

Additional Rules of Thumb

In addition we are including certain rules of thumb from a paper[2] by Dr. Donald R. Woods and Dr. W. K. Kodatsky. The last item is the author's.

Heat Exchanger Approach: 5°C = 40°F

Approx Friction Factor 0.005 (1 velocity head loss every 50–60 length to diameter ratios)

Efficiencies: Pumps 40–60%
Comprs 60–80%
Frac Trays 60%

Keep cooling water below 50°C = 120°F

Latent heat steam about 5 times that for an organic liquid Btu/#°F

Heat Capacity: Btu/lb °F

liq water	1
stm/HC vapors	0.5
air	0.25

Cooling tower 10°F approach to design wet bulb temperature.

Sources

1. Long, R. Dept. of Chemical Engineering, New Mexico State University, "Guide To Design For Chemical Engineers," handout to students for plant designs, 1994.
2. Dr. Woods R. and Kodatsky, W. K., Dept. of Chemical Engineering, McMaster University, Hamilton, Ontario, "Discovering Short Cut Methods of Equipment Sizing and Selection," presented at 1985 Annual Conference (Computer Aided Engineering) of American Society For Engineering Education, Atlanta, Georgia, June 16–20, 1985, Vol. 1.

Appendix 7
Modern Management Initiatives

The first edition's appendix had a section titled "Modern Quality Control." This section discussed modern initiatives in the spirit of quality pioneer W. Edwards Deming. Since then, I have learned of several management initiatives aimed at improvement from various directions. Here is a brief list:

Total Quality Management—discussed in the first edition as Modern Quality Control.

Computer-Integrated Manufacturing—Computers and robotics.

Management by Objective/Theory Z—overall goals.

The Learning Organization—training.

Reengineering—reorganization (usually downsizing).

Virtualization—scattered offices.

Decentralization—making divisions more independent.

Flat Organizations—eliminate middle management.

Critical Path Analysis—studying general procedures.

Sales Force Automation—computerizing the sales force.

Chaordic Organizations—structured "chaos" and "order" combined.

Post-Capitalism/Co-Opetition—cooperation between companies.

Source

Malone, M. S., *Forbes ASAP*, April 7, 1997.

Appendix 8

Process Specification Sheets

WO. No. ____________
PAGE No. ____________
REVISION No. ____________
PREPARED BY ____________
APPROVED BY ____________
DATE ____________

VESSEL DATA SHEET

PLANT	LOCATION	ITEM Nos.	
SERVICE			
SIZE Dia X	S.-S. X (SKIRT) (LEGS)		

CAPACITY	SPECIFIC GRAVITY-CONTENTS		No. REQ'D	
STANDARD REFERENCE DRAWINGS		DESIGN DATA		
STD.C.3. Ladders, Platforms & Handrail	CODE		STAMP REQ'D (Yes) (No)	
STD.C/4.5/Col, Davits & M.H. Hinge & Davit	DESIGN PRESS		psig. DESIGN TEMP.	
STD.C/4.6. Pipe Guides & Supports	OPER. PRES.		psig. OPER. TEMP.	
STD.C/4.7. Vessel Skirt Openings	CORROSION ALLOWANCE			
	WIND LOAD.		mph, SEISMIC FACTOR	
	No. & TYPE OF TRAYS		Furn. By	
REMARKS:	STRESS RELIEVE (Yes) (No)			
	TESTS		X-Ray (Yes) (No)	

ITEM	THK.	MAT'L	COMMENTS
SHELL			
HEADS			TYPE
CLADDING, LINER			
SUPPORT			(Skirt) (Saddles) (legs)
TRAY SUP. RINGS			
GASKETS			
NUTS & BOLTS	—		
LADDER (CAGED)	—		
PLATFORMS #1			Ang. or Len X Wid.
PAINT	—	None	
INSULATION			
INSUL. SUP. RGS.			@ 10′ Spacing (Nom)

NOZZLE SCHEDULE

MK.	SERVICE	No.	SIZE	RATING	FACING	PROJECTION
N-1						
N-2						
N-3						
N-4						
N-5						
N-6						
N-7						
N-8						

No.	REVISION	APP'D	DATE

WO. No. ______
PAGE No. ______
REVISION No. ______
PREPARED BY ______
APPROVED BY ______
DATE ______

SHELL & TUBE EXCHANGER DATA SHEET

PLANT	LOCATION	ITEM No.

5	Service of Unit		Item No.
6	Size	Type	
7			Horiz. - Vert.
8	Surface per Unit sq ft	Shells per Unit : Series : Parallel	: Surface per Shell sq. ft.
9	PERFORMANCE OF ONE UNIT	Shell Side	Tube Side
10	Fluid Circulated		
11	Total Fluid Entering	lb per hr	lb per hr
12	Vapor	lb per hr	lb per hr
13	Liquid	lb per hr	lb per hr
14	Steam	lb per hr	lb per hr
15	Non-Condensables	lb per hr	lb per hr
16	Fluid Vaporized or Condensed	lb per hr	lb per hr
17	Steam Condensed	lb per hr	lb per hr
18	Gravity - Liquid		
19	Density @ Avg. Cond.	lb per cu ft	lb per cu ft
20	Viscosity	Centipoises at °F	Centipoises at °F
21	Viscosity	Centipoises at °F	Centipoises at °F
22	Molecular Wt.		
23	Conductivity		
24	Specific Heat	Btu per lb deg F	Btu per lb deg F
25	Latent Heat - Vapors	Btu per lb	Btu per lb
26	Temperature In	°F	°F
27	Temperature Out	°F	°F
28	Operating Pressure	psig	psig
29	Number of Passes Per Shell		
30	Velocity	Allow. Calc. fps	Allow. Calc. fps
31	Pressure Drop	Allow. Calc. psi	Allow. Calc. psi
32	Heat Exchanged - Btu per hr.	MTD (Corrected)	
33	Transfer Rate - Service	Clean	Fouling Resistance: Shell Tubes
34		CONSTRUCTION	
35	Design Pressure	psig	psig
36	Test Pressure	psig	psig
37	Design. Temp. (Max. Metal Temp.)	°F	°F
38	Tubes	No. O.D.	BWG Length Pitch
39	Shell I.D.; Mat'l		Tube Fins
40	Shell Cover		Floating Head Cover
41	Channel		Channel Cover
42	Tube Sheets—Stationary		Floating
43	Baffles—Cross	Type Spacing	Thickness Hole Dia.
44	Baffle—Long	Type	Thickness Impact Baffle
45	Tube Supports	No.	Thickness
46	Gaskets		Gland Packing
47	Connections—Shell - In	Out	Series Dome
48	Channel - In	Out	Series
49	Corrosion Allowance—Shell Side		Tube Side
50	Weight Code	Customer's Specifications D2-1G	Temp Class
51	Tube to Tube Sheet Joint		
52	S.R. IX-Ray:		
53	Remarks:		
54			
55			

WO. No. ____________
PAGE No. ____________
REVISION No. ____________
PREPARED BY ____________
APPROVED BY ____________
DATE ____________

DOUBLE PIPE (G-FIN)
EXCHANGER
DATA SHEET

PLANT	LOCATION	ITEM No.

No.				
5	Service of Unit		Item No.	
6	Size and Type			
7			Horiz. - Vert.: Conns.	
8	Surface per Unit sq ft	Sections per Unit	Surface per Section	sq. ft.
9	PERFORMANCE OF ONE UNIT	Shell Side	Tube Side	
10	Fluid Circulated			
11	Total Fluid Entering	lb per hr	lb per hr	
12	Vapor	lb per hr	lb per hr	
13	Liquid	lb per hr	lb per hr	
14	Steam	lb per hr	lb per hr	
15	Non-Condensables	lb per hr	lb per hr	
16	Fluid Vaporized or Condensed	lb per hr	lb per hr	
17	Steam Condensed	lb per hr	lb per hr	
18	Gravity - Liquid			
19	Density @ Avg. Cond.	lb per cu ft	lb per cu ft	
20	Viscosity	Centipoises at °F	Centipoises at °F	
21	Viscosity	Centipoises at °F	Centipoises at °F	
22	Molecular Wt.			
23	Conductivity	at °F	at °F	
24	Specific Heat	Btu per lb deg F	Btu per lb deg F	
25	Latent Heat - Vapors	Btu per lb	Btu per lb	
26	Temperature In	°F	°F	
27	Temperature Out	°F	°F	
28	Operating Pressure	psig	psig	
29	Velocity	fps	fps	
30	Pressure Drop	psi	psi	
31	Heat Exchanged - Btu per hr.	MTD (Corrected)		
32	Transfer Rate - Service	Clean		
33	Fouling Resistance: Shell	Tubes : Gross	Net	
34	CONSTRUCTION			
35	Design Pressure	psig	psig	
36	Test Pressure	psig	psig	
37	Design. Temp. (Max. Metal Temp.)	°F	°F	
38	Shell	Tubes		
39	Shell Cover	Fins		
40	Shell Connections	Tube Connections		
41	Shell Connectors	Tube Connectors		
42	Shell Gaskets: Cover	Conn.	Corr. Allow: Shell	Tube
43	Weight	Code	Customer's Specifications	
44	Hook-up–Shell Side	Parallel Banks of	Sections in Series	
45	Tube Side	Parallel Banks of	Sections in Series	
46	Stacking Arrangement:	Wide	High	
47	Length	Width	Height	
48	Ref. Print	Inspection by whom		
49	Based upon operating conditions as given maximum temperature of tubes will not exceed °F			
50	Remarks: All heat transfer data based upon total outside surface of heat transfer elements.			

THESE SECTIONS ARE MANUFACTURERS' STANDARD AND FEATURE METAL TO METAL JOINTS AT UNION AND SHELL CLOSURES.

WO. No. ____________
PAGE No. ____________
REVISION No. ____________
PREPARED BY ____________
APPROVED BY ____________
DATE ____________

AIR COOLED (FIN-FAN) EXCHANGER DATA SHEET

PLANT		LOCATION	ITEM No.
5	Service		
6	Size Type	Induced Forced Draft	No. of Bays
7	Surface/Item - - - External	Bare Tube	Sq Ft
8	Heat Exchanged - - - BTU/Hr		Effective MTD °F
9	Transfer Rate - - - External Surface	Bare Tube Surface	BTU/Hr Sq Ft °F
10			
11		PERFORMANCE DATA	
12		TUBE SIDE	
13	Fluid Circulated		Temperature In °F
14	Total Fluid Entering Lbs/Hr		Temperature Out °F
15	Vapor		Inlet Pressure PSIG
16	Liquid		Gravity—Liquid
17	Steam		Viscosity Cps @ °F
18	Non-Condensables		Viscosity Cps @ °F
19	Vapor Condensed		Molecular Weight
20	Steam Condensed		Specific Heat BTU/Lb °F
21	Density-Vap-Liquid Mixt. Lbs/Cu. Ft.		Latent Heat BTU/Lb
22	Conductivity BTU Ft/Hr. Sq. Ft. °F		Allowable Press. Drop PSI
23	Fouling Resistance 1.S. Hr. Sq. Ft. °F/BTU		Design Pressure Drop PSI
24		AIR SIDE	
25	Air Quantity Lbs/Hr SCFM		Temperature In °F
26	Air Quantity/Fan ACFM		Temperature Out °F
27			Altitude Ft.
28			
29		CONSTRUCTION	
30	Design Pressure PSI	Test Pressure PSI	Design Temperature °F
31	SECTION	HEADER	TUBE
32	Size Rows	Type	Material Seamless Welded
33	No/Bay	Material	OD In. BWG. Avg. Min Wall
34	Arrangement:	No. Passes	No/Section
35	Sections in Parallel in Series	Plug—Design Material	Length Ft
36	Bays in Parallel in Series	Gasket Material	Pitch in. Δ
37		Corrosion Allowance In.	FIN
38	MISC.	Size Inlet Nozzle In.	Material
39	Structure	Size Outlet Nozzle In.	OD
40		Rating	No/In
41		Code—	Type
42			
43		MECHANICAL EQUIPMENT	
44	FAN	DRIVER	SPEED REDUCER
45	Type	Type	Type
46	No/Bay HP/Fan	No/Bay HP/Driver	No/Bay
47	Diameter Ft RPM	RPM	Model
48	No. Blades Pitch ADJUSTABLE AUTO VARIABLE	Enclosure	AGMA HP Rating
49	Blade Material	Volt Phase Cycle	Ratio
50	Hub Material	Mfr	Mfr
51	NOTES:		
52			
53			
54			
55			
56			
57			
58	Plot Area	Proposal Drawing No.	Shipping Weight Lbs

WO. No. ____________
PAGE No. ____________
REVISION No. ____________
PREPARED BY ____________
APPROVED BY ____________
DATE ____________

DIRECT FIRED HEATER
DATA SHEET

PLANT	LOCATION		ITEM No.
SERVICE			
	THERMAL DESIGN		
	INLET		OUTLET
FLUID CIRCULATED			
TOTAL FLUID ENTERING			
VAPOR	=/HR.		=/HR.
LIQUID	=/HR.		=/HR.
GRAVITY-LIQUID			
DENSITY-VAPOR	=/CU. FT.		=/CU. FT.
VISCOSITY-LIQUID	CENTIPOISES		CENTIPOISES
VISCOSITY-VAPOR	CENTIPOISES		CENTIPOISES
MOLECULAR WEIGHT-LIQUID			
MOLECULAR WEIGHT-VAPOR			
SPECIFIC HEAT	B.T.U./=		B.T.U./=
LATENT HEAT	B.T.U./=		B.T.U./=
TEMPERATURE	°F		°F
VAPOR PRESSURE	PSIG		PSIG
OPERATING PRESSURE	PSIG		PSIG
PRESSURE DROP			=/SQ. IN.
TYPE OF FLOW			
VELOCITY			FT./SEC
EFFECTIVE SURFACE	RADIANT SQ. FT.	CONVECTION	SQ. FT.
HEAT ABSORPTION			B.T.U./HR.
RADIANT TRANSFER RATE	AVERAGE	MAXIMUM	B.T.U./SQ. FT./HR.
CONVECTION TRANSFER RATE			B.T.U./SQ. FT./HR.
HEAT RELEASE			B.T.U./SQ. FT./HR.
EFFICIENCY			BASED ON LHV
FLUE GAS TEMP. (APPROX.)		% EXCESS AIR:	°F
TYPE OF FUEL			
LHV OF FUEL			B.T.U./

MECHANICAL DESIGN			
NUMBER OF TUBES: RADIANT			CONVECTION
EFFECTIVE TUBE LENGTH: RADIANT			CONVECTION
RADIANT TUBES: OD "	TUBE WALL "	ASTM SPEC.	
CONVECTION TUBES: OD "	TUBE WALL "	ASTM SPEC.	
RETURN BENDS:			
RETURN BEND SPECIFICATIONS:			
NOZZLES: INLET	OUTLET	ABA	
DESIGN PRESSURE:	PSIG	DESIGN TEMPERATURE:	
CODE:			
HEATER SIZE:		APPROX. WEIGHT:	
REMARKS:			

WO. No. ____
PAGE No. ____
REVISION No. ____
PREPARED BY ____
APPROVED BY ____
DATE ____

CENTRIFUGAL PUMP
Horizontal or Vertical
DATA SHEET

PLANT	LOCATION	ITEM Nos.	
SERVICE	No. REQ'D		
TYPE DRIVE	FURNISHED BY		

OPERATING CONDITIONS

FLUID	GPM Nor., Rated	SPEED RPM
PUMPING TEMP °F	GPM(x) Minimum Cont. Flow	DES. EFF. (x) % Rated
SPEC. GRAV. @P.T.	DISCH. PRES, psig Rated	BHP (x) Rated
VISCOSITY @P.T.	SUCT. PRES, psig MAX Rated	BHP (x) Non-Overload
VAPOR PRES, psia @P.T.	DIFF PRES, psi Rated	BHP (x) Recommended
NPSH, (Ft.) Avail (x) Req'd	DIFF HEAD, FT. Rated	
SUSPENDED SOLIDS		
CORROSIVES		
ENTRAINED GASES		

PUMP SPECIFICATIONS

MODEL	SIZE		ROTATION Facing Cplg.
No. OF STAGES	MAX. WORK PRES.	psig	MAX. WORK TEMP °F
SUCTION: SIZE	RATING	FACING	POSITION
DISCHARGE: SIZE	RATING	FACING	POSITION
IMPELLER DIA. RATED		MAX.	BACK PULL-OUT REQ'D (Yes) (No)

MATERIALS-MAJOR COMPONENTS

PART	MATERIAL/TYPE	PART	MATERIAL/TYPE
SUPPORT	/	MECHANICAL SEAL-TYPE	
CASE	/	METAL PARTS	
SHAFT	/	SEAL FACES	
IMPELLER	/	PACKING No. Rings (x)	TYPE
CASE WEAR RING		LANTERN RING	
IMPELLER WEAR RING		PACKING BOX	
SHAFT SLEEVE		PACK. BOX BUSH	
RADIAL BEARING	/	PACK. GLAND	
HOUSING MAT'L		PRES. RED. BUSH	
COUPLING	/	CASE VENT	
COUPLING GUARD		CASE DRAIN	
PUMP & DRIVE BASE	/	GAUGE CONN.	
LUBRICATION			
		COOLING WATER-SEAL	
		GLAND	
MULTI-STAGE PUMP: IMPELLER HUB SLEEVES		BEARINGS	
SPACER SLEEVES		PEDESTAL	
INTERSTAGE SLEEVES		TOTAL WATER REQUIRED	(x) GPM @ °F
INTERSTAGE DIAPHRAMS			
INTER. DIAPH. BUSHING		SETTING DEPTH (Vert. Pump)	
DIFFUSER		MOTOR DATA	
THRUST BEARING	/	ITEM No.	MFR
HOUSING MAT'L		H.P. S.F.	RPM
BALANCING DRUM		VOLTS	HERTZ PHASE
BALANCING RING		ENCLOSURE	FRAME
		RADIAL BRG.	LUBE
		INSUL.	TEMP RISE
		STARTING TORQ	%, F.L. AMPS
		INNER BRG. CAP. Req'd., JUNCTION BOX - Cast Iron	

REMARKS:

PUMP
Vertical Turbine
Con or Propellor
DATA SHEET

WO. No. ______
PAGE No. ______
REVISION No. ______
PREPARED BY ______
APPROVED BY ______
DATE ______

PLANT		LOCATION	ITEM Nos.	
SERVICE		No. REQ'D		
TYPE DRIVE		FURNISHED BY		

OPERATING CONDITIONS

FLUID	GPM Nor.,	Rated	SPEED RPM
PUMPING TEMP °F	GPM(x)	Minimum Cont. Flow	DES. EFF. (x) % Rated
SPEC. GRAV. @P.T.	DISCH. PRES, psig	Rated	BHP (x) Rated
VISCOSITY @P.T.	SUCT. PRES, psig MAX	Rated	BHP (x) Non-Overload
VAPOR PRES, psia @P.T.	DIFF PRES, psi	Rated	BHP (x) Recommended
NPSH. (Ft.) Avail (x) Req'd	DIFF HEAD, Ft.	Rated	
SUSPENDED SOLIDS			
CORROSIVES			
ENTRAINED GASES			

PUMP SPECIFICATIONS

MODEL	SIZE		ROTATION Facing Cplg.
No. OF STAGES	MAX. WORK PRES. psig		MAX. WORK TEMP °F
SUCTION: SIZE RATING		FACING	POSITION
DISCHARGE: SIZE RATING		FACING	POSITION
IMPELLER DIA. RATED		MAX.	BACK PULL-OUT REQ'D (Yes) (No)

MATERIALS-MAJOR COMPONENTS

PART	MATERIAL/TYPE	PART	MATERIAL/TYPE
DRIVER STAND		MECHANICAL SEAL-TYPE	
DISCHARGE HEAD		METAL PARTS	
COLUME OR CAN, LEN./DIA.	/	SEAL FACES	
MAT'L/TYPE	/	PACKING No. Rings (x)	TYPE
SHAFT		LANTERN RING	
UPPER BRG	/	PACKING BOX	
HOUSING		PACK. BOX BUSH	
INTERMEDIATE BRG.	/	PACK. GLAND	
SPIDER		PRES. RED. BUSH	/
LOWER BRG.	/	CASE VENT	
HOUSING		CASE DRAIN	
BOWLS		GAUGE CONN.	
BOWL WEAR RING			
IMPELLER	/		
IMPELLER WEAR RING		COOLING WATER-SEAL	
SHAFT SLEEVE		GLAND	
SPACER SLEEVE		BEARINGS	
COUPLING	/	PEDESTAL	
LUBRICATION		TOTAL WATER REQUIRED	(x) GPM @ °F
SUCTION MANIFOLD	/		
STRAINER		MOTOR DATA	
		ITEM No.	MFR
		H.P. S.F.	RPM
		VOLTS	HERTZ PHASE
		ENCLOSURE	FRAME
		SHAFT (Hollow) (Solid) Thrust Brg.	
		RADIAL BRG.	LUBE
REQ'D THRUST CAP OF DRIVER	(x) (UP)(DOWN)	INSUL.	TEMP RISE °C
		STARTING TORQ	%, F.L. AMPS
		INNER BRG. CAP. Req'd.,	JUNCTION BOX - Cast Iron

REMARKS:

WO. No. ______
PAGE No. ______
REVISION No. ______
PREPARED BY ______
APPROVED BY ______
DATE ______

TANK DATA SHEET

PLANT	LOCATION	ITEM Nos.	
SERVICE			
SIZE DIA X	HIGH		
CAPACITY	SPECIFIC GRAVITY-CONTENTS	No. REQ'D	
STD. REFERENCE DRAWINGS	DESIGN DATA		
STD.C.3. Ladders, Platforms & Handrail	CODE CONSTRUCTION:		
	STAMP REQUIRED: (YES) (NO)		
	DESIGN PRES. psig. DESIGN TEMP. °F		
	OPER.PRES. psig. OPER. TEMP. °F		
	CORROSION ALLOWANCE:.		
	WIND LOAD mph, SEISMIC FACTOR		
	REFERENCE SPEC.		

REMARKS:		MATERIAL	REMARKS
	SHELL—thk.		
	GIRDER		
	DECK—thk.		(Cone) (Dome) (Float) (Flat)
	BOTTOM—thk.		
	SUMP		
	MANHOLES		
	NOZZLES, CPLGS., PIPE		
	GASKETS		
	NUTS & BOLTS		
	LADDER (Caged)		
	CIRCULAR STAIRWAY		
	LANDING		
	PAINT	None	
	INSUL. TOP—thk.		By Others
	SIDES—thk.		By Others
	INSIDE COATING		
	LINER		

NOZZLE SCHEDULE

MK.	SERVICE	No.	SIZE	RATING	FACING	PROJECTION
N-1	INLET					
N-2	OUTLET					
N-3						
N-4						
N-5	THERMOWELL					
N-6	DRAIN					
N-7	VENT					
N-8	OVERFLOW					
N-9	GAUGE HATCH					
N-10	TANK GAUGE					
N-11	RELIEF VALVE					
N-12						
N-13						
N-14						
N-15						
M-1	SHELL MANHOLE					
M-2	ROOF MANHOLE					

No.	REVISION	APP'D	DATE

WO. No. ____________
PAGE No. ____________
REVISION No. ____________
PREPARED BY ____________
APPROVED BY ____________
DATE ____________

COOLING TOWER DATA SHEET
Page 1 of 3

PLANT	LOCATION	ITEM No.
GENERAL		
Selection		
Tower Model		
Type		
DESIGN & OPERATING CONDITIONS		
• Circulating Water Flow, U.S. GPM		
• Hot (inlet) Water Temp. F.		
• Cold (outlet) Water Temp. F.		
• Wet Bulb Temp. F., Inlet		
Ambient		
Tower Pump Head, Ft.		
Total Fan B.H.P., (Driver Output)		
Drift Loss, % of circulating flow		
Evaporation Loss (at design)		
• Design Wind Load, Lbs./sq. ft.		
Mi./hr.		
• Design Seismic Load, % G		
• Tower Site (ground level, roof, etc.)		
• Elevation Above Sea Level, ft.		
• Tower Exposure		
STRUCTURAL DETAILS		
Number of Cells		
Fans per Cell		
Total Number of Fans		
Nominal Cell Dimen., L x W, ft.		
Overall Tower Dimension, L x W, ft.		
Height-Basin Curb to Fan Deck, ft.		
Fan Stack Height, ft.		
Overall Tower Height, ft.		
Inside Basin Dimensions, ft.		
• Column Extensions, Perimeter, below		
basin curb, ft.		
Internal, below		
curb, ft. (max.)		
Anchorage		
Hot Water Inlet-Number		
Nom. Diameter, in.		
Description		
Height Inlet Above Basin Curb, ft.		
• Access to Top of Tower		
Shipping Weight, lbs.		
Operating Weight, lbs.		

COOLING TOWER DATA SHEET
Page 2 of 3

MATERIALS OF CONSTRUCTION	
Framework Members	
Casing	
Filling	
Support	
Drift Eliminators	
Spacer	
Fan Stacks	
Louvers, Material	
Partitions	
Fan Deck	
Water Distribution—Type	
Material	
Lumber Pre-Treatment	
Type of Treatment	
Items Treated	
Splashers or Spray Nozzles	
Stairway and Handrail	
Structural Connectors	
Ring Joint Connectors	
Bolts, Nuts, Washers	
Anchor Connectors	
Nails	
Mechanical Equipment Support	
• Anchor Bolts—Material	
Furnished by	
• Cold Water Basin—Material	
Furnished by	
• Basin Accessories, by Mfg.	
MECHANICAL EQUIPMENT	
Fans	
Number	
Type or Model	
Manufacturer	
Diameter, ft.	
Number of Blades	
Fan Speed, RPM	
Tip Speed, FPM	
BHP per fan, driver output	
Blade Material	
Hub Material	
Total Static Pressure, in. H_2O	
Velocity Pressure, in. H_2O	
Air Delivery per Fan, ACFM	
Fan Static Efficiency	
Speed Reducer	
Number	
Type	
Model	
Manufacturer	
Reduction Ratio	
AGMA Mechanical H.P. Rating	
Service Factor at Rated H.P. of Driver	
No. of Reductions	

COOLING TOWER DATA SHEET
Page 3 of 3

MECHANICAL EQUIPMENT (Cont'd.)	
Drive Shaft	
Number	
Type	
Model	
Manufacturer	
Rated H.P.	
Drive Shaft Material	
Coupling Material	
Driver	
Number	
• Kind	
• Type	
Manufacturer	
Full Load Speed, RPM	
• Elec. Char.—phase/cycles/volts	
Rated H.P.	
ADDITIONAL DATA	

Source

Branan, C. and Mills, J., *The Process Engineer's Pocket Handbook, Vol. 3,* Houston, Texas: Gulf Publishing Co., 1984, pp. 144–154.

Index